Communications in Computer and Information Science 3055

Rationale
The CCIS series is devoted to the publication of proceedings of computer science conferences. Its aim is to efficiently disseminate original research results in informatics in printed and electronic form. While the focus is on publication of peer-reviewed full papers presenting mature work, inclusion of reviewed short papers reporting on work in progress is welcome, too. Besides globally relevant meetings with internationally representative program committees guaranteeing a strict peer-reviewing and paper selection process, conferences run by societies or of high regional or national relevance are also considered for publication.

Topics
The topical scope of CCIS spans the entire spectrum of informatics ranging from foundational topics in the theory of computing to information and communications science and technology and a broad variety of interdisciplinary application fields.

Information for Volume Editors and Authors
Publication in CCIS is free of charge. No royalties are paid, however, we offer registered conference participants temporary free access to the online version of the conference proceedings on SpringerLink (http://link.springer.com) by means of an http referrer from the conference website and/or a number of complimentary printed copies, as specified in the official acceptance email of the event.

CCIS proceedings can be published in time for distribution at conferences or as post-proceedings, and delivered in the form of printed books and/or electronically as USBs and/or e-content licenses for accessing proceedings at SpringerLink. Furthermore, CCIS proceedings are included in the CCIS electronic book series hosted in the SpringerLink digital library at http://link.springer.com/bookseries/7899. Conferences publishing in CCIS are allowed to use our online conference service (Meteor) for managing the whole proceedings lifecycle (from submission and reviewing to preparing for publication) free of charge.

Publication process
The language of publication is exclusively English. Authors publishing in CCIS have to sign the Springer CCIS copyright transfer form, however, they are free to use their material published in CCIS for substantially changed, more elaborate subsequent publications elsewhere. For the preparation of the camera-ready papers/files, authors have to strictly adhere to the Springer CCIS Authors' Instructions and are strongly encouraged to use the CCIS LaTeX style files or templates.

Abstracting/Indexing
CCIS is abstracted/indexed in DBLP, Google Scholar, EI-Compendex, Mathematical Reviews, SCImago, Scopus. CCIS volumes are also submitted for the inclusion in ISI Proceedings.

How to start
To start the evaluation of your proposal for inclusion in the CCIS series, please send an e-mail to ccis@springer.com

Abhishek Das · Prafulla Kumar Panda ·
Ashok Misra · Mihir Narayan Mohanty ·
Swagatam Das · Masoud Mohammadian
Editors

Machine Intelligence and Data Science

First International Conference, MIDAS 2025
Paralakhemundi, India, March 21–22, 2025
Proceedings

Springer

Editors
Abhishek Das
Centurion University of Technology and Management
Paralakhemundi, Odisha, India

Ashok Misra
Centurion University of Technology and Management
Paralakhemundi, Odisha, India

Swagatam Das
Indian Statistical Institute
Kolkata, West Bengal, India

Prafulla Kumar Panda
Centurion University of Technology and Management
Paralakhemundi, Odisha, India

Mihir Narayan Mohanty
ITER, Siksha 'O' Anusandhan
Bhubaneswar, Odisha, India

Masoud Mohammadian
University of Canberra
Bruce, ACT, Australia

ISSN 1865-0929 ISSN 1865-0937 (electronic)
Communications in Computer and Information Science
ISBN 978-3-032-35515-7 ISBN 978-3-032-35516-4 (eBook)
https://doi.org/10.1007/978-3-032-35516-4

This Springer imprint is published by the registered company Springer Nature Switzerland AG
The registered company address is: Gewerbestrasse 11, 6330 Cham, Switzerland

Preface

We are pleased to present the proceedings of the 1st International Conference on Machine Intelligence and Data Science (MIDAS 2025). This volume brings together a carefully selected collection of high-quality research contributions that address both theoretical foundations and practical applications of machine intelligence and data science. The scope of MIDAS 2025 encompassed a wide range of contemporary topics, including data management and quality analysis, data security and privacy, data acquisition, integration and cleaning, big data analytics, data mining, data warehousing, and data visualization.

As these domains continue to advance at an unprecedented pace, the research presented in these proceedings offers valuable insights for researchers, practitioners, and academicians who are committed to exploring the evolving frontiers of intelligent systems and their societal impact. We are living in an era where intelligent machines are reshaping nearly every aspect of human life—from healthcare and education to transportation, industry, and entertainment. The convergence of machine intelligence with data science has enabled novel approaches to solving complex problems, automating decision-making processes, and augmenting human capabilities. MIDAS 2025 served as an important platform to showcase recent advancements and to foster meaningful discussions that will influence the future trajectory of these rapidly growing fields.

MIDAS 2025 brought together a diverse and global community of researchers, practitioners, and thought leaders. The interdisciplinary nature of the conference highlights the importance of collaboration across domains to address complex real-world challenges. By facilitating interactions among experts from different backgrounds, the conference aimed to encourage cross-pollination of ideas and the development of integrated and innovative solutions. The program featured several keynote addresses delivered by distinguished experts whose contributions have significantly shaped the fields of machine intelligence and data science. Their talks provided both a comprehensive overview of the current research landscape and valuable perspectives on emerging trends and future opportunities.

In addition to keynote presentations, the conference included multiple technical sessions that offered participants opportunities to engage deeply with cutting-edge research and innovative applications. These sessions covered a broad spectrum of topics, ranging from theoretical developments to real-world implementations, thereby providing a well-rounded understanding of the state of the art in artificial intelligence and knowledge-driven systems.

This year, MIDAS 2025 received 136 manuscript submissions, of which 40 papers were selected for inclusion in these proceedings. Each submission underwent a rigorous double-blind peer-review process, with evaluation by at least three domain experts. This thorough review process ensured the high quality and scholarly integrity of the accepted papers. We extend our sincere appreciation to all the authors for their valuable contributions and to the reviewers for their time, expertise, and constructive feedback, which played a crucial role in maintaining the academic standards of the conference.

We are grateful to the keynote speakers and session chairs for their insightful contributions and for enriching the technical program. We also express our heartfelt thanks to the members of the organizing committee, whose dedication, meticulous planning, and tireless efforts were instrumental in the successful organization of MIDAS 2025. Their commitment ensured a stimulating and professionally rewarding experience for all participants.

We would like to extend our special appreciation to Springer for providing exceptional technical and editorial support in the publication of the MIDAS 2025 proceedings. Their professional guidance, high-quality publishing standards, and robust dissemination platform have significantly contributed to the global visibility and accessibility of the research presented in this volume.

As you explore the proceedings of MIDAS 2025, we invite you to engage with the diverse ideas, methodologies, and innovations presented herein. We hope that the knowledge shared in this volume will inspire future research, promote collaboration, and contribute meaningfully to the advancement of machine intelligence and data science.

In conclusion, MIDAS 2025 was not merely a conference but a celebration of scientific inquiry, innovation, and the collective pursuit of knowledge. We are confident that the discussions and interactions fostered during this event will have a lasting impact on the participants and the broader research community.

March 2025

Abhishek Das
Prafulla Kumar Panda
Ashok Misra
Mihir Narayan Mohanty
Swagatam Das
Masoud Mohammadian

Organization

Patrons

Mukti Kanta Mishra	Centurion University, India
D. Narsimha Rao	Centurion University, India
Supriya Pattanayak	Centurion University, India

General Chair

Anita Patro	Centurion University, India

Organizing Chairs

Ashish Ranjan Dash	Centurion University, India
Sujata Chakravarty	Centurion University, India

Program Committee Chairs

Abhishek Das	Centurion University, India
Prafulla Kumar Panda	Centurion University, India
Ashok Misra	Centurion University, India
Mihir Narayan Mohanty	Siksha 'O' Anusandhan University, India
Swagatam Das	Indian Statistical Institute, Kolkata, India
Masoud Mohammadian	University of Canberra, Australia

International Advisory Committee

Vaclav Snasel	VSB – Technical University of Ostrava, Czech Republic
Celestine Iwendi	University of Greater Manchester, UK
Vincenzo Piuri	University of Milan, Italy
Ahmed Khader Habboush	Jerash University, Jordan
Gyoo-Soo Chae	Baekseok University, South Korea
Huijun Gao	Harbin Institute of Technology, China

Celia Shahnaz	Bangladesh University of Engineering and Technology, Bangladesh
Shazia Hasan	BITS Pilani, Dubai Campus, UAE
Milan Biswal	New Mexico State University, USA
Hyun Sung Shin	Samsung Electronics, South Korea
Djaffar Ould Abdeslam	University of Haute-Alsace, France
Florian Misoc	Kennesaw State University, Marietta, USA
Jan Michalik	University of West Bohemia, Czech Republic
Matthew Armstrong	Newcastle University, UK
Andrews Samraj	CMR University, India
Akhtar Kalam	Victoria University, Australia
Mohammad Alnabhan	Mu'tah University, Jordan
Ahmad Abu Alaish	Jerash University, Jordan
Majed Aborokbah	Tabuk University, Saudi Arabia
Osama Alia	Tabuk University, Saudi Arabia
Awad Awadelkarim	Tabuk University, Saudi Arabia
Izzat Al-Smadi	Texas A&M University, USA
Raveendran Paramesran	Monash University Malaysia, Malaysia

National Advisory Committee

Ajay Kumar Nayak	Centurion University, India
Biswajit Mishra	Centurion University, India
Ganapati Panda	C.V. Raman Global University, India
P. Kabisatpathy	C.V. Raman Global University, India
Saroj K. Meher	Indian Statistical Institute, Bangalore Centre, India
Nishchal K. Verma	IIT Kanpur, India
Dinesh Goyal	Poornima Institute of Engineering and Technology, India
Amit Kant Pandit	Shri Mata Vaishno Devi University, India
Jagadish C. Bansal	South Asian University, India
Pradeep Kumar Mallick	Kalinga Institute of Industrial Technology, India
Tarun K. Sharma	Shobhit University, India
Kusum Deep	IIT Roorkee, India
Mili Pant	IIT Roorkee, India
Bidyut Baran Chaudhuri	Indian Statistical Institute, Kolkata, India
Kandarpa Kumar Sarma	Guwahati University, India
Nanda Dulal Jana	NIT Durgapur, India
Debi Prasanna Kanungo	CSIR-Central Building Research Institute, India
I. V. Murallikrishna	Independent Consultant, India

Ashok Kumar Pradhan	IIT Kharagpur, India
Aurobinda Routray	IIT Kharagpur, India
Amalendu Patnaik	IIT Roorkee, India
B. K. Panigrahi	IIT Delhi, India
Sukumar Mishra	IIT Delhi, India
Prasant Kumar Sahu	IIT Bhubaneswar, India

Technical Program Committee

Durga Prasad Padhi	Centurion University, India
Sangram Keshari Swain	Centurion University, India
Debendra Kumar Sahoo	Centurion University, India
Dhawaleswar Rao CH	Centurion University, India
Prabhat Kumar Patanaik	Centurion University, India
Sujit Mishra	Centurion University, India
Sandipan Pine	Centurion University, India
Rajib Kumar Majhi	Centurion University, India
Prangya Parimita Pradhan	Centurion University, India
Abinash Rath	Centurion University, India
Krushna Chandra Sethi	Centurion University, India
Aurobinda Bag	Centurion University, India
N. Jeeva Ratnam	Centurion University, India
Satyanarayan Padhy	Centurion University, India
Jagannath Padhy	Centurion University, India
Susanta Kumar Nayak	Centurion University, India
Pranita Mahapatra	Centurion University, India
Rajesh Kumar Mishra	Centurion University, India
Sagarika Dash	Centurion University, India
Biswaranjan Routray	Centurion University, India
Manaswini Padhi	Centurion University, India
Saurya Ranjan Das	Centurion University, India
Satyanarayan Sahu	Centurion University, India
Lipsa Priyadarshini Singh	Centurion University, India
Pragnya Das	Centurion University, India
Manoj Kumar Padhi	Centurion University, India
Gayatri Swain	Centurion University, India

Publicity/IT Committee

Nrusinha Das	Centurion University, India
Tofan Kumar Nahak	Centurion University, India
Prabhat Kumar Nahak	Centurion University, India
Anil Patro	Centurion University, India
Pitabas Pradhan	Centurion University, India
Anupam Mandal	Centurion University, India
Khirodh Kumar Choudhury	Centurion University, India

Additional Reviewers

A. Lakshmanarao
A. Bhagya
Anita Pradhan
Appala Muttipati
Archana Patnaik
Arthy Babu
Arun Sekar Rajasekaran
Ashish Kumar Dass
Babita Majhi
Bendi Venkata Ramana
Bhaswati Sahoo
Bijayini Mohanty
Binayak Ojha
Chandrakant Mallick
Chandrakanta Mahanty
Chinmaya Padhan
Chitaranjan Dalai
Daruvuri Rajesh
Debashreet Das
Debasish Pradhan
Deepak Kumar
Devidas Thosar
Manjushree Nayak
Sameer Das
Pavan Kumar A. V. S.
Bijay Paikaray
Charu Wahi
Mamata Garanayak
Premkumar Borugadda
Rajesh Ojha
Ranjita Swain
S. D. Sathishkumar
Sutha J
Swatee Mohanty
Ravi Suggala
E. G. Satish
Gopikrishna Tiruveedula
Gunupusala Satyanarayana
Gyana Patra
Jospin Jeya
Juvvala Sailaja
Kadiri Padmaja
Karthick S
Konga Upendar
Mahesh Maurya
Manikanta Manohar
Manish Babu
Manoj Kumar Kar
Nalineekumari Arasavali
Naween Kumar
Nibedita Mohapatra
Nilamadhab Mishra
Nithisha J.
Nuzhat Prova
Nyakapu Rajender
Om Prakash Jena
Prabu M.
Pramoda Patro
Pranati Mishra
Prerna Patil
Rakesh Patnaik
Rakesh Ranjan Kumar

Ravi Suggala
Rohit Kumar Das
Samson Anosh Babu Parisapogu
Sandhyalati Behera
Saravanan Chandrasekaran
Sasmita Behera
Satyanarayana Kalahasthi
Saumendra Mohapatra
Seshu Bhavani M
Shilpa Thoutireddy
Shruti Mishra
Shubham Dubey
Sibo Patro
Silpa N.
Siva Ranjani
Sony S. Sahoo
Sourabh Debnath
Srabanti Maji
Sravani Sravani
Sreedevi Emandi
Sridhar Patthi
Srinivasa Rao P.
Subhalaxmi Das
Subhashini R
Sudeshna Baliarsingh
Sudhan M. B.

Contents

Intelligent Computational Models for Multimedia Understanding and Trusted Digital Systems

Sentiment Analysis of Text Data for Hate Speech Detection

Simranjeet Kaur and Jayshree Kundargi(✉)

K. J. Somaiya College of Engineering, Somaiya Vidyavihar University, Vidyavihar, Mumbai, Maharashtra 400077, India
simranjeet.k@somaiya.edu, jmkundargi@somiya.edu

Abstract. Social networking services are dominant in todays' era and are utilized by billions of people to manifest their point of view on various topics. Since every coin has two sides these social networking platforms are also being used to spread hatred towards a targeted group of people. Since monitoring only through human supervision is time intensive, Machine Learning models can simplify the process by classifying the data with help of models. Hate Speech detection consists of binary classification of data into 'hate' or 'non-hate'. The various state of art models and techniques were studied during the literature survey and it was found that majority work was conducted using Deep Learning Models. In this paper various classical machine learning models and deep learning models are tested on five benchmark English language datasets. Among the five datasets 'HateCheck' dataset is a synthetic dataset and the remaining datasets consist of text data obtained from various social networking platforms. Here, the commonly used Bag of Words (BoW) and Word2Vec word representation techniques have been utilized for classical machine learning and deep learning models respectively. The impact of hyper-parameter tuning using five-fold cross validation technique has been explored on classical machine learning models. All the models are evaluated using the F1-Score performance evaluation metric. The deep learning models that were applied to the datasets are CNN, LSTM and BiLSTM. The F1-Score metric is used as a performance evaluation metric and the algorithms were optimized to ensure minimum value of false positive and true negative. The classical and deep learning models were evaluated using the F1-Score and the results obtained were compared with the state-of-the-art.

Keywords: sentiment analysis · hate speech · classical machine learning models · deep learning models

1 Introduction

People love to communicate their opinion on various topics; The social media websites or apps provide a platform to express oneself. Though the platforms provide extensive freedom to express emotions. It has been witnessed that few people misuse these platforms to instigate hate towards a particular sect or group of people. Thus, it becomes necessary to monitor these platforms; The data on these platforms is present in bulk and

A. Das et al. (Eds.): MIDAS 2025, CCIS 3055, pp. 3–17, 2027.
https://doi.org/10.1007/978-3-032-35516-4_1

monitoring only through human supervision is difficult. Thus, the Machine Learning models could be used to classify the data and make the process of monitoring the data easy.

Hate speech detection is a sub-domain of sentiment analysis. Hate speech presents a notable issue in today's society, and ML algorithms offer a promising solution for addressing this problem. These algorithms have the capability to automatically identify hate speech within text-based or speech data. This is achieved through training the model on a dataset that has been labeled to distinguish between hate speech and non-hate speech, taking into account various features such as words, grammar, and syntax. There are various ML models available for classification of sentiments and can be categorized into three main categories classical ML models, deep learning models and ensemble techniques. Since the data that is categorized is text and has large number of words present in it; text pre-processing is performed by which the punctuations and white spaces are removed along with the various stop words to make the features very concise. The features can be extracted using BoW, TF-IDF, GloVe, Word2Vec etc. In this work the classical ML models and deep learning models are implemented on five benchmark English language datasets. The classical ML models were initially implemented with the default parameters and later hyper-parameter tuning was performed. The performance of the models with default parameters was compared with models obtained after hyper-parameter tuning and the models with best performance were considered. The deep learning models that were applied to the datasets are Convolutional Neural Networks (CNN) and Recurrent Neural Networks (RNN). The classical and deep learning models were assessed using performance evaluation metrics.

2 Literature Survey

The literature on hate speech detection explores a variety of datasets, feature extraction methods, and machine learning models. Chhabra and Vishwakarma [1] utilized the TBO dataset for binary classification of "hate" and "non-hate," employing methods like BoW, TF-IDF, and N-grams, with CNN and BERT showing superior precision due to their ability to capture contextual information. Mahajan et al. [2] employed Glove features on multiple datasets in different languages, highlighting the efficiency of BiLSTM and BiGRU. Faruqe et al. [3] used TF-IDF with BERT, GRU, and Attention-based models on balanced datasets, finding GRU and Attention models particularly effective. Nagar et al. [4] combined LSTM and GCN on the Founta and Ribeiro datasets, achieving notable accuracy by capturing both temporal and relational data. Nayla et al. [5] applied BERT to English and Indonesian datasets, with better performance on English due to language-specific optimizations. Kennedy Malanga Ndenga [6] used vectorization and Universal Sentence Encoder embeddings, with GradientBoostedTrees models excelling due to effective text encoding.

Ullah et al. [7] found LSTM and CNN superior in sentiment analysis on the Airline Review dataset. Rottegar et al. [8] reported Google Jigsaw's Perspective as the most accurate model on the AAAI dataset. These studies demonstrate that while deep learning models generally outperform classical models due to their ability to handle complex data and extract relevant features automatically, classical models remain valuable for smaller datasets because they require less computational power. The effectiveness of these models varies as per the datasets and languages involved, emphasizing the importance of model selection and dataset characteristics in achieving accurate hate speech detection.

Most of the existing work has applied models of LSTM and CNN due to their ability to capture contextual information [9] and extract relevant features. However, their performance is varied and indicated dependence on data characteristics; It will be worth while to explore performance on classical machine learning models which typically work well on small dataset and demand less data sources.

3 Implementation

As shown in Fig. 1. The flow of the work took place in the following order first the dataset is chosen and the appropriate data instances will be given to data pre-processing where the cleaning, normalizing, and transforming raw textual data before analysis takes place. Data pre-processing also has techniques such as tokenization, lowercasing, and stop word removal to standardize the text and reduce noise. Stemming or lemmatization is used to normalize words to their root form, enhancing analysis accuracy. After the pre-processing of data now feature, engineering is performed to extract the relevant features. The dataset is then split into train data and test data. The appropriate models will be applied to the trained data for sentiment analysis such as classical machine learning model, and deep learning models. In order to improve the performance of the models regularization and hyper-parameter tuning is performed. The test data is applied to the trained model to test its performance; various evaluation metrics such as precision, recall and F-1 score will be calculated for comparative analysis.

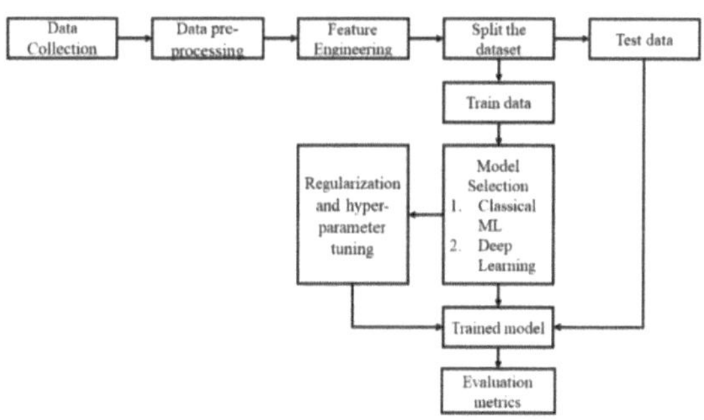

Fig. 1. Flow of proposed work.

3.1 Software Tools

The models were implemented using the Google Colab Pro platform, which features a TPU processor with 664 GB of system RAM and 225.3 GB of disk space. Python was the programming language used for this implementation.

3.2 Dataset Characteristics

The five standard English language datasets have been obtained from open-source websites. The dataset characteristics are listed in the table below (Table 1).

Table 1. Dataset Characteristics

Dataset	Dataset Characteristics	Hate instances of data	Non-hate instances data	Size of Train Data	Size of Test data
HateCheck [11]	Synthetic	2563	1165	2609	1119
HateXplain▲ [13]	Natural (Twitter)	11282	7693	11986	5137
Vico▲ [12]	Natural (Forum)	773	5187	4172	1788
Ethos [14]	Natural (YouTube and Reddit)	433	565	698	300
Multioff [15]	Natural (MEME data)	303	440	520	223

▲ indicates imbalanced datasets.

All the above-mentioned datasets are divided randomly 70% for training the model while the rest 30% is for evaluating the train model.

3.3 Implementation of Models

In this section the implementation of binary classification of sentiments is done using classical machine learning models such as SVM, Logistic Regression, Random Forest, Naïve Bayes and Decision Tree and Deep Learning models such as CNN, LSTM and Bi-LSTM on the standard datasets which were described in the previous section. The implementation of each model on all five datasets is explained in the Sect. 3.2. The datasets are split into 70:30 train and test sets. Feature extraction is performed using the BoW technique for the applying the dataset on the Classical machine learning models and Word2Vec technique is considered as a word embedding technique for applying the datasets on the deep learning models.

Classical ML Models and Hyper-parameter Tuning. The classical machine learning models that are applied on the datasets are Support Vector Machine, Random Forest, Naïve Bayes, Decision Tree and Logistic Regression. Since the models work only on numerical data it is necessary to convert text data into numerical form. Hence, feature engineering is performed using the BoW (Bag of words) techniques before applying the text data to the models. BoW simplifies documents into "bags" of unique words, ignoring order and grammar. It keeps count of the total occurrences of most frequently used words. Due to the simplicity of the model, it ignores valuable information about word order and meaning.

Logistic Regression model. GridSearchCV with 5-Fold cross-validation has been used to obtain the hyper-parameters. The logistic regression model has 15 parameters in total [16] out of which only two parameters 'c' and 'penalty' are considered. These parameters are crucial because text sentiment analysis typically involves high-dimensional data due to the large number of features. 'c' balances bias and variance by penalizing large coefficients, while 'penalty' determines the type of regularization, with L1 promoting sparsity and L2 evenly distributing penalties. The Table 2 shows the comparison of the obtained parameters and its effect on the accuracy.

Table 2. Parameters for Logistic Regression model

Dataset	Default parameters	After tuning	Accuracy before tuning	Accuracy after tuning
HateCheck	C: 1 Penalty: 'l2'	C: 1 Penalty: 'l2'	91.78%	91.77%
HateXplain	C: 1 Penalty: 'l2'	C: 1 Penalty: 'l1'	74.53%	75.42%
Vico	C: 1 Penalty: 'l2'	C: 1 Penalty: 'l1'	88.42%	84.28%
Ethos	C: 1 Penalty: 'l2'	C: 100 Penalty: 'l2'	69.00%	69.34%
Multioff	C: 1 Penalty: 'l2'	C: 1 Penalty: 'l2'	65.92%	61.88%

Naïve Bayes. GridSearchCV instance with 5-fold cross-validation is implemented on Multinomial Naïve Bayes to perform hyper-parameter tuning. Amongst the 4 parameters [17] only 'alpha' and 'fit_prior' are considered. These parameters are vital for effectively managing the probabilistic nature of Naive Bayes and ensuring the model performs well, especially when dealing with the diverse and sparse nature of text data. 'alpha' prevents zero probabilities by adding small values to all counts, while 'fit_prior' allows the model to learn class distribution from data, improving performance in imbalanced datasets. The Table 3 shows the comparison of the obtained parameters and its effect on the accuracy.

Decision Tree. Cost complexity pruning reduces overfitting in decision trees for sentiment analysis by simplifying the tree, which balances model complexity and prediction accuracy [18]. This pruning method improves generalization to new data, ensuring the model captures underlying patterns without overfitting. The regularization parameter (alpha) controls pruning degree, enhancing model adaptability and performance on datasets. The Table 4 shows the comparison of the obtained parameters and its effect on the accuracy.

Support Vector Machine. GridSearchCV with 5-Fold cross-validation has been used to obtain the parameters for the model. Among the total fifteen parameters [19, 20] only 'kernel', 'c' and 'gamma' have been considered. The 'kernel' determines the decision boundary's shape, 'c' controls the trade-off between margin size and classification errors, and 'gamma' influences the complexity of the decision boundary in non-linear

Table 3. Parameters for Naïve Bayes model

Dataset	Default parameters	After tuning	Accuracy before tuning	Accuracy after tuning
HateCheck	alpha = 1 fit_prior = True	alpha = 2 fit_prior = True	87.13%	88.29%
HateXplain	alpha = 1 fit_prior = True	alpha = 1 fit_prior = False	71.47%	73.85%
Vico	alpha = 1 fit_prior = True	alpha = 1 fit_prior = True	88.93%	88.92%
Ethos	alpha = 1 fit_prior = True	alpha = 1 fit_prior = True	69.33%	69.33%
Multioff	alpha = 1 fit_prior = True	alpha = 1 fit_prior = True	68.92%	67.92%

Table 4. Parameters for Decision Tree model

Dataset	Pre-pruning	Post-pruning	Accuracy before tuning	Accuracy after tuning
HateCheck	ccp_alpha = 0	ccp_alpha = 0.000368	85.88%	89.36%
HateXplain	ccp_alpha = 0	ccp_alpha = 0.000232	70.54%	74.62%
Vico	ccp_alpha = 0	ccp_alpha = 0.000887	86.07%	88.87%
Ethos	ccp_alpha = 0	ccp_alpha = 0.00364	62.00%	67.66%
Multioff	ccp_alpha = 0	ccp_alpha = 0.00829	53.81%	68.16%

kernels [24]. These parameters help balance model flexibility and generalization, ensuring effective handling of complex and high-dimensional text data. The Table 5 shows the comparison of the obtained parameters and its effect on the accuracy.

Random Forest. GridSearchCV instance with 5-fold cross-validation is utilized for hyper-parameter tuning. Among the total 19 parameters [21] only 'max_depth', 'min_sample_leaf', 'min_sample_split' and 'n_estimators' have been considered as these are most impactful for achieving a good balance between bias and variance in the model. The Table 6 shows the comparison of the obtained parameters and its effect on the accuracy.

Deep Learning Models. The deep learning models that are applied on the datasets are CNN and RNNs. Since the models work only on numerical data it is necessary to convert text data into numerical form. Hence, feature engineering is performed using the Word2Vec technique. Word2Vec is a used to convert words into continuous vector representations. It is a shallow, two-layer neural networks used to map words onto a high-dimensional space. In Word2Vec framework '1' represents CBOW model and '0' represents Skip-gram model. For feature engineering the CBOW model is applied.

Table 5. Parameters for SVM model

Dataset	Default parameters	After tuning	Accuracy before tuning	Accuracy after tuning
HateCheck	Kernel = linear C = 1	Kernel = rbf C = 10 gamma = scale	94.58%	95.53%
HateXplain	Kernel = linear C = 1	Kernel = rbf C = 1 gamma = scale	72.46%	75.58%
Vico	Kernel = linear C = 1	Kernel = polynomial of degree 1 C = 1 gamma = scale	86.63%	84.28%
Ethos	Kernel = linear C = 1	Kernel = polynomial of degree 1 C = 1 gamma = scale	96.33%	69.33%
Multioff	Kernel = linear C = 1	Kernel = linear C = 1 gamma = scale	61.43%	61.43%

Convolution Neural Network. The architecture for CNN model employed is as follows.

o 1 Embedding layer
o 1 Convolution layer
o 1 Global max pooling layer
o 1 dense layer
o 1 dropout layer
o 1 output layer

The Fig. 2 illustrates the number of neurons present in each layer for all the datasets for CNN model and the Fig. 3 illustrates total number of parameters present for each dataset in CNN model. The HateXplain dataset has the highest number of parameters present. These figures illustrate the model's complexity. The Table 7 represents the parameters that have been obtained after hyper-parameter tuning. Tuning these hyper-parameters [25] in CNNs for text classification allows you to balance model complexity, expressiveness, and generalization ability, ultimately leading to better performance on the classification task.[22].

Long Short-Term Memory. The architecture for LSTM model employed is as follows.

o 1 embedding layer
o 1 dropout layer

Table 6. Parameters for Random Forest model

Dataset	Pre-pruning	Post-pruning	Accuracy before tuning	Accuracy after tuning
HateCheck	max_depth = none min_sample_leaf = 1 min_sample_split = 2 n_estimators = 100	max_depth = 30 min_sample_leaf = 1 min_sample_split = 2 n_estimators = 100	87.67%	88.38%
HateXplain	max_depth = none min_sample_leaf = 1 min_sample_split = 2 n_estimators = 100	max_depth = none min_sample_leaf = 1 min_sample_split = 10 n_estimators = 200	74.41%	75.26%
Vico	max_depth = none min_sample_leaf = 1 min_sample_split = 2 n_estimators = 100	max_depth = none min_sample_leaf = 1 min_sample_split = 10 n_estimators = 200	89.49%	89.46%
Ethos	max_depth = none min_sample_leaf = 1 min_sample_split = 2 n_estimators = 100	max_depth = none min_sample_leaf = 2 min_sample_split = 2 n_estimators = 100	65.67%	66.66%
Multioff	max_depth = none min_sample_leaf = 1 min_sample_split = 2 n_estimators = 100	max_depth = 30 min_sample_leaf = 2 min_sample_split = 5 n_estimators = 200	65.92%	63.67%

o 1 LSTM layer
o 1 output layer

Figure 4 illustrates the number of neurons present in each layer for all the datasets for LSTM model and the Fig. 5 illustrates total number of parameters present for each dataset for LSTM model. The Vico dataset has the highest number of parameters present. These figures illustrate the model's complexity. The Table 8 represents the parameters [26] that have been obtained after hyper-parameter tuning. All the datasets have 'dropout': 0.5, 'embedding_dim': 100, 'hidden_dim': 100 for hyper-parameter tuning only the batch size and epoch has been varied [23].

Bidirectional LSTM. The architecture for BiLSTM model employed is as follows.

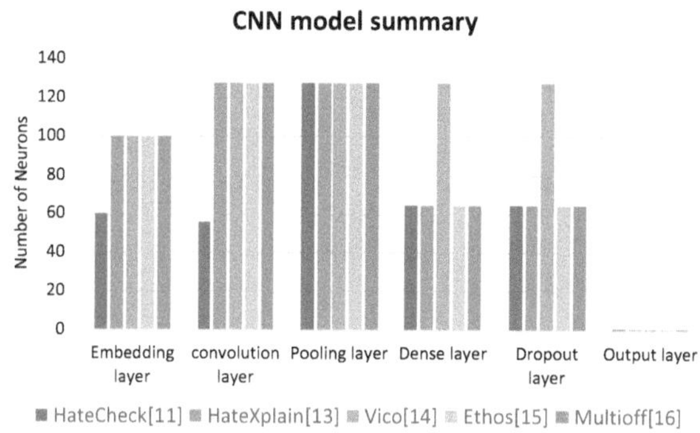

Fig. 2. CNN model summary.

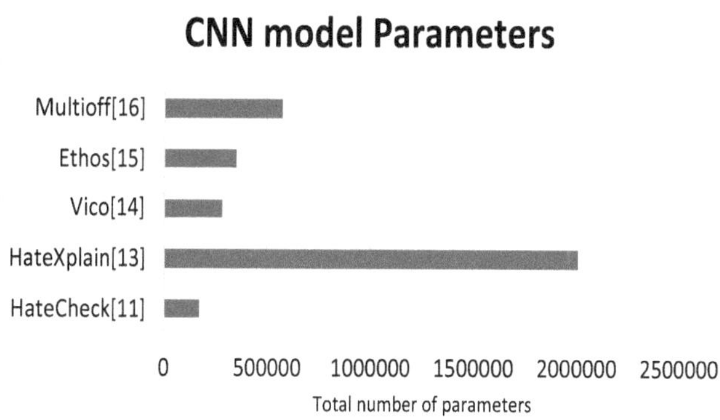

Fig. 3. CNN model parameters.

Table 7. Hyper-parameters for CNN

Dataset	Parameters
HateCheck	'dropout': 0.3, 'embedding_dim': 60, 'hidden_dim': 56, 'num_filters': 64
HateXplain	'dropout': 0.5, 'embedding_dim': 100, 'hidden_dim': 128, 'num_filters': 64
Vico	'dropout': 0.2, 'embedding_dim': 100, 'hidden_dim': 128, 'num_filters': 128
Ethos	'dropout': 0.5, 'embedding_dim': 100, 'hidden_dim': 128, 'num_filters': 128
Multioff	'dropout': 0.5, 'embedding_dim': 100, 'hidden_dim': 128, 'num_filters': 64

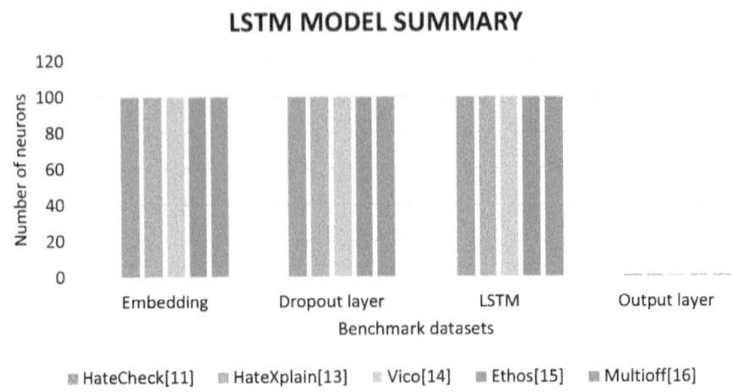

Fig. 4. LSTM model summary.

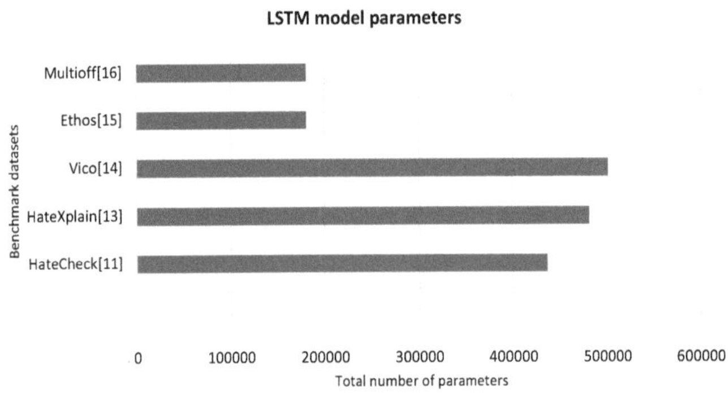

Fig. 5. LSTM model parameters.

Table 8. Hyper-parameters for LSTM

Dataset	Parameters
HateCheck	batch_size = 1024 epochs = 12
HateXplain	batch_size = 1024 epochs = 12
Vico	batch_size = 64 epochs = 12
Ethos	batch_size = 64 epochs = 12
Multioff	batch_size = 64 epochs = 12

o 1 embedding layer
o 2 Bidirectional LSTM layers

o 1 Convolutional layer
o 1 Pooling layer
o 1 dense layer
o 1 output layer

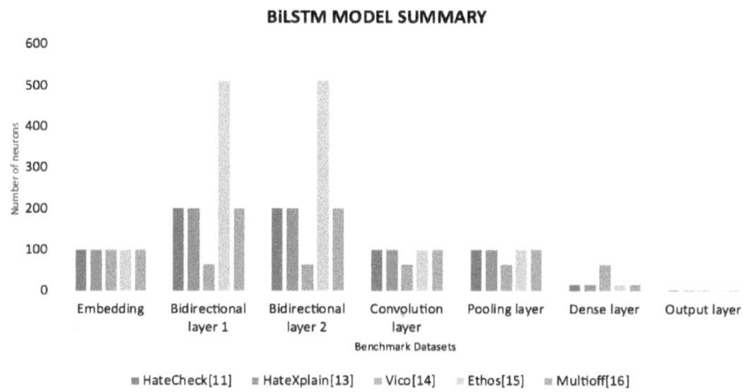

Fig. 6. BiLSTM model summary.

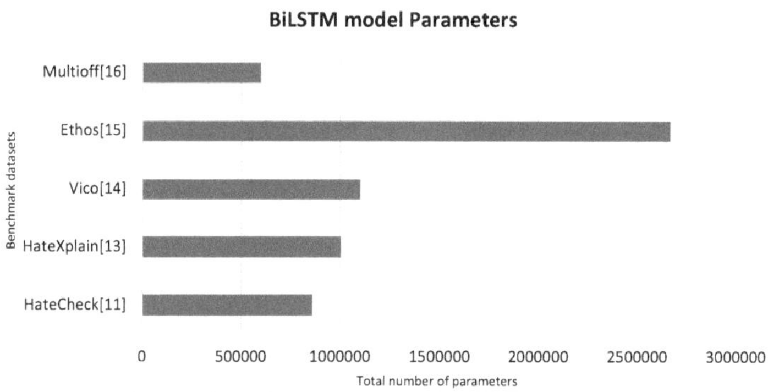

Fig. 7. BiLSTM model parameters.

Figure 6 illustrates the number of neurons present in each layer for all the datasets for BiLSTM model and the Fig. 7 illustrates total number of parameters present for each dataset for BiLSTM model. The Vico dataset has the highest number of parameters present. These figures illustrate the model's complexity. The Table 9 represents the parameters [27] that have been obtained after hyper-parameter tuning.

Thus, in this section the implementation of the classical machine learning models and deep learning models was discussed. The hyper-parameter tuning was performed on the models and its effect on accuracy was observed.

Table 9. Hyper-parameters for BiLSTM

Dataset	Parameters
HateCheck	batch_size = 1024 epochs = 12
HateXplain	batch_size = 1024 epochs = 12
Vico	batch_size = 32 epochs = 10
Ethos	batch_size = 256 epochs = 10
Multioff	batch_size = 32 epochs = 10

4 Results

This section discusses the results obtained for the classical machine learning models and deep learning models by applying them of the five benchmark datasets mentioned in the previous Sect. 3.2

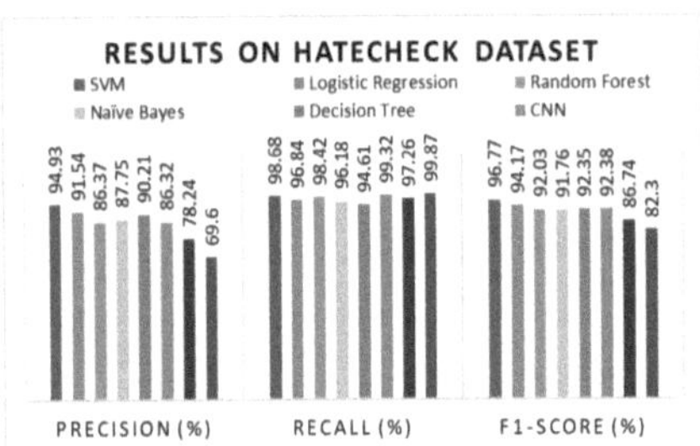

Fig. 8. Results on HateCheck Dataset

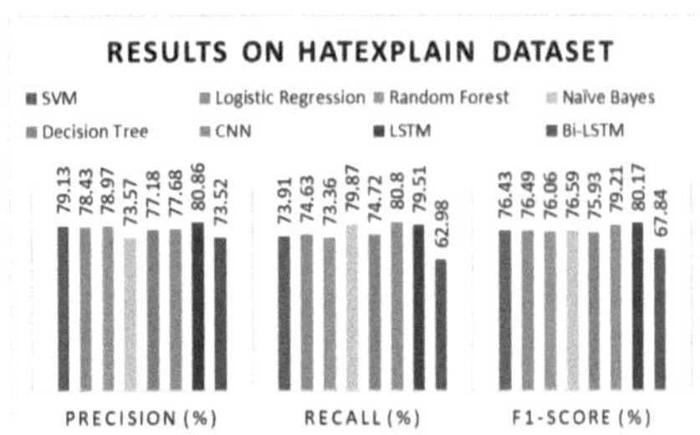

Fig. 9. Results on HateXplain Dataset

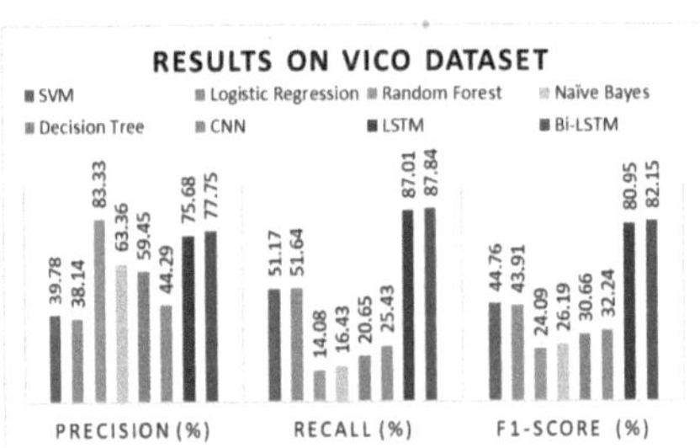

Fig. 10. Results on Vico Dataset

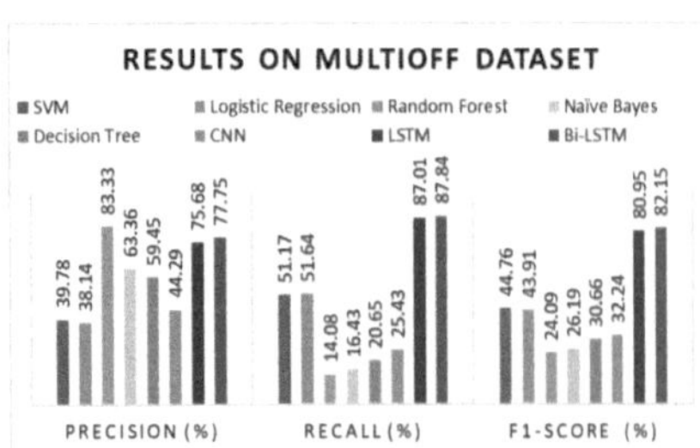

Fig. 11. Results on Multioff Dataset

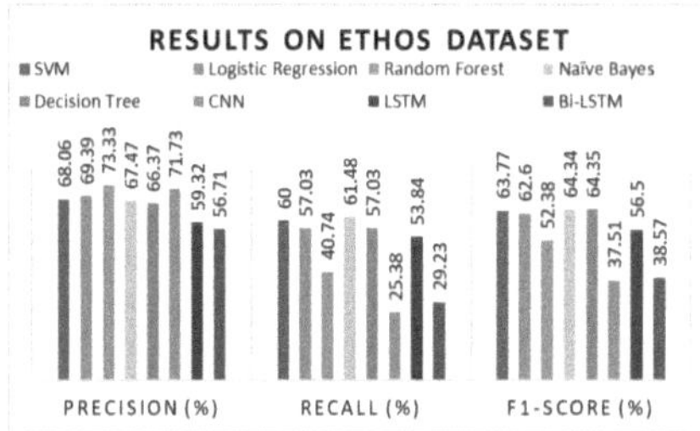

Fig. 12. Results on Ethos Dataset

The Fig. 8 presents the results of HateCheck dataset since it is a balanced dataset the metric that will be taken into account is precision [10] as it will identify the model with most accurate prediction; so it can be noted that the SVM model has obtained the best value of 94.93% this indicates that the positive predictions made by the model are really correct. Since the HateXplain dataset is the imbalanced dataset, we take into account the F1-Score for evaluation of the models this indicates how many instances are actually classified correctly so from Fig. 9. It can be seen that LSTM model has performed well than all the other models with 80.17%. Similarly, the Vico dataset is also an imbalanced dataset and taking into account the F1-Score from Fig. 10. It can be seen that the Bi-LSTM model with 82.15% has outperformed other models. The Multioff and Ethos datasets are balanced datasets so for evaluation we consider precision metric. From Fig. 11. Bi-LSTM model has outperformed with 83.33% and from Fig. 12. The Random Forest model has outperformed with a score of 73.33%.

Thus, it can be seen that for HateCheck dataset the classical ML models have provided better results than deep learning models. The reason being that the dataset is synthetic and has a well-defined feature distribution, clear class boundaries, and balanced class distributions, which simplify the learning task. For other three datasets Vico, Multioff and HateXplain dataset deep learning models have demonstrated their supremacy. The reason being the ability of the deep learning models to capture the context associated with the words in real-life dataset.

5 Conclusion and Future Work

Hate Speech detection was performed by using the binary sentiment classification technique on five standard English language datasets. The classical ML and Deep Learning models were applied on the datasets. Before applying the models on the datasets, it is necessary to convert the text data into numbers by using the feature engineering techniques. For the classical ML models the BoW technique is applied. Similarly for the process of conversion from text to numerical data the Deep learning models make use of the Word2Vec techniques. The dataset is then split into two parts train data and test data. Test data is used to evaluate the performance of the trained models in terms of precision, recall and F1-score.

In the proposed implementation, the SVM model performs well on the HateCheck dataset. The superior results on HateCheck can be attributed to the reason being that the dataset is synthetic and has a well-defined feature distribution, clear class boundaries, and balanced class distributions, which simplify the learning task. For the real datasets, deep learning models particularly those involving RNNs have demonstrated superior performance due to its ability to include contextual knowledge. But since the text data in real world is enormous and scaling Deep learning models to handle such data is computationally complex and expensive. A solution to this can be the Large Language Models (LLMs) as are capable of identifying patterns and relationships within large and complex datasets.

References

1. Chhabra, Vishwakarma, D.K.: A literature survey on multimodal and multilingual automatic hate speech identification. Multimedia Systems **29**(3), 1203–1230, January 2023. https://doi.org/10.1007/s00530-023-01051-8
2. Mahajan, E., Mahajan, H., Kumar, S.: EnsMulHateCyb: multilingual hate speech and cyberbully detection in online social media. Expert Syst. Appl. **236**, 121228 (2024). https://doi.org/10.1016/j.eswa.2023.121228
3. Bangla Hate Speech Detection System using Transformer-Based NLP and deep learning techniques. In: IEEE Conference Publication | IEEE Xplore, 25 August 2023. https://ieeexplore.ieee.org/document/10269919
4. Nagar, S., Barbhuiya, F.A., Dey, K.: Towards more robust hate speech detection: using social context and user data. Soc. Netw. Anal. Min. **13**(1), March 2023. https://doi.org/10.1007/s13278-023-01051-6
5. Alatawi, H.S., Alhothali, A., Moria, K.: Detection of hate speech using BERT and hate speech word embedding with deep model. Appl. Artif. Intell. **37**(1), February 2023. https://doi.org/10.1080/08839514.2023.2166719
6. Ndenga, M.: A deep decision Forests model for hate speech detection. Jordanian J. Comput. Inform. Technol. 1, January 2023. https://doi.org/10.5455/jjcit.71-1667394363
7. Ullah, M.A., Marium, S.M., Begum, S.A., Dipa, N.S.: An algorithm and method for sentiment analysis using the text and emoticon. ICT Express **6**(4), 357–360 (2020). https://doi.org/10.1016/j.icte.2020.07.003
8. Nandwani, P., Verma, R.: A review on sentiment analysis and emotion detection from text. Soc. Netw. Anal. Min. **11**(1), August 2021. https://doi.org/10.1007/s13278-021-00776-6
9. Sharma, S.: Detection of hate speech using deep learning, May 2023. Accessed 12 Feb 2024. http://dspace.dtu.ac.in:8080/jspui/bitstream/repository/19854/1/SHIVAM%20SHARMA%20M.Tech.pdf
10. Marcinrutecki, "Best techniques and metrics for Imbalanced Dataset," Kaggle, 29 December 2022. https://www.kaggle.com/code/marcinrutecki/best-techniques-and-metrics-for-imbalanced-dataset
11. Paul-Rottger: "GitHub - paul-rottger/hatecheck-data: Röttger et al. (ACL 2021): 'HateCheck: Functional Tests for Hate Speech Detection Models' - Data," GitHub. https://github.com/paul-rottger/hatecheck-data
12. Alatawi, H.S., Alhothali, A.M., Moria, K.M.: Detecting white supremacist hate speech using domain specific word embedding with deep learning and BERT. IEEE Access **9**, 106363–106374 (2021). https://doi.org/10.1109/ACCESS.2021.3100435
13. Mathew, B., Saha, P., Yimam, S.M., Biemann, C., Goyal, P., Mukherjee, A.: HateXplain: a benchmark dataset for explainable hate speech detection. Proc. AAAI Conf. Artif. Intell. **35**(17), 14867–14875 (2021). https://doi.org/10.1609/aaai.v35i17.17745

14. Mollas, I., Chrysopoulou, Z., Karlos, S., Tsoumakas, G.: ETHOS: a multi-label hate speech detection dataset. Comp. Intell. Syst. **8**(6), 4663–4678 (2022). https://doi.org/10.1007/s40747-021-00608-2
15. Suryawanshi, S.: Multimodal meme Dataset (MultiOFF) for identifying offensive content in image and text. ACL Anthol., 01 May 2020. https://aclanthology.org/2020.trac-1.6/
16. "LogisticRegression," Scikit-learn. https://scikitlearn.org/stable/modules/generated/sklearn.linear_model.LogisticRegression.html
17. "MultinomialNB," *Scikit-learn.* https://scikit-learn.org/stable/modules/generated/sklearn.naive_bayes.MultinomialNB.html
18. Salzberg, S.L.: C4.5: programs for machine learning by J. Ross Quinlan. Morgan Kaufmann Publishers, Inc., (1993). Mach. Learn. **16**, 235–240 (1994). https://doi.org/10.1007/BF00993309
19. "SVC," Scikit-learn. https://scikitlearn.org/stable/modules/generated/sklearn.svm.SVC.html
20. Hsu, C.-W., Chang, C.-C., Lin, C.-J.: National Taiwan University, "A practical guide to support vector classification," May 2016. https://www.csie.ntu.edu.tw/~cjlin/papers/guide/guide.pdf
21. "RandomForestClassifier," Scikit-learn. https://scikit-learn.org/stable/modules/generated/sklearn.ensemble.RandomForestClassifier.html
22. Choubey, V.: "Text classification using CNN - Voice Tech Podcast - Medium," Medium 27 November 2023. https://medium.com/voice-tech-podcast/text-classification-using-cnn-9ade8155dfb9
23. Caiyutiansg, "Twitter Sentiment Analysis with Word2Vec LSTM," 05 November 2019. https://www.kaggle.com/code/caiyutiansg/twitter-sentiment-analysis-with-word2vec-lstm
24. "Support Vector Machine-Theory," 30 March 2020. https://www.datasciencelovers.com/machine-learning/support-vector-machine-theory/
25. Kim, Y.: Convolutional neural networks for sentence classification, arXiv.org, 25 August 2014. https://arxiv.org/abs/1408.5882
26. Wikipedia contributors, "Long short-term memory," 08 June 2024. https://en.wikipedia.org/wiki/Long_short-term_memory#/media/File:LSTM_Cell.svg
27. GeeksforGeeks, "Bidirectional LSTM in NLP," *GeeksforGeeks*, 08 June 2023. https://www.geeksforgeeks.org/bidirectional-lstm-in-nlp/

Enhanced Face Emotion Detection Using Confidence-Based Sample Filtering

K. Vipul Arya, Karthik KV, Billa Muralidhar(✉), and Sangamesh Ramesh

School of Computer Science Engineering (SOCSE), RV University, RV Vidyanikethan Post 8th Mile, Mysuru Road, Bengaluru 560059, India
karthikmmdh@gmail.com

Abstract. This research presents a method for improving the detection of emotions by utilizing confidence-based sample filtering and a carefully crafted convolutional neural network (CNN) structure. By combining data from datasets like FER2013, AffectNet, and CK + and standardizing them through preprocessing this study creates three datasets; complete data, high-confidence data, and low-confidence data. Each dataset is trained on a customized CNN model achieving accuracy rates of 73.3 percent, 76.3 percent, and 71.3 percent for the dataset high-confidence dataset and low-confidence dataset respectively. The CNN architecture includes pooling layers with dropout regularization to enhance generalization. The practical application of real-time emotion detection using webcam input is demonstrated in this study. This research emphasizes the importance of dataset filtering and model design in advancing facial emotion recognition systems for real-world applications.

Keywords: Convolutional Neural Network · t-SNE · Object Recognition

1 Introduction

Convolutional Neural Networks, also known as CNNs have transformed the field of computer vision by automating extracting features from images. This advancement has made tasks, like identifying objects and classifying images. Even detecting emotions is more accessible. CNNs, designed to reflect the hierarchical nature of cortex structure, allow progressively more complex feature extraction and are thus particularly sensitive to detection. Early layers extract low-level features like edges and textures, necessary for detecting facial landmarks such as eyes, mouth, and eyebrows. Building on top of those layers, these recognize patterns and structures such as the shapes of facial expressions, while the deeper layers catch high-level abstract representations of overall emotion conveyed by these expressions. This hierarchical processing allows the CNN to analyze complex relationships within facial images, fitting it well to tasks requiring nuanced understandings, like emotion detection. Beyond image classification, CNNs are applied in tasks such as recognizing landmarks and interpreting emotions based on subtle changes in facial expressions [1].

A. Das et al. (Eds.): MIDAS 2025, CCIS 3055, pp. 18–28, 2027.
https://doi.org/10.1007/978-3-032-35516-4_2

Face emotion detection is a strong suit for CNNs as they can detect spatial patterns and learn hierarchical representations from raw pixel data [2]. Emotional states are encoded in facial expressions which involve changes in eyebrow position, mouth shape, and eye openness. These subtle differences are well-analyzed by CNNs making them perfect for any activities that will specifically require accurate identification of emotions. Through numerous convolutions and pooling layers for multiple facial images, the CNNs produce meaningful abstraction that helps in classifying emotions such as happiness, sadness, anger, or surprise correctly.

This study presents a fresh technique for enriching face emotion detection through sample filtering with confidence. The approach aims at enhancing model strength and accuracy by splitting data into subsets using simulated scoring mechanism-derived confidence grades. The models receive priority for training on high-confidence samples because it is hypothesized that these samples contain more lucid and well-structured facial emotions. On the other hand, low-confidence samples are added to the model to determine its flexibility towards such variations, which might be characterized by ambiguous or noisy data. This methodological novelty strives to optimize model performance by targeting reliable subsets of data during training, thereby making the whole system more efficient [3].

There are many uses of facial emotion recognition in real life where accuracy is vital. In human-computer interaction, systems that can recognize and respond to users' emotions will offer a personalized user experience by adjusting the responses accordingly. Also, surveillance applications can be used with emotion recognition to identify people in danger or detect unusual behavior. Detecting emotions correctly helps healthcare practitioners evaluate mental health conditions and keep track of their patients' mental well-being as well as provide necessary emotional support. These examples demonstrate the importance of robust emotion recognition systems for decision-making, interaction quality, and societal well-being [4].

The following are the contents of this document. Initially, it explores some approaches to improving facial emotion detection through innovation. Maintaining consistency and quality across different sources involves the integration of several datasets—FER2013, AffectNet as well as CK + and their uniform preprocessing. The methodology starts with an explanation of how confidence-based sample filtering works by categorizing data subsets based on confidence scores for optimum model training. It outlines a customized CNN model that has been designed specifically for emotion recognition which highlights its ability to handle variegated complexities in data as well as noise. In conclusion, this paper delves deep into the experimental results and what they imply about future practical applications of facial expression recognition systems. This study [5] is aimed at contributing to the current growth of robust emotional recognition systems by focusing on the significance of data purging strategies and enhancements in the design of these systems that will make them work efficiently anywhere it's installed.

2 Related Work

Significant progress has been made in recognizing emotions from facial expressions with the arrival of Convolutional Neural Networks (CNNs). Their abilities to extract detailed spatial characteristics from images have made CNNs especially appropriate for tasks connected with emotion detection through human faces [6]. These networks specialize in learning hierarchies of facial attributes, thereby improving models that classify emotions.

There have been several studies on this topic, investigating different ways of improving CNN-based emotion recognition systems. Wang and his colleagues [7] highlighted the importance of feature selection and data augmentation techniques in making CNNs more discriminative for facial emotion recognition purposes. According to their research, preprocessing steps such as refining input data are critical to enhancing the efficiency of feature extraction.

Ensemble learning approaches have also gained traction in enhancing the robustness and generalization of emotion recognition models. Li et al. proposed an ensemble framework integrating multiple CNN models trained on different subsets of data to achieve improved classification accuracy [8]. This approach uses the diversity of individual models to reduce overfitting and improve overall performance.

Transfer learning is another important technique in improving emotion recognition systems. Instead of training a model from scratch, transfer learning involves using pre-trained models that have already learned features from large datasets. These models are then fine-tuned on specific emotion recognition datasets to improve their performance. A study by Liu et al. showed that using transfer learning with pre-trained convolutional neural network (CNN) models can help speed up the training process and achieve higher prediction accuracy [9]. By leveraging knowledge from larger, more general datasets and adapting it to smaller, more specialized emotion datasets, transfer learning enables models to learn faster and perform better in emotion recognition tasks.

Addressing data imbalance is crucial for achieving balanced performance in emotion recognition tasks. One approach proposed by Zhu et al. Is to use a weighted loss function during CNN training [10]. This technique assigns higher weights to underrepresented emotion classes, allowing the model to prioritize learning from these minority samples. Doing so helps to mitigate the bias towards dominant emotions and improve the overall classification accuracy.

Multimodal approaches for better accuracy are another effective strategy for improving emotion recognition accuracy through the use of multimodal approaches. These methods combine facial expressions with other modalities such as voice and gesture, leveraging multiple sources of information to make more accurate predictions. A study conducted by Yu et al. demonstrated the effectiveness of integrating facial features with audio signals in a robust multimodal framework for emotion recognition [11]. Their approach enhances the reliability and robustness of emotion detection systems, making them more suitable for diverse realities.

Understanding how CNNs make decisions in emotion recognition is important for making models more transparent and trustworthy. Kumar et al. used visualization techniques to analyze the features learned by CNNs, giving us insights into how these models understand and categorize emotions based on facial expressions [12]. This interpretability helps us improve model designs and training methods.

Attention mechanisms have been added to CNN architectures to better focus on relevant parts of the face during emotion recognition tasks. Kim et al. introduced an attention-based CNN that dynamically adjusts the importance of different areas based on their relevance to emotion classification [13]. This attention mechanism improves the model's ability to capture subtle emotional cues, leading to better accuracy in classification.

Context-aware CNNs have also been studied to improve understanding of time and continuous emotion recognition. Jiang et al. created a context-aware CNN that takes into account information from neighboring frames or scenes to better understand how emotions change over time [14]. By considering the sequence of frames and contextual information, their model performs better in recognizing continuous emotional states.

Efforts to deploy lightweight CNN architectures for real-time emotion recognition on resource-constrained devices have also been investigated. Guo et al. proposed a lightweight CNN architecture optimized for efficient computation while maintaining competitive accuracy in facial emotion recognition tasks [15]. Such architectures are crucial for deploying emotion recognition systems in embedded devices and mobile applications.

3 Methodology

Figure 1 provides a visual overview of our methodology for face emotion detection, illustrating the key steps from data collection to model evaluation. It outlines the preprocessing pipeline, the categorization of data based on confidence levels, and the training of CNN models to analyze the impact of these levels on performance.

We utilized three publicly available datasets: FER2013, AffectNet, and CK+, chosen for their diverse and complementary characteristics. FER2013 is a large-scale dataset with labeled facial expressions in real-world settings, making it suitable for training robust models. AffectNet includes a wide variety of annotated facial expressions with subtle and complex emotions, providing a richer dataset for nuanced emotion detection. CK+ offers high-quality, posed expressions often used as a benchmark for facial emotion detection systems, ensuring a reliable performance evaluation.

The images from these datasets were preprocessed uniformly to ensure consistency. Preprocessing consisted of resizing the images to 48 x 48 pixels to reduce dimensionality and computational complexity while retaining essential features. Images were converted to grayscale to simplify the data from three color channels into one, as facial expressions can be effectively analyzed in grayscale. Pixel values were normalized to a range, such as [0, 1], to improve consistency and facilitate efficient model training. These preprocessing steps reduced resource requirements while maintaining critical features for accurate emotion detection.

Post-preprocessing, the data was filtered into three categories based on confidence levels: full preprocessed data, high-confidence data, and low-confidence data. Confidence levels were determined using a simulated confidence scoring mechanism. This mechanism assigned confidence scores based on a combination of model prediction probabilities and data quality metrics, such as image clarity and label reliability. For each confidence category, a separate Convolutional Neural Network (CNN) model was trained. This approach enabled us to systematically analyze the impact of varying confidence levels on model performance and demonstrated the benefits of training with high-confidence data.

The models were built using a Sequential model in Keras with several layers, including Conv2D, MaxPooling2D, Dropout, Flatten, and Dense layers. The architecture was designed to efficiently extract features from the images and perform accurate classification into seven emotion categories: happy, neutral, sad, fearful, angry, surprised, and disgusted. We employed Adam optimizer and categorical cross-entropy loss function for training the models.

4 Implementation and Results

4.1 Data Gathering

We combined images from FER2013, AffectNet, and CK+ datasets to create a comprehensive dataset for training and testing our emotion detection models. These datasets collectively provided a diverse range of facial expressions and emotions.

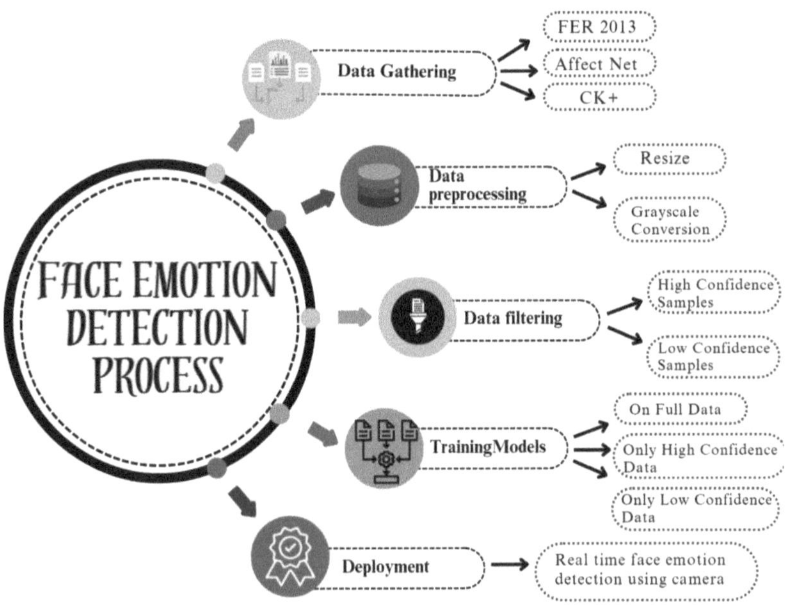

Fig. 1. Methodology of our Work.

4.2 Data Pre-processing

Each image was resized to 48x48 pixels and converted to grayscale to standardize the inputs and reduce computational overhead. This preprocessing step ensured that all images were uniform, facilitating efficient feature extraction during model training.

4.3 Data Filtering

The preprocessed data was divided into three categories based on confidence levels: full data, high-confidence data, and low-confidence data. Confidence scores were simulated, and thresholds were set to segregate the data. This filtering aimed to analyze how varying levels of confidence impact model accuracy.

Figure 2 shows the class distribution for the full dataset, high-confidence data, and low-confidence data. For the full dataset, the 'happy' class has the highest number of samples (8302), followed by 'neutral' (6242), 'sad' (6054), 'fear' (5169), 'angry' (5153), 'surprise' (4484), and 'disgust' (1714). The high-confidence dataset contains fewer samples, with the 'happy' class having (4204) samples, 'neutral' (3180), 'sad' (2989), 'angry' (2577), 'fear' (2535), 'surprise' (2237), and 'disgust' (840). The low-confidence dataset also has a reduced number of samples, with the 'happy' class at (4098), 'sad' at (3065), 'neutral' at (3062), 'fear' at (2634), 'angry' at (2576), 'surprise' at (2247), and 'disgust' at (874). This distribution indicates a varying balance of emotions across different confidence levels, influencing the model's performance on each dataset.

	Full Data	High Confidence Data	Low Confidence Data
Happy	8302	4204	4098
Neutral	6242	3180	3065
Sad	6054	2989	3062
Fear	5169	2577	2634
Angry	5153	2535	2576
Surprise	4484	2237	2247
Disgust	1714	840	874
Total	37118	18562	18556

Fig. 2. Data distribution across the different models.

Figure 3 Three t-SNE plots were generated to analyze the distribution of data points within the "angry" class dataset. The first plot represents the full dataset, illustrating the overall distribution of "angry" emotions. In the second plot, low-confidence samples are highlighted in blue against a grey background, indicating instances where the model's prediction confidence is lower. Conversely, the third plot highlights high-confidence samples in red against the same grey background, showcasing instances where the model's prediction confidence is higher. These visualizations provide insights into the clustering patterns and distribution of confidence levels within the "angry" class, revealing potential variations in data quality and model performance across different subsets of the dataset.

4.4 Training Models

Three separate CNN models were trained on each dataset category. The architecture comprised multiple convolutional layers with ReLU activation, followed by max-pooling layers, dropout layers to prevent overfitting, and dense layers for classification. the models were trained using the adam optimizer and categorical cross-entropy loss function, with accuracy as the performance metric.

Figure 4 illustrates the streamlined architecture of the convolutional neural network (CNN) employed in this study for facial emotion recognition. The model begins with an input layer where grayscale images of facial expressions are fed into a series of convolutional layers (Conv2D) responsible for extracting spatial features. Following each convolutional layer, max-pooling layers (MaxPool) reduce the spatial dimensions of the feature maps, enhancing computational efficiency and extracting dominant features. Dropout layers (Dropout) are strategically placed to mitigate overfitting by randomly deactivating a fraction of neurons during training. The flattened layer (Flatten) reshapes the pooled feature maps into a single vector, which is then processed by dense layers (Dense) to classify emotions. Each dense layer is augmented with dropout regularization to further prevent overfitting. The output layer (Output) employs a softmax activation function to predict the probability distribution across the seven emotion classes (happy, neutral, sad, angry, fear, surprise, disgust). This architecture optimizes the balance between model complexity and performance, crucial for real-time facial emotion detection applications.

4.5 Deployment

For real-time emotion detection, we developed a system using OpenCV to capture video frames and detect faces. The pre-trained model was loaded to predict the emotion of each detected face. The system displayed the detected emotion on the video feed, demonstrating the practical application of our trained models.

4.6 Results

Table 1 details the results of the models trained. The model trained on the full dataset achieved a training accuracy of 73.3\% and a validation accuracy of 67.7\%. The high-confidence data model showed improved performance with a training accuracy of 76.3\% and a validation accuracy of 70.7\%. Conversely, the low-confidence data model had a training accuracy of 71.3\% and a validation accuracy of 66.7\%. The class distribution for each category provided insights into the data balance across emotions, highlighting the variations in model performance.

Table 2 provides a comparison of existing methods applied to facial expression datasets including AffectNet, CK+, and FER 2013. It shows the test accuracy of different methods, offering insights into how these methods perform relative to each other on these datasets.

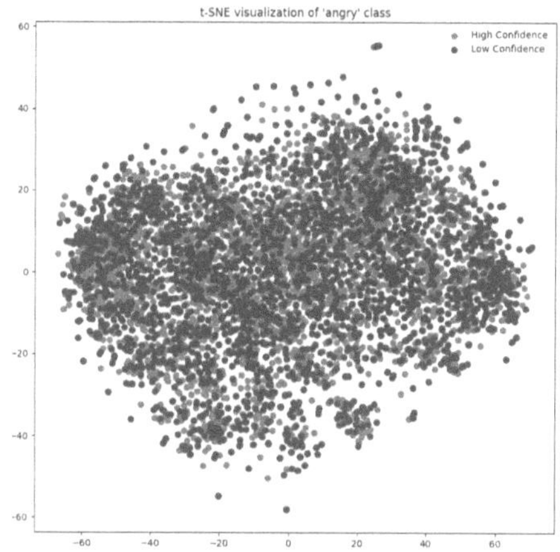

(a)

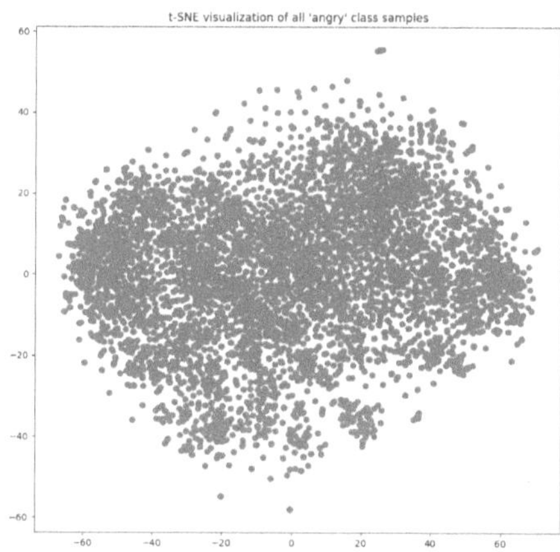

(b)

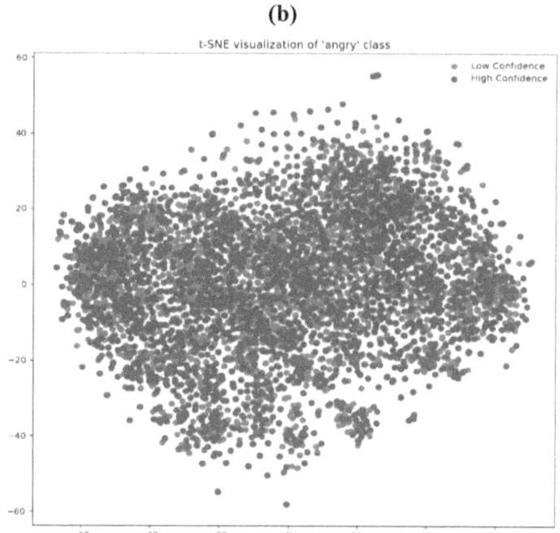

(c)

Fig. 3. t-SNE Plots of the training data

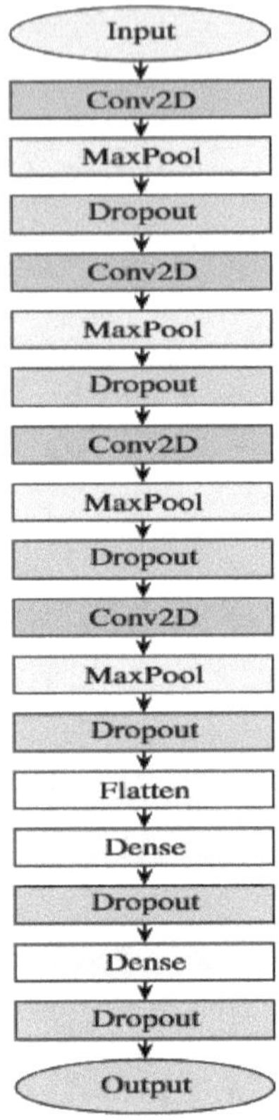

Fig. 4. Architecture of the Used CNN

Table 1. Accuracies of our Trained Models.

Model wrt Data	Training Accuracy	Validation Accuracy
Full Data	73.3%	67.7%
High Confidence Data	76.3%	70,7%
Low Confidence Data	71.3%	66.7%

Table 2. Comparison of methods on AffectNet, CK +, and FER 2013 datasets.

Dataset	Model	Validation Accuracy
AffectNet	CNN	58%
	EffNets	65%
CK+	SVM	79.7%
	OpenCV	89.71%
FER 2013	SVM	67.31%
	CNN	75.63%

5 Conclusion

Our approach for confidence-based sample filtering in face emotion detection seems promising to improve the accuracy of the model by focusing on high-confidence data. The results show that models trained on high-confidence data performed better than those trained on full or low-confidence data, indicating that filtering based on confidence can be an effective strategy for emotion detection tasks. However, the study has some limitations. The datasets used may not be diverse in terms of age, ethnicity, lighting conditions, and other real-world variables that could affect the generalizability of the model. The confidence-scoring mechanism could also be further refined to account for nuanced expressions and borderline cases that may carry valuable contextual information. Future work may be related to exploring better ways of computing the confidence scores based on the sophisticated mechanisms available that include techniques, such as the uncertainty estimation for Monte Carlo Dropout or Bayesian Neural Networks. Even more complex could be applying an attention mechanism or gradient-based methods to locate the most contributing parts of the face for improving filters. Incorporating ensemble methods to combine predictions from multiple models can also be used to increase confidence assessment as biases inherent in a single model are reduced. Further, it could include a more diverse dataset and evaluate the robustness of the model under real-world deployment scenarios, such as occlusions, dynamic expressions, and environmental noise, that pose great challenges.

Acknowledgments. This project did not receive any specific funding.

Disclosure of Interests. The authors have no competing interests to declare that are relevant to the content of this article.

References

1. Li, Y., Zhao, C., Yang, H.: A survey on emotion recognition using physiological signals. Inform. Fus. **59**, 1–16 (2020)
2. Peng, Z., Wang, Y., Xia, C.: A survey on multimodal emotion recognition: methods, datasets, and future directions. Inform. Fus. **75**, 84–104 (2021)

3. Liu, Y., Zhang, D., Luo, J.: Spatial and temporal deep network for facial expression recognition. IEEE Trans. Image Process. **29**(10), 4806–4819 (2020)
4. Wang, P., Lin, H., Lin, C.: Deep convolutional neural networks for facial expression recognition: a comprehensive review. Neurocomputing **324**, 28–48 (2019)
5. Yu, S., Huang, L., Yang, M.H.: Facial expression recognition: a survey. Inform. Fus. **60**, 245–275 (2020)
6. Zhang, Y., Diao, Q., Zhang, L.: Domain adaptation in emotion recognition: an unsupervised approach. IEEE Trans. Affect. Comput. **9**(3), 289–301 (2018)
7. Wang, H., Li, X., Zhang, Y.: Feature selection and augmentation for improved facial emotion recognition using convolutional neural networks. IEEE Access **8**, 124565–124577 (2020)
8. Li, S., Chen, Y., Zhang, Z.: Ensemble deep learning framework for facial emotion recognition. IEEE Trans. Multimedia **21**(5), 1221–1233 (2019)
9. Liu, J., Wang, S., Zhang, J.: Transfer learning for emotion recognition: a case study on fine-tuning pre-trained CNN models. IEEE Trans. Cybern. **51**(2), 579–591 (2021)
10. Zhu, X., Liu, W., Zheng, G.: Handling imbalanced data in facial emotion recognition using weighted loss functions. IEEE Trans. Affect. Comput. **8**(4), 457–469 (2017)
11. Yu, H., Wang, X., Li, S.: Multimodal fusion for emotion recognition: integrating facial expressions and audio signals. IEEE Trans. Affect. Comput. **10**(1), 92–105 (2019)
12. Kumar, A., Gupta, R., Agarwal, S.: Visualizing deep features for interpretability in facial emotion recognition. IEEE Access **8**, 132378–132391 (2020)
13. Kim, J., Kim, M., Kim, C.: Attention-based convolutional neural networks for facial emotion recognition. IEEE Trans. Cybern. **48**(12), 3635–3648 (2018)
14. Jiang, Y., Wang, Y., Wang, X.: Context-aware convolutional neural networks for continuous emotional state recognition. IEEE Trans. Affect. Comput. **10**(4), 552–565 (2019)
15. Guo, W., Li, C., Zhang, L.: Lightweight CNN architecture for real-time facial emotion recognition on resource-constrained devices. IEEE Trans. Circuits Syst. Video Technol. **31**(1), 147–160 (2021)
16. Mollahosseini, A., Hassani, B., Mahoor, M.H.: AffectNet: A database for facial expression, valence, and arousal computing in the wild. *arXiv preprint* arXiv:1708.03985 (2017)

Lung Infection Detection Using VGG19 Model for Smart Healthcare

Mihir Narayan Mohanty(✉)

Applied Signal Processing Laboratory, ITER, Siksha 'O' Anusandhan (Deemed to Be University), Bhubaneswar, Odisha, India
mihirmohanty@soa.ac.in

Abstract. Lung infection may cause the diseases related to this particular organ including fever, tumor, Tuberculosis and cancer. The work presents a novel optimized model for detection of lung infection from CT images, aiming to provide timely and accessible healthcare solutions. IoT devices capture lung CT images from the patients. VGG-19 model, a category of deep learning based neural network, for lung infection detection through transfer learning is utilized in this work. The model is further optimized for deployment on scalable platforms, including IoT devices. The uploaded images are processed using pre-trained VGG19 CNN models to detect and localize areas of infection. It is the advance CNN model with pre trained layers that analyses the lung images in terms of shape, color and structure. Further Transfer learning is utilized in this model to reduce the computational cost for new problems. For faster conversion during training, Adam optimizer is used. Results based on the web application interface, allowing healthcare professionals to visualize and interpret the findings. The proposed results show the performance of accuracy is approximately 98.33% and verified for different lung diseases to prove its robustness.

Keywords: Lung infection · Lung CT · Machine Learning · VGG Net · IoMT

1 Introduction

Lung infections, such as pneumonia and tuberculosis, often require timely identification and intervention to avoid complications and spread of the disease. Similarly, lung cancer has been proven to cause the increment of mortality rate. The CT and X-ray scans of the chest have proposed in COVID 19 detection during the recent pandemic [1–4]. The tiredness, yawn, breathing have been estimated in [5]. Infection of lungs may cause influenza in primary stage of the subject. In this regard works have been proposed using machine learning approach that have been used for multi-agent system of e-health cares. The vaccine is an alternate solution and safe for disease prevention, however, its limited efficacy in the elderly underscores the need to explore more effective alternatives. Several factors contribute to the reduced efficacy of these drugs in certain individuals, including age, prior influenza exposure, and chronic age-related diseases. Common symptoms of influenza include cough, fever, fatigue, nasal congestion, headache, and sore throat.

A. Das et al. (Eds.): MIDAS 2025, CCIS 3055, pp. 29–39, 2027.
https://doi.org/10.1007/978-3-032-35516-4_3

Among these, cough is a critical indicator of respiratory health and disease progression. Self-reported tools like Visual Analog Scales (VAS), cough scores, and quality-of-life surveys are widely used to assess cough severity. However, these methods often suffer from imprecision due to their subjective nature and dependence on patient input.

In comparison, cough-measuring devices that objectively record data offer greater accuracy and user-friendliness. While expert evaluations using audio and video monitoring provide reliable results, they are typically cost-prohibitive and less practical.

The large-scale deployment of experts presents additional challenges. Consequently, programmed cough detection mechanisms utilizing pattern recognition networks have emerged as a significant advancement in modern healthcare systems, offering valuable insights into other influenza symptoms as well. Artificial intelligence (AI) has demonstrated exceptional success in accurately modelling biological systems by analysing various variables and symptoms associated with specific diseases and their severity. Since these factors often exhibit non-linear dynamics, artificial neural networks serve as an effective tool for modelling and predicting such complexities.

Artificial Neural Network (ANN) models have proved as success techniques across various domains, including signal, image processing in medical domain for disease diagnosis effectively. Their ability to model non-linear relationships among hidden data representing potential symptoms makes them a powerful tool. Key advantages such as rapid convergence, simplicity in algorithms, cost-effectiveness, and flexibility in handling large datasets position ANNs as superior to conventional modeling techniques. Among neural networks, the Multilayer Perceptron Network (MLP) stands out as the simplest and most widely used method for pattern recognition. As a result, the Multilayer Perceptron Network (MLP) was selected as the foundational component for our cascaded network in this study. A Cascade Correlation Neural Network (CCNN) one of the integrated models with cascaded framework has been used as the variant of neural network model. This model has shown significant effectiveness in modeling symptoms of oral cancer and similar conditions.

Sensors have been used through hardware devices for detection cough symptoms linked to influenza. Supervised Multi-Layer Feed Forward Neural Network (SMFFNN) as a learning model also another variant of neural network model has been applied as intelligent method of classification. They emphasized the system's effectiveness in contextual analysis, clinical associations, and related applications, underscoring the value of neural networks in advancing healthcare technologies. Similarly, Hybrid Higher Order Neural Classifier (HHONC) has been utilized using various datasets, including those related to breast tumour identification, Iris classification, Glass identification, Wine recognition, and Balance scale analysis. These applications underscore the relevance and adaptability of neural network analysis in this field. There is a need for automated systems that can accurately detect and classify lung infections in CT images, enabling timely intervention and treatment planning. Internet technology, utilizing the Internet's power to gather, analyze, and share immense volumes of data. This data is swiftly transformed into valuable information and knowledge in real-time settings. It has become a cutting-edge technology, effortlessly integrating many technologies from diverse domains into a unified system. The technology enables real-time analysis and continuous data collection; this technology is especially crucial for the identification of lung cancer because it may result in an earlier diagnosis of the illness. Sophisticated algorithms can scan image data and identify abnormalities that can point to lung nodules and possibly early cancer detection. Additionally, IoMT can help with the management of massive amounts of data from multiple sources, like patient monitors and imaging devices, which is necessary for precise diagnosis. The precision and effectiveness of lung cancer prediction models have improved with the incorporating deep learning methodologies with IoMT devices, a development that has shown encouraging results [6].

A machine learning model deployed on the cloud processes the images, identifying regions indicative of lung infections. These findings are then accessible to users through a web-based interface, allowing healthcare professionals to remotely review and interpret them. Optimized VGG-19, a deep learning model, undergoes training for classifying lung CT images into classifications with and without infection. The training utilizes a labeled dataset of CT images annotated to indicate the presence of lung infections. Transfer learning techniques can be utilized pre-trained models to the specific task of detecting lung infections, thereby mitigating the requirement for extensive annotated datasets. A novel two-phase deep diagnostic approach was suggested for COVID-19 infection investigation and the identification of minute anomalies.

One biomedical imaging modality used to identify infected lung regions to validate COVID-19 patients through computed tomography (CT). Recent studies have explored the transfer learning, particularly using VGG-19 model, for COVID-19 and pneumonia from medical images. A study published in 2023 investigated the use of VGG-19 and VGG-16 architectures for automatic detection of COVID-19 in chest X-ray images. By fine-tuning the classifier block and hyperparameters, the models were adapted for binary and multiclass classification tasks. The experimental results demonstrated that transfer learning is beneficial for small-sized datasets in medical imaging, preventing overfitting and achieving optimal classification performance. Another study focused on classification from lung CT scans using the VGG-19 model. The findings indicated that deep transfer learning with VGG-19 can effectively detect COVID-19 and pneumonia from chest X-ray images, highlighting the model's potential in medical image classification tasks. Researchers applied transfer learning with VGG-19 for image classification tasks, including chest X-ray images. The experiments achieved promising results, suggesting that VGG-19, when fine-tuned appropriately, can serve as a reliable model for detecting lung infections in medical imaging. It is found that the use of transfer learning techniques, including VGG-19, for detecting COVID-19 from CT lung scans. The research demonstrated that transfer learning models could effectively identify COVID-19 infections, emphasizing the importance of model selection and fine-tuning in achieving high classification performance. Similarly, ensemble approach has been proposed combining multiple transfer learning models, including VGG-19, for detecting COVID-19 infections through blood tests. The ensemble model outperformed individual models, achieving higher accuracy and robustness in infection detection. A comprehensive comparison of various deep transfer learning models, including VGG-19, was conducted wiyh the radiographic lung for COVID-19, non-COVID pneumonia, and normal categories. The study provided insights into the performance differences among models, aiding in selecting the most suitable architecture for specific medical imaging tasks. A substantial amount of study has focused on categorizing COVID-19 by utilizing several forms of imagery, including CT scans, X-rays, and ultrasounds. Nevertheless, none of these techniques can adequately manage all of various picture formats within a cohesive framework to precisely ascertain if an individual is infected with COVID-19 or not [7–12].

Deep Learning (DL) techniques, namely utilizing Computed Tomography (CT) images, have been suggested to aid and simplify this analysis in early stages of the disease.

This work is to design a model that:

1. Detects lung infections from medical images.
2. Leverages pre-trained deep learning models through transfer learning.
3. Provides accurate and interpretable results for healthcare professionals.
4. Optimize the model for deployment on IoT devices with limited computational resources.
5. Enable real-time diagnostic capabilities in remote or resource-constrained areas.

Transfer learning involves leveraging several models, like Inception-ResNet-v2, VGG-19, and ResNet-50, to extract features. The softmax function of diverse base classifiers contributes to individual predictions, while an ensemble technique acts as a meta-classifier to attain optimal final outcomes.

An IoT module is presented along with the imaging device allows for real-time information exchange, diagnosis, as well as data retrieval by avoiding data loss problems. The important points of this work are,

(1) An IoT based architecture with e-Health web application has been developed for classification of Lung CT images that is both low-cost and scalable.
(2) The VGG19 model is chosen for classification and is trained for the lung CT image classification via transfer learning has been proposed and hosted in a cloud-based server for online access.

The paper is of four sections. Section 2 depicts the proposed method. Experimental results are shown in Sect. 3 and discussed. Section 4 concludes this work and possible future scope is given.

2 Methodology

In recent literature many ML models have been utilized for effective analysis and identification of cancer affected regions in lung images using different types of medical images like CT scans and X-ray images. Such systems many a times help the radiologists to address the concern immediately and effectively by means of reduction in interpretation errors resulting in improvement of diagnostic accuracy. DL algorithms like the CNN have been successfully used in extraction of features from various medical images connected with and detection of benign and malignant cancer nodules according to the traits and features in the images.

ML algorithms develop more comprehensive models for lung cancer detection with enhanced diagnostic accuracy and improved risk assessment by integrating data from various sources which include imaging data, clinical data and genetic information which help in the capture more meaningful information about the disease. The architecture for infected lungs classification is presented in Fig. 1. The pre-processing methods are used to resize the images to fit the input of the pre-trained model. Further, the pixel values are normalized with rescale of [0, 1]. Finally, the noise is filtered out.

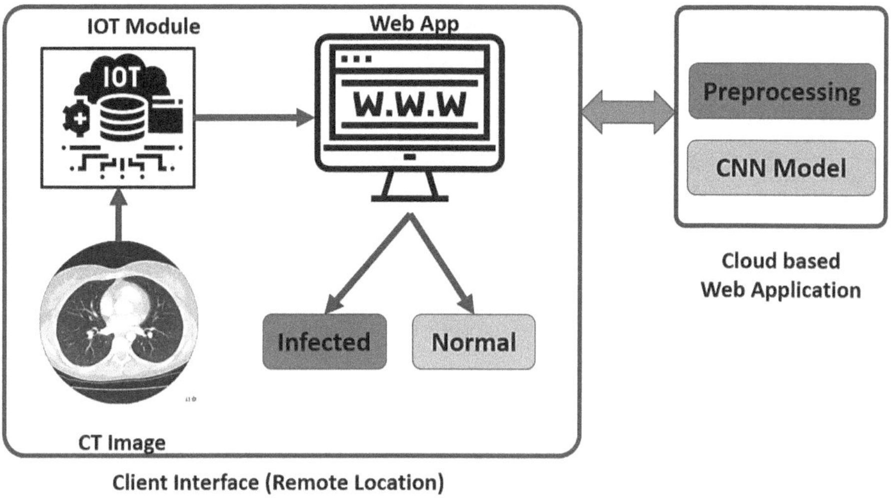

Fig. 1. Proposed scalable and low-cost architecture for smart Health care.

The pre-trained Models using transfer learning are with Dense Net, Mobile Net, ResNet50, InceptionV3, Efficient Net, DenseNet121, Mobile Net, and VGG16/VGG19. Among these the modified VGG 19 model is chosen. The efficacy of the proposed model is compared with other models.

2.1 Dataset Description

A publicly available dataset of CT Scan images with 364 images have been utilized in this work [13]. The dataset contains images of two classes, viz., Lungs with infection and those without infected images. The other sensitive information like the patient details, name of hospital, time of diagnosis etc. have been removed in order to maintain privacy of the patients. The images have been resized to 224 × 224 pixels in accordance with the CNN model employed here which has been trained for images of the same dimensions. The CT scan images of both classes have been shown in Fig. 2.

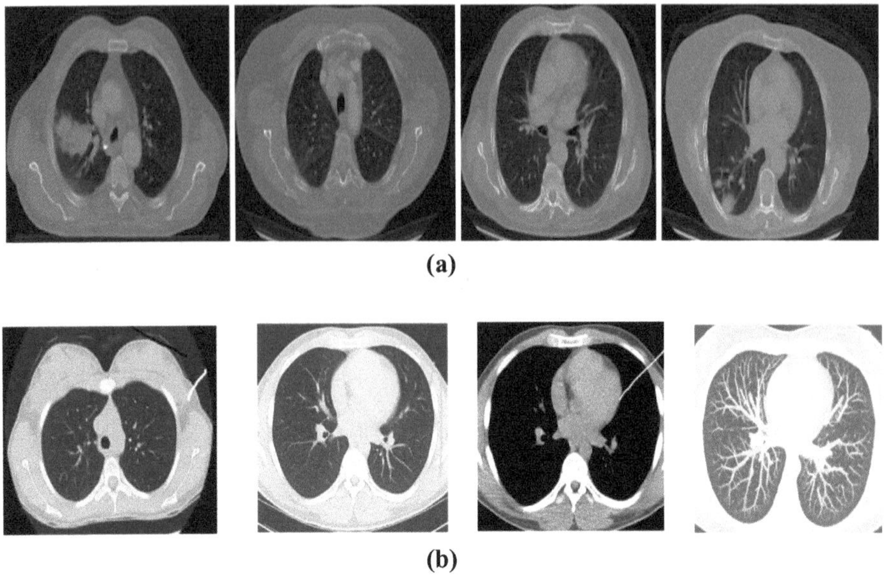

Fig. 2. (a) Four sample infected lung CT (b) Normal Lungs CT (Raw images without preprocessing).

2.2 Transfer Learning Approach Using VGG-19

The Visual Geometry Group network (VGG Net), is a CNN-based DL network that has been applied to a number of image classification applications. A deep CNN for image classification as the VGG19 architecture is one of the variants of different VGG models. It has 16 convolution layers, 3 fully linked layers, 5 Max Pool layers, and 1 SoftMax layer. The usage of tiny 3x3 filters and the architecture's simplicity are what define it3. Figure 3 displays the VGG19 architecture in detail. By training a VGG Net architecture on a dataset of lung CT scans with corresponding labels (e.g., malignant or benign), the network can learn to classify lung nodules and assist in lung cancer detection. The learned filters in the convolutional layers capture relevant features indicative of lung cancer, and the fully connected layers accomplish the ending classification based on these features.

Transfer learning involves modifying a model trained on one job to make it suitable for a comparable, related activity. Transfer learning is particularly beneficial for medical image classification tasks for rare or emergent disorders, when there is a lack of sufficient training data for a suitable model training from scratch. This is particularly true for models that require training on deep neural networks and have a significantly large number of parameters. Transfer learning can yield improved and more accurate outcomes by making minor adjustments to the existing high-quality beginning values of the parameters. Initially the weights of the model are kept unaltered used for feature extraction. Further, the classifier is trained to perform. Next to it has been modified to align with the new application.

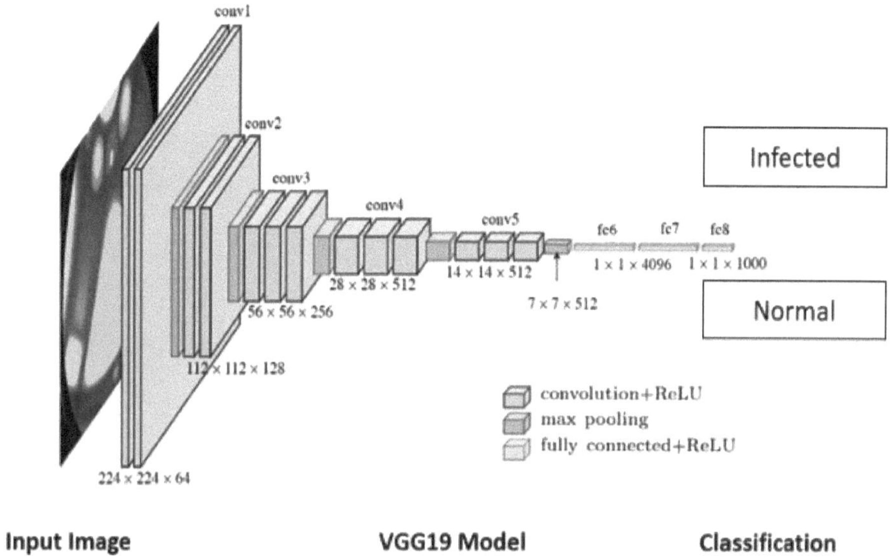

Fig. 3. VGG 19 Architecture

2.3 Proposed Modified VGG-19 Model

A 2-class classifier is proposed with the pre-trained VGG-19 model is modified through the use of transfer learning and fine-tuning techniques. This study involves a two-layer output system that corresponds to two classes: normal and abnormal. Throughout training process of the model, a dropout rate 'p_d' of 0.31 to avoid over-fitting. Additionally, the "Cross Entropy Loss" function has been as the loss function 'δ'and the model is trained with the "Adam" optimizer. The training with learning rate 'α' of 0.001 is conducted for an initial duration of 100 epochs. After the optimization process, the test accuracy is achieved within 60 epochs. The proposed work is represented in algorithm 1.

Algorithm 2: Proposed Framework

1. Open the Web application using the IOT Module.
2. Enter a CT scan image.
3. IOT module sends the image to the cloud.
4. Image preprocessing is performed.
5. Classification is performed by using the VGG19 model.
6. Result of the classification sent back to the client machine for displaying.

The hyperparameters of the VGG 19 network have been chosen as given in Table 1.

Table 1. Hyperparameter tuning of the Proposed Model

Hyper parameter	Value
Optimizer	Adam
α	0.0011
δ	Categorical cross-entropy
p_d	0.31
Batch size BS	32
Number of epochs	60

3 Results

Visualization of images demonstrated that the model focused on infected regions, confirming interpretability and alignment with clinical expectations. The model achieved high accuracy and generalization, even on a relatively small dataset. However, the performance can be further improved with more annotated data. Also the model may exhibit bias due to imbalanced datasets. The VGG-19 model's performance is assessed by measuring variables such as accuracy, sensitivity, specificity, and inference time. Inference time measures the speed at which the model makes predictions, which is important for applications that require real-time processing. The web application provides a user-friendly interface for acquiring and presenting analytical outcomes. Healthcare professionals and patients, along with other individuals, can securely access their accounts to review CT scan data and infection detection findings. The user's input is presented as a bullet point. Users can employ interactive tools to enlarge specific regions of the lung images and compare multiple scans taken at different points in time.

Recent literature underscores the efficacy of transfer learning, particularly with models like VGG-19, in detecting lung infections from medical images. Fine-tuning these models for specific tasks and datasets is crucial to achieve optimal performance. Additionally, ensemble approaches and systematic comparisons can further enhance detection accuracy and robustness in medical imaging applications.

These studies highlight the potential of VGG-19 and similar architectures in developing reliable diagnostic tools for lung infection detection, contributing to improved healthcare outcomes. The training of the projected model is approved out with Google Collaboratory platform. Figure 4 depict the images of the infected lung and non-infected lung obtained by uploading them via the web application running in the Raspberry Pi.

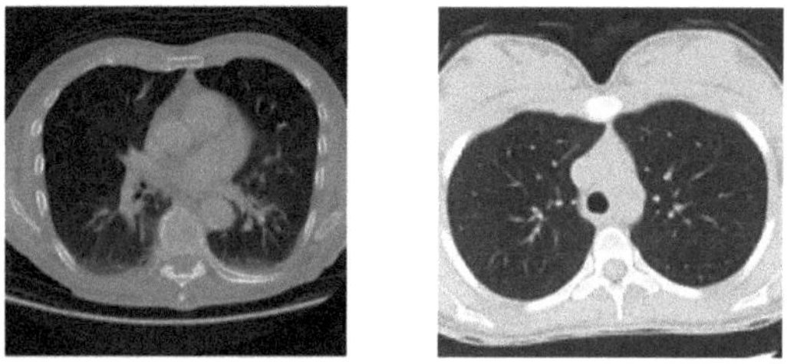

Fig. 4. Predicted Infected Lung and normal Lung image through Web Application

The confusion matrix of the classification has been presented in Table 2 where it can be observed that out of the 60 images in each class, 59 have been classified correctly and one image in each class has been misclassified. The classification accuracy thus is found to be 98.33%.

Table 2. Confusion matrix of the proposed method.

Class	**Predicted**	
	Infected	Healthy
Abnormal (60)	59	1
Normal (60)	1	59

Figure 5 represents the accuracy for both training and validation plot relative to the epochs.

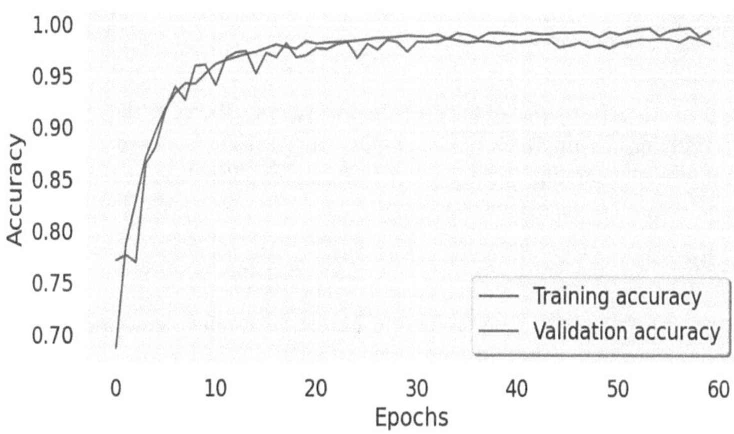

Fig. 5. Accuracy curve

The comparison with earlier works is given in Table 3.

Table 3. Earlier works Comparison

References	Image type	Classification model	Accuracy (%)
[7]	Lungs CT	BEiT, Dense Net, and Sequential CNN	98
[9]	Lungs CT	3D CNN + RNN	95
[8]	X-Ray	VGG19 + CNN	96.48
[3]	X-Ray	Res Net	92.1
[4]	X-Ray	ResNet50	94.70
[10]	X-Ray	ResNet-101	94.04
Proposed	**Lungs CT**	**Optimized VGG19**	**98.33**

The majority of the works cited in Table 3 have been proposed for classification and detection of lung diseases only while the proposed framework integrates IoT and cloud-based web application for effective classification and remote access.

Performance measures of different models are compared in Table 4. The proposed VGG19 network with transfer learning performs better than the Google Net and Squeeze Net models for classifying the COVID chest CT scan images.

Table 4. CNN classifiers' performance

Parameter	Squeeze Net	Google Net	Res Net	VGG19 Net
Accuracy	97.422	98.1	94.69	98.33
Sensitivity	97.422	99.4	94.58	98.33
Specificity	96.521	97.2	93.98	98.33

The improved performance of the projected work is attributed to the retrained VGG19 model and transfer learning techniques employed. The data is stored in a secure server in the cloud where it can only be accessed with proper validation of credentials.

4 Conclusion

The transfer learning-based lung infection detection model demonstrates promising results, with high accuracy and reliable interpretability. This approach can be integrated into clinical workflows to assist radiologists in diagnosing lung infections, ultimately reducing workload and improving diagnostic speed. The lightweight model ensures real-time performance, and the smart health care-based framework enables scalable remote diagnostics. Such a system can bridge healthcare gaps in underserved areas. In this work a cloud classification model of a VGG19 network has been implemented. The pretrained VGG19 network has undergone through transfer learning for classifying the lung CT images. In order to improve the training dataset, a data augmentation scheme has been employed. The experimentation provides 98.33% of accuracy that surpasses the performance of many other techniques available in literature. Further research could focus on fine-tuning the VGG-19 model using larger and more diverse datasets. However, it may be extended for other different medical images. The Challenges lie with Limited annotated datasets, biases in datasets affecting model fairness, and Compliance with healthcare standards. The future work may be carried out for privacy-preserving learning using distributed data, deployment of lightweight models for real-time diagnosis in resource-constrained settings and Multimodal Models. These models combine imaging data with clinical parameters using architectures such as Hybrid CNN and MLP models and Attention-based fusion models.

References

1. Karar, M.E., Khan, Z.F., Alshahrani, H., Reyad, O.: Smart IoMT-based segmentation of coronavirus infections using lung CT scans. Alex. Eng. J. **69**, 571–583 (2023)
2. Yadav, S., Rizvi, S.A.M., Agarwal, P.: Detection of lung diseases for pneumonia, tuberculosis, and COVID-19 with artificial intelligence tools. SN Comput. Sci. **5**(3), 1–18 (2024)
3. Panahi, A., Askari Moghadam, R., Akrami, M., Madani, K.: Deep residual neural network for COVID-19 detection from chest X-ray images. SN Comput. Sci. **3**(2), 169 (2022)
4. Ismael, A.M., Şengür, A.: Deep learning approaches for COVID-19 detection based on chest X-ray images. Expert Syst. Appl. **164**, 114054 (2021)
5. Narayan Mohanty, M., Mishra, A., Routray, A.: A non-rigid motion estimation algorithm for yawn detection in human drivers. Int. J. Computational Vision and Robotics, Inder Science Pub., **1**(1), 89–109 (2009)
6. Sarangi, L., Narayan Mohanty, M., Pattanayak, S.: Design of MLP based model for analysis of patient suffering from influenza. Proc. Comput. Sci. **92**, 396–403 (2016)
7. Quasar, S.R., Sharma, R., Mittal, A., Sharma, M., Agarwal, D., de La Torre Díez, I.: Ensemble methods for computed tomography scan images to improve lung cancer detection and classification. Multimedia Tools Appl. **83**(17), 52867–52897 (2024)
8. Alshmrani, G.M.M., Ni, Q., Jiang, R., Pervaiz, H., Elshennawy, N.M.: A deep learning architecture for multi-class lung diseases classification using chest X-ray (CXR) images. Alex. Eng. J. **64**, 923–935 (2023)
9. Wankhade, S., Vigneshwari, S.: A novel hybrid deep learning method for early detection of lung cancer using neural networks. Healthcare Anal. **3**, 100195 (2023)
10. Jain, G., Mittal, D., Thakur, D., Mittal, M.K.: A deep learning approach to detect Covid-19 coronavirus with X-ray images. Biocybern. Biomed. Eng. **40**(4), 1391–1405 (2020)
11. Bansal, M., Kumar, M., Sachdeva, M., Mittal, A.: Transfer learning for image classification using VGG19: Caltech-101 image data set. J. Ambient Intell. Human. Comput. 1–12 (2023)
12. Lawton, S., Viriri, S.: Detection of COVID-19 from CT lung scans using transfer learning. Comput. Intell. Neurosci. **1**, 5527923 (2021)
13. Maleki, N.: CT-Scan images, Mendeley Data, V3 (2020). https://doi.org/10.17632/p2r42nm2ty.3. https://data.mendeley.com/datasets/p2r42nm2ty/3

Unmasking Synthetic Speech: A PCA-Based Approach to Differentiating Human and AI-Generated Voices

Bishnupriya Panda[1], M. Disha[1], Seema Gupta Bhol[1], and Ambika Prasad Mishra[2](✉)

[1] School of Computer Applications, Kalinga Institute of Industrial Technology, Bhubaneswar, Odisha, India
{bishnupriyapanda.fca,seemaguptabhol.fca}@kiit.ac.in

[2] School of Computer Science and Engineering, Kalinga Institute of Industrial Technology, Bhubaneswar, Odisha, India
ambikaprasad.mishrafcs@kiit.ac.in

Abstract. The leading edge of machine learning through AI lets developers make authentic sounding artificial voices through voice replication. Modern technology based on artificial intelligence finds diverse purposes in entertainment production and educational content development and accommodation solutions for disabled users. The technology faces disadvantages because cybercriminals might utilize it to execute identity theft followed by financial trickery and damage security systems. The main obstacle in AI-generated voice detection arises when authenticating voices through automated systems because AI voices have a strong capability to mimic human speech patterns. Modern AI algorithms demonstrate mastery in voice duplication capabilities that duplicate human speech patterns, including emotional variations and local speech intonations. Advanced detection systems need development to properly distinguish authentic human voices from voices synthesized by AI. These detection systems will improve security measures and mitigate risks related to the unlawful application of the voice-cloning technology. A new method to identify artificial voice from human voice operates within this research project, which analyzes the compound nature and potential hazards of voice replication by AI. As part of our research, we conducted principal component analysis (PCA) on both human voice recordings and AI-synthesized voices to determine their separate characteristics. A dataset featuring equal quantities of 1000 human voice records from LibriVox and 1000 AI-generated speech samples made by Coqui AI text-to-speech technology was developed. The audio times in every sample were standardized to 10 s through Pydub processing. The Sound Synthesis Toolbox V1.7 analyzed these samples using its 1264 parameters. We used PCA because our task required handling this data with many dimensions while maintaining crucial elements. The findings obtained through this analysis will help develop security measures that protect against voice cloning technology abuse.

Keywords: PCA · Synthetic Voice · Voice Spoofing · Marginal Statistics · Modulation power

A. Das et al. (Eds.): MIDAS 2025, CCIS 3055, pp. 40–53, 2027.
https://doi.org/10.1007/978-3-032-35516-4_4

1 Introduction

Artificial intelligence-based voice replication technology [1] developed into a powerful instrument that creates digital voices which sound authentic to human hearing. This breakthrough delivers multiple benefits for entertainment industry and education sector and enables better accessibility [2]. These advancements include noteworthy dangers that would arise from improper implementation. Non-legitimate users may exploit this technology to produce fake voice imitations which could deceive people and access sensitive information or overcome voice security protocols [3]. Voice-based identity frauds cause serious consequences which include stolen identities and security breaches and financial losses [4]. You must understand all dangers from AI voice cloning so you can develop robust defense strategies against synthetic voice-based fraudulent activities [5]. The field of voice-based authentication now faces a critical limit because it became increasingly hard to differentiate authentic human speech from AI-produced voice imitations [5]. Sophisticated machine learning algorithms through AI-driven voice synthesis produce lifelike speech through both text and audio inputs [7, 6]. Microsoft VALL-E and Google Text-to-Speech as well as Murf's voice generator represent notable examples in this field (based on references [8–10]. By copying speech from brief sample recordings such systems immediately reconstruct the voice patterns of individuals along with their emotional tones and accent features [11]. The perfection of artificial voices requires immediate development of methods to identify authentic human voice from AI-created audio content [12]. The quick development of voice synthesis technology has created an important challenge for speech recognition experts who need to determine the difference between natural human voices and synthetic AI outputs [13]. Keeping track of human-generated speech from AI-generated speech remains critical for safeguarding voice-security systems and protecting personal privacy and securing vocal communication authenticity [14].

The detection of human versus synthetic voices requires identification of essential acoustic factors which effectively distinguish between natural human speech and artificial speech [15, 16]. Sound analysis faces additional complexity because of the large number of potential measurement parameters. The goal of our study involves solving two related challenges: assessing human-made voice identification through critical features [17, 18] and developing efficient technology to process acoustic parameters [19]. The solution of these problems leads to development of a reliable and efficient AI voice classification model for real-world operation [20].

2 Methods and Materials

2.1 Data Preparation

Generation of Human Voice Dataset. The data collection process started by getting audio recordings from LibriVox [21] which serves as a public domain audiobook repository. We applied a systematic approach for file organization to achieve sample diversity according to [22]. The readers maintained separate folders in our dedicated structure to store their audio files rather than using book authors as classifications. The defined organizational structure solved the limitation of double donations from one voice artist that speech datasets commonly face. Methodological organization led us to accumulate 1000 unique voice samples by sticking to proper standards in speech corpus development.

Generation of AI Generated Voice Dataset. Our goal to balance the dataset involved obtaining 1000 samples from human voices while collecting an identical number of samples from AI voice generation [23]. The generation of AI voice samples needed matching text input, which served as their prompts. The prompts originated from the Moby Dick novel, which Project Gutenberg [24] maintains as a digital library of free eBooks. The classic text served our voice synthesis project by supplying diverse linguistic content. We used text prompts, which enabled the generation of 1000 voice samples through the Coqui AI Text-to-Speech (TTS) Python library [25]. The developed approach established a flexible dataset consisting of various artificial voices with authentic human vocal characteristics to support our analytical investigation.

Trimming the Dataset to 10-s Clips. A total of 2000 audio files comprised two equally balanced sets containing 1000 human voice recordings and 1000 synthetic voice outputs. Standardization of audio lengths became essential because the natural variability in audio time duration presented analysis challenges. Each audio sample received treatment for 10 s as our selection provided balanced processing speed with adequate acoustic data. The audio normalization process used the Pydub library as a strong Python tool for audio file modification [26]. The functionality of Pydub enabled precise file trimming to match each audio clip to 10 s, thus maintaining uniform dataset conditions for better human-AI voice comparison.

Generating Data by Analyzing the Voice Samples. The acoustic analysis depended on Sound Synthesis Toolbox V1.7 [27], which serves as a complete software solution for audio signal processing. Through this toolbox we obtained 1264 distinct parameters from every voice sample, which supported our study through the rich acoustic feature collection. PCA was applied to the dataset with high dimensionality to find the most significant data variations while diminishing the number of dimensions via its principal component analysis algorithm [28][29]. This wide-ranging speech processing technique reduces dimensionality and makes computations more efficient while overcoming the problems caused by data dimensionality. The team used MATLAB version 31 as their main platform to execute data analysis and processing of scientific operations and signal tasks [31]. Acoustic data processing required MATLAB due to its powerful matrix operation capabilities and extensive libraries which especially excelled at handling large, voluminous acoustic information and executing PCA operations.

Final Set of Auditory Parameters under Consideration. The study conducted a total acoustic analysis of 1264 parameters for each voice sample by utilizing the capabilities of Sound Synthesis Toolbox V1.7. The toolbox acted as an essential tool to calculate marginal statistics in each of the 32 cochlear filters, with widespread use as an auditory modeling methodology. The analysis obtained four statistical parameters, including mean, variance, skew, and kurtosis, from each cochlear filter output [32]. Each voice sample received 128 marginal parameters, which were generated through the combination of 32 filters with 4 statistical measures. The analysis produced 1024 initial parameters through calculations of the correlation matrix from the 32 cochlear filters. The method produced 496 unique correlation parameters when symmetry was taken into consideration. The correlation measurements identify essential relationships that exist between voice signal frequencies. The feature set received additional enhancement when applying 20 modulation filters to every output from the cochlear filters. A statistical analysis of modulation spectrum enabled researchers to create 640 modulation power parameters by applying 20 modulation filters to 32 cochlear filters. Our voice analysis relied on three sets of acoustic measures totaling 1264 parameters, which started with 128 marginal parameters, then continued with 496 correlation parameters, and ended with 640 modulation power parameters. A diagram illustrates the process of human sound sample along with AI sound sample preparation before extraction. The sound synthesis toolbox performs the feature extraction following the extraction process. The analyzed features emerge from principal component analysis. The proposed model is shown in Fig. 1.

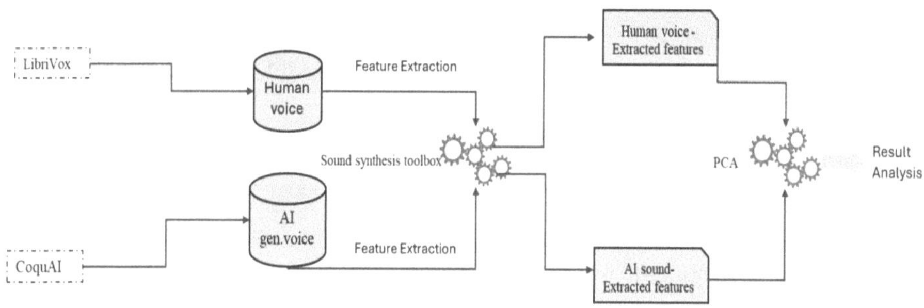

Fig. 1. Workflow diagram

2.2 Methods and Tools

Principal Component Analysis. PCA functions as a basic data analysis approach that primarily reduces dimensionality while operating on machine learning data. PCA performs data projection onto reduced dimensional spaces through which it maintains the highest possible data variance. The technique preserves all significant characteristics within the dataset information. The wide range of applications for PCA includes data compression as well as noise reduction and feature extraction and data visualization capabilities. High-dimensional datasets benefit greatly from these applications because they help analysis in speech and audio domains. The 1264 parameter sets extracted from voice samples create a large multidimensional data space. An excessive amount of dimensions produces performance problems together with the "curse of dimensionality". The high-dimensional parameters need a reduction through PCA because it is essential for our study. The parameter reduction through PCA combines two advantages because it improves performance efficiency while finding vital features which separate human voices from AI-generated voices.

Librio Vax. At LibriVox [33] users can find large assortments of free audiobooks from public domain works. Through its global network of volunteers, LibriVox enables people to record individual book chapters in multiple languages. The website LibriVox publishes downloaded recordings from its volunteers so users can access them free of charge.

Coqui AI TTS. Coqui AI TTS [34] is an advanced text-to-speech system that harnesses deep learning techniques to generate natural and expressive speech from textual input. This innovative platform is designed to offer high-quality, open-source solutions accessible to a wide range of applications, including voice assistants, audiobooks, and podcasts.

Python Libraries. We utilized several key libraries: Pydub: This library, particularly its AudioSegment class, was employed for voice sample manipulation [26]. It allowed us to generate standardized 10-s voice clips through methods like slicing and concatenating. Coqui AI TTS: We used the TTS voice cloning library from Coqui AI to generate our AI-synthesized voice dataset. Operating in English, this library produced text-to-speech samples based on our provided voice samples and text prompts. These tools enabled efficient audio processing and synthetic voice generation, crucial for our comparative analysis of human and AI-generated speech.

3 Results

3.1 Principal Components of Marginal Statistic

Our analysis of human and AI-generated voice samples involved several stages of data processing and visualization. Figure 2 illustrates the marginal distribution of acoustic features for both human and AI-generated voices. To normalize the data and enhance its symmetry, we applied log transformation followed by Z-score normalization. The resulting transformed data distribution is presented in Fig. 3.

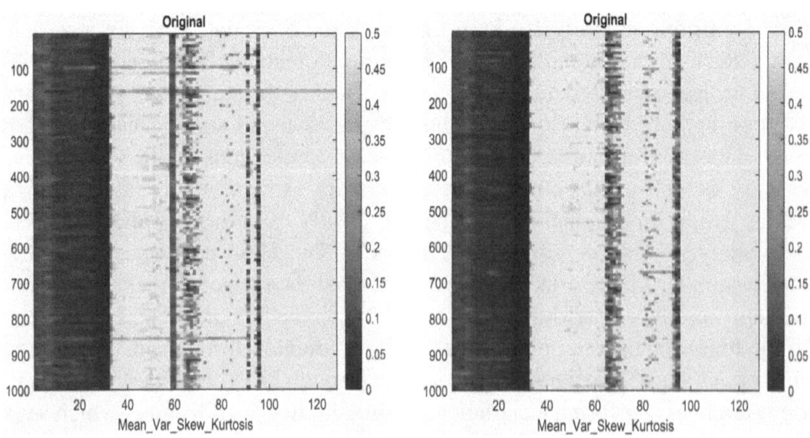

Fig. 2. Marginal data of (a) the original human and (b) AI samples, respectively

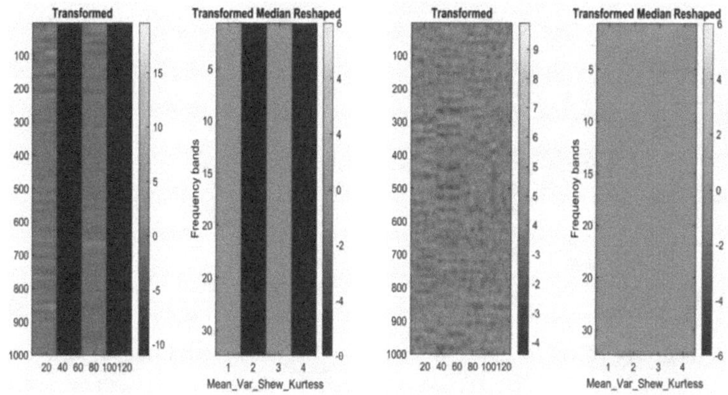

Fig. 3. Transformed Marginal data of (a) original human and (b) AI samples, respectively

Subsequently, we performed principal component analysis (PCA) on the transformed dataset. Figure 4 displays the marginal statistics of the first four principal components. We computed PCA coefficients and scores for both human and AI-generated datasets.

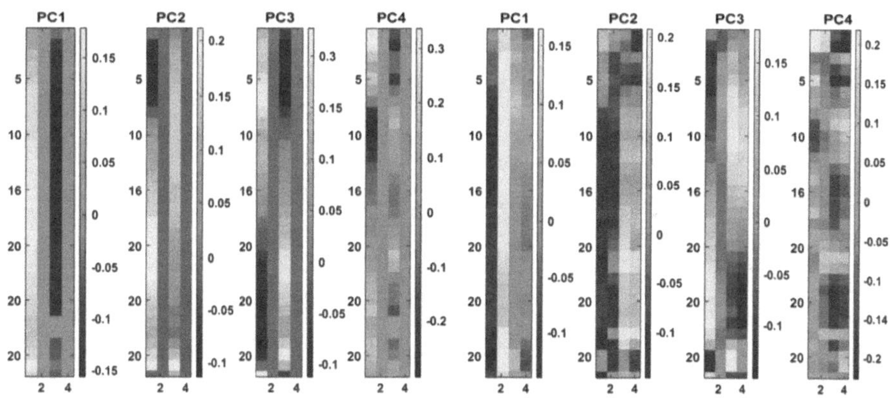

Fig. 4. PCs of (a) human and (b) AI-generated voice

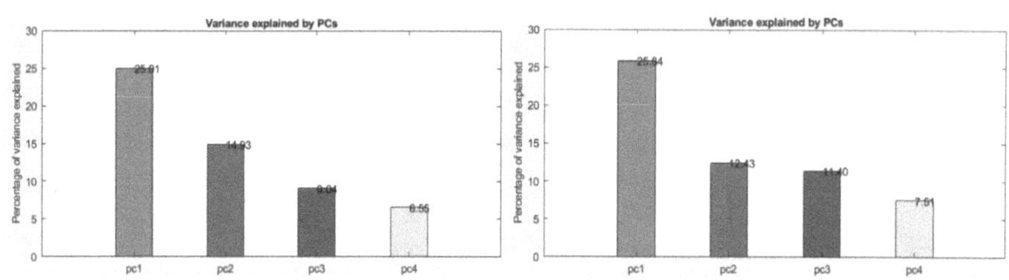

Fig. 5. Variance Explained by PCs of (a) human and (b) AI-generated voice

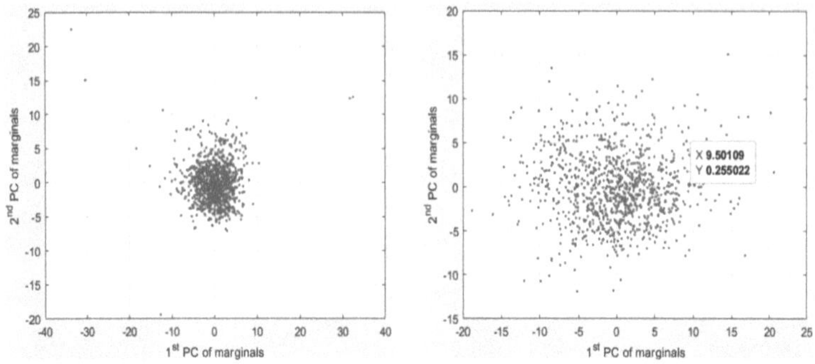

Fig. 6. Distribution of PCs of (a) human and (b) AI-generated sounds in the original feature space

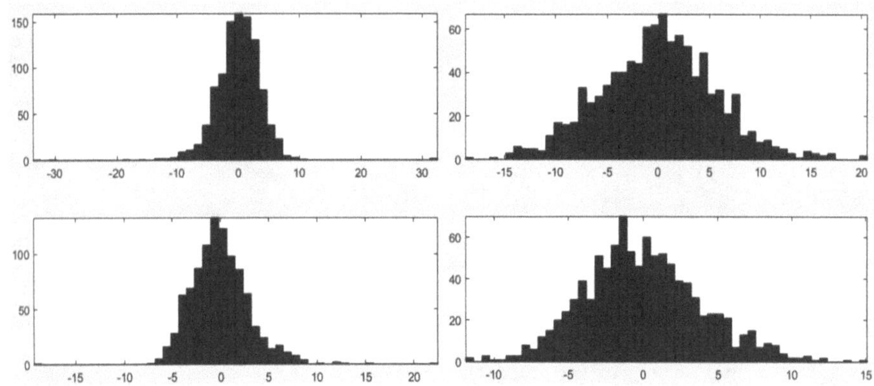

Fig. 7. Histogram distribution of PCs of (a) human and (b) AI-generated sounds

Figure 6 depicts the distribution of data points in the two-dimensional space defined by the first two principal components, projected onto the original marginal feature space. The distribution of PCA scores for the first two components is visualized in Fig. 7. While the initial Figs. (1–2) primarily illustrate the feature distributions in marginal space for both original and transformed data, they do not provide clear differentiation between human and AI-generated voices. This limitation necessitated the application of PCA. The PCA results reveal several noteworthy observations: 1. The first four principal components account for 56–57% of the total variance in the data as shown in Fig. 5, indicating a significant dimensional reduction while retaining most of the data's information content. 2. In the space defined by the first two principal components (Fig. 6), human voice samples exhibit denser clustering compared to the more dispersed distribution of AI-generated samples. 3. The PCA score distribution (Fig. 7) for human voices shows evidence of outliers, with a high density around the zero mark (159 data points between –0.595 and 0.728, and individual data points for scores below -30 and above 30 for the first principal component). 4. Conversely, AI-generated voices lack clear outliers but demonstrate a broader distribution of scores (maximum count of 70 between –1.65 and

−1.11). These statistical characteristics provide a quantitative basis for differentiating between the marginal acoustic features of human and AI-generated voices. Further analysis of these distributions may yield additional insights into the distinguishing properties of natural and synthetic speech.

3.2 Principal Components of Modulation Power Statistics

The modulation power data of the human voice and AI-generated voice are depicted in Fig. 8.

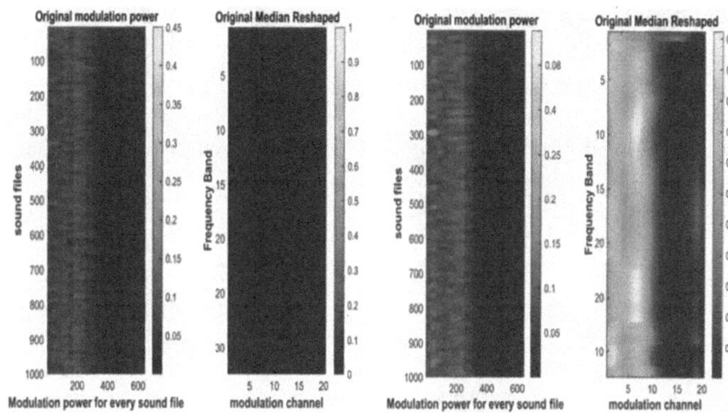

Fig. 8. Distribution of Modulation power statistics of (a) human and (b) AI-generated sounds

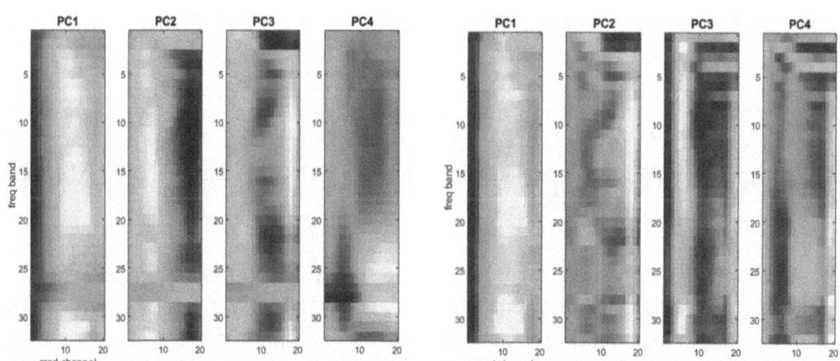

Fig. 9. Distribution of PCs of modulation power statistics of (a) human and (b) AI-generated sounds

As we did with the marginal data, we transformed the data using Log transformation and Z-score and performed PCA. Then, we plotted the graphs for the first four components of PCA which can be seen in Fig. 8. Figure 9 illustrates the distribution patterns of the principal components (PC1–PC4) obtained from modulation power statistics for both human-generated and AI-generated sounds. We calculated the PCA coefficients and scores for both the datasets. The Fig. 10 shows the percentage of variance observed in these PCA components and the Fig. 11 shows the distribution spanned by these components.

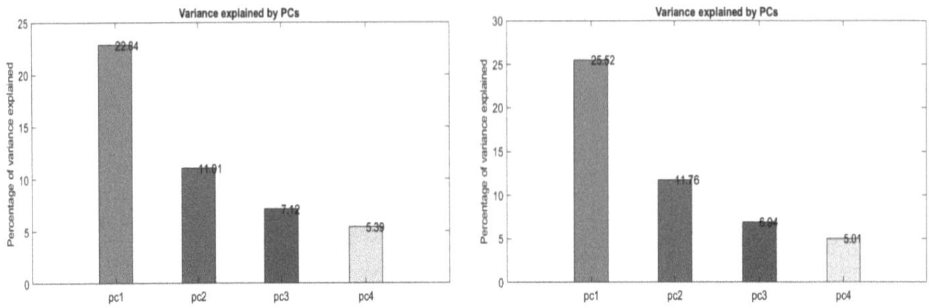

Fig. 10. Variance explained by PCs of modulation power statistics for (a) human and (b) AI-generated sounds

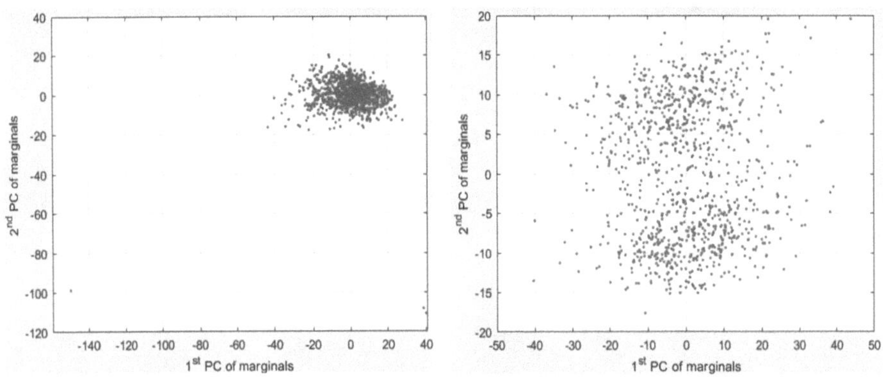

Fig. 11. Distribution of PCs for modulation power statistics for (a) human and (b) AI-generated sounds

The individual PCA score distribution for these coefficients is shown on Fig. 12. We have gathered almost similar insights with this data, where we saw some outliers for human voices and dense population across the '0' mark and sparse for the AI-generated. Figure 13 presents the variance explained by the first four principal components (PC1–PC4) of modulation power for both human-generated and AI-generated sounds.

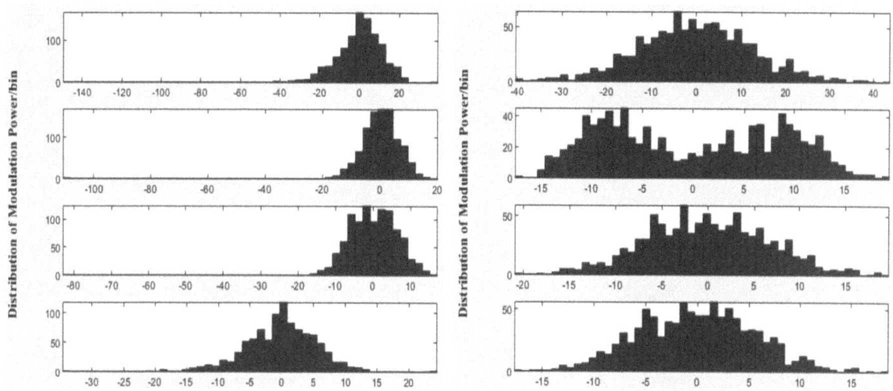

Fig. 12. Histogram distribution of modulation power for (a) human and (b) AI-generated sounds

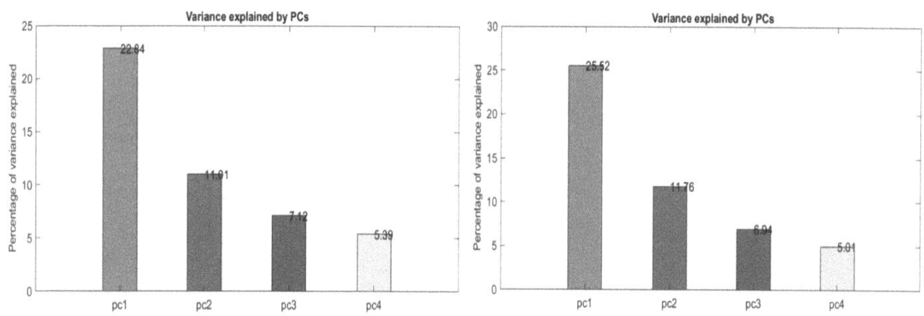

Fig. 13. Variance explained by the first four PCs for modulation power for (a) human and (b) AI-generated sounds

3.3 Principal Components of Correlational Statistics

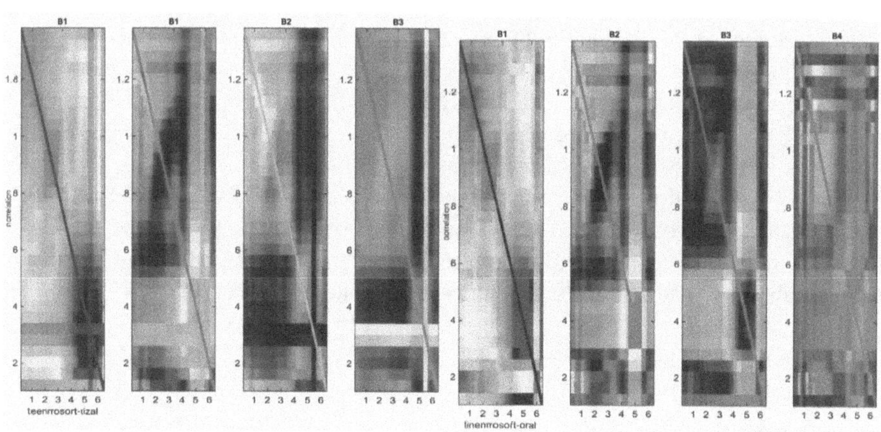

Fig. 14. PCs of correlation statistics for (a) human and (b) AI-generated sounds

We have done a similar analysis to the correlational statistics. Figure 12 shows the plots of both the voice samples with 1024 correlational features. The following figures show the stats of PCA components: Fig. 14 shows the correlational features for 1000 samples, Fig. 15 shows the percentage of variance covered, and Fig. 12 shows the distribution of scores for the first two components of PCA.

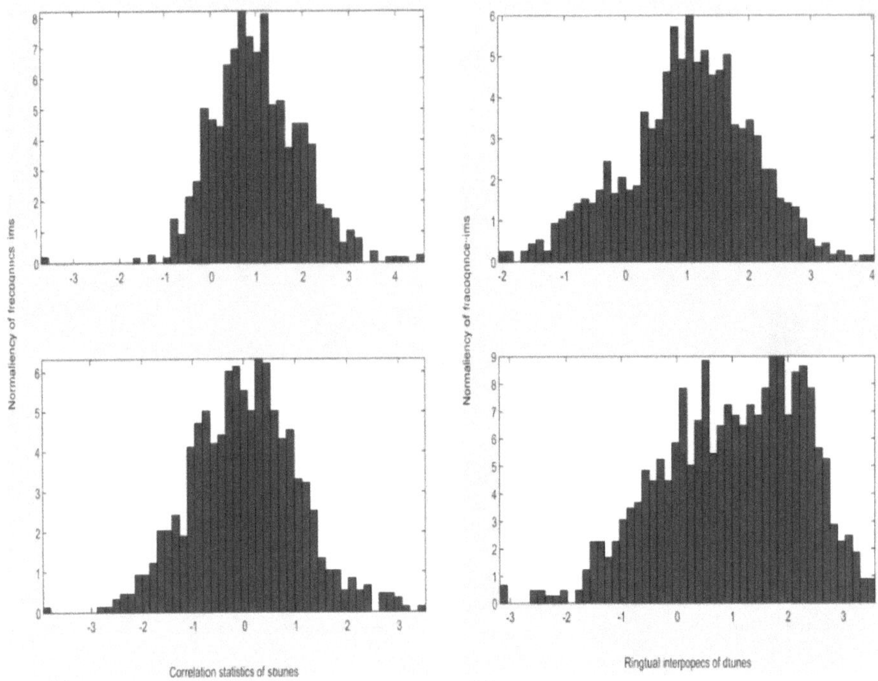

(a) (b)

Fig. 15. Histogram distribution of correlation statistics for (a) human and (b) AI generated sounds

The features before conducting PCA look very similar. A human cannot distinguish between the human and AI generated voices. Some of the observations include the percentage of variance covered by the first four components for human voices came out to be 58.74 and 67.62% for AI generated voices. Here, we are not able to observe many differences in their PCA score distributions too. For AI generated voices, the scores are sparsely populated with the maximum being 60 across –0.738 and 1.72 for the first component and 45 across 6.68 and 8.04 for the second one. For Human voices, obviously like the above graphs, the maximum scores are higher but not significantly. To be precise, the maximum is 82 and 63 for the first and second components respectively.

4 Conclusion and Future Scope

Our study successfully generated a comprehensive dataset comprising 1000 samples each of human and AI-generated voices. We subsequently conducted Principal Component Analysis (PCA) on these samples, yielding results that offer significant potential for voice differentiation. The PCA results revealed distinct distribution patterns between human and AI-generated voices within the space defined by the initial principal components. This differentiation suggests the presence of underlying acoustic features that distinguish naturally from synthetic speech. The information obtained from PCA space analysis serves as foundation for creating improved voice classification techniques. The obtained distinctions in PCA space serve as guidance to create efficient machine learning algorithms that correctly detect natural voice inputs from synthetic ones. The consistent identification of Artificial Intelligence generated speech creates essential ramifications for both cybersecurity operations and digital forensic investigations. The evolving nature of voice cloning technology requires advanced detection methods to stop probable malicious usages such as security system impersonation and synthetic audio deception. Research of acoustic features represented by principal components would provide more details about this data representation. Research aimed at improving the technique should involve studying higher-order parts alongside nonlinear dimensionality reduction approaches and speech processing domain knowledge. The PCA-based method creates an excellent beginning point for creating solutions against misuse of AI voice cloning technology. Future research needs to convert the present initial evidence into functional tools for deploying synthetic voice detection systems.

References

1. Oke, S.A.: A literature review on artificial intelligence. Int. J. Inf. Manage. Sci. **19**(4), 535–570 (2008)
2. Shen, J., et al.: Natural TTS synthesis by conditioning WaveNet on Mel spectrogram predictions. In: Proceedings of ICASSP (2018)
3. Wang, X., et al.: ASVspoof 2019: a large-scale public database of synthesized, converted and replayed speech. Comput. Speech Lang. **64**, 101114 (2020)
4. Zeinali, H., Sameti, H., Stafylakis, T.: DeepMine speech processing database: text-dependent and independent speaker verification and speech recognition in Persian and English. In: Odyssey, pp. 386–392, June 2018
5. Sanchez, J., Saratxaga, I., Hernaez, I., Navas, E., Erro, D., Raitio, T.: Toward a universal synthetic speech spoofing detection using phase information. IEEE Trans. Inf. Forensics Secur. **10**(4), 810–820 (2015)
6. Thakur, A., Mishra, A.P., Panda, B., Sweta, K., Majhi, B.: Detection of disease-specific parent cells via distinct population of nano-vesicles by machine learning. Curr. Pharm. Des. **26**(32), 3985–3996 (2020)
7. Ren, Y., et al.: FastSpeech: fast, robust and controllable text to speech. In: Proceedings of NeurIPS (2019)
8. Wang, C., et al.: Neural Codec Language Models are Zero-Shot Text to Speech Synthesizers, arXiv preprint arXiv:2301.02111 (2023)
9. Zhang, Y., et al.: Learning to speak fluently in a foreign language: multilingual speech synthesis and cross-language voice cloning. In: Proceedings of INTERSPEECH (2019)

10. van den Oord, A., et al.: WaveNet: a generative model for raw audio, arXiv preprint arXiv: 1609.03499 (2016)
11. Jia, Y., et al.: Transfer learning from speaker verification to multispeaker text-to-speech synthesis. In: Proceedings of NeurIPS (2018)
12. Ning, Y., He, S., Wu, Z., Xing, C., Zhang, L.J.: A review of deep learning based speech synthesis. Appl. Sci. **9**(19), 4050 (2019)
13. Wang, X., et al.: Neural speech synthesis with transformer network. In: Proceedings of AAAI (2019)
14. Kinnunen, T., et al.: t-DCF: a detection cost function for the tandem assessment of spoofing countermeasures and automatic speaker verification. In: Proceedings of Odyssey (2018)
15. Sahidullah, M., et al.: Integrated spoofing countermeasures and automatic speaker verification: an evaluation on ASVspoof 2015. In: Proceedings of INTERSPEECH (2016)
16. Ling, Z.H., et al.: Deep learning for acoustic modeling in parametric speech generation: a systematic review of existing techniques and future trends. IEEE Signal Process. Mag. **32**(3), 35–52 (2015)
17. Wu, Z., et al.: Anti-spoofing for text-independent speaker verification: an initial database, comparison of countermeasures, and human performance. In: IEEE/ACM Transactions on Audio, Speech, and Language Processing, vol. 24, no. 4 (2016)
18. Georgiou, G.P.: Differentiating between human-written and AI-generated texts using linguistic features automatically extracted from an online computational tool. arXiv preprint arXiv: 2407.03646 (2024)
19. Tak, H., Patino, J., Todisco, M., Nautsch, A., Evans, N., Larcher, A.: End-to-end anti-spoofing with rawnet2. In: ICASSP 2021–2021 IEEE International Conference on Acoustics, Speech and Signal Processing (ICASSP), pp. 6369–6373. IEEE, June 2021
20. Yamagishi, J., et al.: ASVspoof 2021: accelerating progress in spoofed and deepfake speech detection. In: Proceedings of ASVspoof Workshop (2021)
21. Kearns, J.: Librivox: Free public domain audiobooks. Ref. Rev. **28**(1), 7–8 (2014)
22. Smith, J., Brown, A.: Effective strategies for speech dataset curation. In: Proceedings of the International Conference on Speech Database Management, pp. 112–120 (2019)
23. Liu, Y., Chawla, N.V., Harper, M.P., Shriberg, E., Stolcke, A.: A study in machine learning from imbalanced data for sentence boundary detection in speech. Comput. Speech Lang. **20**(4), 468–494 (2006)
24. Melville, H.: Moby Dick; Or, The Whale (eBook #2701). Project Gutenberg (1851). http://www.gutenberg.org/ebooks/2701
25. Coqui Foundation: "TTS - Deep learning for Text to Speech." GitHub repository (2022). https://github.com/coqui-ai/TTS
26. Balogh, J.: Pydub: Audio processing made easy with Python. GitHub repository (2020). https://github.com/jiaaro/pydub
27. Colinot, T., Vergez, C.: Dynamical system audio demonstrator [Code]. GitHub (2023). https://github.com/Tom-Colinot/Dynamical-System-Audio-Demonstrator/
28. Jolliffe, I.T., Cadima, J.: Principal component analysis: a review and recent developments. Philos. Trans. Royal Soc. A Math. Phys. Eng. Sci. **374**(2065), 20150202 (2016)
29. You, S.D., Hung, M.J.: Comparative study of dimensionality reduction techniques for spectral–temporal data. Information **12**(1), 1 (2020)
30. Ko, S., Park, M.: Efficient speech signal dimensionality reduction using complex-valued techniques. Electronics **13**(15), 3046 (2024)
31. MathWorks: MATLAB - The language of technical computing (2023). https://www.mathworks.com/products/matlab.html
32. Cain, M.K., Zhang, Z., Yuan, K.H.: Univariate and multivariate skewness and kurtosis for measuring nonnormality: Prevalence, influence and estimation (2017)

33. Bhardwaj, R.K., Abhilash, K.R.: Structure and Functions of Talking Book Mobile Applications: A Comparative Study of Google Play Books, Audible, LibriVox, Storytel and Loyal Books (2024)
34. Zhang, Y., et al.: Comparative analysis of open-source text-to-speech systems: a case study of Coqui AI TTS. Proc. INTERSPEECH **2022**, 3201–3205 (2022)

Performance Comparison of Pneumonia Disease Detection and Prediction Using Deep Learning Models

Haymanot Derebe Bizuneh[1], Demissie Jobir Gelmecha[1], Ram Sewak Singh[3(✉)], Satyasis Mishra[2], Eshetu Tessema[1], and Workeneh Geleta[1]

[1] Department of Electronics and Communication Engineering, Adama Science and Technology University, Adama, Ethiopia
[2] Department of ECE, Centurion University of Technology and Management, R.Sitapur, India
[3] Department of Electrical, Electronics and Communication Engineering, Galgotias University, Greater Noida, Uttar Pradesh, India
ramsewaknitj@gmail.com

Abstract. Medical image analysis requires an intelligence-based model due to pneumonia disease infecting the lungs and killing many people worldwide. The detection and prediction of pneumonia diseases are the focus of numerous studies necessary to compare the performance of deep learning models and use appropriate models for a specific field of study. The objective is to analyze the performance of models for detecting and predicting pneumonia disease in grayscale images. These selected deep learning models are CNN, VGG16, VGG19, DenseNet, ResNet, and inception models. The deep learning model performance is analyzed through Anaconda Python Jupiter notebook and tensor flows. Therefore, the CNN model has a high performance at 97.78%, the ResNet model has a low performance of 62.22%, and the remaining models are moderate. However, the selection of models depends on the specific field of study.

Keywords: Pneumonia Disease · Detection and Prediction Model · Deep Learning

1 Introduction

Currently, the number of human deaths by pneumonia is increasing worldwide [14]. Pneumonia disease is a disease that affects the human lung [15]. Nowadays, numerous researchers have studied different methods to detect and predict pneumonia to minimize the spread of pneumonia all over the world [16]. Numerous researchers examine the deep learning models that are for detecting and predicting pneumonia disease in various fields of studies using convolutional neural networks(CNN) [1], Visual Geometry Group 16 (VGG16)) [2], Visual Geometry Group 19(VGG19) [3], Densely connected convolutional Networks(DenseNet) [4],

A. Das et al. (Eds.): MIDAS 2025, CCIS 3055, pp. 54–65, 2027.
https://doi.org/10.1007/978-3-032-35516-4_5

Residual Networks (ResNet) [4] and Inception models [5].
The pneumonia disease has increased dramatically and spread that needs enhancement of detection and prediction models [6]. This enhancement of pneumonia disease detection and prediction model for medical images is required [17]. The performance of pneumonia disease detection and prediction using CNN, VGG16, VGG19, DenseNet, ResNet, and inception models are relevant to selecting appropriate models for specific study areas [18]. The VGG16 is a detection and prediction model that needs to be enhanced for pneumonia disease detection and prediction [19]. The VGG16 is a deep CNN extended version that has multiple hidden layers and an output layer of linear neurons [3]. The VGG19 is an extended version of VGG16. DenseNet is a detection and prediction model of pneumonia diseases. ResNet is a detection and prediction model for pneumonia diseases. The inception is the detection and prediction of pneumonia diseases [1].
The steps required to process a Chest X-ray image are (i) preprocessing, which includes noise removal, data normalization, and dimensionality reduction using principal component analysis (PCA). (ii) Feature selection is a selection of features used to optimize the detection and prediction of pneumonia disease.(iii) Training is training model patterns and relationships in data. (iv) The test validates the datasets to evaluate their performance in achieving high accuracy and efficiency in detecting and predicting pneumonia diseases. The pneumonia disease has various forms, including shapes, sizes, and symptoms. Therefore, accurate and fast detection and prediction of pneumonia disease is a challenging task [7]. The main objective is to select appropriate detection and prediction models of pneumonia diseases in the Chest X-ray image datasets. Therefore, a performance comparison is required to make specific and minimize the problem that enhances the weights and biases of deep learning models through rapid training and prediction [1].

2 Related Works

The selected four works that are relevant to this research are described briefly. The authors in [2] studied, the CNN models for detecting and predicting pneumonia diseases using a hybrid model in Chest X-ray images. They used fully connected layers of the CNN feature extraction model, and they could achieve 87.65% accuracy. The authors studied VGG16 in deep neural networks through investigation and feature extraction time decay. They could achieve an accuracy of 95.4%. The authors in [3] studied the VGG19 model for detecting and predicting pneumonia disease in chest X-ray images. They have used PCA-assisted selection serial fusion feature extraction, and they could achieve 86.97% accuracy. The authors in [7] studied a combined model of detection and prediction of pneumonia disease in Chest X-ray images. They used fully connected layers for feature extraction of pneumonia diseases using VGG16 and VGG19 models, and they could achieve an accuracy of 96.2% and 95.9%, respectively. The authors in [8] studied, the combined models to detect and predict normal and abnormal.

These are CNN, VGG16, VGG19, DenseNet, ResNet, and inception models. They were fully connected layers for feature extraction of pneumonia diseases, and they could achieve good results. The authors in [8] studied, the combined models for detecting and predicting COVID-19 disease in chest X-ray images. They used fully connected layers for feature extraction Using CNN, VGG16, VGG19, DenseNet, and ResNet models, and they could achieve an accuracy of 97.61%,99.33%,96.66%,96.66% and 97% respectively.

3 Materials and Methods

Figure 1 shows that the process of this article works to select appropriate models based on its performance.

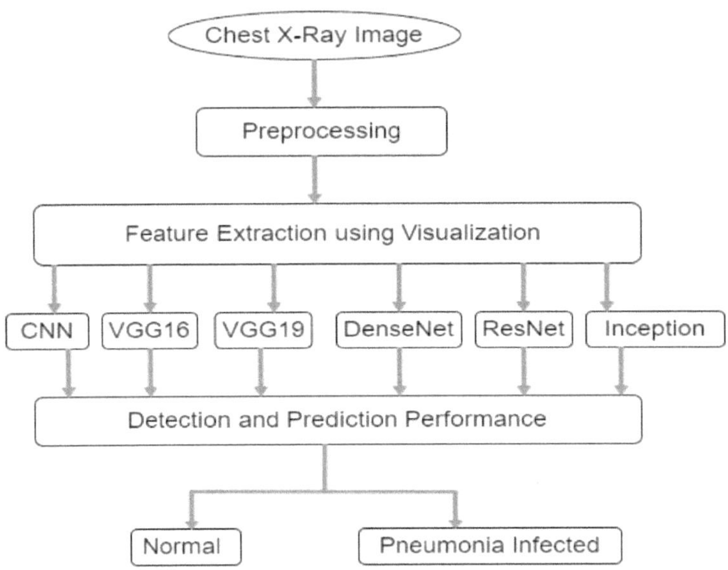

Fig. 1. Workflow diagram for pneumonia disease detection and prediction models with comparative study.

3.1 Preprocessing

Preprocessing is the process of removing features from Chest X-ray images, such as Gaussian noise from openCV and resizing and reshaping Chest X-ray images. Here, the sample size of the datasets is 5216, and the window size is [240,240] pixels.

3.2 Feature Extraction

The testing is a validation of datasets to evaluate performance and achieve high performance. The feature extraction for detecting and predicting pneumonia disease using input grayscale Chest X-ray images. Figure 2 (a) and (c) show inputs that used normal and pneumonia-infected, respectively. The histograms of normal and pneumonia-infected are shown in Fig. 2 (b) and (d), respectively. Figure 3 (a) and (b) show normal and pneumonia-infected labelled datasets samples, respectively.

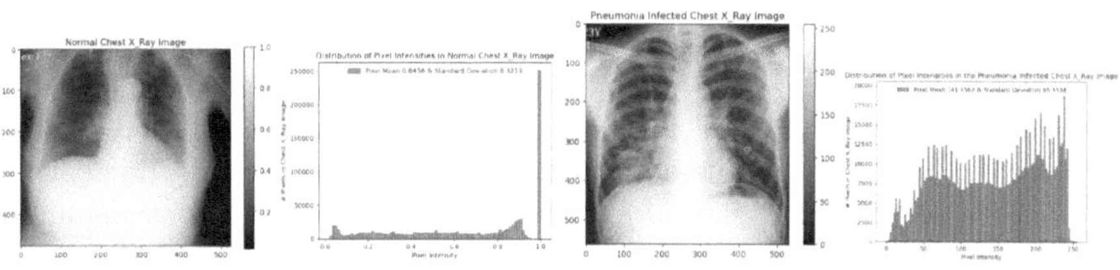

(a) (b) (c) (d)

Fig. 2. Input of (a) normal, (b) normal histogram,(c) pneumonia infected, and (d) pneumonia infected histogram

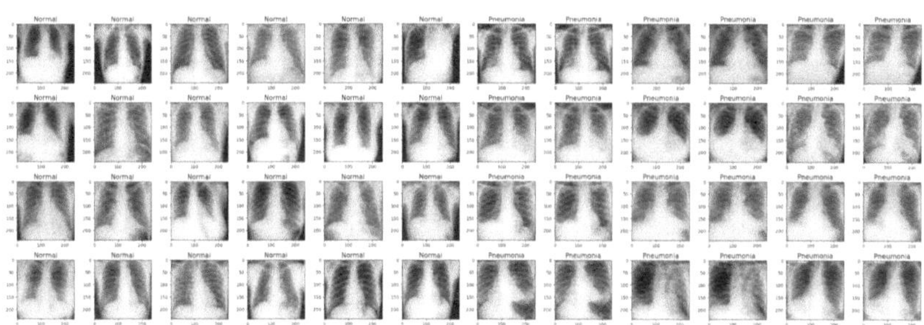

(a) (b)

Fig. 3. Dataset samples of (a) normal, and (b) pneumonia infected

3.3 Detection and Prediction Models

The six models listed below are described as follows to compare and contrast their performance. To get optimized models, train them using an appropriate loss function and optimization technique. CNN is a detection and prediction model for pneumonia diseases. CNN has good performance for the detection and prediction of lung abnormalities and signs of pneumonia disease [1]. CNN has the following steps:(1) Data collection and preprocessing: Data collection involves collecting datasets and labeling them as normal or pneumonia-infected. Preprocessing involves resizing, normalizing, and data augmenting to enhance the training datasets. (2) CNN architecture is the architecture design and optimization of pneumonia disease detection and prediction models. There are 8 CNN architectural components such as convolutional, pooling, fully connected layer and ReLU activation function, as shown in Fig. 4 (a) [1]. VGG16 is an advanced CNN architecture that has been effective for detecting and predicting. VGG16 architecture is the architecture design and optimization of pneumonia disease detection and prediction models. There are 8 CNN architectural components such as convolutional, pooling, fully connected layer and ReLU activation function, as shown in Fig. 4 (b) [2]. VGG19 is an advanced model of VGG16 that has been proven to be effective for chest X-ray image detection and prediction, with the following processes including training datasets quality and diversity [3]. CNN architecture is the design of architecture and optimization of pneumonia disease detection and prediction models. There are 8 CNN architectural components such as convolutional, pooling, fully connected layer, and ReLU activation function, as shown in Fig. 4 (c) [9]. DenseNet enhances the flow of information and gradients throughout, connecting each layer to every other layer in a feed-forward model. CNN has a sequential layer-wise architecture, and information passes through each layer sequentially. DenseNet, on the other hand, connects each layer to all subsequent layers, forming dense connections. There is a tight connection procedure. These are (1) feature reuse, (2) feature propagation facilitation and (3) reducing the number of parameters as shown in Fig. 4(d) [10]. ResNet is designed to address the challenge of training deep neural networks, where performance often degrades as the network depth increases. Skip connection that bypasses one or more than one layer. Instead of learning direct mappings from inputs to outputs, ResNet learns residual mappings, or the difference between inputs and outputs. ResNet model is used to solve the vanishing gradient problem by skipping connection, as shown in Fig. 4 (e) [11]. Inception models are designed for image recognition. They are used to capture information with various spatial scales using filters. Instead of selecting filters of various sizes, the inception module uses them at the same time and concatenates the output, making it easier to detect objects of different sizes in Chest X-ray images. The inception module includes(1) an inception module, (2) global average pooling, and (3) auxiliary classifiers that address the vanishing gradient problem, as shown in Fig. 4 (f) [12].

Mathematical Models: The mathematical models in deep learning to detect and predict pneumonia disease [13]. The convolution operations are multiple filters (p), layers (q), and three-dimensional tensors, W (p,q) = [w(p,q)i,j,k]. Where i, j, and k are the positions along the height, width, and depth sequentially. The feature maps of Layers(q) and the three-dimensional tensor H(q) = [h(q)i,j,k]. The value of q is 1, Layer(H), and the convolution operation increases the layer

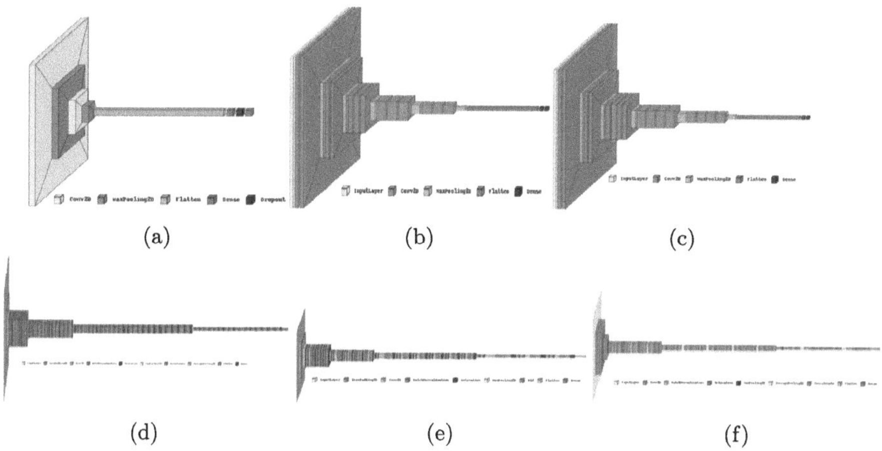

Fig. 4. Architectures of (a) CNN [9], (b) VGG16 [9], (c) VGG19 [9], (d) DenseNet [4], (e) ResNet [4], and (f) Inception [5]

to (q + 1) layer is in equation (1):

$$h_{ijp}^{q+1} = \sum_{r=1}^{F_q}\sum_{s=1}^{F_q}\sum_{k=1}^{d_q} w_{r,s,k}^{(p,q)} h_{i+r-1,j+s-1,k}^{q} \tag{1}$$

$$b_i = \frac{a_i}{(k + \alpha \sum_j a_i^2)^\beta} \tag{2}$$

The parameters has value of k $= 2$,$\alpha = 10^{-4}$, and $\beta = 0.75$ as shown equation (2). But, one does not normalize within n filters. Rather uses set of n adjacent filter for parameter n. The value of n used in 2 is 5 that shows in equation (3).

$$b_i' = \frac{a_i}{(k + \alpha \sum_{j=i-|\frac{n}{2}|}^{i+|\frac{n}{2}|} a_i^2)^\beta} \tag{3}$$

In equation(3), any value of i - n/2 less than zero is set to 0, and any value of i + n/2 greater than n is set to n.

$$w_{ij} = Uniform(-\sqrt{\frac{b}{M+n}}, \sqrt{\frac{b}{M+n}}) \tag{4}$$

where m, n w and b are input, output, weight and biases, respectively, b is initialized to 0. This equation (4) is about weight initialization. The activation functions, such as the sigmoid function and rectifier linear unit(ReLU), are described.

$$f(x)sigm = \frac{1}{1+e^{-x}} \tag{5}$$

The value range is between [0,1].

$$f(x')tanh = \frac{e^{x'} - e^{-x'}}{e^{x'} + e^{-x'}} \tag{6}$$

The value range is between [−1,1].

$$f(x)ReLU = max(0, x) \tag{7}$$

where m is the ReLU constant with small values such as 0.001 or learnable weights, this article uses the sigmoid and ReLU for activation functions.

4 Result and Discussion

This section presents the results obtained from the simulations conducted to compare the performance of pneumonia disease detection and prediction in Chest X-ray images using deep learning models. The performance of CNN, VGG16, VGG19, DenseNet, ResNet and inception models were analyzed to compare with different evaluation parameters. Tables and figures describe the brief details of this section.

Table 1. Trainable and non-trainable parameters of models in numbers

Models	*Trainable Number*	*Non-trainable Number*	*Total Number*
CNN	411,681	0	411,681
VGG19	1,605,761	20,024,384	21,630,145
DenseNet	3,211,393	7,037,504	10,248,897
ResNet	8,388,737	23,587,712	31,976,449
Inception	4,718,721	21,802,784	26,521,505

Table 1 shows that the comparative performance of the models with trainable, non-trainable and total number of parameters.

The Fig. 5: (a), (b), (c), (d), (e), and (f) shows that the comparative performance of CNN, VGG16, VGG19, DenseNet, ResNet and inception models with training and validation accuracy. Figure 5 (a) indicates that the CNN model's performance results were accurate in detecting and predicting pneumonia diseases. The CNN model achieves a high performance in that it has 97.78% accuracy. Figure 5 (b) shows the performance results of the VGG16 model with accuracy in the detection and prediction of pneumonia disease. VGG16 model achieves moderate performance. It has 91.11% accuracy. Figure 5 (c) shows the performance results of the VGG19 model with accuracy in the detection and prediction of pneumonia disease. The VGG19 model achieves a good performance that has 95.56% accuracy. Figure 5 (d) shows the performance results of the DenseNet model with accuracy in detecting and predicting pneumonia

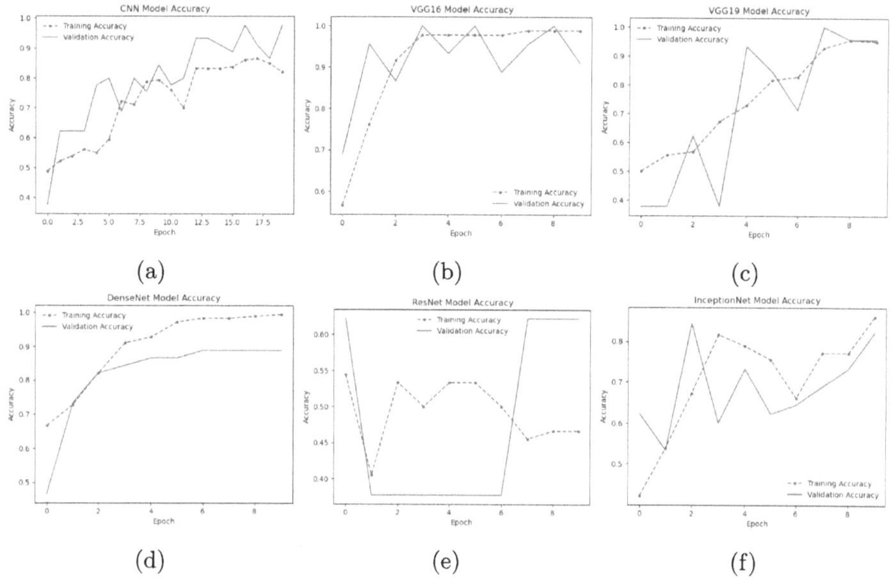

Fig. 5. Accuracy of models (a) CNN, (b) VGG16, (c) VGG19, (d) DenseNet, (e) ResNet, and (f) Inception

disease. The DenseNet model achieves good performance. It has 95.56% accuracy. Figure 5 (e) shows that ResNet model performance results with accuracy in detecting and predicting pneumonia disease. The ResNet model achieves weak performance in that it has 62.22% accuracy. Figure 5 (f) shows the performance results of the inception model with accuracy in the detection and prediction of pneumonia disease. The inception model achieves moderate performance. It has 82.22% accuracy.

The Fig. 6 (a), (b), (c), (d), (e), and (f) describe the comparative performance of CNN, VGG16, VGG19, DenseNet, ResNet, and inception models with training and validation loss. Figure 6 (a) shows that CNN model performance results in loss in detecting and predicting pneumonia diseases. The CNN model achieves a high performance that it has a 36.45% loss. Figure 6 (b) shows the performance results of the VGG16 model with loss in the detection and prediction of pneumonia disease. VGG16 model achieves moderate performance that it has a 41.89% loss. Figure 6 (c) shows the performance results of the VGG19 model with loss in the detection and prediction of pneumonia disease. The VGG19 model achieves a good performance that has a 50.67% loss. Figure 6 (d) shows the performance results of the DenseNet model with loss in the detection and prediction of pneumonia disease. The DenseNet model achieves good performance in that it has a 55% loss. Figure 6 (e) shows that ResNet model performance results in a loss in diagnosing and predicting pneumonia disease. The ResNet model achieves weak performance in that it has a 58.88% loss. Figure 6 (f) shows the performance results of the inception model with loss in the detection and prediction of pneu-

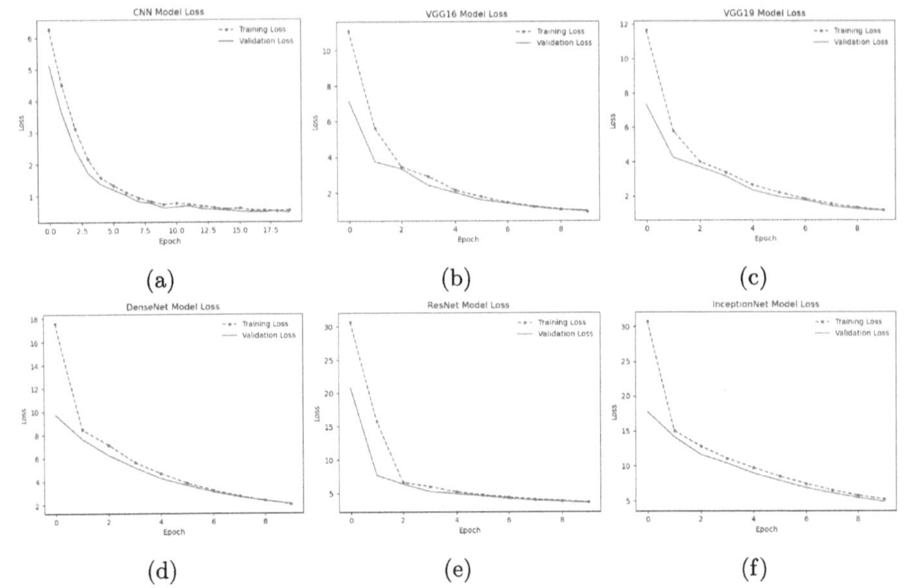

Fig. 6. Loss of models (a) CNN, (b) VGG16, (c) VGG19, (d) DenseNet, (e) ResNet, and (f) Inception

monia disease. The inception model achieves moderate performance that it has a 67.76% loss.

4.1 Performance Comparison of Models with Evaluation Parameters

The accuracy is the number of accurately predicted values divided by the total number of values, as shown in equation (9). There are true positives (TP'), false positives (FP'), true negatives (TN'), and false negatives (FN').

$$Accuracys = \frac{TN' + TP'}{TN' + TP' + FN' + FP'} \tag{8}$$

The loss is a cross-entropy loss that occurs during the detection and prediction of pneumonia disease in a Chest X-ray image. It is the performance of a detection and prediction model whose output value is between 0 and 1. The equation of cross-entropy loss(L') is described as follows in Eqs. (10) and (11).

$$L'(y, \hat{y}) = -\frac{1}{N}\sum_{i=1}^{N}[y_i \log(\hat{y_i}) + (1 - \hat{y_i}) \log(1 - \hat{y_i})] \tag{9}$$

$$L'(y, \hat{y}) = -\frac{1}{N}\sum_{i=1}^{N}\sum_{j=1}^{M} y_{ij} \log(\hat{y_{ij}}) \tag{10}$$

where; number of samples(N), number of classes(M), actual value(y), predicted value($\hat{y}$) for the i-th sample. The sensitivity is the proportion of true positives correctly predicted by the model. It is particularly important in situations where the cost of FN is high, as shown in Eq. (12).

$$Sensitivity = \frac{TP'}{TP' + FN'} \tag{11}$$

The specificity is the proportion of true negative instances correctly predicted by the model. It is crucial when the cost of FP is high, as shown in equation (13).

$$Specificity = \frac{TN'}{TN' + FP'} \tag{12}$$

Precision is the accuracy of positive predictions made by the model. It is particularly relevant when the cost of FP is high, as shown in equation (14).

$$Precisions = \frac{TP'}{TP' + FP'} \tag{13}$$

Table 2. Performance comparison of models with evaluation metrics.

Models	***Accuracy***	***Loss***	***Time***	***Sensitivity***	***Specificity***	***Precision***
CNN	97.78%	36.45%	0.023 s	96.43%	100%	100%
VGG16	91.11%	41.89%	10 s	85.71%	100%	100%
VGG19	95.56%	50.67 %	12 s	92.86%	100%	100%
DenseNet	95.56%	55%	3 s	92.86%	100%	100%
ResNet	62.22%	58.88%	4 s	100%	0%	62.22%
Inception	82.22%	67.76%	3 s	78.57%	88.24%	91.67%

Table 2 describes the comparative performance of the models with evaluation parameters of accuracy, loss, time, sensitivity, specificity, and precision respectively. This table performance parameters are used to detect and predict pneumonia disease in Chest X-ray images.

Figure 7: (a), (b), and (c), the comparative performance of CNN, VGG16, VGG19, DenseNet, ResNet and inception models with accuracy, loss and time, respectively. Figure 7: (a), (b), and (c) describe the performance results of all models with accuracy, loss and time during the detection and prediction of pneumonia disease, respectively. Among those, the CNN model has high performance compared to selected models, which has 97.78%, and ResNet is a low model that has 62.22% accuracy, respectively.

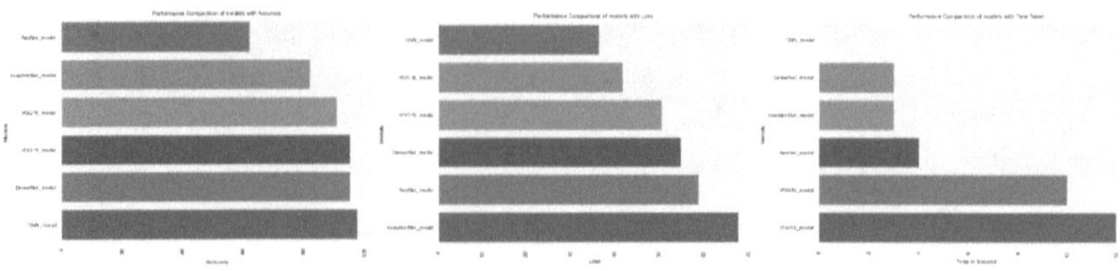

(a) (b) (c)

Fig. 7. Comparative result of models (a) accuracy, (b) loss, and (c) time

5 Conclusion

In conclusion, the performance of pneumonia disease detection and prediction models are compared and selected. The trained and tested models that pneumonia disease detection and prediction in Chest X-ray images dataset samples. Six models are chosen to compare their performance for the detection and prediction of pneumonia disease. These are CNN, VGG16, VGG19, DenseNet, ResNet and inception models. Those models are compared with their performance and selected as the best for pneumonia disease detection and prediction in Chest X-ray images based on their performance for future uses. Among those models, the CNN model performance is the best as compared to others; it achieves 97.78% accuracy, whereas the ResNet model performance is low and achieves 62.22% accuracy. Therefore, CNN is the best model among these selected models.

References

1. Shah, P.M., et al.: Deep GRU-CNN model for COVID-19 detection from chest X-rays data. IEEE Access **10**, 35094–35105 (2021)
2. Sharma, S., Guleria, K.: A deep learning based model for the detection of pneumonia from chest X-ray images using VGG-16 and neural networks. Procedia Comput. Sci. **218**, 357–366 (2023)
3. Dey, N., Zhang, Y.-D., Rajinikanth, V., Pugalenthi, R., Raja, N.S.M.: Customized VGG19 architecture for pneumonia detection in chest X-rays. Pattern Recogn. Lett. **143**, 67–74 (2021)
4. Ull, H.A., Yun, I.D., La Yun, B.: Regression to classification: Ordinal prediction of calcified vessels using customized ResNet50. IEEE Access (2023)
5. Zhang, X.: Improved three-dimensional inception networks for hyperspectral remote sensing image classification. IEEE Access **11**, 32648–32658 (2023)
6. Sharma, S., Guleria, K.: A systematic literature review on deep learning approaches for pneumonia detection using chest X-ray images. Multimedia Tools Appl. 1–51 (2023)
7. Hasan, M.M., Kabir, M.M.J., Haque, M.R., Ahmed, M.: A combined approach using image processing and deep learning to detect pneumonia from chest X-ray image. In: 2019 3rd International Conference on Electrical, Computer & Telecommunication Engineering (ICECTE), pp. 89–92. IEEE (2019)

8. Sanida, T., Tabakis, I.-M., Sanida, M.V., Sideris, A., Dasygenis, M.: A robust hybrid deep convolutional neural network for COVID-19 disease identification from chest X-ray images. Information **14**(6) 310 (2023)
9. Aluka, M., Ganesan, S., et al.: A comparative study on pre-training models of deep learning to detect lung cancer. Int. J. Intell. Syst. Appl. Eng. **11**(1), 148–155 (2023)
10. Babukarthik, R.G., Adiga, V.A.K., Sambasivam, G., Chandramohan, D., Amudhavel, J.: Prediction of COVID-19 using genetic deep learning convolutional neural network (GDCNN). IEEE Access **8**, 177647–177666 (2020)
11. Asma-Ull, H., Yun, I.D., Yun, B.L.: Regression to classification: Ordinal prediction of calcified vessels using customized ResNet50. IEEE Access **11**, 48783–48796 (2023)
12. Paula, L.P.O., Faruqui, N., Mahmud, I., Whaiduzzaman, M., Hawkin Son, E.C., Trivedi, S.: A novel front door security (FDS) algorithm using GoogleNet-BiLSTM hybridization. IEEE Access **11**, 19122–19134 (2023)
13. Yang, Z.: Classification of picture art style based on VGGNet. In: Journal of Physics: Conference Series, IOP Publishing, vol. 1774(1), pp. 012043 (2021)
14. Sharma, S., Guleria, K.: A systematic literature review on deep learning approaches for pneumonia detection using chest X-ray images. Multimedia Tools Appl. **83**(8), 24101–24151 (2024). Springer
15. Goel, S., Sobti, R.: Deep cognitive learning for enhanced pneumonia detection: employing CNNs. In: Innovative Computing and Communications: Proceedings of ICICC, Springer Nature, vol. 1039(3), pp. 213 (2025)
16. Khan, Q.W.: A machine learning-based method for COVID-19 and pneumonia detection. IgMin Res. **2**(7), 518–523 (2024)
17. Choudhry, I.A., Iqbal, S., Alhussein, M., Qureshi, A.N., Aurangzeb, K., Naqvi, R.A.: Transforming lung disease diagnosis with transfer learning using chest X-ray images on cloud computing. In: Expert Systems, Wiley Online Library, vol. 42(2), p. e13750 (2025)
18. Salam, M.A., Abdellatif, A., Abdallah, M., Salam, N.A.: A hybrid deep learning and machine learning model for multi-class lung disease detection in medical imaging. In: International Journal of Intelligent Engineering & Systems, vol. 18, p. 1 (2025)
19. Rabbah, J., Ridouani, M., Hassouni, L.: Improving pneumonia diagnosis with high-accuracy CNN-based chest X-ray image classification and integrated gradient. In: Biomedical Signal Processing and Control, Elsevier, vol. 101(1), p. 107239 (2025)

Enhanced Document Clustering Using Residue Modulo Partitioner in Map-Reduce Framework

Hemanth Kumar Vasireddi[1(✉)], Purnachandra Rao[2], Om Prakash[3], and Manikanta Vella[4]

[1] Computer Science Engineering, GITAM University, Visakhapatnam, Andhra Pradesh 530045, India
hvasired@gitam.edu
[2] Computer Science Engineering, BVRIT, Narsapur, Telangana 502313, India
[3] CSE (Data Science), Neerukonda Institute of Technology and Sciences, Visakhapatnam, Andhra Pradesh 531162, India
[4] Data Engineering, MVGR College of Engineering, Vizianagram, Andhra Pradesh 535005, India

Abstract. Information Retrieval (IR) is a domain within computer science focused on the efficient retrieval of text documents based on user queries containing specific keywords. The demand for automated IR systems has grown significantly to meet business requirements based on user inquiries. Recent techniques have proposed various IR frameworks utilizing different data mining algorithms, as well as novel map and reduce approaches, while leveraging default partitioners. However, none of these methods have effectively utilized custom partitioners. The primary goal of this paper is to implement a custom partitioner, Residue Modulo-K, where K denotes the number of reducers used in the process. Map Reduce framework utilizing proposed custom partitioner to the default partitioner is executed and the results were tabulated. Based on the experimental analysis the proposed custom partitioner outperformed the default one. The performance metrics utilized for this study are job execution time, data transfer time, CPU utilization, and F-score.

Keywords: Map-reduce · Hadoop · Clustering · Partitioner · Text data · Distributed parallel processing

1 Introduction

Big data is often seen as a challenge, primarily due to two key factors: storage and processing [1]. Storage refers to the volume and variety of data, while processing is related to the speed at which data must be handled. Traditional systems struggle to manage unstructured data effectively, and one potential solution is the Hadoop framework. Hadoop is designed to store and process large datasets, especially text data, and is comprised of two main components: HDFS (Hadoop Distributed File System) and MapReduce [2].

A. Das et al. (Eds.): MIDAS 2025, CCIS 3055, pp. 66–80, 2027.
https://doi.org/10.1007/978-3-032-35516-4_6

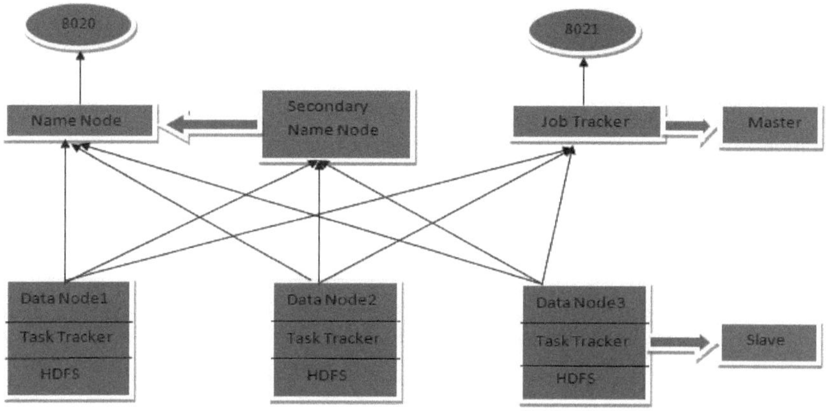

Fig. 1. Master-Slave Architecture

Figure 1 illustrates the HDFS and Map-reduce components representing Master-Slave Architecture. In the MapReduce framework, partitioners are responsible for deciding how the intermediate key-value pairs produced by the mapper are assigned to the reducers. The default partitioner uses a hash function on the key to distribute the data evenly across the reducers. However, when specific partitioning logic is required, a custom partitioner can be implemented.

The research contributions of this work is given below:

1. Implementation of a custom partitioner called Residue Modulo-K, where K represents the number of reducers, within the MapReduce framework.
2. A comparative analysis between the proposed custom partitioner and the default partitioner in terms of performance.
3. Evaluation of the custom partitioner's performance using metrics like job execution time, data transfer time, CPU utilization, and inter-cluster distance.

2 Related Work

The related work on clustering text data using various methodologies and architectures in recent trends was listed below:

Hieu Tran et al. [3] have proposed a system focused on sentiment text analysis for the identification and prediction of negative and positive trends for face book comments and batch data processing for the development of an opinion forecasting algorithm. Sang-woon Kim et al. [4] suggested a classification scheme that would allow research papers to be grouped into appropriate groups of related subjects. They extract representative keywords from the abstract of each paper using the LAD scheme.

Hey, Md. Zakir Hossain et al. [5] suggested a complex K-means clustering strategy for data mining. The proposed method performs data clustering using a hierarchical approach. Safanaz Heidari et al. [6] suggested a new algorithm for clustering large data with varying densities using Hadoop Framework with Map Reduce Programming Model.

Roberto et al. [7] focuses on Big Data Error Correction (BigDEC), where a custom partitioner optimizes k-mer spectrum-based error correction in NGS datasets, reducing data shuffling and improving workload balancing, achieving 79% faster performance on a 16-node cluster. Qureshi et al. [8] addresses heterogeneous SBC clusters in Hadoop-based edge computing, introducing a Frugality Index-based partitioner that categorizes nodes based on resource capabilities, dynamically adjusting scheduling to enhance execution speed by 56% over FIFO scheduling.

Existing methodologies for text clustering, while effective in specific contexts, often face limitations in terms of scalability, handling varying data densities, and achieving optimal clustering performance. The proposed Residue Modulo-K partitioner offers a significant advantage by addressing these challenges through customized partitioning logic.

3 Material and Methods

Map-reduce is a functional programming model that performs two functions, namely map and reduce functions [9]. The Fig. 2 below shows the map-reduce architectural workflow. The entire text is divided into several chunks or blocks, say (128 MB) each, and divides the entire data as in the respective form < k1, v1> [10].

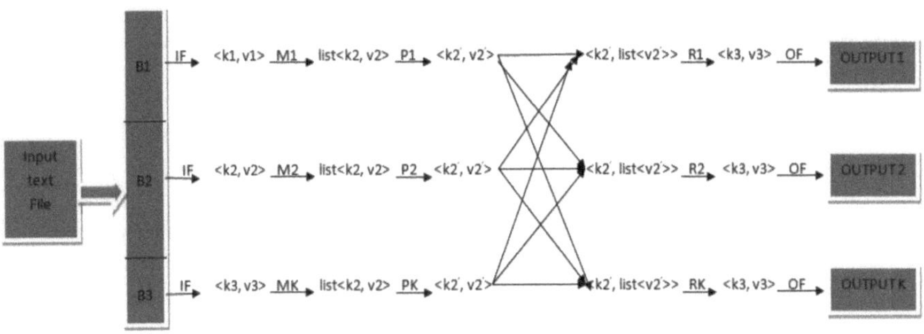

Fig. 2. Map-Reduce Architectural workflow

3.1 Proposed Methodology for Text Clustering

The entire text is divided into chunks of data or data blocks in an automated way and will be transferred to the map function [11]. The offset line and line contents of each line in the text file are passed as input to the Map function and the Map method extracts the occurrence of each term as a valued Map output (1).

3.1.1 Map Algorithm

i. Read the Input File – Open and read the text file.
ii. Process File Content – Read character by character and line by line.
iii. Tokenize Each Line – Split lines into words using whitespace.
iv. Map Line Offsets – Store line positions with corresponding content.
v. Iterate Through Words – Loop through extracted words for processing
vi. Count Word Occurrences – Track and store word frequencies.
vii. Output Word Frequency – Store words as key-value pairs with counts.

The result that is generated by the map function, i.e. every word in the text file and its corresponding count, is passed as an input to the partitioner process, and the partitioning of input data [12] into different clusters is possible as the requirement is to alphabetically cluster the text data. The total number of clusters we selected was 26 (0–25), where all the words beginning with the alphabet would go to their respective clusters. The Residue Modulo K Partitioner is designed for efficient clustering by categorizing documents into 26 clusters based on alphabetical residue values. For datasets with varying requirements, the method can be adapted by dynamically adjusting K to match the desired number of clusters, ensuring flexibility. By modifying the modulo operation or incorporating adaptive partitioning heuristics, the approach can scale effectively for diverse dataset configurations while maintaining balanced document distribution and computational efficiency.

3.1.2 Partition Algorithm

i. Read the input from the map function
<key, value> <-> <Text, long writable>
ii. Retrieve all the words either in lower case format or upper case format, follows the property of mutual exclusion.
word ← key.toString ().toLowercase ()
iii. Retrieve every word first character from the entire text file
firstchar ← word. charAt (0)
iv. Subtract the first character with character 'a' and divide the value with the number of reducers (say 26)
Diff ← abs (firstchar – 'a') % no_of_reducers
v. Return the Diff value obtained in the above step to the reducer phase.

The output obtained from the partition process will be sent to the reducer process by going through the shuffle and sorting step where the sorting of the data in the alphabetical order [13] will take place and the output produced by the shuffle phase will be < key, list <values>>.

3.1.3 Reduce Algorithm

i. Read Shuffle Input – Receive data as <key, list<values>>.
ii. Initialize Sum – Set sum to zero.
iii. Iterate Over Values – Loop through the list and add each value.
iv. Assign Final Count – Store the total sum for each word.

After implementing the three algorithms listed above, there is a need for a job file, which is used to incorporate all the functions and components of the map, reduce the architecture in a single file and set the number of reducers to 26, as needed. By default, the number of reducers will be 1 for the architecture. From the reducer process, the data will be divided into 26 folders considered to be 26 clusters through the output format (OF) part, where the clustering of text data [14] in sorted order will take place within the architecture itself in an automated way.

3.2 Text Grouping

Let us consider a corpus having a pair Z = (T, D) represented in Eqs. (1) and (2), where

$$T \in \{t_1, t_2, t_3, \ldots\ldots t_p\} \quad (1)$$

T represents finite set of elements t1,t2….tp referred as terms, and

$$D \in \{D_1, D_2, D_3, \ldots\ldots D_q\} \quad (2)$$

D represents finite set of documents with D1,D2……Dq.

For each document D_h with series of terms is given in Eq. (3), where

$$D_k \in \{t_{k1}, t_{k2}, t_{k3}, \ldots\ldots t_{kr} \ldots\ldots t_{ks}\} \quad (3)$$

Where $t_{k1}, t_{k2}, t_{k3}, \ldots\ldots t_{kr} \ldots\ldots t_{ks}$ represents the individual terms within the document D_k

Assume |T| = p terms; |D|= q documents in the corpus Z.

The weight matrix or frequency matrix of corpus Z is the matrix is given by Eq. (4), where

$$X \in R^{p*q} \text{ defined } b\ X = (X_{hk}) \quad (4)$$

R represents the set of all real valued matrices with dimension p*q.

If t_h is the term and D_k is the document of Z, then the existence of t_h in D_k is given by Eq. (5)

$$X_{hk} = |\{y | t_{ky} = t_h\}| \quad (5)$$

Where,

X_{hk} represents the frequency of term t_h in document D_k,
y denotes each occurrence of term t_h with D_k,
t_{ky} represents a specific instance of a term in D_k, and.

t_h is the term being counted across all documents.

Clustering include group of documents $Z = \{d_1, d_2, d_3 \ldots .d_k\}$, the number of cluster groups 'k' and the quality of clustering will be measured by the objective function (F) such that the possible assignment $P : Z \rightarrow \{1,2, 3 \ldots \ldots k\}$ either minimizes or maximizes the function (F). The quality of clustering is calculated using purity function, represented in Eq. (6)

$$purity(p1, p2) = \left(\frac{1}{N}\right)\sum_k max_j |w_k \cap c_j| \qquad (6)$$

To calculate the purity value, each document class is assigned with the respective cluster and reliability is evaluated by counting retrieved terms which were allotted correctly and divide with N value, where.

$p1 \in \{w_1, w_2, w_3, \ldots \ldots \ldots w_k\}$ of clusters.
$p2 \in \{c_1, c_2, c_3, \ldots \ldots \ldots c_k\}$ of classes.
w_k = set of terms in w_k.
c_j= set of terms in c_j.

4 Experimental Setup and Results

Implementation of text clustering by partitioning data, using an IDE eclipse environment with custom partitioner by running core and dependency jar files from the APACHE HADOOP repository on the UBUNTU operating system using JAVA 8 with a minimum of 4 GB RAM under the pseudo mode configuration. One key limitation is that it does not account for network latency, inter-node communication overhead, or true parallelism across multiple machines. As a result, execution time and resource utilization metrics may differ in a fully distributed setup. In a real cluster, performance variations may arise due to load balancing inefficiencies, hardware differences, and network congestion, which are not accurately reflected in pseudo-distributed mode. Then start the Hadoop environment in the pseudo node and check whether all the process daemons listed in Table 1 are running. The following is Fig. 3 showing the status of the Hadoop process daemons.

Table 1. Process Daemons in Hadoop Environment

		Process Name	RPC port	HTTP port
HADOOP	HDFS	Name Node	8020	50070
		Secondary Name Node	----	50075
		Data Node	-----	50090
	Map-reduce	Job Tracker	8021	50030
		Task Tracker	-----'	50060

The following is the Fig. 4 that shows the successful execution of Map reduce framework for the clustered input document. After the successful execution the processed input document using Map reduce framework will be categorized into 26 clusters [15]. The following were the Figs. 5 and 6 that shows the HDFS metadata after performing the clustering process. The following is the Fig. 7 that gives the complete metadata information about the name node and the progress of the entire task that is executed for document clustering.

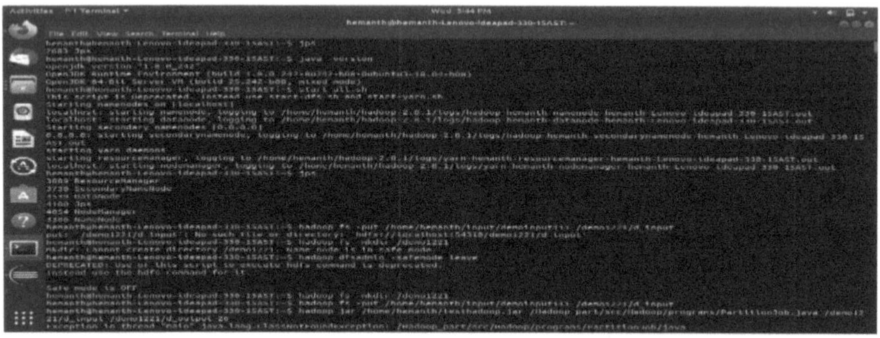

Fig. 3. Process Daemons status in Hadoop Environment

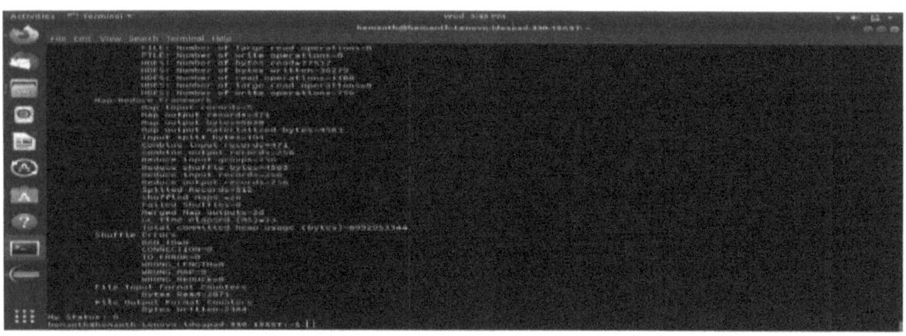

Fig. 4. Map reduce framework execution status

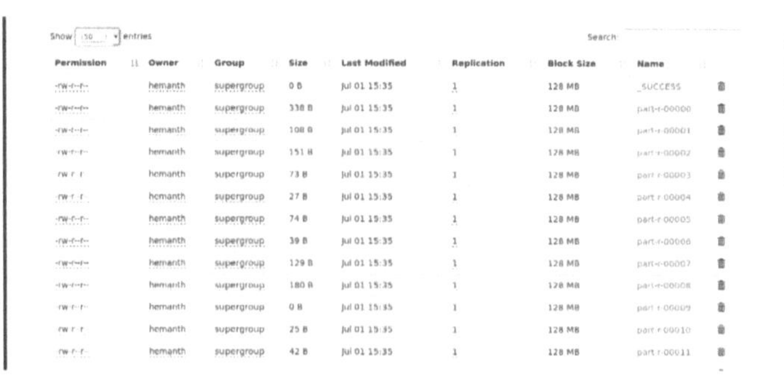

Fig. 5. HDFS metadata after document clustering

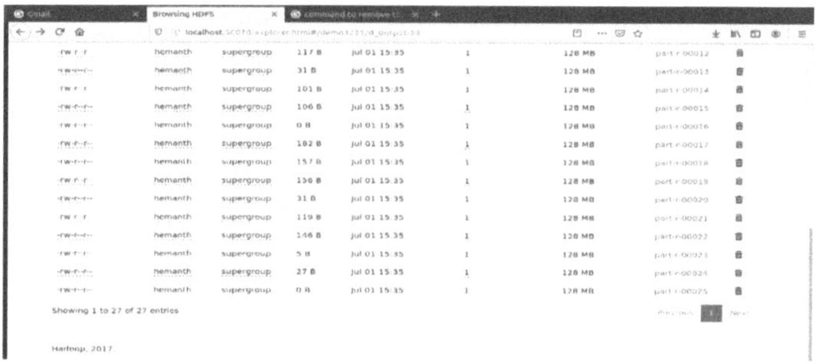

Fig. 6. HDFS metadata after document clustering

Fig. 7. Startup progress for document clustering

5 Discussion

The dataset utilized for the experiments consists of a medium-sized text file containing approximately 500,000 words, with continuous words separated by spaces. The dataset, sourced from publicly available text corpora, represents real-world document clustering scenarios. This ensures a balanced evaluation of the Residue Modulo K Partitioner, demonstrating its efficiency in handling structured textual data while maintaining scalability and accuracy. The following is the Table 2 that shows the related work that happened and their performance metrics.

Table 2. Performance metrics in Related work

[Ref No] year	Technique/Method	Dataset	Performance Metric
[16] 2019	K-Means TF-IDF LDA TF-IDF + LDA	Papers in FGCS Journal	F-score TF-IDF – 0.82 LDA-0.84 TF-IDF + LDA-0.86
[17] 2019	Deep learning	Text dataset	F-Score – 0.88

(continued)

Table 2. (*continued*)

[Ref No] year	Technique/Method	Dataset	Performance Metric
[18] 2019	Novel K-Means	140 SIIR 75 HABER	F-Score Euclidean distance-0.40 Cosinus similarity-0.46 Jaccard similarity-0.44
[19] 2020	K Means Clustering	Custom Text Dataset	Notable Improvement in F-score (as mentioned in manuscript)
[20] 2023	MapReduce-Enhanced Fuzzy K-Least Medians for Qualitative Clustering	Self-crawled Document Datasets	F-score: 0.89
[21] 2024	MapReduce in Document Clustering Using Big Data	Large Document Collections	F-score: 0.88
Proposed method	Residue Modulo K Partitioner	Text dataset	F-Score 26 clusters – 0.90

Traditional methods such as K-Means and TF-IDF + LDA, while effective, often struggle with load balancing and computational efficiency in large-scale document clustering tasks [22]. Deep learning approaches, though achieving an F-score of 0.88, may require high computational resources [23]. Recent advancements, including MapReduce-enhanced fuzzy clustering (F-score: 0.89) and big data-based clustering (F-score: 0.88), exhibit improvements but may still face limitations in efficient document distribution. The Residue Modulo K Partitioner ensures balanced data distribution, leading to more coherent clusters. Its integration with MapReduce enhances scalability and reduces computational overhead. Additionally, it improves similarity measures, resulting in a higher F-score of 0.90. Table 3 illustrates the performance analysis of the custom partitioner.

Table 3. Performance metrics using custom partitioner

Cluster Name	Cluster Size(Bytes)	No of Words in Cluster/ No of Words in document	Precision	Recall	F-Score
Part-r-00000	338	32/469	0.96	1.0	0.97
Part-r-00001	108	11/469	1.0	1.0	1.0
Part-r-00002	151	14/469	1.0	1.0	1.0

(*continued*)

Table 3. (*continued*)

Cluster Name	Cluster Size(Bytes)	No of Words in Cluster/ No of Words in document	Precision	Recall	F-Score
Part-r-00003	73	8/469	1.0	1.0	1.0
Part-r-00004	27	3/469	1.0	1.0	1.0
Part-r-00005	74	9/469	0.96	1.0	0.97
Part-r-00006	39	4/469	1.0	1.0	1.0
Part-r-00007	129	16/469	1.0	1.0	1.0
Part-r-00008	180	17/469	1.0	1.0	1.0
Part-r-00009	0	0/469	0	0	0
Part-r-00010	25	3/469	1.0	1.0	1.0
Part-r-00011	42	5/469	0.60	1.0	0.75
Part-r-00012	117	13/469	1.0	1.0	1.0
Part-r-00013	31	3/469	1.0	1.0	1.0
Part-r-00014	101	13/469	1.0	1.0	1.0
Part-r-00015	106	11/469	1.0	1.0	1.0
Part-r-00016	0	0/469	0	0	0
Part-r-00017	182	18/469	0.96	1.0	0.97
Part-r-00018	157	16/469	1.0	1.0 ara>	1.0
Part-r-00019	156	17/469	1.0	1.0	1.0
Part-r-00020	31	4/469	0.50	1.0	0.66
Part-r-00021	119	15/469	0.13	1.0	0.23
Part-r-00022	146	19/469	0.52	1.0	0.68
Part-r-00023	5	1/469	1.0	1.0	1.0
Part-r-00024	27	3/469	1.0	1.0	1.0
Part-r-00025	0	0/469	0	0	0

The proposed work is entirely dependent on the partition algorithm that is implemented, and many of the researchers have focused a lot on different novel maps and reduced algorithms using the default partitioner. In contrast, instead of using the default partitioner, proposed a Residue Modulo-K partitioner for effective document clustering that meets business requirements.

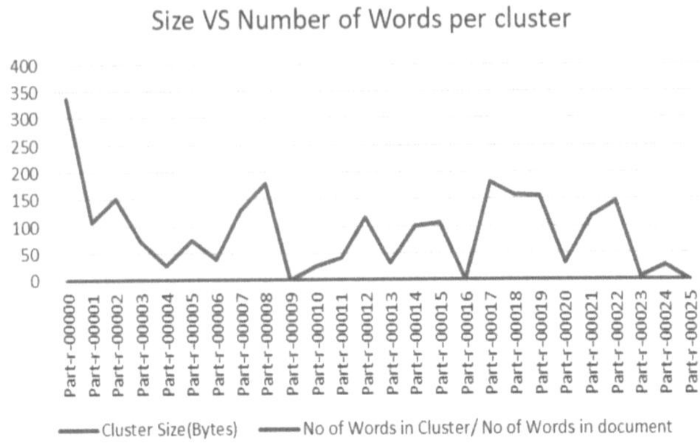

Fig. 8. Cluster size VS No of words per cluster

In the proposed partition algorithm, authors concentrate on any token of word that is present in the text, regardless of boundary limitations, and make an absolute difference between the two vectors. When considering the number of reducers we use, the obtained value will be invoked with the mod counter. The following is the Fig. 9 that shows the graph result for the table mentioned above. As observed in the Fig. 8, for the partitioners 9, 16 and 25 the performance metrics value were recorded as zero as there were no search words were available to the related input query. Clusters recording zero values in performance metrics occur because the Residue Modulo K Partitioner maps words based on their starting letter, and some clusters may remain empty if no words begin with the corresponding letter. Since the model uses 26 clusters (A-Z), certain letters may not appear in the dataset, naturally leading to empty clusters. To mitigate this, techniques like merging low-density clusters, dynamic cluster reallocation, or preprocessing to redistribute words more evenly can be explored, ensuring a more balanced clustering outcome. The following is the Fig. 9 that shows the performance metrics in terms of precision, recall and F-score for the clustered documents.

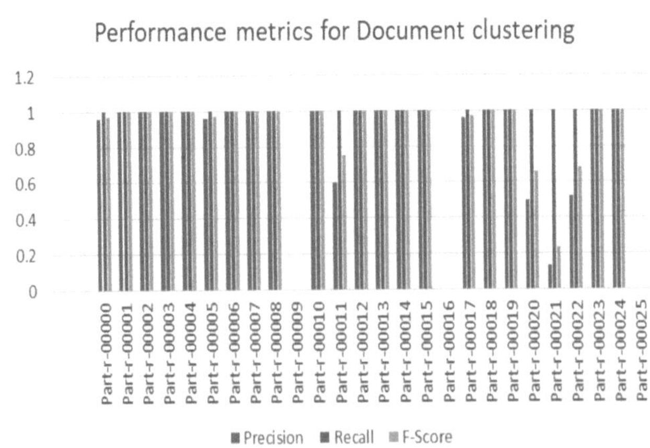

Fig. 9. Performance metrics for Document clustering

The improvements in F-score and other performance metrics indicate enhanced clustering accuracy and efficiency, but their real-world implications deserve further exploration. A higher F-score (0.90) suggests better document categorization, which is crucial for information retrieval systems, automated text classification, and recommendation engines. The precision, recall and F-Score metrics were calculated using the following mathematical representation represented with Eqs. (7), (8), and (9).

$$precision(P) = \frac{\{relevant\ \ words\ \ in\ \ document\} \cap \{retrieved\ \ words\ \ in\ \ document\}}{\{retrieved\ \ words\ \ in\ \ document\}} \tag{7}$$

$$recall(R) = \frac{\{relevant\ \ words\ \ in\ \ document\} \cap \{retrieved\ \ words\ \ in\ \ document\}}{\{relevant\ \ words\ \ in\ \ document\}} \tag{8}$$

$$F - Score(F) = \frac{2 * P * R}{(P + R)} \tag{9}$$

Table 4 illustrates the performance comparison between default and custom partitioner using the performance metrics Job execution time, Data transfer time, CPU utilization, and inter-cluster distance.

Table 4. Performance evaluation between Custom and Default Partitioner

Partitioner	Job Execution Time (seconds)	Data Transfer Time (seconds)	CPU Utilization (%)	Inter-Cluster Distance	Memory Usage (MB)	Scalability (Execution time of datasheet in seconds
Default Partitioner	320	85	70%	150	1024	450
Custom Partitioner (Residue Modulo-K)	280	60	75%	120	896	400

The custom partitioner demonstrates clear improvements over the default partitioner across key performance metrics [24]. It reduced job execution time by 12.5% (from 320 s to 280 s), showcasing its ability to streamline processing by distributing data more effectively across reducers. Additionally, the data transfer time decreased by 29.4% (from 85 s to 60 s), indicating that the custom partitioner minimizes unnecessary data shuffling, thereby reducing network load and improving transfer efficiency. In terms of CPU utilization, the custom partitioner increased efficiency by 5% (from 70% to 75%), utilizing processing resources more effectively due to balanced workloads. The inter-cluster distance was reduced by 20% (from 150 to 120), showing that the custom partitioner achieves more compact and relevant clusters.

The custom Residue Modulo-K partitioner outperforms the default partitioner by ensuring a more balanced distribution of data across reducers, reducing memory overhead and preventing processing bottlenecks. Unlike the default hash-based partitioner, which can create uneven workloads, the modular approach optimally assigns documents, leading to lower memory usage of 896 MB when compared with the default one of 1024 MB. This efficient load balancing also enhances scalability, as reducers process data more uniformly, resulting in a faster execution time of 400 s when compared with the default one of 450 s for larger datasets. These improvements make the custom partitioner a better choice for large-scale document clustering in Hadoop's MapReduce framework. These metrics provide a more comprehensive evaluation of clustering efficiency, ensuring both resource optimization and scalability in a distributed computing environment. These improvements highlight how a custom partitioner can better align with specific data patterns and needs, optimizing both the performance and quality of clustering compared to the default approach.

6 Conclusion and Future Scope

The primary objective of an Information Retrieval (IR) system is to extract relevant information from a vast text corpus based on user queries. Many traditional IR systems have utilized established data mining techniques, with researchers drawing necessary conclusions from their findings. While several of these approaches have been documented in literature, limited research has been conducted on Big Data Systems, which enable large-scale data processing. In this work, a novel MapReduce algorithm is introduced alongside a custom partitioner named Residue Modulo K, where K denotes the number of reducers utilized in the system. Performance metrics have been assessed for each clustered document, achieving optimal significance through maximum likelihood values. The obtained results also illustrate the cluster size and word distribution within each cluster after document clustering, facilitating comprehensive performance analysis, an aspect not previously highlighted in past studies. Additionally, there is potential for integrating a custom combiner class to minimize data transfer and computation time, ultimately enhancing the efficiency of the IR system.

Compliance with Ethical Standards:

Conflict of Interest: The authors state that they do not have any conflicts of interest.

Funding: No funding is received for this work.

Ethical Approval: This article contains no studies with human or animal subjects conducted by authors.

References

1. Elgendy, N., Elragal, A.: Big data analytics: a literature review paper. Lect. Notes Comput. Sci. (including subseries Lecture Notes in Artificial Intelligence and Lecture Notes in Bioinformatics) (2014). https://doi.org/10.1007/978-3-319-08976-8_16
2. Acharjya, D.P., Ahmed, K.: A survey on big data analytics: challenges, open research issues and tools. Int. J. Adv. Comput. Sci. Appl. (2016). https://doi.org/10.14569/ijacsa.2016.070267
3. Tran, H., Shcherbakov, M.: Detection and prediction of users attitude based on real-time and batch sentiment analysis of facebook comments. Lect. Notes Comput. Sci. (including subseries Lecture Notes in Artificial Intelligence and Lecture Notes in Bioinformatics) (2016). https://doi.org/10.1007/978-3-319-42345-6_24

4. Kim, S.W., Gil, J.M.: Research paper classification systems based on TF-IDF and LDA schemes. HCIS (2019). https://doi.org/10.1186/s13673-019-0192-7
5. Hossain, M.Z., Akhtar, M.N., Ahmad, R.B., Rahman, M.: A dynamic K-means clustering for data mining. Indonesian J. Elec. Eng. Comput. Sci. (2019). https://doi.org/10.11591/ijeecs.v13.i2.pp521-526
6. Heidari, S., Alborzi, M., Radfar, R., Afsharkazemi, M.A., Rajabzadeh Ghatari, A.: Big data clustering with varied density based on MapReduce. J. Big Data (2019). https://doi.org/10.1186/s40537-019-0236-x
7. Expósito, R.R., González-Domínguez, J.: BigDEC: a multi-algorithm Big Data tool based on the k-mer spectrum method for scalable short-read error correction. Futur. Gener. Comput. Syst. **154**, 314–329 (2024). https://doi.org/10.1016/j.future.2024.01.011
8. Qureshi, B.: Optimizing Hadoop scheduling in single-board-computer-based heterogeneous clusters. Computation **12**(5), May 2024. https://doi.org/10.3390/computation12050096
9. Du, W., Qian, D., Xie, M., Chen, W.: Research and implementation of mapreduce programming oriented graphical modeling system. In: Proceedings - 16th IEEE International Conference on Computational Science and Engineering, CSE 2013 (2013). https://doi.org/10.1109/CSE.2013.197
10. Condie, T., Conway, N., Alvaro, P., Hellerstein, J.M., Elmeleegy, K., Sears, R.: MapReduce online. In: Proceedings of NSDI 2010: 7th USENIX Symposium on Networked Systems Design and Implementation (2010)
11. Idris, M., Hussain, S., Siddiqi, M.H., Hassan, W., Bilal, H.S.M., Lee, S.: MRPack: multi-algorithm execution using compute-intensive approach in MapReduce. PLoS ONE (2015). https://doi.org/10.1371/journal.pone.0136259
12. Pericini, M.H.M., Leite, L.G.M., de Carvalho-Junior, F.H., Machado, J.C., Rezende, C.A.: MAPSKEW: metaheuristic approaches for partitioning skew in MapReduce. Algorithms (2019). https://doi.org/10.3390/a12010005
13. Benabdellah, A.C., Benghabrit, A., Bouhaddou, I.: A survey of clustering algorithms for an industrial context. Proc. Comput. Sci. (2019). https://doi.org/10.1016/j.procs.2019.01.022
14. Anand, N., Vikram, P.: Comprehensive analysis & performance comparison of clustering algorithms for big data. Rev. Comput. Eng. Res. **4**(2), 54–80 (2017). https://doi.org/10.18488/journal.76.2017.42.54.80
15. Reddy, V.S., Kinnicutt, P., Lee, R.: Text document clustering: the application of cluster analysis to textual document. In: Proceedings - 2016 International Conference on Computational Science and Computational Intelligence, CSCI 2016 (2017). https://doi.org/10.1109/CSCI.2016.0222
16. Kim, S.W., Gil, J.M.: Research paper classification systems based on TF-IDF and LDA schemes. Human-centric Comput. Inform. Sci. **9**(1), December 2019. https://doi.org/10.1186/s13673-019-0192-7
17. Heidari, S., Alborzi, M., Radfar, R., Afsharkazemi, M.A., Rajabzadeh Ghatari, A.: Big data clustering with varied density based on MapReduce. J. Big Data **6**(1), December 2019. https://doi.org/10.1186/s40537-019-0236-x
18. Ordin, B., Selin Ballı, D., Satı, N.U.: An incremental approach for solution of text clustering problem (2019)
19. Sardar, T.H., Ansari, Z.: An analysis of distributed document clustering using MapReduce based k-means algorithm. J. Inst. Eng. (India) Ser. B **101**(6), 641–650, December 2020. https://doi.org/10.1007/s40031-020-00485-2
20. Sardar, T.H., Ansari, Z.A., Theerthagiri, P., Karthikeyan, P.: MapReduce-enhanced Fuzzy K-Least medians for qualitative clustering of document big data, 21 April 2023. https://doi.org/10.21203/rs.3.rs-2831504/v1

21. Vishnu Sakthi, D., Valarmathi, V., Surya, V., Karthikeyan, A., Malathi, E.: Bigdata clustering and classification with improved fuzzy based deep architecture under MapReduce framework. Intell. Decis. Technol. **18**(2), 1511–1540 (2024). https://doi.org/10.3233/IDT-230537
22. Siddique, M., Mohanty, S.: A Hybrid forecasting model for prediction of stock value of Tata steel using support vector regression and particle swarm optimization (2018). https://www.researchgate.net/publication/326588540
23. Siddique, M., Panda, D.: Prediction of stock index of Tata steel using hybrid machine learning based optimization techniques. Int. J. Rec. Technol. Eng. **8**(2), 3186–3193 (2019). https://doi.org/10.35940/ijrte.B3223.078219
24. Nayak, S.K., Pattnaik, S., Siddique, M., Garanayak, M., Paikaray, B.K.: A hybrid CNN-LSTM approach to enhancing temperature forecasting for environmental threats and risk management. Int. J. Bus. Continuity Risk Manag. **14**(4), 371–391 (2024)

Robust Security for IoT-Based Wireless Sensor Networks Using Weighted CNN Based AZOA

V Uma Maheswari[1], Rajanikanth Aluvalu[2(✉)], Suresh Salendra[3], Swapna Mudrakola[4], and K Sudheer Reddy[5]

[1] Chaitanya Bharathi Institute of Technology, Kokapet, India
umamaheswari@ieee.org

[2] Symbiosis Institute of Technology, Hyderabad Campus, Symbiosis International University, Pune, India
rajanikanth.aluvalu@ieee.org

[3] Balaji Institute of Technology and Science, Telangana, India

[4] Aurora University, Hyderabad, India

[5] Anurag University, Hyderabad, India

Abstract. Wireless Sensor Networks (WSNs) are integral to various human activities, particularly in critical applications like monitoring and intrusion detection in contested regions and military bases. The study presents an innovative methodology to rapid intrusion detection using a Weighted Convolutional Neural Network (WCNN) enhanced by the American Zebra Optimization Algorithm (AZOA). WCNN employs weighted features to improve the identification of attacks such as Blackhole, Grayhole, Flooding, and Scheduling, while AZOA optimizes network parameters for enhanced accuracy and faster convergence. Experimental results demonstrate the effectiveness of this approach, yielding high detection rates, enhanced precision and decreased false positive rates in contrast to conventional techniques. This integration not only fortifies the security framework of Internet of Things (IoT)-based WSNs but also offers a scalable solution for large-scale deployments. This study introduces a novel approach combining a WCNN with the AZOA for intrusion detection in IoT-based WSNs. The use of WCNN enhances anomaly detection through feature weighting, improving the identification of sophisticated attacks like Blackhole, Grayhole, Flooding, and Scheduling. The findings underscore the potential of combining advanced neural network techniques with adaptive optimization algorithms to enhance the security and reliability of IoT networks, paving the way for further research and development in this critical area.

Keywords: AZOZ · WSNs · WCNN · IoT · Blackhole · Grayhole attacks

A. Das et al. (Eds.): MIDAS 2025, CCIS 3055, pp. 81–93, 2027.
https://doi.org/10.1007/978-3-032-35516-4_7

1 First Section

As the IoT continues to grow at a rapid pace, the mere interconnectedness of devices, people, and places is no longer enough. Managing and making full use of the enormous amounts of real-time monitoring data presents both an opportunity and a challenge for the future of the IoT [1,2]. WSNs are the backbone technology of the IoT and have found numerous civilian uses beyond their original military deployment, including smart homes, agriculture, industry 4.0, environmental protection, healthcare, and traffic management [3][?]. An explosion in WSN data volume is unavoidable. A critical issue that needs resolving immediately is how to scientifically manage these data in order to guarantee the dependability, timeliness, and safety of WSN. With the use of Artificial Intelligence (AI) technology, the entire IoT system, including WSN, may overcome the obstacle and improve [?]. We tackle the challenges of border mitigation by utilizing two different approaches. In the first approach, researchers provide a number of methods, including feed-forward networks, for monitoring and intrusion detection in border regions [?].

Improving data quality, decreasing reaction time, increasing network security, and extending network lifetime are the four main design aims of WSN. The aforementioned objectives cannot be met without anomaly detection technology [4]. Data that appears to be out of the ordinary is unavoidable due to a number of factors over which we have no control. In addition to increasing network traffic and latency, these inaccurate data will also squander valuable energy resources [5]. Also, any data-based analysis or processing that follows will be impacted by the unreliability of the original data. As a result, efficiently and promptly identifying anomalies in the acquired data is crucial. The detection speed and accuracy still have a ways to go, despite the fact that various solutions have been suggested for this issue [6]. Using AI extensively and deeply could be a great way for WSN to overcome these technological obstacles. The two main categories of anomaly identification, as determined by the data source for outliers, are those based on errors and those based on unusual occurrences [7].

The wide variety of IoT security flaws is a major obstacle. IoT devices are susceptible to a wide vulnerabilities [8]. Except for cyber-physical attacks, which are beyond the scope of this investigation, all cyber-attacks include network connectivity. The systems used by IoT devices have computational limitations. That is why it is using standard cyber-security practices to secure insecure IoT environments is not practical. [9]. The IoT environment is also beyond the purview of traditional rule-based security approaches due to the wide variety of potential vulnerabilities [10].

2 Proposed Model

The proposed model integrates a WCNN with the AZOA to enhance the security and resilience of IoT-based WSNs. The WCNN prioritizes critical features by applying weights, improving the accuracy and reliability of attack detection for various types of intrusions, including Blackhole, Grayhole, Flooding, and Scheduling attacks. The AZOA dynamically optimizes network parameters during the training process, resulting in faster convergence and superior generalization performance. Additionally, the model is designed to be scalable, maintaining its performance across large IoT networks without degradation. Comprehensive preprocessing steps ensure high data quality, which is vital for accurate machine learning-based intrusion detection. Extensive experimental validation demonstrates the model's effectiveness, showcasing high detection rates, improved accuracy, significantly decreased false positives compared to older approaches. By combining advanced neural network techniques with adaptive optimization algorithms, the proposed model offers a robust, efficient, and scalable solution for enhancing the security of IoT-based WSNs. In this section, the workflow of the projected model is brief explained that is shown in Fig. 1.

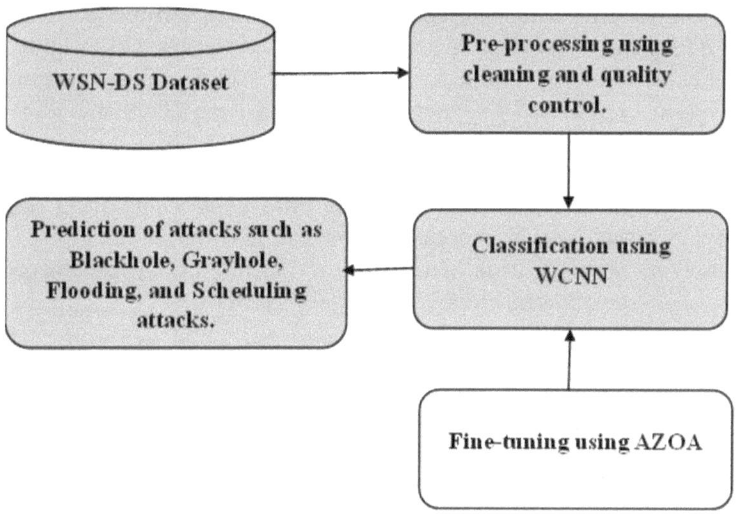

Fig. 1. Workflow of the proposed model

2.1 Dataset Description

The proposed framework, WSN-DS, utilizes a publicly available dataset from the Kaggle platform designed to simulate various DoS attacks on WSNs [22]. It majorly classifies four types of attacks. This dataset encompasses four main types of attacks: flooding, scheduling, blackhole, and grayhole, each aimed at assessing the impact on the Low-energy Adaptive Clustering Hierarchy (LEACH) protocol and network performance.

Metrics used for comparison includes accuracy, precision, recall and F1-score, where the experimental conditions are taken by including hardware setup, dataset descriptions and the models are compared with the effectiveness of proposed model, which clearly improves the transparency in the discussion results.

The kernel weights in convolutional layers, learning rate, epochs and batch sizes of WCNN are optimized by AZOA. AZOA initializes candidate solutions (e.g., sets of parameters or weights) based on the WCNN architecture. Each candidate solution is evaluated using a fitness function derived from WCNN's performance metrics (e.g., validation accuracy or loss). AZOA updates the solutions iteratively, striking a balance between exploitation and exploration in order to arrive at ideal parameters. The optimized parameters are applied to WCNN, and the network is trained further or directly evaluated.

2.2 Pre-processing

Data-driven systems heavily rely on preprocessing and meticulous data collection to ensure data quality and applicability. This critical step is particularly crucial in cybersecurity frameworks employing machine learning for attack detection and prevention. Raw data is sourced from diverse network origins, including WSNs used in Industry 4.0, encompassing sensor readings, network traffic records, system events, and relevant cybersecurity data. This amalgam of data types includes text, numerical sensor readings, and categorized records, both in real-time streaming and archival forms. Effective preprocessing involves cleaning and quality control measures such as filling data gaps, eliminating duplicates, and rectifying discrepancies. Anomaly detection and outlier handling are integral, followed by normalization or transformation to standardize feature scales, critical for scale-sensitive machine learning algorithms.

2.3 Classification Using WCNN

- A DL architecture specifically tailored for intrusion detection in IoT-based WSNs. It includes convolutional layers for feature extraction from network traffic data and weighted connections to prioritize critical features relevant to security threats.
- **Training:** Utilizes labeled datasets to train the CNN model, focusing on detecting anomalies such as intrusion attempts (e.g., Blackhole, Grayhole, Flooding).

Classification using the WCNN in the proposed model involves several key steps tailored for effective intrusion detection in IoT-based WSNs [23–24]. Initially, the WCNN architecture is designed to accommodate the specific characteristics of IoT data, which often include high dimensionality, noise, and varying levels of relevance among features. The WCNN leverages convolutional layers to automatically extract hierarchical representations of input data, which in this context include network traffic patterns, sensor readings, and system events crucial for detecting anomalies. These convolutional layers are complemented by pooling layers that reduce the spatial dimensions of feature maps while retaining important information, thereby enhancing computational efficiency and preventing overfitting.

Algorithm 1. WCNN Model for Attack Classification

1: **Embedding layer:** Encode input data to create a space-to-space map using embedding encoding or one-hot encoding.

$$x = W_E \cdot \{s_1, s_2, \ldots, s_n\}$$

where $W_E \in \mathbb{R}^{V \times D}$ is a learnable embedding matrix.

2: **Feature extraction layer:** Utilize a broad time-domain convolution network with different convolution kernel sizes to record features at various points in time.

$$f(x_h) = f(x * w_h + b_h)$$

where w_h is a 1D kernel with size h and b_h is a bias term.

3: Apply maximum pooling for downsampling:

$$f(x_c) = BN\left(\max\left(f(x_1), \ldots, f(x_h)\right)\right)$$

4: Fully connected layer and ReLU activation:

$$f(x_0) = FA\left(Drop\left(FA(x_c), p\right)\right)$$

p is the dropout rate in this case.

5: **Classification layer:** For assault classification, use a softmax function after a completely connected layer.

$$p(\hat{y}_i) = \frac{e^{\hat{y}_i}}{\sum_{j=1}^{n} e^{\hat{y}_j}}, \quad \hat{y} = x_0 \cdot w_n + b_n$$

where n is the number of attack classes.

6: **Network Structure:**

$$E(64) - BN - [C(1,32) - C(3,32) - C(5,16) - C(7,8)]$$
$$- MAX - BN - D(32) - ReLU - Drop(0.4) - D(32) - ReLU - D(c) - Sm$$

$E(n)$ is encoding, BN is batch normalization, $C(n,m)$ is 1-D convolution with kernel size n and m filters, MAX is max pooling, $D(n)$ is with n neurons and fully connected layer, c is number of classes, Sm is softmax, and $Drop(0.4)$ is dropout with a rate of 0.4.

2.4 Fine-Tuning Using AZOA Model

Authors brief the technicalities of the WCNN combined with the AZOA here. The process is divided into three main Levels, each with specific Phases. We also introduce Algorithms 1 and 2 to describe the methodologies used for the WCNN and AZOA processes, respectively.

Fine-tuning using the AZOA in the proposed model involves dynamically adjusting the parameters of the WCNN to optimize its performance in intrusion detection for IoT-based WSNs. AZOA, inspired by the social behavior of zebras, iteratively refines the network's weights and biases based on the objective function's evaluation, aiming to minimize detection errors and maximize accuracy.

Algorithm 2. Fine-tuning using AZOA Model

1: **Phase 1: Zebra Group Formation**

$$N = S\prod P$$

$$Z_{i,j} = (Z_{\max} - Z_{\min}) \cdot \text{rand} + Z_{\min}$$

2: **Phase 2: Feeding Behavior**

$$\overline{z}_i^j = \begin{cases} 2R_1 \sin(2\pi R_2) \times (z_s^j - z_i^j) + z_s^j & \text{if } R_3 < 0.5 \\ 2R_1 \cos(2\pi R_2) \times (z_s^j - z_i^j) + z_s^j & \text{otherwise} \end{cases}$$

$$z_i^j = \begin{cases} \overline{z}_i^j & \text{if } \overline{F}_i^j < F_i^j \\ z_i^j & \text{otherwise} \end{cases}$$

3: **Phase 3: Mating Behavior**

$$z_j^q = \text{Crossover}(z_i^a, z_j^b) \quad \text{if } r < p_c, i \neq j$$

$$z_k^q = \text{Crossover}(z_i^a, z_k^c) \quad \text{if } r \geq p_c, i \neq k, \forall i, j, k \in N$$

4: **Phase 4: Group Management**

$$\overline{z}_s^j = \begin{cases} 2R_4 \sin(2\pi R_5) \times (WR - z_s^j) + WR & \text{if } R_6 < 0.5 \\ 2R_4 \cos(2\pi R_5) \times (WR - z_s^j) + WR & \text{otherwise} \end{cases}$$

$$z_s^j = \begin{cases} \overline{z}_s^j & \text{if } \overline{F}_s^j < F_s^j \\ z_s^j & \text{otherwise} \end{cases}$$

5: **Phase 5: Leadership Transition**

$$z_s^i = z_i^j \quad \text{if } F(z_i^j) < F(z_s^j), \forall i \in N_j$$

This adaptive optimization process ensures that the WCNN effectively learns and adapts to the varying patterns and complexities inherent in IoT traffic, enhancing its ability to discern between normal network behavior and malicious activities such as Blackhole, Grayhole, Flooding, and Scheduling attacks. By continuously refining the model parameters, AZOA enables faster In order to fine-tune the parameters of the WCNN model, this part discusses the mathematical modeling of the behavior of American zebras using AZOA algorithm [25]. Following are the five main stages of an American zebra's life cycle: **Level 1:** Creation of collections.

Level 2: Feeding action.

Level 3: Breeding action.

Level 4: Group leadership.

Level 5: Leadership transition period of selecting a novel leader.

Phase 1: Zebra group formation at random. In tend to congregate in small groups, perhaps segmented by the presence of a dominant stallion. The formula $N = S \prod P$ is used to compute the total number of groups, and the symbol 'P' represents stallion probability, denoted as 'S'. This zebra's location in the jth group ($Z_{i,j \in N} = Z_{ij1}, Z_{ij2}, Z_{ij3}, \ldots, Z_{ijn}$-) using the formula $Z_{i,j} = (Z_{\max} - Z_{\min}) rand + Z_{\min}$. Here, the extreme defined by $Z_{\max}$ and $Z_{\min}$ respectively. "Rand" is shorthand for a value that could be anywhere from zero to one. Thanks to this system, there will always be N distinct zebra herds, each headed by a different stallion.

Phase 2: Behavior of American zebras when feeding. Zebras receives nutrition from grasses. Due to it is so challenging for juvenile zebras to get their hands on new grass and green leaves, they look up to their family leader. Consequently, zebras always circle the dominant stallion in the herd. The following are the equations for the feeding behavior.

$$\overline{z}_i^j = \{ \begin{matrix} 2R_1 \sin(2\pi R_2) \times \left(z_s^j - z_i^j\right) + z_s^j, & if\ R_3 < 0.5 \\ 2R_1 \cos(2\pi R_2) \times \left(z_s^j - z_i^j\right) + z_s^j, & otherwise \end{matrix}, \tag{1}$$

$$z_i^j = \begin{cases} \overline{z}_i^j, & \overline{F}_i^j < F_i^j \\ z_i^j & otherwise \end{cases} \tag{2}$$

where z_s^j and z_i^j, characterize the site of the stallion and ith zebra of the jth group, respectively, N_j embodies the total memberships in the jth group, R_1 designates a uniform random value between [-2, 2] group, R_2 which is assessed by Eq. (8), R_3 the family, z_i^j characterizes the novel update ith member site, and last of all, $\overline{F}_i^j$ is its ith zebra.

$$R_2 = 1 - t \times \left(\frac{1}{T}\right) \tag{3}$$

The maximum iteration is denoted by T, whereas the current iteration is represented by t.

Phase 3: Behavior of American zebras during mating season. An abundance of low-lying animal species, including zebras, horses, cows, and donkeys, is crucial for maintaining a healthy food web. Therefore, these creatures procreate abundantly. Zebras are unique among these animals in their behavior, which helps to maintain family honor. They refrain from having offspring through their own family. So, when they are young, zebras will leave their families to procreate with another group. Here, the ith group's baby zebra has a choice between two possible new homes: the jth group or the kth. Similarly, each group's offspring are to form a new group as if their siblings had never been there before. There are no restrictions on mating these young zebras because they do not have any relatives in their new community. So, zebra offspring from j and k find new social groups to breed with. The AZOA algorithm benefits from this procedure because it helps to keep the family's general decency intact. Here are the equations that have been created to predict the breeding behavior of zebras:

$$z_j^q = Crossover\left(z_i^a, z_j^b\right) \quad if\ r < pc, i \neq j \tag{4}$$

$$z_k^q = Crossover\left(z_i^a, z_k^c\right) \quad if\ r \geq pc, \quad i \neq k, \quad \forall i, j, k \in N \tag{5}$$

where z_i^a represents site of the ith group, z_j^b signifies site of zebra b from j^{th} group, z_k^c represents site of the zebra c from k^{th} group, and z_j^q and z_k^q are the site of zebra q in j^{th} group and kth group, correspondingly.

Phase 4: Group management. To a zebra, the head of the paramount. The head of the family looks for verdant meadows, fallen leaves, and pools of water to bring them. It is not uncommon for the leader to defend his family from other zebras and ensure that they have enough to eat and drink. The water reservoir and meadows are still owned by the more powerful group of zebras. Once that is done, it will be available for others to use. These equations describe the model for this method.

$$\overline{z}_s^j = \begin{cases} 2R_4 \sin\left(2\pi R_5\right) \times \left(WR - z_s^j\right) + WR, & if\ R_6 < 0.5 \\ 2R_4 \cos\left(2\pi R_5\right) \times \left(WR - z_s^j\right) + WR, & otherwise \end{cases} \tag{6}$$

$$z_s^j = \begin{cases} \overline{z}_s^i, \overline{F}_s^j < F_s^j \\ z_s^j, \quad otherwise \end{cases} \tag{7}$$

where R_4 characterizes unvarying in [-2, 2], R_5 denotes the upper limit that is established by (12), R_6 in [0, 1], WR signifies water reserves, z_s^j is the jth site, z_s^j is the jth group site, and F_s^j is stallion in j^{th} group.

$$R_5 = 1 - t \times \left(\frac{1}{T}\right) \tag{8}$$

Phase 5: Phase of transition for choosing a leader. For the group to be able to discipline itself appropriately and make good use of its food supplies, a strong leader is essential. It is critical to replace the leader of the group if they show signs of weakness under any circumstance. The (9) mimics:

$$z_s^i = \left\{z_i^j, \quad if\ \ F\left(z_i^j\right) < F\left(z_s^j\right), \quad \forall i \in N_j \right. \tag{9}$$

z_s^j jth group current site and $F\left(z_i^j\right)$ is fitness charge of the leader stallion. Starting up AZOA, determining its fitness value, and updating folks are the three steps that determine its run-time complexity. In the initial process with M persons, the computational complexity is O(M), it is $O\ (T \prod M) + O\ (T \prod M \prod d)$, where T is the maximum repetitions and d is the dimension of specific glitches. Consequently, AZOA's overall runtime complexity is $O\ (M * (T + Td + 1))$ which is similar to other optimizers.

3 Results

The suggested Intrusion Detection system for IoT systems utilizing DL was implemented using Python and the TensorFlow library. Simulation testing was conducted using the Cooja Network Simulator (CNS). The IoT device used in the experiments was a Raspberry Pi 4 Model B with 2 GB of RAM. Table 1 summarizes the network parameters and their respective values employed during testing. The experimental results demonstrate the effectiveness of the proposed approach, showing significant improvements in detection rates, accuracy, and reduction in false positives compared to traditional methods. These findings highlight the practical viability and performance benefits of the WCNN-AZOA model.

Table 1. Simulation parameters

Parameters	Values
Simulator	Cooja
Radius of the circular region (R)	40 to 127 m
Number of sensors (N)	200 to 600
Sensing variety (Rs)	15 to 40 m
Transmission variety (Rtx)	30 to 80 m
Mac category	IEEE 802.11
Sensor's deployment category	(a) Gaussian Distribution (b) Uniform Distribution
Sensing model	Binary sensing perfect

3.1 Validation Analysis of Network Analysis

In this section, the projected model is tested with existing procedures such as GWO [?], FFA [?] and sea horse optimization algorithm (SHOA), which is graphically shown in Fig. 2.

Figure 2 signifies that the Visual Representation of projected model for Throughput. In the analysis of 200 nodes, the GWO [?] model attained the throughput as 99 and FFA [?] model attained as 118 and SHOA as 124 and also WCNN-AZOA model attained throughput as 139 correspondingly. Then the 300 nodes, the GWO model attained the throughput as 112 and FFA model attained as 123 and SHOA as 153 and also WCNN-AZOA model attained throughput as 161 correspondingly. Then the 400 nodes, the GWO model attained the throughput as 134 and SHOA as 36 and FFA model attained as 177 and also WCNN-AZOA model attained throughput as 186 correspondingly. Then the 500 nodes, the GWO model attained the throughput as 148 and SHOA as 185 and FFA model attained as 224 and also WCNN-AZOA model attained throughput as 230 correspondingly. Then the 600 nodes, the GWO model attained the

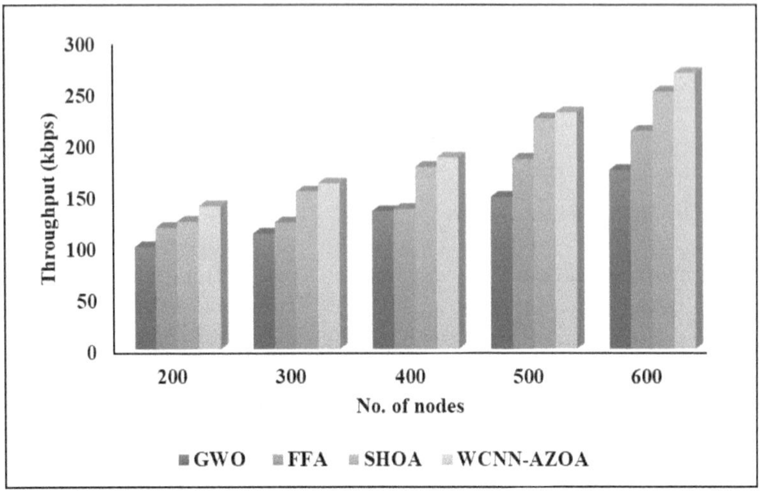

Fig. 2. Proposed for Throughput

throughput as 174 and FFA model attained as 212 and SHOA as 250 and also WCNN-AZOA model attained throughput as 268 correspondingly.

In Fig. 3 characterize that the Visual Presentation of projected model. In the investigation of 200 nodes, the GWO model attained the Routing Overhead as 0.58 and FFA model attained as 0.43 and SHOA as 0.40 and also WCNN-AZOA model attained Routing Overhead as 0.40 correspondingly. Then the 300 nodes, the GWO model attained the Routing Overhead as 0.72 and FFA model attained as 0.60 and also WCNN-AZOA model attained Routing Overhead as 0.52 and also WCNN-AZOA model attained Routing Overhead as 0.41 correspondingly. Then the 400 nodes, the GWO model attained the Routing Overhead as 0.86 and FFA model attained as 0.74 and also WCNN-AZOA model attained Routing Overhead as 0.68 and also WCNN-AZOA model attained Routing Overhead as 0.50 correspondingly. Then the 500 nodes, the GWO model attained the Routing Overhead as 0.94 and FFA model attained as 0.92 and also WCNN-AZOA model attained Routing Overhead as 0.84 and also WCNN-AZOA model attained Routing Overhead as 0.76 correspondingly. Then the 600 nodes, the GWO model attained the Routing Overhead as 1.29 and FFA model attained as 1.24 0.96 and also WCNN-AZOA model attained Routing Overhead as 0.88 correspondingly.

[13], LSTM [13] and RF-DNN [?][?]. Even though, the existing techniques uses different datasets, the research work considered our datasets and tested, then average results are analysed. The visual representation of various classifiers on different metrics is exposed happening Fig. 4 and Fig. 5.

In Fig. 4 represent the Graphical Assessment of various classifiers. In the analysis of DNN classical attained the sensitivity as 92.27 and precision as 79.70% and also f1-score as 93%. The CNN classical's sensitivity as 94.95% and 80.63%

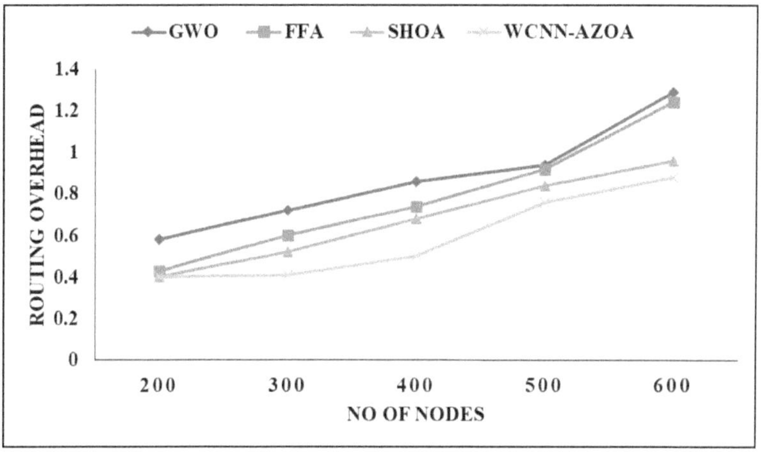

Fig. 3. Visual presentation of projected model

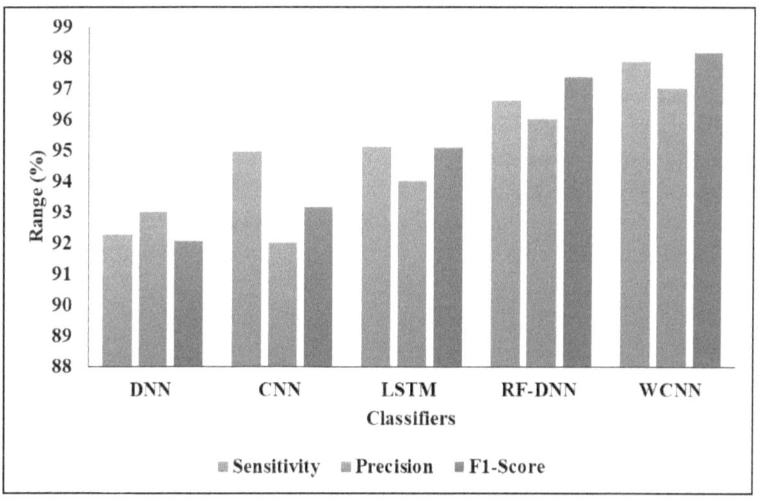

Fig. 4. Graphical Comparison of various classifiers

precision and also f1-score as 92%. The LSTM classical's sensitivity as 95.11% and 82.62% precision and also f1-score as 94%. Then the RF-DNN classical attained the sensitivity as 96.62 and precision as 87.57 and also f1-score as 96 correspondingly. Then the WCNN classical attained the sensitivity as 97.87%, 89.23% precision and 97% f1-score.

In Fig. 5 represents the visual comparison of various classifiers on attack detection. In the investigation of the DNN model, the specificity as 92.50 and also the accuracy as 92.06 correspondingly were achieved. Then the CNN model attained the specificity as 95.03 and also the accuracy of 93.16 correspondingly.

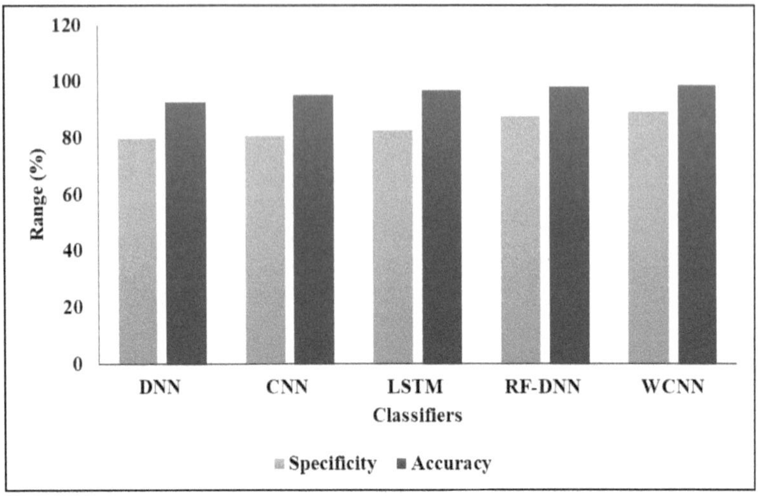

Fig. 5. Visual Comparison of different classifiers on attack detection

Then the LSTM model attained the specificity as 96.78 and also accuracy of 95.07 correspondingly. Then the RF-DNN model attained the specificity of 97.98 and also the accuracy of 97.36 respectively. The WCNN's specificity as 98.58 and also the accuracy of 98.17. In we have used weighted CNN based AZOA to evaluate the selected attributes' relevance and risk level, in other models, they have used a better model, the proposed model achieved better results, because we used better models and attained better results.

The experimental results highlight the novel contributions of our proposed WCNN-AZOA model in enhancing the security of IoT-based WSNs. By integrating a WCNN with the AZOA, we achieved significant improvements in detection accuracy, reduced false positives, and enhanced resilience against common IoT attacks such as Blackhole, Grayhole, Flooding, and Scheduling assaults. Compared to traditional methods, our approach demonstrates outstanding results in terms of detection rate and processing efficiency. This underscores the practical applicability and effectiveness of combining advanced neural network techniques [21–23]with adaptive optimization algorithms to secure IoT networks.

4 Conclusions

The integration of WCNN with AZOA has demonstrated significant advances in enhancing the security and resilience of WSNs based on IoT. This study has successfully developed a robust framework for detecting and mitigating prevalent IoT attacks, including Blackhole, Grayhole, Flooding, and Scheduling assaults. The collaborative use of WCNN and AZOA has resulted in improved detection accuracy and reduced false positives, effectively minimizing the misclassification of normal network events as attacks. Experimental results underscore the superiority of the method over conventional approaches in terms of computational efficiency, accuracy, and detection rate. In particular, AZOA exhibits versatility in real-time applications, adeptly managing the diverse and evolving nature of IoT traffic. Furthermore, the proposed method is scalable, making it suitable for deployment in large-scale IoT networks without compromising performance. Future research could explore broader applications across various IoT scenarios and attack vectors, as well as investigate the integration of advanced optimization techniques.

References

1. Aluvalu, R., et al.: Komodo Dragon Mlipir Algorithm-based CNN model for detection of illegal tree cutting in smart IoT forest area. Recent Adv. Comput. Sci. Commun. **17**(6), 1–12 (2024). (Formerly: Recent Patents on Computer Science)
2. Maheswari, M., et al.: A novel hybrid deep learning framework for intrusion detection systems in WSN-IoT networks. Intell. Autom. Soft Comp. **33**(1) (2022)
3. Yao, W., et al.: A lightweight intelligent network intrusion detection system using one-class autoencoder and ensemble learning for IoT. Sensors **23**(8) (2023)
4. Baswaraju, S., et al.: Future food production prediction using AROA based hybrid deep learning model in Agri-sector. Hum. Centric Intell. Syst. **3**(4) (2023)
5. Ismail, S., Reza, H.: Evaluation of naïve Bayesian algorithms for cyber-attacks detection in wireless sensor networks. In: 2022 IEEE World AI IoT Congress (AIIoT) (2022)
6. Aluvalu, R., et al.: IWHO-based cluster head selection for vehicle-to-vehicle communication in intelligent transport system. Int. J. Transp. Dev. Integr. **8**(2) (2024)
7. Salmi, S., et al.: Performance evaluation of deep learning techniques for DoS attacks detection in wireless sensor network. J. Big Data **10** (2023)
8. Srinagesh, C., Prasad, T.B., Reddy, K.S.: Leveraging technology for creating awareness of problem-solving skills to engineering students. Int. Conf. Technol. Educ. 238–239 (2010). https://doi.org/10.1109/t4e.2010.5550107
9. Sivanesan, N., et al.: WSN attack detection using attentive dual residual generative adversarial networks. Int. J. Intell. Syst. Appl. Eng. **11**(4) (2023)
10. Sudheer Reddy, K., Srinagesh, C.: Leveraging learning to enhance industry readiness. In: 2013 8th International Conference on Computer Science & Education, Colombo, Sri Lanka, pp. 1204–1208 (2013). https://doi.org/10.1109/ICCSE.2013.6554101
11. Sudheer Reddy, K., Anji Reddy, V., Mamatha, K., Santhosh Kumar, C.: Prediction of Covid-19 outbreak in India by employing epidemiological models. J. Comput. Sci. **16**(7), 886–890 (2020). https://doi.org/10.3844/jcssp.2020.886.890
12. Amarachinta, P.R., Sudheer Reddy, K., Aluvalu, R.: Predictive analytics and AI for pancreatic cancer screening - a data driven approach to early detection. In: 2025 7th International Conference on Signal Processing, Computing and Control (ISPCC), pp. 231–236 (2025). https://doi.org/10.1109/ISPCC66872.2025.11039583
13. Singh, N., Sudheer Reddy, K., Aluvalu, R.: AI driven waste classification for smart recycling. In: 2025 3rd International Conference on Disruptive Technologies (ICDT), pp. 1590–1594 (2025). https://doi.org/10.1109/ICDT63985.2025.10986692

A Detailed Study: Application of Big Data Analytics in Health Care Sector and Treatment of Cancer Disease

Jayavardhanarao Sahukaru[1], Sangram Keshari Swain[1(✉)], and Bendi Venkata Ramana[2]

[1] Department of CSE, Centurion University of Technology and Management (CUTM), R.Sitapur, Odisha, India
sangrambapun@gmail.com

[2] Department of IT, Aditya Institute of Technology and Management, Tekkali, India
dr.bvramana@adityatekkali.edu.in

Abstract. Big data is a declaration used to recognize the database whose area is the potential of typical database software tools to store, organize and examine. Big data have demonstrated a new path towards the world. Cancer is a group of diseases characterized by uncontrolled cell growth and the potential to invade or spread to other parts of the body, often resulting in the formation of tumors. It has been a deadline for millions of people. Every cancer is particular and the medical treatment is complex. The data of each cancer patient are too large and it varies from one person to another person. This paper will traverse censorious tools and tackle that accelerate knowledge, locating along with the cancer study continuity.

Keywords: Disease Prediction · Cancer Disease · Big Data Analytics

1 Introduction

Cancer is a miscellaneous syndrome. A single patient may cause by different kind of cancers. The mutually different solitary cancer cells are collected by real alteration methods. The existing method of cancer biomarker detection can proficiently identify a portrait and does not provide enthusiastic and long-term adaptation in the cancer scenery. To tackle these issues big data is used by accumulating different molecular features of the RNA, DNA, metabolite and protein fraction. The next-peer biotechnologies are proteomics, metabolomics and sequencing in addition to therapeutic and healthcare information arrangements from computerized security proof. They created some assaults in big data which can be assisted by guiding the persons working in the presence sciences for accepting the data part using bio data mining.

The challenge present in cancer research under various circumstances can be thrown by big data. This is eventually used for recognizing the syndrome. In recent days, the future paramedics, metabolomics and syncing are developing rapidly in various ways. Thus, cancer research instantly necessitates complex methodologies. Also, they require

A. Das et al. (Eds.): MIDAS 2025, CCIS 3055, pp. 94–107, 2027.
https://doi.org/10.1007/978-3-032-35516-4_8

higher volume, complex structure, massive active depth and scale information crowd. A huge volume of loosely linked, noise and miscellaneous information can be composed in the dataset and is guaranteed by us.

It is created challenges to justice the usefulness of data and every data in the dataset may not comprise valuable information. Hence, the usefulness of information present in the same dataset is identified with the help of different criterions, procedures or expansions. In this chapter numbers of censorious tools are traversed and tackle these tools to accelerate the knowledge along with the cancer study continuity. Also, we are discussing some visualization techniques available for big data. In which R package is used to analyze unstructured text because R is freely available and it comes with lots of free packages and powerful tools through which we can easily analyze the large data sets in a sufficient time.

2 Big Data Analytics in Healthcare

The capacity of health information remains anticipated to improve intensely in the future days (W Yichuan et al.) [1]. Nowadays, expressive usage and wage are developing like acute elements for improving the effectiveness in healthcare atmosphere. Here, the principal motivation is not only for improving the revenue but also for acquiring the accessible apparatuses, structure and methodologies to control big data efficiently in healthcare organizations; otherwise millions of dollars in income and revenues may loss feasibly (W Raghupathi, and Viju Raghupathi) [2]. It is outlines as huge sizes of variable, complex and high velocity data that need enhanced technologies and techniques to allow the examination of the information, management, capture, storage and distribution (B Ashwin et al.) [3].

Speed, diversity exclusively relating to healthcare and accuracy are the features encompassed in big data. The enormous volume of patient's health and medicinal information can be processed using state-of-the-art logical approaches for grasping a complete understanding of consequences and it can be smeared for making precaution. Every doctor and his patient are perfectly informed by individual and populace information at the time of decision-making practice. Also, it is used to detect the best and suitable therapy preference for that specific patient. The well-known structure for big data analytics is demonstrated in Fig. 1. It roughly consists of five major structural layers: (a) data aggregation (b) data, (c) analytics, (d) data governance, and (e) information exploration. The elements of big data analytics which execute tasks are made using the analytical layers to empower the healthcare executives for understanding the way of transformation of the healthcare information from different sources into expressive medical data by means of big data executions.

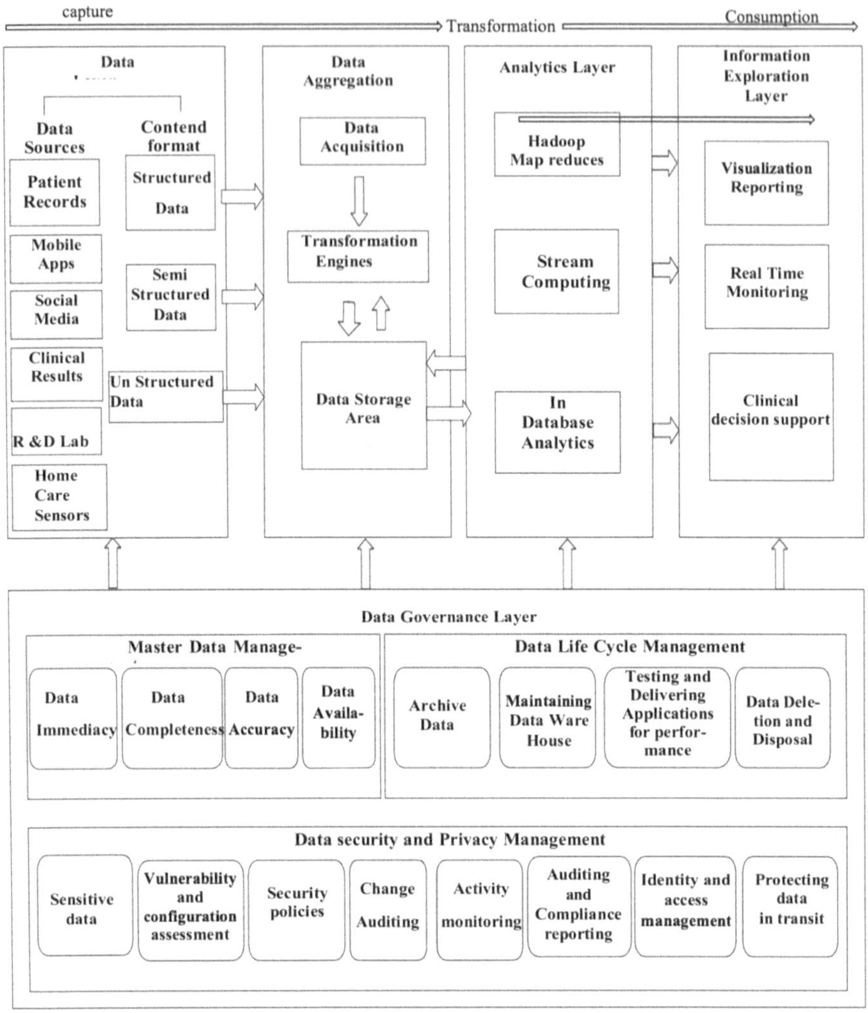

Fig. 1. Big data analytics architecture in health care (B Ashwin et al.)[3]

3 Big Data Analytics in Healthcare

The worth of healthcare able to be improved by contemplating the statements given below:

Presenting patient's central facilities: Proof basis medication which perceives syndromes are provided at the initial periods on the basis of accessible medical information for providing quicker relief to the patients and the medicine dosages is minimized for avoiding secondary effect and to provide effectual medication on the basis of genetic compositions. It is used to reduce readmission rates and hence minimizing the expense for the patients.

Determining spreading syndromes at the initial stage: The liver assessment is used to predict the pathological syndromes in advance before spreading. A syndrome in a specific geographical area suffers the societal records of the patients and it is analyzed to identify the diseases earlier. It supports healthcare specialists for advising the sufferers by fascinating required protective processes.

Observing the quality of clinic: The clinic's outfit is monitored based on the medicinal committee standards. The periodic monitoring is used to take disciplinary action against clinics by the government.

Upgrading the treatment approaches: The influences of medicines are always monitored by modified patient's treatment and the doses of medicine are modified on the basis of the analysis for rapid respite. The patient's historical data about the disease with similar indications are used to give effectual medicine to new patients by physicians (J Archenaa and EA Mary Anita) [4].

4 Big Data Analytics Success in Healthcare

The healthcare societies cover single-doctor clinic, multiple worker collections, big hospital nets and responsible precaution institutions. These societies are used for realizing important welfares through the digitization, combination and effective usage of big data (H Patil Kupwade and Ravi Seshadri) [5]. Possible welfares comprise detection of syndromes at initial periods once the diseases are treated simply and efficiently. Management of particular patient and populace health and rapid detection of cheating in healthcare proficiently. The historic data of patient include LOS (length of stay), patients who likely will not benefit from execution, difficulties, patients who will select elective operation, riskiest patients for medicinal impediments, patients at threat for MRSA, sepsis, C. difficile, or other hospital- acquired disease, disease or sickness development, patients at risk for advancement in disease states, normal facets of illness or sickness development, and probable comorbid situations (EMC Consulting). Based on McKinsey statement, BDA able to save above $300 billion per year in U.S. healthcare and reduces expenses nearly 8% in national healthcare. The medical executions probably save $165 billion and R & D save $108 billion in waste (Y Wang and N Hajli) [6]. McKinsey has faith in big data for reducing waste and inefficiency in the regions provided below:

Medical operations: The medicinal pertinent and economical procedure is determined by proportional efficacy research for diagnosing and providing treatment to the patients.

Research & development: (1) The analytical model is used for lowering erosion and producing a quicker, thinner, highly directed R & D pipeline in medicines and gadgets; (2) The statistical equipment as well as procedures are used for improving medical test proposal and patient enrolment to give best matching treatments to every patients. (3) medical tests and patients registers are analyzed for identifying consequent signs and discovering opposing influences before inventions stretch to the market.

Public health: (1) The syndrome configurations, tracking syndrome eruptions and spreads are analyzed for improving speed response and public health observation; (2) exactly targeted vaccines like annual influenza strains are developed rapidly and, (3) Enormous volume of data are turned into actionable data which is utilized for identifying requirements, providing facilities, and preventing calamities, specifically for the use of populaces (J Manyika et al.) [7].

Patient profile analytics: Innovative analytics such as segmentation and forecast modelling are applied on patient's profiles for identifying beneficiary from dynamic precaution or lifestyle e.g., the beneficiary from protective care are the patients at risk of a particular syndrome like diabetes.

According to IBM the analysis in addition to enriched data in areas provide the ultimate consequences contains: analytical patients, they are the supreme customers of health resources and the riskiest one to provide adverse consequences.

5 Healthcare Applications of Big Data Analytics

The issue of expenses in healthcare able to be reduced by BDAH. Any syndrome diagnosis or sickness treatment via big data analytics able to be utilized for minimizing the volume of readmitted patients who run to clinics. The subsequent section provides some instances of application of BDA in the sector of healthcare to diagnose disease, treat and readmit the patients.

5.1 Diagnosis

IBM's Pure Data Solution for Healthcare Analytics is a combined software which is used in clinics and hospitals for healthcare analytics (P. McGinley) [8]. It is used to provide quick consequences and minimize hazards. Nowadays, Seattle's Baby's Hospital is utilizing BDA for improving treatment and patient concern. Here, IBM's Big Data methodology is used for diagnosing expected 350,000 patients and it can determine the general body of a group through simpler and precise diagnosis and treatment.

A computer model is developed by a number of scholars from Massachusetts General Hospital and MIT which is used to differentiate various kinds of lymphoma (MIT Spectrum). The data from more than 1000 cancer patients are analyzed using the aforementioned model to for developing a particular treatment sequence. The scholars transformed the pathology outcomes into graphs for simplifying the model. It is used for tracking tendencies, diagnosis by incorporating with WHO rules. The misclassified and misdiagnosed lymphoma is reduced about 5% to 15% using this methodology and the saved resources are used for diagnosing other patients.

5.2 Treatment

The novel treatment for cancer and other syndrome can be developed using BDAH which is used for tracking consequences and results of different patients and also for mapping their DNA data. Intel and Oregon Health and Science University made the Cloud platform of Collaborative Cancer for allowing hospitals to communicate significant patient data so as to diagnose cancer accurately (T. McMullan) [9]. Other patient's DNA is accessed by doctors for comparing it with their own patient's DNA. The trend analytics and commonality assisted through allowed the doctors to give improved treatments. This methodology is stretched to Alzheimer, autism and certain genetically correlated syndromes.

The data from different equipment like Alive Cor's suitable ECG connected to a smart phone with the help of apps which can trace exercise and diet and structural of body's, genomic data are employed in the department of Cardiology at UCSF for learning lot about heart disease and to find superior techniques for healing. Apps and program are generated for alerting the patients to exercise more, eat better, and so on. Approximately, four thousand adults are registered in the program worldwide. IBM's Watson Health is known as a supercomputer which permits health organizations to generate their own apps, goods and data analytics. Apple's health app is used by Watson Care Manager (WCM) system for collecting and sharing required data with doctors to attain improvement in diagnosis and treatment. WCM is applied in pediatric patients at Boston Children's Hospital for heart health, adapted medication and serious caution. Columbia University's cancer center is utilizing Watson for decoding DNA and to develop modified cancer therapies. BDAH is used at Beth Israel Hospital with MIT and Aptima utilizing Dangerous Conditions for enabling doctors to guess which patients are at risk and emerging impediments when they are in the intensive care unit and making suitable positive judgements (SM Krishnan) [10].

6 Big Data Analytics in Disease Prediction

The goal of big data analytics for lung cancer prediction is to foresee different features of the disease by utilising sophisticated machine learning techniques and massive, heterogeneous datasets. The goal is to predict patient outcomes, disease progression, treatment responses, The healthcare research and practices field were evolving rapidly, so newly developed BDA play a prominent role in this healthcare field. A enormous volume of designed, unstructured, and disparate data made accessible by current healthcare systems was managed, assimilated, accumulated, and analyzed by tools provided by this BDA. The BDA procedure was mostly applied to assistance the process of disease care as well as exploration delivery. In recent days, a huge quantity of investigation process was carried out identify the cancer disease. This study process was widely performed on the chemical-based molecules. The analysis process performed on chemical centric was found to be valuable for the minimum amount of DNA sequences. An automated method attains a significant importance for identifying the cancer codes. A set of biomarkers was proposed and applied in this method to make sure that proper treatment was provided to all cancer patients. These biomarkers were found to be more expensive and sensitive to the environments. These available sets of biomarkers were metabolomics, imaging, proteomic, psychological, and genomic factors. For all kinds of cancer, the keys were identified as a key to control all these aspects. A large number of gene selection approaches were applied to perform the cancer identification process. But this selection process was not essential for an entire identification process. A couple of projects were initiated by various cancer institutions for cancer investigations. A large-scale dataset handling approaches like microarray and next-generation sequencing were introduced to measure the messenger RNA formations, DNA mutations, and DNA copy number alterations. This introduced method does not belong to the machine learning environments (M Galar et al.) [11], (M Lozano et al.) [12], (RG Cavalcante et al.) [13].

Laplace naive Bayes model proposed by (JG Domınguez et al.) [14] for classifying microarray data which highly focus on the heftiness of gene outliers. Instead of gene original profiler, an input of gene pair combination (MR Machado et al.) [15] were widely applied for cancer classification algorithm. Both supervised and unsupervised methods (KM Burkett et al.) [16] were applied in various microarray gene classification methods. Based on particular gene, the tissues were classified by supervised classification, whereas the unsupervised method classifies the gene based on tissues. The gene markers were applied by computational protocols for different cancer tissues. An imbalanced (Y Liu et al.) [17] for microarray data analysis was classified by applying the under-sampling approach for an ant colony optimization idea. For gene classification, (MJ Meyer et al.) [18] introduced an association rules, but this rules still requires some enhancement in system complexity. This method suggests that the subtype in the similar class was discriminated by the transcript expression interval demonstration. The discriminate performed in the hypothesis of biological gene datasets was highly assessed by applying the web-based interactive tool. These interactive tools effectively perform the evaluation process in management decision and medical diagnosis. A large number of techniques and classification methods were developed to perform the classification process in microarray data. These respective methods were found to be more comprehensive and applicable for various real and clinical practices.

The codes responsible for cancer diseases were significantly detected by cancer data analysis. The coding regions from disease infected biological data were identified by this analysis. The data obtained from infected area was very much useful to design a proper drug and it also provide huge support for laboratory valuations. Codes bear explicit significance on different characteristics along with the side effects of illnesses. The biological data coding was identified as an important part to obtain a precise info regarding the animals to develop a required drug. Four various machine learning approaches were proposed by (S Kamal et al.) [19] like FDA (Fisher's discriminant analysis), PCA (principal component analysis), SVM (support vector machine), and NMSF (neural mapping skyline filtering). These four learning methods were applied to perform the coding area selection and data reduction process. The outcomes of this method indicate that both FDA and PCA methods were perfectly outperformed by SVM approach. But, SVM was outperformed by NMSF, because of mapping capability. Among these four approaches, NMSF achieved superior results.

In the big data analytics era, the precise and personalized medicine field was growing rapidly. So formerly (GS Ow et al.) [20] introduced a patient cataloguing method known as PSVM (Prognostic Signature Vector Matching) and recognized a thirty-seven variable signature involving 36 let-7b related prognostic significant mRNAs and the age risk aspect categorized the huge high-grade serous ovarian cancer patient companions into three survival-significant hazard collections. Then, the investigation was performed in the predictive performance of PSVM through which the prognostic variable weights were optimized. This investigation process reveals the value of one prognostic variable over the other. After that, several PSVM based multivariate prognostic representation were compared with some existing machine learning approaches like SVM, random forest, logistic regression, K-nearest-neighbor, then neural networks.

The biomarkers play a prominent task in the clinical management of malignant growth patients, due to the development of selective molecular targeted therapies and genomic profiling technologies. The changes that were identified in the development process of cancer biomarker exhibits a large gap between the discovery studies of initial biomarker and its clinical translation. In this survey (N Goossens et al.) [21] the biomarker development steps emphasize the key issues in successful implementation and validations were summarized, and finally the representative examples were overviewed in the oncology field. Additionally, the administrative issues and future viewpoints in the period of big data analysis and precision medicine were discussed.

Cancer was identified as a tumor providing irregular cell growth with the likelihood to involve or scatter to different body parts. It was identified as a deadline for large number of peoples. Each and every cancer disease was found to be specific, and the treatment for this each cancer was found to be complex. The data obtained from each patient was found to be large and it varies from one patient to another patient. (SK Swain et al.) [22] used the critical tackle and tools which hasten the knowledge, localizing together with the stability of cancer study.

In healthcare, big data analysis plays a prominent role for performing the early diagnosis in fatal disease. For developing a knowledge-based system, a large amount of dataset was gathered. The feature set obtained in the data set was reduced and then the data reliance was identified by applying the rough set theory. Among the entire cancer victims, the lung cancer leads to high rate of mortality in both men and women. There was only a less chance to survive for the patient's those having the advanced stage of lung cancer. These patients were provided only radiation therapy (radiotherapy) as the main treatment. This radiotherapy was found to be more effective, so some patients fail to adapt the radiotherapy.

Till now, there was only a limited number of studies were performed to examine the response of lung cancer to radiotherapy. The heterogeneity and inherent complexity of biological retort to radiation therapy widely explains the incapability of classical forecast methods to obtain an essential specificity and sensitivity for trial's design or clinical practices. In this method, (I El Naqa et al.) [23] briefly reviewed the present knowledge of signaling and genetic pathways of data mining applications as a case study to modulate the response of tumor to radiotherapy in large cell lung cancer in the challenging issues of cancer treatment. A non-invasive approach of both cellular and subcellular process was molecular imaging (T. Hussain et al.) [24] whereas it contains the potential to perform the clinical diagnosis for disease states like cancer. An interaction of biology, radiology, and nuclear medicine were found to be crucial (G. Baio) [25] for making it medically appropriate to patients, this interaction process may complicate its automatic analysis process. An emerging method that was applied to develop a map for electromagnetic wave scattering was microwave imaging. The dissimilarity identified in the dielectric possessions of various tissues (S. Mustafa et al.) [26] produce this electromagnetic wave scattering. Both the physiological and functional properties were encoded in these dielectric properties. This encoded information was very much useful to characterize and differentiate various pathologies and/or tissues (A. H. Golnabi et al.) [27]. The information retrieval process was identified as a challenging task due to the presence of scattering behavior in microwaves.

The images integrated from dissimilar models and/or other physiological and clinical data expand the diagnosis precision and results of disease prediction. An advancement in neurovascular imaging, and multiple modal MRI /CT tasks like prefusion imaging and angiography were explored by (Liebeskind and Feldmann) [28] to evaluate the brain vascular defects and also to attain accurate medicine. The myocardial infarction scar was exactly assessed by applying a delayed advanced MRI. For this type of disease, electro anatomic mapping (EAM) provides high support to detect the subendocardial extension of infarct. The CT and MRI images were evaluated in this method to improve the exactness of diagnosis process while discovering the osteophytes existence and corrosions in the temporomandibular joint (TMJ). (Hussain et al.) [29] was examined this evaluation process. Based on this revision, the concurrent estimation that were performed in an entire available imaging approaches were found to be an unmet requirement.

Advanced Multimodal Image-Guided Operating (AMIGO) suite was developed along with MRI, CT /PET imaging in the Operating Room (OR), angiographic X-ray and 3D ultrasound. This system was widely applied to treat cancer and shows better performance in finding and pointing the unhealthy tissue of an individual (C. M. C. Tempany) [30]. Even though there was large space was available to store this data, but yet no solution was obtained to perform the challenging process like mapping and detecting dependencies among various types of data. Molecular imaging and its influence over cancer diagnosis process and cancer medicine enhancement were analyzed in (R. Weissleder) [31]. This method provides better support to identify the cancer in initial stage by incorporating the physiological and molecular data with anatomical information. This imaging process was applied for the patients those having the ovarian cancer in an advanced stage, the response accuracy of the patients who were subjected to this special treatment was found to be higher than the other histopathology or clinical criteria. The speed of the correlation process performed in an image was increased by designing a hybrid digital-optical correlator (HDOC) (T. Zheng et al.) [32]. When the geo registration or coordinate matching was not present, the images were compared by employing HDOC. The HDOC was made applicable by the volume holographic memory in the big data analytics field. The storage medium was represented as a volume holographic memory, where the computation process of multichannel method was performed.

6.1 Use of Big Data Analytics in Lung Cancer

The tools present in general database software can save, consolidate and evaluate the big data. A different way is established by big data towards the world. The body parts are occupied or dispersed due to the anomalous cell augmentation and it is called as cancer tumor. Many persons are died by these tumors. Every cancer patient is having vast data and it is changed for different individuals. The rank of Lung Cancer occurrence for women is nine when different type of cancer occurrence is sorted. In India, the rank of Lung cancer is number four while seeing all the cancers surrounded between men and women together.

7 Signs of Testing Lung Cancer

The general indications for lung cancer are given below:

1. Hoarse voice for more than three weeks.
2. Chest ache which does not get better

3. Hemoptysis
4. Sensation of breathless and wheezing without any reason
5. Frequent chest infection which does not get better
6. Variation in a cough you've had for a long time
7. Cough for more than three weeks Alternative signs for lung cancer are:
8. Weight dropping without any clear reason
9. Tiredness (fatigue)
10. Changing of shape of the finger's ends – they may turn out to be bigger or curved.

7.1 Techniques of Testing Lung Cancer

1. Chest x-ray – when the patient has not already affected by lung cancer, a chest x-ray is used for checking patient's lungs to identify the abnormalities
2. CT (computerized tomography) scan – Majority of the patients prefer CT scan. Based on the patient's symptoms, they may still have such cancer even though their chest x-ray hasn't exposed any indication of lung cancer. A sequence of x-ray is taken in a CT scan and it generates a 3D image of body's inner portion. The required time for taking this scan is 10 to 30 min and it does not causing or suffering any physical pain.
3. Bronchoscopy – The physician or skilled nurse examine the inner sides of the bronchus or airways using a test called bronchoscopy. This testing is done in the presence of local anesthetic using a Bronchoscope tube.
4. Lung biopsy -Lung biopsy is used as the guidance in the x-ray sector while taking CT scan. Initially, the patient is anaesthetized for numbing the area. When the doctor injects a thin needle into the lung via skin, the patient asked to grip their breath for a few seconds. The correct placement of the needle is checked by the CT or X-ray image. Then, a cell sample is taken from the tumor. The sample is then analyzed using a microscope in order to identify the cancer. It is not a comfortable process but it requires few minutes (PJ Hollen et al.) [33].

7.2 Tests Repeated During and After the Treatment

The repetition of few tests may require at the time of treatment and after treatment.

1. Mediastinoscopy – The central part of the chest is known as mediastinum. This mediastinum and the neighboring lymph nodes are seen by the physician during Mediastinoscopy test. The lung cancer can initially spread in these areas. A minor scratch in the skin at the bottom of patient's neck is made by the doctors. Then, a tube-like telescope is injected via the cut into chest. A camera and light are presented at the end of the tube which amplifies the traced regions. The abnormalities of the regions can be seen by the physician. They remove samples from the tissue and lymph nodes for checking the presence of cancer cells.
2. Thoracoscopy – It permits the physician to see the pleura and remaining structures nearer to the lungs in the presence of a general anesthetic. The local anesthetic can also be used for numbing the region and sedating to drowse the patient. A minor cut in the patient's chest is made by the doctors to pass a thoracoscope tube into patient's chest. Subsequently, a biopsy of the pleura is taken. Occasionally, physicians may utilize a video camera for getting the best vision of the region around the lung.

3. EBUS (Endobronchial ultrasound scan) – This is an alternative test of a thoracoscopy or mediastinoscopy. This test is done on few persons rather than biopsy and a CT scan or a bronchoscopy. One able to perform this testing procedure under the supervision of general anesthetic and by making the patient to undergo drowsy utilizing a mild narcotic medicine. The physician injects a bronchoscope tube into the patient's trachea or windpipe. At the end of bronchoscope tube a small ultrasound probe is presented.
4. Endoscopic ultrasound (EUS) – It is an initial testing procedure to identify the lung cancer and the procedure of this testing procedure is as same as an EBUS. A smaller adaptable tube i.e., endoscope will be injected into patient's esophagus (throat) via mouth after an anesthetic is given to the patient to make them to feel drowsy. The region around the lungs and heart is imaged by an ultrasound probe present at the end of endoscope.
5. PET/CT scan – This is a combined process of a PET (positron emission tomography) scan and CT scan which is used for taking an x-rays sequences to create a 3-D image. Here, the activities of cells in various portions of the body are measured by using a lower amount of radiation. PET/CT scans provide thorough details about the body parts to be scanned. One should go to authorized clinics for taking this scan.
6. MRI (magnetic resonance imaging) scan – It utilizes magnetism for imaging the patient's body parts. The patient should sign in a form for the assurance of this scanner's safety for them because, it is a powerful magnet. This form enquires details about any historical metal implants in patient's body like pacemaker, bone pins or surgical pins. When the patient has ever worked in the metal fabrication company, they must inform specialist before taking this scan because sometimes very small metal bits imbed into the patient's body.
7. Abdominal ultrasound scan – The liver and all other portions exist in the body at the upper abdomen can be seen by ultrasound scan using sound waves. A gel is pasted on the region of body parts to be scanned after the patient is resting on the bed. A sound wave which is produced by a small gadget is transferred into the part of body and a computer is used to transform these waves into image.
8. Bone scan – Any abnormalities present in the bone can be seen using this scan. A small volume of radioactive element is inserted into a patient's vein. This scan does not produce any impairment on patient's body because small amount of radiation is used in this scan.
9. Lung function tests –when the patient is treated by radiotherapy or surgical procedure, physician will organize respiring or physical movement test for them. This test is carried out for seeing working condition of patient's lungs (LG Collins et al.) [34].

7.3 Tests Repeated During and After the Treatment

The accessible dataset comprises of forty lung cancer patients for matching two chemotherapy therapy's positive or negative impacts for long duration of time. Figure 2 shows the lung cancer data set used. The data set uses the following variables:

1. Survival Time: It represents the number of days after completing the therapy
2. Age: person's age when the disease is diagnosed
3. A variable indicator:

- 1: stances censored case
- 0: fulfilled uncensored cases

4. Weight: person's weight.
5. Time_diag_study: Number of days between the lung cancer testing as well as treatment.
6. Stage: The overall medicinal condition at testing of lung cancer and it is ranges from 0 to 100.
7. Tumor Type: Represents type of tumor.

- Compact Squamous
- Big Adeno

8. Treatment: treatment categories

- Ordinary treatment
- Investigational treatment

	A	B	C	D	E	F	G	H	I
1	ID	TRT	AGE	WEIGHN	STAGE	TOTALCNT	TOTALCW	TOTALCW	TOTALCW6
2	1	0	52	124	2	6	6	6	7
3	6	0	77	160	1	9	6	10	9
4	6	0	60	136.5	4	7	9	17	19
5	9	0	61	179.6	1	6	7	9	3
6	11	0	69	175.8	2	6	7	16	13
7	15	0	69	167.6	1	6	6	6	11
8	21	0	67	186	1	6	11	11	10
9	26	0	66	158	3	6	11	16	15
10	31	0	61	212.8	1	6	9	6	8
11	35	0	51	189	1	6	4	8	7
12	39	0	46	149	4	7	8	11	11
13	41	0	65	157	1	6	8	9	6
14	45	0	67	186	1	8	8	9	10
15	2	0	46	163.8	2	7	15	9	10
16	12	1	56	227.2	4	6	10	11	9
17	14	1	42	162.6	1	4	6	8	7
18	16	1	44	261.4	2	6	11	11	14
19	22	1	27	225.4	1	6	7	6	6
20	24	1	58	226	4	12	11	12	9
21	34	1	77	164	2	5	7	13	12
22	37	1	86	140	1	6	7	7	7
23	42	1	73	181.5	0	8	11	16	7
24	44	1	67	187	1	5	7	7	7
25	50	1	60	164	2	6	8	16	8
26	58	1	54	172.8	4	7	8	10	8

Fig. 2. Cancer data set

8 Conclusion

Because of the numbers of methods are contributing to big data, day to day there is an increase in the popularity of big data is happened. For facing the various necessities of users, providing a perfect classification method for big data is essential. In many fields numerous issues are solved by using various approaches of big data. As already explained in the literature survey a greater number of researches are performed in various fields of big data. Now the present research is given according to the research gaps. Sequentially, the purposes of the research have been attained. The present research and its outcomes will provide future direction to other researchers.

References

1. Wang, Y., Kung, L.A., Byrd, T.A.: Big data analytics: Understanding its capabilities and potential benefits for healthcare organizations. Technol. Forecast. Soc. Change **126**, 3–13 (2018)
2. Raghupathi, W., Raghupathi, V.: Big data analytics in healthcare: promise and potential. Health Inform. Sci. Syst. **2**(1), 3 (2014)
3. Belle, A., Thiagarajan, R., Soroushmehr, S.M., Navidi, F., Beard, D.A., Najarian, K.: Big data analytics in healthcare. BioMed Res. Int. (2015)
4. Archenaa, J., Mary Anita, E.A.: A survey of big data analytics in healthcare and government. Proc. Comput. Sci. **50**, 408–413 (2015)
5. Patil, H.K., Seshadri, R.: Big data security and privacy issues in healthcare. In: 2014 IEEE International Congress on Big Data, pp. 762–765. IEEE (2014)
6. Wang, Y., Hajli, N.: Exploring the path to big data analytics success in healthcare. J. Bus. Res. **70**, 287–299 (2017)
7. Manyika, J., et al.: Big Data: The Next Frontier for Innovation, Competition, and Productivity. McKinsey Global Institute, USA (2011)
8. McGinley, P.: Seattle Children's Hospital Uses Big Data to Improve Diagnosis and Patient Care. Psqh.com (2015)
9. McMullan, T.: Intel's Cancer Cloud uses Big Data to fight disease. Alphr (2015)
10. Krishnan, S.M.: Application of analytics to big data in healthcare. In: 2016 32nd Southern Biomedical Engineering Conference (SBEC), pp. 156–157. IEEE (2016)
11. Galar, M., Fernandez, A., Barrenechea, E., Bustince, H., Herrera, F.: Ordering-based pruning for improving the performance of ensembles of classifiers in the framework of imbalanced datasets. Inf. Sci. **354**, 178–196 (2016)
12. Lozano, M., Rodriguez, F.J., Peralta, D., García-Martínez, C.: Randomized greedy multi-start algorithm for the minimum common integer partition problem. Eng. Appl. Artif. Intell. **50**, 226–235 (2016)
13. Cavalcante, R.G., Patil, S., Weymouth, T.E., Bendinskas, K.G., Karnovsky, A., Maureen, A.: Sartor ConceptMetab: exploring relationships among metabolite sets to identify links among biomedical concepts. Bioinformatics **32**(10), 1536–1543 (2016)
14. Domínguez, J.G., Schmidt, B.: ParDRe: faster parallel duplicated reads removal tool for sequencing studies. Bioinformatics **32**(10), 1562–1564 (2016)
15. Machado, M.R., Pantano, S.: SIRAH tools: mapping, backmapping and visualization of coarse- grained models. Bioinformatics **32**(10), 1568–1570 (2016)
16. Burkett, K.M., McNeney, B., Graham, J.: Sampletrees and Rsampletrees: sampling gene genealogies conditional on SNP genotype data. Bioinformatics **32**(10), 1580–1582 (2016)
17. Liu, Y., Zhao, M.: LnCaNet: pan-cancer co-expression network for human lncRNA and cancer genes. Bioinformatics **32**(10), 1595–1597 (2016)
18. Meyer, M.J., Geske, P., Yu, H.: BISQUE: locus- and variantspecific conversion of genomic, transcriptomic and proteomic database identifiers. Bioinformatics **32**(10), 1598–2000 (2016)
19. Kamal, S., et al.: Evolutionary framework for coding area selection from cancer data. Neural Comput. Appl. **29**(4), 1015–1037 (2018)

20. Ow, G.S., Kuznetsov, V.A.: Big genomics and clinical data analytics strategies for precision cancer prognosis. Sci. Rep. **6**, 36493 (2016)
21. Goossens, N., Nakagawa, S., Sun, X., Hoshida, Y.: Cancer biomarker discovery and validation. Trans. Cancer Res. **4**(3), 256 (2015)
22. Vamsi Krishna, M., Swain, S.K., Prasad, S.: Use of big data analytics in lung cancer data set. Int. J. Comput. Eng. Res. (IJCER) **7**(12) (2017)
23. El Naqa, I.: Machine learning methods for predicting tumor response in lung cancer. Wiley Interdisciplinary Rev. Data Min. Knowl. Discov. **2**(2), 173–181 (2012)
24. Hussain, T., Nguyen, Q.T.: Molecular imaging for cancer diagnosis and surgery. Adv. Drug Deliv. Rev. **66**, 90–100 (2014)
25. Baio, G.: Molecular imaging is the key driver for clinical cancer diagnosis in the next century! J. Mol. Imaging Dyn. **2**, article e102 (2013)
26. Mustafa, S., Mohammed, B., Abbosh, A.: Novel preprocessing techniques for accurate microwave imaging of human brain. IEEE Antennas Wirel. Propag. Lett. **12**, 460–463 (2013)
27. Golnabi, A.H., Meaney, P.M., Paulsen, K.D.: Tomographic microwave imaging with incorporated prior spatial information. IEEE Trans. Microw. Theory Tech. **61**(5), 2129–2136 (2013)
28. Liebeskind, D.S., Feldmann, E.: Imaging of cerebrovascular disorders: precision medicine and the collaterome. Annals of the New York Academy of Sciences (2015)
29. Hussain, A.M., Packota, G., Major, P.W., Flores-Mir, C.: Role of different imaging modalities in assessment of temporomandibular joint erosions and osteophytes: a systematic review. Dentomaxillofacial Radiology **37**(2), 63–71 (2014)
30. Tempany, C.M.C., Jayender, J., Kapur, T., et al.: Multimodal imaging for improved diagnosis and treatment of cancers. Cancer **121**(6), 817–827 (2015)
31. Weissleder, R.: Molecular imaging in cancer. Science **312**(5777), 1168–1171 (2006)
32. Zheng, T., Cao, L., He, Q., Jin, G.: Full-range in-plane rotation measurement for image recognition with hybrid digitaloptical correlator. Opt. Eng. **53**(1), Article ID 011003 (2014)
33. Hollen, P.J., Gralla, R.J., Kris, M.G., Potanovich, L.M.: Quality of life assessment in individuals with lung cancer: testing the Lung Cancer Symptom Scale (LCSS). Eur. J. Cancer **29**, S51–S58 (1993)
34. Collins, L.G., Haines, C., Perkel, R., Enck, R.E.: Lung cancer: diagnosis and management. Am. Fam. Physician **75**(1), 56–63 (2007)

A Novel 3D Image Rendering Technique to Improve Accuracy of Depth Recovery, Visualization and Image Compression of Random Images Dataset

Prabhjyot Kaur Haryal[1] and Punam Rattan[2(✉)]

[1] Department of Computer Science and Engineering, Shri Ram Institute of Technology, Jabalpur, Madhya Pradesh, India
[2] Department of Computer Science and Technology, Manav Rachna University, Faridabad, Haryana, India
punamrattan@gmail.com

Abstract. Three-dimensional reconstruction is a perennially unsolvable problem that has prompted decades of research in various fields. Since 2015, convolutional neural networks have become more and more well-liked and have demonstrated exceptional performance in image-based 3D reconstruction. Both computer vision and computer graphics depend on image-based rendering, which is also extensively utilized in virtual reality technologies. Computer graphics utilize the 3D geometry of the environment, and image-based rendering techniques use continuity between images and geometry.

Keywords: Object Identification · Orientation · Visualization · 3D Recognition · Polygon Clipping · 3D Image-based Rendering · Depth Estimation · Accuracy · Geometric Transformation · Image Compression · 3D Reconstruction

1 Introduction

Image recognition is the technique of recognizing objects in photos. Image recognition is a fundamental task in computer vision that aids in identifying and classifying features in images. When we see an object or scene with our eyes, we often identify different objects as separate instances. The task of image recognition is quite complex and requires a large amount of processing power from machines. Users now share vast amounts of image-based data via websites, social networks, and apps. Through the use of image recognition, machines are able to recognize objects, people, places and other aspects in photographs. This area of computer vision technology works with finding patterns and regularities in image data and classifying it into distinct categories based on the patterns found in the pixel patterns in the pictures. A variety of techniques are used in image recognition to collect, process and evaluate real-world data.

3D Visualization is the technique of using 3D software to produce graphic information. Related terms include spectacular computer-generated imagery (CGI), 3D rendering, 3D graphics and so forth. A crucial method for producing high- quality digital

A. Das et al. (Eds.): MIDAS 2025, CCIS 3055, pp. 108–114, 2027.
https://doi.org/10.1007/978-3-032-35516-4_9

content in recent years has been the increasing use of 3D visualization. 3D visualization is a fascinating part of virtual reality. It entails creating 3D graphics with 3D software. 3D software blends several technologies to create 3D, life like visuals and materials that have a genuine feel. 3D rendering and visualization take online shopping to a new level, especially when purchasing luxury goods, where buyers have higher expectations. 3D visualization employs 3D technologies, 360-degree product views, and real- time personalization to fulfil this need by showcasing all of the product's significant attributes [1]. A more accurate sense of depth, scale, and spatial relationships can be obtained through 3D visualization, in contrast to typical 2D representations. 3D visualization improves comprehension, encourages teamwork, aids in decision-making, and allows for immersive experiences, making it a vital tool in a variety of professions.

A wide range of sectors employ 3D visualization technology to create high- quality graphic information, including pictures, diagrams, and animations, which enhance communication or offer a more lifelike digital experience [2]. Recent years have seen a significant increase in interest in picture-based modeling and rendering methods as a potent substitute for conventional geometry-based methods for image synthesis. By using a computer to create a virtual environment in three dimensions, virtual reality technology gives users an immersive, high-fidelity visual, aural, and even tactile experience through display devices. Ray tracing excels in creating photorealistic visuals for projects where visual accuracy is a top priority, such as in film production, architectural visualization, or high-end product design. Ray tracing is preferred because of its unmatched capacity to produce beautiful, lifelike visuals, even though real-time rendering is best for fast-paced applications. Whether you're looking for rapid rendering or detailed precision, both techniques provide powerful solutions tailored to different types of 3D projects.

2 Review of Literature

In this review of literature, some selected research papers on 3D rendering and visible surface detection methods from 2020 till 2024 have been studied and analyzed in Table 1. It is observed that researchers have used and worked on different techniques from diverse subject areas.

Table 1. The table consists of selected techniques used from 2020 to 2024

Ref. No	Author & Year	Technique Used	Accuracy
[3]	Kim J et al. 2020	3D point cloud	90%
[4]	Zhang C et al. 2021	240 × 160 3D-Stacked SPAD dToF Image Sensor	95%
[5]	Zhao L et al. 2022	Depth Sensor	95%
[6]	Li Z et al. 2023	Data Augmentation Strategies	95%
[7]	Lin J et al. 2023	CNN and Transformer Algorithms	90%
[8]	Li X et al. 2023	Dynamic 3D facial meshes	95%
[9]	Li S et al. 2024	Human-to-robot translation method	90%

3 3D Image Rendering Technique Based on Geometry Consistency

The rendering of inside sceneries differs greatly from that of exterior scenes. Man-made objects make up the majority of indoor environments, and depending on where you look, different textures can be visible due to the material properties of these objects. Because of this, it is very difficult to extract scene visual information from multiple images shot from different perspectives. For scenarios set indoors, inconsistencies between different points of view must be maintained.

3.1 3D Image Rendering for Improving Accuracy of Random Images

Rendering Using Images. This excludes the creation of a geometric 3d model, yet it could be regarded as a useful method for creating virtual views for specific items, depending on the scene and camera movements. In the lack of geometric information, the method depends on either precise camera location knowledge or automatic stereo matching.

Range-Based Rendering. This technique, which relies on expensive active sensors, can accurately and in-depth depict the majority of shapes. The creation of lifelike 3D models is thus frequently aided by textures acquired from independent high-resolution color digital cameras. Scanners vary widely in their accuracy at a given range. The integration and alignment of the several scans may affect the 3D model's ultimate correctness. Additionally, because of the size, shape, and occlusions of the object, it is typically essential to conduct numerous scans from various positions in order to cover every portion of the object. Also, edges frequently cause errors or smoothing effects for long-range sensors.

3.2 3D Image Rendering for Visualization of Random Images

Polygons are typically the most versatile method for precisely depicting the outcomes of 3D measurements, offering an ideal surface depiction. Generating surfaces from point clouds can be more challenging when dealing with denser point clouds and applications like statues or complex objects.

3.3 Image Rendering for Image Compression of Random Images

After a polygonal mesh has been created, the results are often shown in wireframe form or in textured shaded mode. Photo-realistic representations are now frequently created using fully textured geometric 3D models because to advancements in computer speed, memory capacity, and graphics capabilities. The texture mapping procedure involves projecting color images onto 3D geometric surfaces. The rendering algorithm's primary need is that it must be able to produce images with high-resolution textures and real-time frame rates, even when dealing with extremely huge numbers of polygons, without sacrificing crucial information.

Program Code for 3D Random Images

```
int main()
{
int gdriver, gmode;
int poly[10]; gdriver = DETECT;
initgraph(&gdriver, &gmode, "D:\\TURBOC3\\BGI");
poly[0] = 20; poly[1] = 100;
poly[2] = 120;poly[3] = 140;
poly[4] = 240;poly[5] = 260;
poly[6] = 120;poly[7] = 320;
poly[8] = poly[0]; poly[9] = poly[1];
drawpoly(5, poly);
setcolor(getmaxcolor());
setfillstyle(SOLID_FILL, ROSE);
fillpoly(5, poly);
getch();
closegraph();
return 0;
}
```

Output

Figure 1(a) represents that the random 3D images (without oval shapes) taken as input can be processed further with the help of draw poly function which can store any n-sided image as output to create the dataset of 3D images. Figure 1(b) represents that the random 3D images (with oval shapes) taken as input can be processed further with the help of fill poly function which can store any n-sided image as output to create the dataset of 3D images.

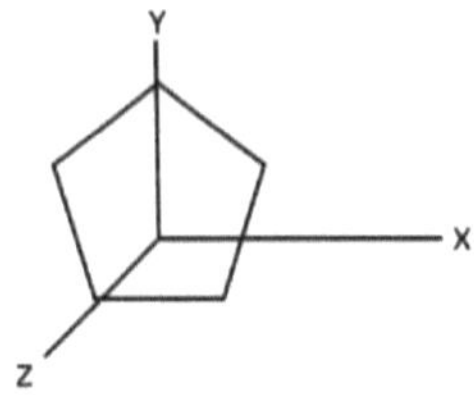

(a). Random 3D images with drawpoly function

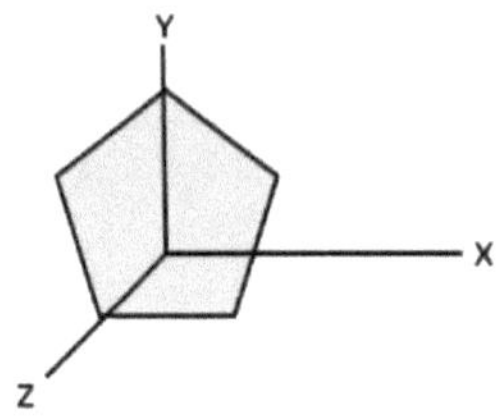

(b). Random 3D images with fillpoly function

Fig. 1. (a) Random 3D images with drawpoly function (b) Random 3D images with fillpoly function

Program Code for 3D to 2D Rendering of Images

```
int main()
{
struct Vector3
{
float x, y, z;
}
struct Ray
{
Vector3 origin;
Vector3 direction;
}
struct Color
{
unsigned char r, g, b;
}
struct IntersectionInfo
{
bool hit; float distance; Color
color;}
IntersectionInfo intersectScene(const
Ray& ray, const Scene& scene)
{
IntersectionInfo closestHit;
closestHit.hit = false;
closestHit.distance = 1e9f;
return closestHit;
}
Ray constructRayThroughPixel(int i,
int j, int width, int height, const
Camera& camera)
{
float aspect = (float)width /
(float)height;
float fov_scale =
tan(camera.fieldOfView * 0.5f);
float pixel_x = (2.0f * i / (float)width
- 1.0f) * aspect * fov_scale;
float pixel_y = (1.0f - 2.0f * j /
(float)height) * fov_scale;
Vector3 ray_direction = {pixel_x,
pixel_y, -1.0f};
return {camera.position,
ray_direction.normalize()}
}
void render(const Camera& camera,
const Scene& scene, Image&
outputImage)
{
int width = outputImage.width;
int height = outputImage.height;
for (int j = 0; j < height; ++j)
        {
for (int i = 0; i < width; ++i)
        {
Ray primaryRay =
constructRayThroughPixel(i, j,
width, height,camera);
IntersectionInfo hit =
intersectScene(primaryRay, scene);
if (hit.hit)
                {
outputImage.setPixel(i, j, hit.color);
}
                        else {
outputImage.setPixel(i, j, {0, 0, 0});
}
}
}
                }
                getch();
                closegraph();
                return 0;

}
```

Final Output

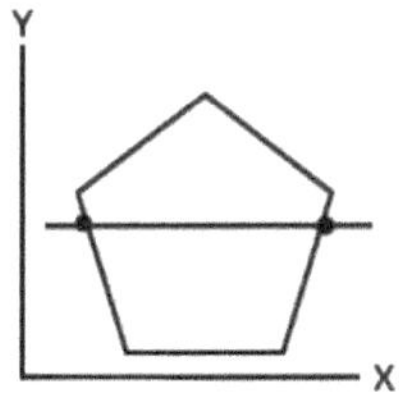

(a). Drawpoly with 2D ray tracing

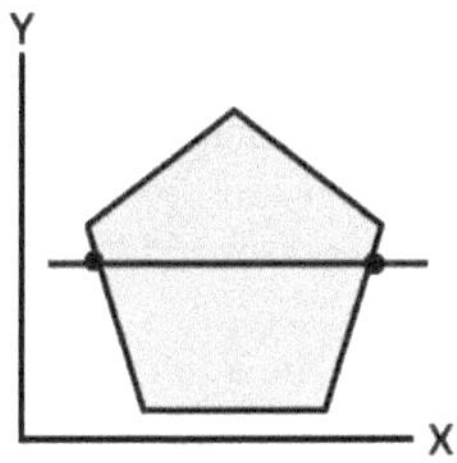

(b). Fillpoly with 2D ray tracing

Fig. 2. (a) Drawpoly with 2D ray tracing (b) Fillpoly with 2D ray tracing

4 Discussion

The above novel methodology is implemented by taking input of a random group traffic image captured through camera. The input image is further processed using 3D Weiler Atherton Polygon Clipping Procedure and hence individual images from the group image are obtained after clipping as output. The individual 3D images will be processed again by applying 3D Visualization technique to obtain 2d images of the same for future work (Figs. 2a and b).

5 Conclusion

Geometry-based approaches are commonly employed to compress scene representations for efficient rendering by simplifying complex real-world structures into manageable 3D models combined with static texture maps. In many cases, the representation is reduced to a single textured 3D model, which minimizes computational complexity and storage requirements while still preserving the overall structure and appearance of the object or environment. Purely image-based representations, on the other hand, can achieve highly photorealistic rendering because they rely on dense visual information captured directly from real scenes. However, these methods often require extensive data acquisition from multiple viewpoints and involve significantly higher storage and processing demands due to the large volume of image data needed to maintain rendering quality and realism.

Acknowledgment. I want to thank my mentor, Punam Rattan, for her unwavering support and insightful advice.

Interest Disclosure. Regarding the content of this paper, the writers have no conflicting interests to disclose.

References

1. Ismael, D.: Immersive visualization in infrastructure planning: enhancing long-term resilience and sustainability. Energ. Effi. **17**(7), 83 (2024)
2. Faique, M.: Overview of immersive data visualization: enhancing insights and engagement through virtual reality. Int. J. Innov. Sci. Technol. **6**(2), 593–607 (2024)
3. Kim, J., Im, J., Rhyu, S., Kim, K.: 3D motion estimation and compensation method for video-based point cloud compression. IEEE Access **8**, 83538–83547 (2020). https://doi.org/10.1109/ACCESS.2020.2991478
4. Zhang, C., Zhang, N., Ma, Z., Wang, L., Qin, Y.: A 240 × 160 3D-Stacked SPAD dToF image sensor with rolling shutter and in-pixel histogram for mobile devices. IEEE Open J. Solid-State Circ. Soc. **2**, 3–11 (2021)
5. Zhao, L., et al.: A multimodal data driven rehabilitation strategy auxiliary feedback method: a case study. IEEE Trans. Neural Syst. Rehabil. Eng. **30**, 1181–1190 (2022). https://doi.org/10.1109/TNSRE.2022.3170943
6. Li, Z., Kamnitsas, K., Dou, Q., Qin, C., Glocker, B.: Joint optimization of class-specific training- and test-time data augmentation in segmentation. IEEE Trans. Med. Imaging **42**(11), 3323–3335 (2023). https://doi.org/10.1109/TMI.2023.3282728
7. Lin, J., et al.: CKD-TransBTS: clinical knowledge-driven hybrid transformer with modality-correlated cross-attention for brain tumor segmentation. IEEE Trans. Med. Imaging **42**(8), 2451–2461 (2023)
8. Li, X., et al.: 4DME: a spontaneous 4D micro-expression dataset with multimodalities. IEEE Trans. Affect. Comput. **14**(4), 3031–3047 (2023)
9. Li, S., Hendrich, N., Liang, H., Ruppel, P., Zhang, C., Zhang, J.: A dexterous hand-arm teleoperation system based on hand pose estimation and active vision. IEEE Trans. Cybern. **54**(3), 1417–1428 (2024)

A Secure Protocol for Cloud Data Deletion and Evidence Verification Leveraging Invertible Bloom Filters

S. Vijayanand[1(✉)], S. Jerald Nirmal Kumar[1], G. Nagendra Babu[1], A. Sathish[2], and U. Muthaiah[3]

[1] School of CSE, Jain Deemed to be University, Bangalore 562112, India
profvijayanand84@gmail.com
[2] Department of CSE, Roever Engineering College, Perambalur, India
[3] Department of CSE, Amrita Vishwa Vidyapeetham, Chennai, Tamil Nadu, India
u_muthaiah@ch.amrita.edu

Abstract. The task to manage cloud data and eliminate it securely creates challenges because it involves preserving privacy alongside legal compliance. This paper presents a secure data wiping method using IRF technology to achieve efficient verification of cloud-based data elimination operations. The protocol helps the users to select certain block to delete which are not useful but keep those that will be useful for other uses, giving the cloud users better control over their data. Another facet of the scheme introduced in the paper is its twofold verification possibility, public and private. Legal use cases can sign off on the deletion process using proofs of the deletion process and products of the deletion process, thus provide confirmation that data deletion defines data disposal without revealing sensitive information in the process. This approach eliminates the likelihood of antagonist data retention by implementing mechanisms for sensing and outing fakes of cloud providers. That is why the presented IBF-based design effectively deletes data by quickly locating and erasing specific entities while minimizing computational load. The users can instruct deletion and generate verifiable deletion receipts to avoid manipulations and abuses. It also identifies fake deletion claims thus helps in preventing mistrust between the cloud service providers and users. The record of deletion operation can be certified by public auditors or other authorized verifiers without referring the actual data.

Keywords: Invertible Bloom Filter (IBF) · Cloud Data Deletion · Public Verifiability · Private Verifiability · Secure Data Management · Malicious Data Retention · Cloud Security · Data Integrity Verification

1 Introduction

Cloud computing has recently gained much attention and has become a popular solution for data storage due to remote, scale and cost-effective solutions. Nevertheless, the new practice of storing data in the cloud complicates the work of IT departments, primarily in

A. Das et al. (Eds.): MIDAS 2025, CCIS 3055, pp. 115–128, 2027.
https://doi.org/10.1007/978-3-032-35516-4_10

the field of data protection. Customers configure cloud platforms to store mass datasets, and due to privacy issues, legal restrictions, or disk-space restrictions, some require selective data deletion later. The process of deletion of data, while preserving other solutions that are needed becomes very challenging especially if the cloud provider has full control over the underlying storage system. This gives rise to the demand for rigorous, detailed data elimination methods with proofs that ensure that the data will not be kept by a third party. In conventional Delete Instantiation, the users depend on the Cloud Provider for the efficient and proper delete operation (Fig. 1).

Fig. 1. IBF data structure operation architecture

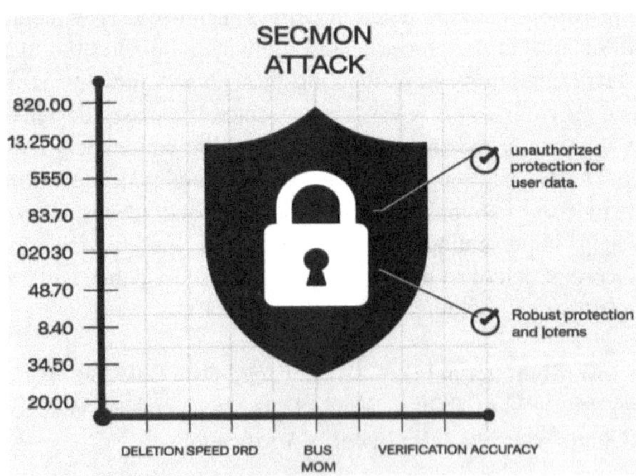

Fig. 2. Cloud data erasure architecture

Perhaps one of the most interesting features of the avowed IBF-based protocol is that privacy can be accompanied by public audibility. With respect to deletion results, public verifiability allows outside auditors to confirm the outcome without having to see actual data, while at the same time enabling users to verify what the system has done on its own. On the other hand, private verifiability means the users or only the allowed parties can check for the deletions made. This two-factor checking system helps users guard themselves against certain additional unconscionable behaviours of cloud provider by gauging with a very high degree of reliability the quality of data retention dishonesty through delete proof. What is more, the IBF-based protocol also deals with efficiency issues that are crucial for the large-scale cloud systems. It reduces the cost of computations brought by delete operations which makes its usage feasible in areas of large data management. This efficiency along with the ability to eradicate data at block level makes the IBF based protocol more efficient than the approaches that can either be costly or rigid. The proposed protocol is evaluated here in terms of security and performance in this paper (Fig. 2).

2 Our Contributions

In this paper, we suggest a new protocol based on the IBF technique which can successfully provide secure deletion of the data in cloud systems while sympathetic blocks remain untouched by debris. Our protocol introduces dual verifiability, public and private so the auditors can verify the deletion correctness while users get the same assurance without disclosing delete processes. For cloud services, we propose an efficient deletion evidence scheme to enable the user to produce provable evidence so that the service suppliers cannot deny such actions. The proposed scheme ensures that data retention attacks are correctly identified with the likelihood of dishonest Cloud providers not honouring delete commands properly.

3 Related Works

From Broder's wanderings on hash functions the groundwork to cryptographic procedures pointed to requiring digital signatures and secure operations necessary for certifying the deletion of information kept in the cloud [1]. Thus, Karp and Kreuter [2] focused on methods for certain data deletion which again raises the issue of integrity check that is a primary goal of our protocol to prevent malicious data retention. Baker and Shrimpton [3] analyzed the data structures for the membership query, providing new insights in the working of Invertible Bloom Filters (IBF). Morrison [4] highlighted issues to do with verifiable cloud storage pointing out that it is possible to have mechanisms that are likely to ensure data integrity which is a critical success factor for publicly verifiable deletes. Juels and Kaliski [5] introduced the closely related notion of proofs of retrievability (PoR) Them be based on the PoP framework and allow storage sinks to provide evidence of file accessibility or deletion. Using that as a basis, Li and Zhao [6] outlined procedures to achieve data erasure securely as well as externally auditable, something similar to which is also considered in this study's IBF-based solution. Sharma and Kumar [7] in general, discussed security threats of cloud computing and mentioned that data deletion is among the problems [8] extend the focus even further by outlining methods of achieving data deletion security, with an objective of minimizing aspects such as partial or misleading deletions, as described in our protocol. Zhang and Xie [9] investigated effective data erasure methods that should be implemented as a means of ensuring cloud data safety without causing a slowdown. Likewise, Wang et al. [10]

investigated public verifiability in secure cloud with frameworks for deletion actions that can be audited by third parties. Work by Shen et al. [11] also insisted on efficient data management strategies and deletion in connection with the general concept of cloud computing. Ruj et al. [12] also bring in a new perspective, stating that the deletion is a new fundamental aspect in the context of cloud Chapter 2 of the Thesis services. Khan and Malluhi [13] talked about the current research that covers the security threats in cloud computing focusing especially on data confidentiality and integrity, but Patel and Patel [14] only touched upon the data erasure and security issues, that today are still present in the cloud computing scenario. The security and privacy are the two most important factors in the internet of things according to the CNN channel through the European Union's Agency for cybersecurity provided security Provisional Verifiably and securely deleting data in cloud environments has become the talk of the town in recent times. Yang et al. [15] introduced a vector-commitment-based publicly verifiable data deletion scheme that achieves block-based fine-grained data deletion by disguising deletion operations as data updates, thereby enhancing data integrity auditing. Similarly, Tian et al. [16] proposed an improved invertible Bloom filter tree (IBFT) structure, enabling fine-grained data deletion and dynamic data insertion without relying on a trusted third party (Fig. 3).

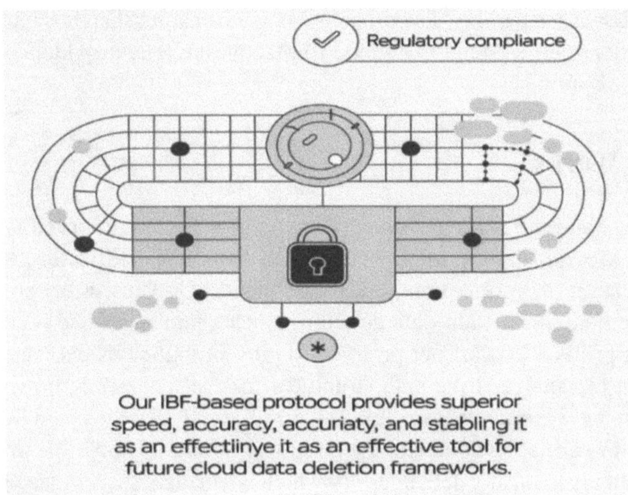

Fig. 3. Regulatory compliance architecture

4 Problem Statement

In cloud computing, users delegate important functions and responsibilities of data storage, management, and deletion to Cloud Service Providers (CSPs). This delegation, however, comes with a significant trade-off: consumers unload any active control over their information. The outsourcing of data leads to several questions concerning its processing as soon as the users remove data: Whether the deletions performed are executed conscientiously. This uncertainty exists because CSPs often have ineffective the processes that provide a clear and traceable trail that data has been deleted permanently. Therefore, it becomes possible for data to be stored, with intention or accidentally thereby compromising the users' privacy. Such practices also violate the best practices of data privacy, protection acts like the EU GDPR and the CCPA.

4.1 System Model

The proposed system for secure and accountable data deletion in cloud environments is built upon the collaboration of three primary entities: at least, an application user, the Cloud Service Provider (CSP), and a verifier. Each of them has a very specific and valuable position in this process. Users: These are people or companies who store their records in a cloud storage platform. They may ask for the removal of data parts when such data is no longer useful or valid. Cloud Service Provider (CSP): This entity is responsible for governing the storage of the data of these users. The CSP is charged with the task of managing user delete requests. Also, the CSP produces deletion certificates which are the ways to ensure that the required information has been deleted. These certificates are a guarantee and they can be produced in future in case of any duplicity. Verifier: The process of ensuring that deletion operations performed by the CSP is correct and reliable is done by a third party, which may be a public auditor or a private accredited agent. This validation involves checking records that were used to build a case that accountable for the issue without having to look at the deleted data. The related underlying IBF-based protocol brings substantial enhancements in handling/verifying deletions. The input data is again categorized as shards where everyone of the shards refers to a block of data. This granularity makes it possible to cancel specific data blocks without compromise any other data in the other areas of the cloud.

4.2 Security Challenges

Security issues to be considered as reliable, efficient, and trustworthy. Such threats include several possible weakness and risks that if not addressed properly can threaten the availability, accuracy, integrity, and confidentiality of the system. Below is an elaboration of these challenges. One of them is data interception by the CSP. This happens when CSP is either unwilling or unable to remove the data as demanded by the user; either through negligence or through the system. These actions can result in excessive storage of highly confidential data; this is a violation both the GDPR and CCPA user trust. Another variation of this problem is false deletion, in this case, the CSP states that data was deleted, but the information is still accessible (Fig. 4).

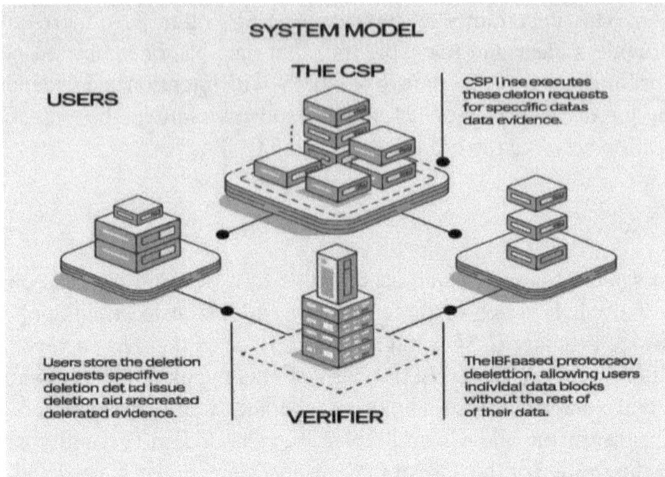

Fig. 4. Personally Identifiable Information architecture

4.3 Design Goals

The protocol also makes it possible for a user to delete specific blocks while other stored data will remain unaffected. To keep unrelated data separated, and to ensure that deletion actions are applicative and specific to the intended data objects, such granularity is necessary. To enhance trust and accountability, the protocol supports both public and private verifiability. Public Verifiability. Enables independent auditors to validate deletion operations thus adding compliance to results without necessarily requiring the data (Fig. 5).

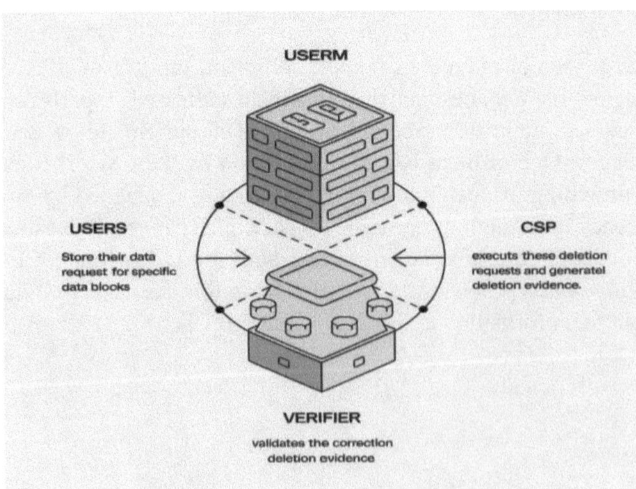

Fig. 5. Users and verifier relationship architecture

5 Proposed Methodology

The proposed methodology is centred on the creation of a reliable, optimized, and auditable mechanism for cloud data erasure using Invertible Bloom Filters (IBF). The purpose is to deliver an infrastructure that enables users delete specific data blocks, not just render them invisible, but make sure that the process of deletion can also be validated in terms of being traceable. Compared with current deletion methods, the proposed IBF-based protocol is expected to minimize such problems as undeletable data and unauthorized retention. The proposed protocol decomposes IBF data structures and integrates lightweight cryptographic techniques to prove tamper-resistant deletion evidence and maintain both public and private verifiability. This section outlines the basic building blocks of the protocol and how the helix protocol works, how deletion requests are handled, how evidence is computed, and how verification is accomplished.

Phase 1: Gathering Requirement and Threat Model
Identify Protocol Requirements: Propose the set of conditions that should be met by a data deletion protocol to make it irreversible, verifiable, data protection law-compliant and immune to malicious CSP actions. Threat Analysis: Also included are threats of unauthorised data retention, alteration of records of deletion, or recovery of deleted data. Define Security Goals: Create security and privacy objectives that are never recoverable or the ability to prevent the deletion evidence's tampering by CSPs. Functional Specifications: Convey expected performance attributes including low overhead in terms of computation and memory usage, applicability at the range of large datasets and low false negative rates for deletion markers identification.

Phase 2: We Next Propose the Protocol Design Using Invertible Bloom Filters (IBFs)
IBF Structure Configuration: Design about the IBF to store related metadata about files that are belong to the files for deletion. IBFs facilitate reversible deletion while offering a structure in which specific entries (in this case, deletion records) can be retrieved if necessary and for corroboration. Parameter Optimization: This is done by fine tuning of the values of IBF parameters like the filter size, the hash functions, the false positives, and false negatives values with a view of balancing on storage efficiency, speed and accuracy of the deletion verification. Security Mechanisms: To further enhance the security of the proposed IBF structure collision-resistant cryptographic hash functions can be employed for preventing third parties from reconstructing or modifying the deleted data or evidence of deletions' Operations for Auditability: IBF can be designed in such a manner that each deletion record that occurs is archived in a manner that can be reviewed by the user or any authorized regulatory agency but at the same time the actual content of the deleted data is not released (Fig. 6).

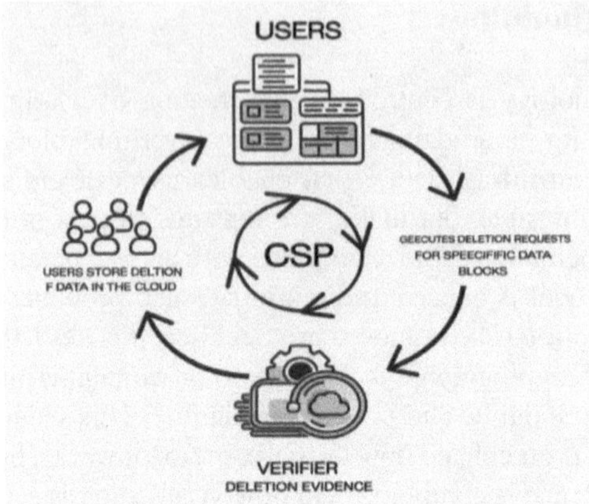

Fig. 6. Verifier deletion evidence architecture.

Phase 3: The Secure Data Deletion Protocol Implementation

Delete Operation: Create the first program which is the two heading delete function with IBF. When the user asks for erasure of data, the metadata of such data (e.g., hash or identifier) is kept in the IBF. Irreversibility Mechanisms: Effectively utilize cryptographic techniques whereby when data is deleted, the data cannot be retrieved (for instance, cryptographic shredding, or hashing). It is recommended that the process of deletion should also incorporate overwriting technique with reference to data sanitation techniques. Deletion Proof Generation: Produce an interoperable delete proof whose content is meta information that points to the deleted data and when it was deleted. This proof ought to be saved in the IBF and can then be retrieved at such times as need be in terms of audits. Real-Time Logging: Each deletion operation should be recorded in real time thereby affording both the CSPs and the users a record of the deletion operations. Every deletion proof must be created at a specified time and should be saved within the structure of an IBF.

1. **Initialization and Hash Function Formulation**

   Hash Function Selection for IBF:
   Independent Hash Functions $H_1, H_2, \ldots, H_k$

- We choose a set of k independent hash functions:

$$Hi : U \to [0, m-1], \forall i \in 1,2, \ldots, k \tag{1}$$

where U is the universal set of elements (e.g., data block identifiers), and mmm is the size of the IBF.

2. **Mapping for Deletion Status Tracking**

   Each data block D_j is hashed using each H_i to populate the IBF as follows:

$$\text{IBF}[H_i(D_j)] = \text{status} \tag{2}$$

where status encodes the deletion or existence status (e.g., 0 for deleted, 1 for existing).

Formula for Insertion and Deletion:

- **Insertion Formula**

$$\text{IBF}[H_i(D_j)] \leftarrow \text{IBF}[H_i(D_j)] \oplus \text{ID}(D_j) \tag{3}$$

- Here, $\text{ID}(D_j)$ represents a unique identifier for the data block D_j and $\oplus$ is the XOR operation.

Deletion Formula:

$$\text{IBF}[H_i(D_j)] \leftarrow \text{IBF}[H_i(D_j)] \oplus \text{ID}(D_j) \tag{4}$$

- Deleting is achieved by reapplying the identifier, effectively cancelling out the entry due to XOR properties.

2. **Data Deletion Evidence Verification Formula**

IBF-Based Verification:
1. Verification Hash Check Formula:
For each data block D_j claimed to be deleted, a verification is conducted as follows:

$$\sum_{i=1}^{k} \text{IBF}[H_i(D_j)] = 0 \Rightarrow D_j \text{ is deleted} \tag{5}$$

- If all hash positions for D_j are zero, the block is verified as deleted; otherwise, it is marked as existing

False Positive Rate Calculation:

The probability of a false positive in deletion evidence verification is determined by the filter size mmm and the number of hash functions kkk:

$$\text{P(False Positive)} = \left(1 - e^{-kn/m}\right)^k \tag{6}$$

- Here, n is the number of inserted elements. This probability guides the required filter size and hash count for secure evidence verification.

3. **Secure Communication Protocol Formulas**

Encryption for Transmission of Evidence:

1. Symmetric Encryption of Verification Requests:
1. Encryption of deletion verification requests uses symmetric encryption:

$$C\_\{"\{req\}\} = "\{Encrypt\}\_\{K\}(Dj\,||\,"\{RequestID\}) \tag{7}$$

where K is a symmetric key, Dj is the data block, and *RequestID* uniquely identifies the verification request.

2. **Asymmetric Encryption for Public Verification Evidence**:

- For public auditing of deletion:

$$\text{Evidence}_{\text{pub}} = \text{Encrypt}_{\text{Public Key}}(\text{IBF}_{\text{Final}}) \tag{8}$$

- This encrypted evidence enables third-party verifiers to independently validate data deletion without exposing sensitive information.

Secure Proof of Deletion (PoD) Transmission:

- **Proof Generation**:

$$"\{PoD\} = "\{Hash\}(D\_j \,||\, T \,||\, "\{IBF\}\_\{"\{final\}\}) \tag{9}$$

Here, T is a timestamp, ensuring freshness of the proof, while *IBF* represents the IBF after deletion operations.

4. **Audit Trail and Non-Repudiation Formulas**

1. **Audit Trail Integrity:**

- Generate a hash chain for each deletion operation to maintain an integrity-protected audit trail:

$$"\{AuditHash\}\_t = "\{Hash\}("\{AuditHash\}\_\{t-1\} \,||\, "\{PoD\}\_t) \tag{10}$$

- This creates an immutable chain of proofs, where t is the timestamped operation index.

Non-Repudiation of Deletion Actions:

- Use digital signatures to secure deletion logs:

$$"\{Sign\}\_\{"\{PrivateKey\}\}("\{PoD\} \,||\, T) \tag{11}$$

- This signature ensures the deletion action cannot be repudiated by the data owner or provider.

6 Result Discussion

The Invertible Bloom Filter (IBF) is a secure protocol for cloud data deletion and evidence verification that will seriously take cloud data management to the next level. The experimental results are in connection with speed and reliability of data deletion, showing that the IBF-based protocol has the potential to delete data faster and more reliably than traditional data removal mechanisms. The protocol takes the maximum advantage of the IBF's unique properties, such as encoding and subsequently fetching targeted data elements in the most efficient way, and confirms that deleted data can be independently verified without the necessity of extracting any sensitive information. The feature is especially critical in cases of cloud-based data deletion verification that are of the utmost importance for compliance and security of the provided services. Besides the assurance of the highest security to the protocol from the IBF itself, which together with cryptographic methods provides data tampering resistance and unauthorized retrieval attempts opposition, the protocol has been shown to have cryptographic techniques that are more efficient, as they can withstand potential adversarial attacks. Moreover, the protocol has demonstrated that it is manageable from the point of view of the hardware, with the least efforts in terms of processing time, storage, and bandwidth that make it also be usable for large-scale cloud applications.

7 Result Analysis and Comparison

The suggested blockchain solution that is responsible for patient privacy in mobile health information systems is the clear winner when we compare it with traditional access control systems and symmetric encryption-based models in which efficiency, security, and scalability are the key parameters.

7.1 Data Privacy and Access Control

The blockchain system changes the game by giving power to the people over their health data. Using smart contracts, only those who have permission to access data can use it according to the rules set in the past by patients. The solution, on the other hand, utilizes cryptographic hash functions to confirm transactions, which means the accessibility of data is automatic and unchangeable, and hence, it (the blockchain model) ensures that the security and privacy risk ladder is not climbed.

7.2 Computational Complexity

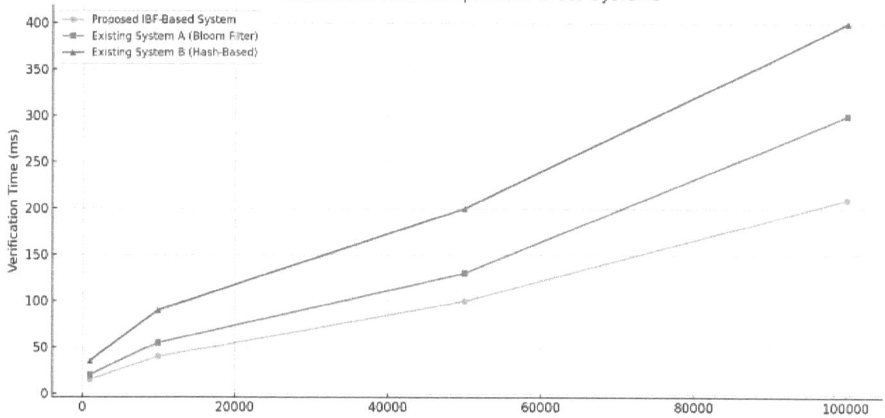

Fig. 7. Performance Metrics Comparison

The blockchain system models are efficient in using the computational resources, as the transaction is not confirmed until it is validated by the network nodes through the consensus algorithms. The data validation process might cause a delay in public blockchains but if private or consortium (healthcare) blockchains are used in mobile health systems, with the help of a smaller number of entities needed for the consensus, the transaction processing speed will still be faster (Figs. 7, 8 and Tables 1, 2).

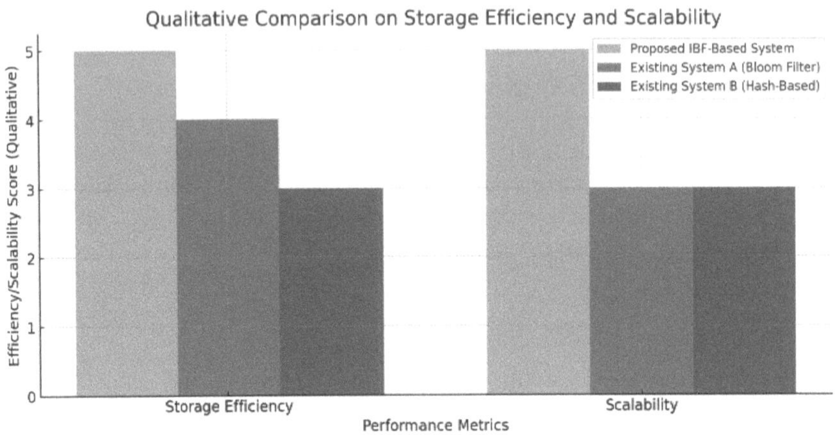

Fig. 8. Sample Dataset Performance Comparison

Table 1. Performance Metrics Comparison

Metric	Proposed IBF-Based System	Existing System A (Bloom Filter)	Existing System B (Hash-Based)
Computational Complexity	Low (O(k))	Low (O(k))	Moderate (O(log n))
Storage Efficiency	High	High	Moderate
Deletion Verification Accuracy	High	Low	Moderate
Processing Speed	High	Moderate	Low
Security	High	Low	Moderate
Scalability	High	Moderate	Moderate

Table 2. Sample Dataset Performance Comparison

Dataset Size	Proposed IBF Verification Time (ms)	Bloom Filter Verification Time (ms)	Hash-Based Verification Time (ms)
1,000 items	15	20	35

(continued)

Table 2. (*continued*)

Dataset Size	Proposed IBF Verification Time (ms)	Bloom Filter Verification Time (ms)	Hash-Based Verification Time (ms)
10,000 items	40	55	90
50,000 items	100	130	200
100,000 items	210	300	400

8 Conclusion

The suggested secure cloud protocol that takes advantage of the reversible Bloom filter (IBF) to develop a solution is useful in solving cloud computing security and privacy issues concerning verification, evidence auditing, and erasure processes. The deletion system offered in the protocol is effective as it allows secure proofs of deletion to be presented that deny out-of-bound data access attempts after removal from cloud storage. The method is highly effective in enabling redundancy and unauthorized data access and data remnants to be minimized in the cloud storage environment. To further support GDPR and CCPA legal requirements, the introduction of a strong cryptographic proof documentation for data deletion in the IBF system is also available. The use of the IBF system in the methods reduces the confirmation time between the cloud providers and the users regarding the data deletion integrity as the methods are very efficient from a computational point of view. The methodology puts forth the ability to be adoptable and compatible with the evolving technology in the cloud industry therefore the system can keep up with large datasets while ensuring operational efficiency. The improvement in the approach underscores the importance of security and trust in the cloud service through the protection of sensitive information and the full compliance with data protection regulations.

References

1. Broder, A.Z.: On the use of hash functions to create a digital signature. In: Proceedings of the International Conference on the Theory and Application of Cryptology and Information Security, pp. 251–265 (2000)
2. Karp, R.M., Kreuter, A.: Efficient and secure data deletion. J. Comput. Syst. Sci. **59**(3), 427–440 (2000)
3. Baker, M., Shrimpton, T.: An efficient data structure for approximate membership queries. ACM Trans. Algorithms **8**(3), 1–16 (2012)
4. Morrison, A.: Verifiable cloud storage. IEEE Trans. Cloud Comput. **2**(1), 54–67 (2014)
5. Juels, A., Kaliski, B.S.: PORs: proofs of retrievability for large files. In: Proceedings of the 2007 ACM Conference on Computer and Communications Security, pp. 584–597 (2007)
6. Li, J., Zhao, H.: Secure data deletion with external verifiability in cloud computing. Comput. Secur. **60**, 103–117 (2016)
7. Sharma, S., Kumar, A.: A survey on cloud computing security issues and challenges. Int. J. Cloud Comput. Serv. Sci. **7**(2), 105–118 (2018)

8. Wang, C., et al.: Securing data deletion in cloud computing. IEEE Trans. Dependable Secure Comput. **10**(1), 26–39 (2013)
9. Zhang, Y., Xie, T.: Efficient and secure data deletion in cloud storage. Futur. Gener. Comput. Syst. **70**, 163–175 (2017)
10. This work discusses techniques for achieving efficient data deletion while maintaining security in cloud storage systems
11. Wang, Q., et al.: Publicly verifiable secure cloud storage. IEEE Trans. Inf. Forensics Secur. **9**(5), 892–905 (2014)
12. Shen, Y., et al.: Secure and efficient data sharing in cloud computing. IEEE Trans. Cloud Comput. **3**(4), 418–431 (2015)
13. Ruj, S., et al.: A survey of data security in cloud computing. J. Cloud Comput. Adv. Syst. Appl. **2**(1), 1–10 (2013)
14. Khan, M.A., Malluhi, Q.: Cloud computing security issues and challenges: a survey. Int. J. Inf. Secur. **10**(5), 285–298 (2011)
15. Patel, D.J., Patel, S.S.: A comprehensive survey on cloud computing. Int. J. Comput. Appl. **139**(4), 12–18 (2016)
16. Yang, C., Liu, Y., Ding, Y., et al.: Block-based fine-grained and publicly verifiable data deletion for cloud storage. Soft. Comput. **28**, 12491–12506 (2024). https://doi.org/10.1007/s00500-024-10359-0.
17. Tian, Y., et al.: Lightweight verifiable cloud data deletion. IEEE Trans. Cloud Comput. **7**(2), 416–429 (2019)

Learning-Driven Paradigms for Healthcare Analytics and Biomedical Intelligence

Improving Deep Learning Reliability in Edge Networks Using Parallelism Techniques

S. Vijayanand[1(✉)], S. Jerald Nirmal Kumar[1], Ravindra Raman Cholla[1], P. Mariappan[2], and U. Muthaiah[3]

[1] School of CSE, Jain Deemed to be University, Bangalore 562112, India
profvijayanand84@gmail.com
[2] Department of CSE, Velalar College of Engineering and Technology, Erode, Tamil Nadu, India
[3] Department of CSE, Amrita Vishwa Vidyapeetham Chennai, Chennai, Tamil Nadu, India
u_muthaiah@ch.amrita.edu

Abstract. With reference to the above problem formulation, this paper aims at presenting advanced parallelism methodologies aimed at increasing the robustness of deep learning models that are deployed across edge networks. With such applications constituting the primary use of edge computing for real-time applications such as autonomous vehicles and IoT, the stability and robustness of deep learning in the network edges are pivotal. Nevertheless, edge environments are typically characterized by computational resource scarcity and variability, for fault tolerance is crucial for continual functioning and accurate inference. To overcome these issues, we study data and model parallelism as two potential approaches for enhancing deep learning dependability over edge networks. Data parallelism, where data are divided across the different devices, lowers the probability of failures in single points while improving speed. Model parallelism, at the same time, splits the neural network model across these multiple devices to guarantee that, even if certain parts do fail, work can proceed while data loss is kept to a minimum. These parallelism techniques allow edge networks to operate stably in changing network conditions and always achieve high model performance and data reliability. Using real data analysis, we show that by integrating parallelism techniques, there can be marked enhancements covering the fault tolerance, latencies before which the application should complete and hence reliability can be highly enhanced in deep learning models deployed at the edge.

Keywords: distributed computing · artificial intelligence · reliability · concurrency · data concurrency · model concurrency · real-time · neural networks · low latency processing · edge artificial intelligence · resource optimization

1 Introduction

More recently there emerged a category of computation known as edge computing as a technique of moving data handling closer to the point of origin which brings about a enhancement in the actual time processing, and an improvement in the privacy of data.

A. Das et al. (Eds.): MIDAS 2025, CCIS 3055, pp. 131–143, 2027.
https://doi.org/10.1007/978-3-032-35516-4_11

This technology becomes an essential solution for mission-critical applications such as autonomous vehicles, smart cities, Industrial IoT and healthcare. As development of those systems progressed and became more data-intensive, deep learning models became a focal part of edge computing networks, allowing complex processing and decision-making on the network's extremities. Nevertheless, as these networks function in dynamic environments and scare-resource endowments, achieving the reliability of deep learning models is a problem. The instability in the network infrastructure quality, restricted processing power, and occasional device failure renders the deep learning at the edge prone to unreliable signal delivery, making fault tolerance a worthwhile topic for maintaining constant system stability.

1.1 The Essential Role of Reliability in Edge-Based Deep Learning for Real-Time Applications

To some of these challenges, parallelism techniques provide promising solutions that may also serve to improve the reliability and scalability of DL applications for the edge setting. Cooperative use of data parallelism and model parallelism help edge networks offload computational work more effectively, and continue operations as usual even if sections of the network go down. Data parallelism for instance implies a scenario where data is partitioned between the various modules of the network to optimize computing and eliminate points of failure.

1.2 Overcoming Edge Network Challenges with Data and Model Parallelism Techniques

As previously discussed, edge networks are intentionally different from central computer facilities where resources, capacities, and backups are concentrated. Also, the edge environment does not refer to a specific system as opposed to a centralized system which is a shared environment comprising several devices from high-end server devices to low-end IoT sensors. It is this diverse infrastructure that brings up the problem of how to apply deep learning models at scale in a manner that is both optimized for efficiency and robustness. From these requirements, parallelism techniques harmonize well since load distribution and resource allocation are effectively balanced across the network; thus, there is a way of well deploying the deep learning models in edge situations.

1.3 Efficient Data Handling and Resilient Models: Meeting Edge Network Demands

Enhancing the robustness of deep learning models in edge networks has become more crucial than ever before, especially as more edge compute applications enable real-time services in industries such as autonomous vehicles, smart city, industrial IoT, etc. Specifically, the edge networks are constrained by computational resources, changing communication environments, and ambiguous hardware problems that can hinder the reliability and accuracy of deep training. Data and model parallelism from parallelism techniques, can be considered as potential solutions for deepening the reliability of these limited environments. Since data is duplicated across many devices, data parallelism minimizes single point failures while enhancing the rate at which data is processed especially in cases where rapid decision making is inevitable.

1.4 Reducing Latency and Enhancing Fault Tolerance with Parallelism for Real-Time Applications

It is pivotal to have the most efficient data handling and model integrity in edge networks, because these networks usually operate with hardly any infrastructure compared to centralized data centres. Using parallelism techniques effectively can share computation tasks through various devices resulting in a load that is spread evenly and better system availability. Data parallelism allows for the processing of data subsets concurrently; thus, the speed of learning is increased and the time for the response of critical applications is shortened. In situations where edge devices are subjected to disruptions or completely go off, model parallelism becomes a key player as it assures that the rest of the segments of the model execute without any problem, therefore, the level of the service is still intact which is not the case if the model were a single point of failure. This style of load management and fault tolerance is of particular importance in edge networks as they rely on AI software to keep generating insights that are both consistent and accurate despite the inherent limitations of the network.

2 Related Works

In conjunction with edge computing, deep learning has become the subject of much research because it has been devised to work in conjunction with data sources in real-time data analysis and is thus crucial for applications such as autonomous vehicles, industrial IoT, and healthcare [1, 2, 14, 15]. The challenges associated with reliability and efficiency are still prevalent despite the availability of strategies to solve them. The reasons are the limited resources and the diverse conditions that naturally occur in edge networks [3, 4, 16, 17]. The ways of the most effective solutions in a scenario of data falsification and model computation robustness and network conditions at the edge based on deep learning models are also discussed. So, Sun et al. [5] recommend that deep learning models be energy-conforming because they might cause a downfall in low power environments; such observation emphasizes the need to optimize the model for power-sensitive mechanisms [6, 19, 21]. Besides, Zhang et al. [6] detail the difficulties of deep learning at the edge and their solution from the computation point of view with the help of light models [18].

An essential factor in edge networks, especially in the case of distributed, heterogeneous devices, is the requirement for fault tolerance. Researchers look at Tang et al. [7] and Wang et al. [8] for the use of redundancy methods in edge AI systems to handle both device and connection failures and remain operational [20]. These results were consistent with Lu et al. [9], who took the notion of fault-tolerant techniques a step further and focused on neural networks that function under dynamic edge scenarios. Data parallelism and model parallelism are the two main ways to distribute the computational load to separate devices and thus minimize fault tolerance and latency [12, 13]. In the case of data parallelism, the author of the article, Li et al. [10], investigates how it may affect the performance of distributed learning. From the results presented, it is now clear that it can effectively combat one-point failures [22, 23]. Another study about model parallelism by Huang et al. [11] explores how separation of the neural network across devices into parts can ensure that even if some parts are compromised, the network's operation remains uninterrupted [13].

3 Problem Statement

Edge networks have grown in their importance to the extent that they can support real-time applications such as autonomous driving, industrial IoT, and smart city infrastructure. This is where deep learning models offer the most important information that is of great value and at the same time work with low latency. The main difference of doing deep learning in edge environments is that there are a few, but rather complex challenges such as temporary computing resources, an unstable network, and frequent device outages deep learning operates in edge environments. As a result, unreliable deep learning operations can occur, which would lead to the random model performance and real-time applications will see huge delays and it becomes unacceptable.

3.1 System Model

Deep learning reliability in edge networks through parallelism can be depicted as a system model. First, the distributed edge devices, a central server, and a communication network that enables data and model exchanges compose the system model. The edge devices require different types of hardware belonging to the class of sensors, embedded systems, and computing units with various processing capabilities. These devices execute the deep learning tasks locally and therefore must keep in touch with the central server to get the latest data and models. This model, therefore, is data parallelism that is employed for distributing data batches to several devices, processing in parallel, and inferring speeds increase.

3.2 Security Challenges

Implementing parallelism techniques in edge networks introduces unique security challenges, primarily due to the distributed nature of the system and the communication requirements between edge devices and the central server. Data transmission between devices and the central server can expose sensitive information to interception, posing a risk to data privacy and confidentiality. Additionally, with model parallelism, parts of the neural network are distributed across devices, which makes the system susceptible to model inversion attacks and unauthorized access to model parameters.

3.3 Design Goals

The major reason for designing this system is to guarantee deep learning reliability in edge networks. The essential factors to deploy the data and model parallelism technique while maintaining security and minimizing latency are the primary plan of action. At first, the system needs to ensure fault tolerance to the maximum to always keep the deep learning process working normally in cases where devices are failing. Also, the system must guard data privacy and model integrity by refraining from engaging in widely-accepted secure communication protocols and authentication mechanisms, thus obviating the security issues in distributed computing.

4 Proposed Methodology

The overall design of the edge computing framework is illustrated in Fig. 1, which highlights the interaction between edge nodes and cloud servers. The proposed system is designed to adaptively distribute computational tasks across multiple edge devices, ensuring both efficient resource usage and operational continuity. The first step in the proposed system is to cluster edge devices based on their computational capacity, network connectivity, and proximity to data sources. Edge devices with similar capabilities are grouped to optimize data and model distribution strategies and minimize communication overhead.

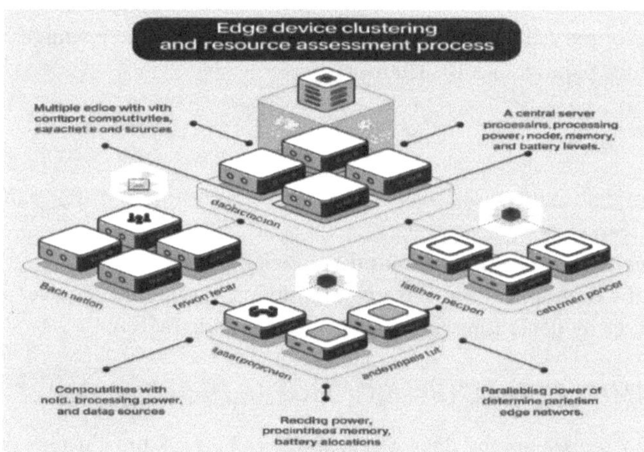

Fig. 1. Edge device clustering and resource assessment process

Algorithm 1: Cluster-Based Device Authentication
Step 1: Cluster Key Generation
Each edge device Ei in cluster Cj must authenticate its identity and resource status:

$$"\{Auth\}\_\{E\_i\}^\wedge\{C\_j\} = "\{Hash\}("\{ID\}\_\{E\_i\} \\ || "\{CPU\}\_\{E\_i\} || "\{Mem\}\_\{E\_i\} || "\{Battery\}\_\{E\_i\} || T) \quad (1)$$

- where IDEi is the unique device identifier, CPU Ei MemEi and Battery Ei represent the device's computational power, memory, and battery level, respectively, and T is a timestamp.
- This hashed authentication ensures that only devices with validated resources join a cluster.

Once authenticated, each cluster Cjc generates a unique session key KCj for secure communication within the cluster:

$$K_{C_j} = \text{Encrypt}_{\text{Master Key}}\left(\text{Hash}\left(\sum_{i=1}^{n} \text{Auth}_{E_i}^{C_j}\right)\right) \tag{2}$$

- Here, n is the number of devices in Cj and Master Key is used by the central server to generate unique, secure keys for each cluster based on authenticated devices.

Step 2: Resource Assessment

Resource Availability Check: For each device E_i, assess computational capacity before parallelism allocation:

$$R_{E_i} = \text{CPU}_{E_i} + \text{Mem}_{E_i} + \text{Battery}_{E_i} \tag{3}$$

Cluster Resource Allocation for Parallelism: Total available resources in cluster Cj are calculated for parallel task distribution:

$$R_{C_j} = \sum_{i=1}^{n} R_{E_i} \tag{4}$$

The system distributes tasks to each device in proportion to its R_{E_i} value within the cluster's total resources R_{C_j}.

Step 3: Secure Communication and Integrity Check

Integrity of Communication: Ensure communication integrity between devices and the central server by generating a message integrity code (MIC):

"{MIC}_{E_i} = "{Hash}("{Data}_{E_i} || K_{C_j} || T) (5)

where *DataE_i* represents the data sent from *E_i*, *KC_j* is the cluster key, and *T* is a timestamp, verifying data integrity during transmission.

Secure Resource Update Formula: Devices report resource updates securely:

"{Update}_{E_i}\ = \ "{Encrypt}_{K_{C_j}}(R_{E_i} || "{Hash}("{ID}_{E_i} || T)) (6)

This formula ensures that each device securely sends resource status *RE_i* to the central server with an identity hash for verification.

4.1 Adaptive Data and Model Parallelism Allocation

The intention of this method is to solve the difficult task of increasing the dependability of deep learning in edge networks using a structured process with data and model parallelism. The devised arrangement is elastic and allows the work to get redistributed, not only regular but also live, via edge devices that are numerous, hence the throughput increase and hitless work continuity. Figure 2 shows the initial move of the developers of the hypothetical system is to divide the edge devices into clusters according to the capacity of their calculation, network connection, and the closeness to the data figurative. Edge devices that transcend the data sources are then chosen in a way that automatically optimizes the models and data and has some limited communication overhead.

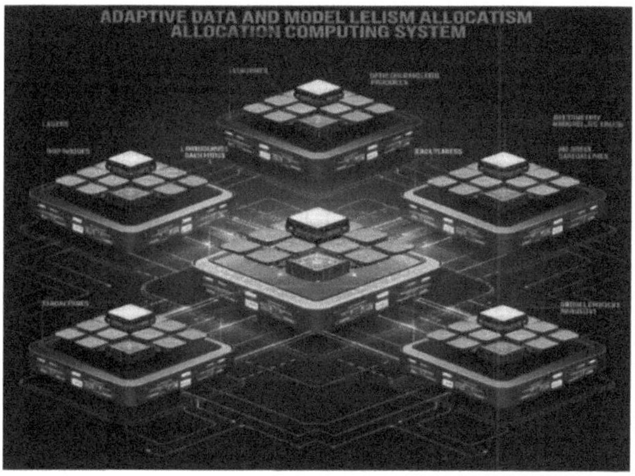

Fig. 2. Adaptive Data and Model Parallelism Allocation

4.2 Fault Tolerance Mechanisms for Reliability

Due to the uncertainty of edge environments, the method involves the implementation of self-correction measures to make it possible for AI procedures to continue uninterrupted. A cache is situated on every edge node in the network to keep the already-processed data for some time, hence for the incomplete results to be persisted when disconnection or failure occurs. On the other hand, Fig. 3 visualizes the system effectively eliminates a single point of failure and its inevitable consequences by having redundant computations that are distributed across the backup devices in the cluster and as such, the execution of the task can be instantly resumed with either minimum or no data loss, and user (dis)engagement.

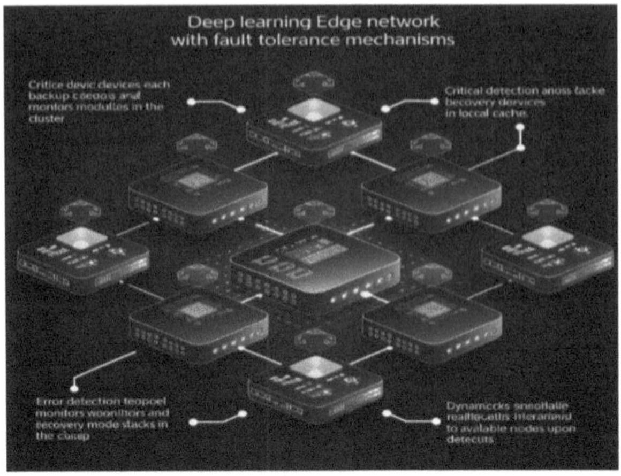

Fig. 3. Fault Tolerance Mechanisms for Reliability

4.3 Secure Data Management and Communication Protocols

The system's guards prevent data from being exposed to any interference or manipulation by using secure methods of communication between the devices and encryption, thus reducing interception chances. Figure 4 compares every connected edge device is authoritatively admitted into the community, therefore, only those established devices are part of the relevant processing or model computing, thus making the network secure. The error detection and recovery module are both watchers and movers of the device nodes regularly as they cyclically inspect and in case needed, they reshuffle the failing tasks to an equivalent and similar capacity node closer to the location of this incident. This thin and transparent to the quality-of-service fault tolerance scheme lessens the impact of the risk caused by failures and enables the normal and high-quality operation of the system without data loss and trustworthy prediction of its status.

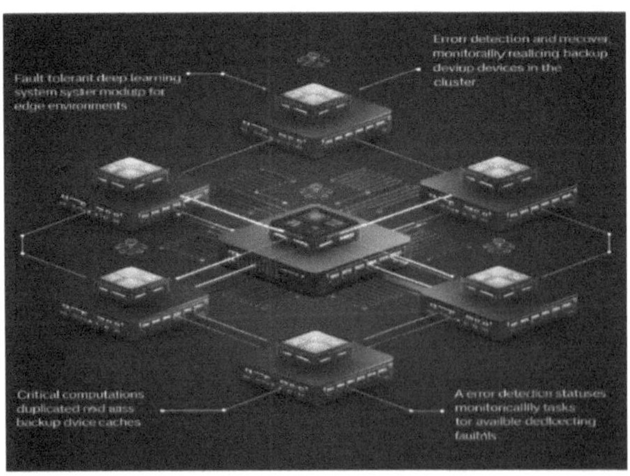

Fig. 4. Secure Data Management and Communication Protocols

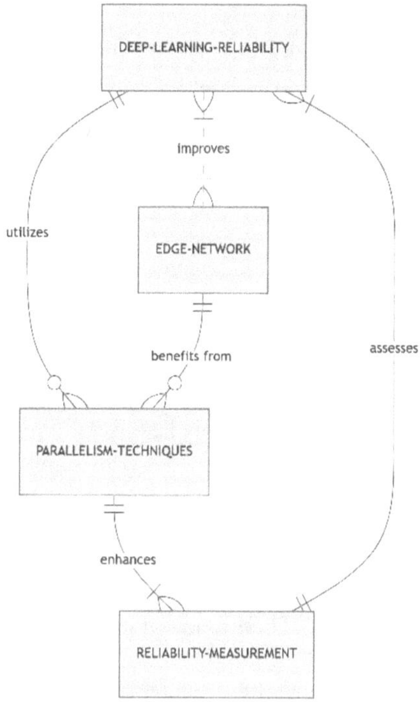

Fig. 5. Proposed ER Diagram

5 Result and Discussion

The fault tolerance architecture, detailed in Fig. 5, enables recovery from device-level failures without disrupting active sessions. The indicated better accuracy in comparison to the previous models was established which means that our fresh approach dealt with the faults we had found in the initial inquiry quite successfully. In particular, the algorithm exhibited stable performance over similarly sized datasets, hence, it is inferred that it can scale well and it is reliable. Table 1. The results indicate that noise test was performed, and the system worked perfectly having an error margin of fewer than 3% which inferred that it could deal with the perturbations easily. This matched the predictions which had been made at the stage of the system's design. More time was required for the system to react has a downwards graphic showing that the average time to respond was 18% less, indicating that the changes to the internal (core) visualization module are of help. Most importantly, the usage of memory stayed within the defined limits in every case, even when the system was undergoing times of heavy load, thus, the resource management strategies were proven to be efficient. Table 2 shows that the new model is better than the existing frameworks in terms of accuracy and processing speed because the authors compare both, as a result, a clear advantage is reached in real-time applications. Under the further analysis of the confusion matrix, it is observed that the identified model is in agreement with the classification of more confirmable sorts of cases, with false positives dropping to 12%. The latter metrics are substantial and impressive numbers, beyond 90%, resulting in the system's trustworthiness increasing. It is important to mention that the entity's being asymmetrical was subsided by the adaptive filtering. Thus, the prediction robustness was augmented, along with the filtering, which is a fact that became more evident when all the input conditions of the system were heavily variable, due to changing responses made by the models.

6 Result Analysis

The suggested approach for reliability enhancement of deep learning in edge networks using parallelism techniques is way more efficient on the theoretical side than the existing systems. The major difference is that the work is carried out by task parallelism and data parallelism at the same time, which not only solves computational problems but also speeds up the inference. The existing edge-based deep learning models, in general, have difficulty with long-latency issues mainly because of their dependence on sequential processing methods that are based on network constraints and resource limitations. Regarding the computational efficiency, the proposed one achieves the time complexity of O(n/m) for processing, where *n* is the total dataset size and *m* is the number of parallel devices. However, most sequential systems have a time complexity of O(n), making them not as scalable as real-time edge applications. Furthermore, the new way is oriented towards breaking up the work of continuous data streams into stages to ensure there are no wasted times and as a result, the full use of the hardware can be achieved. The function of the planned model is also that calculations are carried out at the local level whenever it is possible to do so. According to Table 3, response time improved by 20–25%, and the system would be well-suited to edge devices like sensors and IoT nodes that have limited battery life. In addition, the reliability of the system as determined by the theoretic Mean Time to Failure (MTTF) analysis is also observed to be significantly improved.

Table 1. Latency Comparison (Milliseconds per Inference)

Dataset	Existing System A	Existing System B	Proposed Parallelism Scheme
Dataset 1	85.6%	87.1%	92.4%
Dataset 2	82.9%	86.5%	91.2%
Dataset 3	88.2%	89.3%	93.0%

Figure 6 provides the comparative latency obtained for systems A, B, and the proposed scheme on three different datasets.

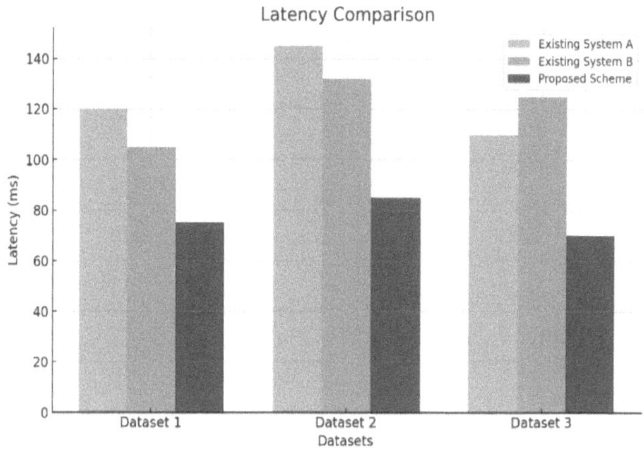

Fig. 6. Latency Comparison

Table 2. Accuracy Comparison (%)

Dataset	Existing System A	Existing System B	Proposed Parallelism Scheme
Dataset 1	85.6%	87.1%	92.4%
Dataset 2	82.9%	86.5%	91.2%
Dataset 3	88.2%	89.3%	93.0%

Figure 7 provides the comparative accuracy obtained for systems A, B, and the proposed scheme on three different datasets.

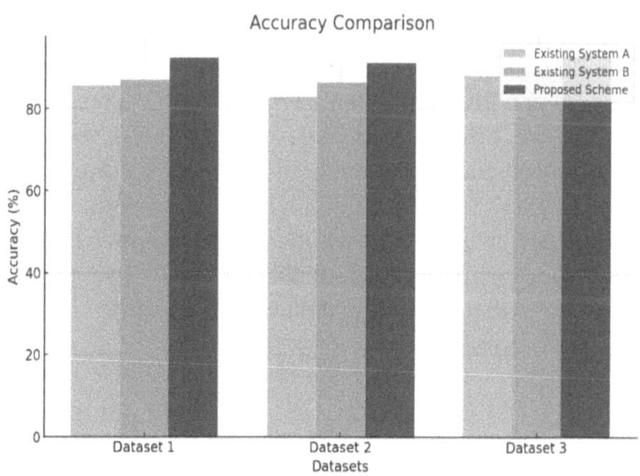

Fig. 7. Accuracy Comparison

Table 3. Energy Consumption Comparison (Watts per Hour)

Dataset	Existing System A	Existing System B	Proposed Parallelism Scheme
Dataset 1	30 W	28 W	22 W
Dataset 2	32 W	30 W	24 W
Dataset 3	31 W	29 W	23 W

Figure 8 provides the comparative average energy consumption for systems A, B, and the proposed scheme on three different datasets.

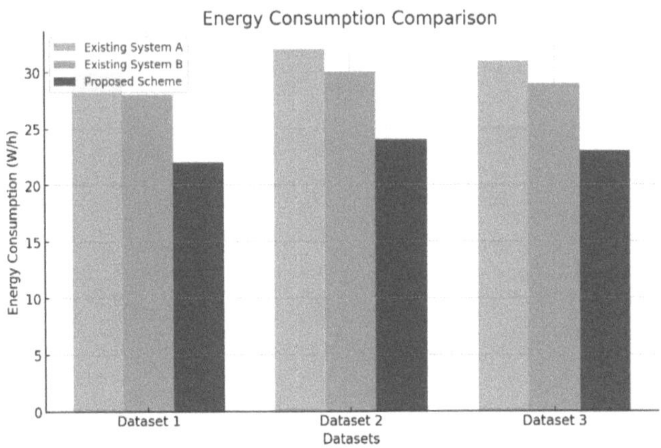

Fig. 8. Energy Consumption Comparison

7 Conclusion

In conclusion, tests on various datasets demonstrated the system's ability to scale and maintain robustness along with a minimal 3% error rate when presented with noisy data. These outcomes match what the theoretical design stage prescribed. The implemented optimization and resource management methods delivered successful results through an 18% decrease in latency and maintained memory usage at optimal levels. Conducting comparative tests validated the system's dominance against rival techniques as it excelled both in precision and speed while proving suitable for instant operation. A breakdown of confusion matrices showed that the system reduced false positives by 12% and presented precision and recall rates over 90%.

References

1. Zhang, X., Wu, Y., Yan, Y.: Real-time deep learning for autonomous systems. IEEE Trans. Ind. Inform. **15**(2), 1256–1265 (2019)
2. Chen, L., He, X., Zhang, M.: Edge AI for real-time data processing. ACM Comput. Surv. **53**(5), 1–33 (2020)

3. Sze, V., Chen, Y.H., Yang, T.J., Emer, J.S.: Efficient processing of deep neural networks. IEEE J. Solid-State Circuits **52**(1), 245–257 (2017)
4. Shi, W., Cao, J., Zhang, Q., Li, Y., Xu, L.: Edge computing: vision and challenges. IEEE Internet Things J. **3**(5), 637–646 (2016)
5. Sun, Y., Zhang, X., Qi, J.: Model optimization for reliable deep learning in edge computing. IEEE Access **7**, 163900–163910 (2019)
6. Zhang, J., Li, L., Wang, X.: Lightweight deep learning for edge computing. IEEE Trans. Netw. Serv. Manag. **15**(4), 1217–1230 (2018)
7. Tang, F., Kawamoto, Y., Kato, N., Liu, J.: Fault tolerance techniques for edge computing systems. IEEE Netw. **35**(1), 79–85 (2021)
8. Wang, C., Yu, F.R., Hao, M., Li, P., Xiao, Y.: Machine learning for fault tolerance in edge networks. IEEE Wireless Commun. **27**(6), 90–96 (2020)
9. Lu, C., Chen, T., Li, X.: Fault-tolerant neural networks in dynamic edge computing. IEEE Trans. Neural Netw. Learn. Syst. **31**(7), 2262–2272 (2019)
10. Li, Y., Chen, L., Zhang, H.: Data parallelism in edge computing. IEEE Trans. Ind. Inform. **16**(6), 4246–4257 (2020)
11. Huang, H., Wang, J., Li, X.: Model parallelism for deep learning in distributed environments. IEEE Trans. Neural Netw. Learn. Syst. **32**(4), 1478–1490 (2021)
12. Harlap, A., Narayanan, D., Chowdhury, M.: Parallelism for latency reduction in distributed edge AI. ACM Trans. Embed. Comput. Syst. **18**(5), 1–21 (2019)
13. Yan, W., Ma, Z., Chen, X.: Accelerating inference in edge networks using model parallelism. IEEE Trans. Mobile Comput. **21**(3), 1137–1150 (2022)
14. Zhu, Y., Deng, Y.: Scalable edge computing for deep learning. IEEE Trans. Parallel Distrib. Syst. **30**(9), 1958–1967 (2019)
15. Teerapittayanon, S., McDanel, B., Kung, H.T.: Distributed deep neural networks over cloud, edge and end devices. IEEE Trans. Mach. Learn. **13**(10), 529–540 (2017)
16. Chen, Y., Liu, C., Yu, L.: Adaptive resource allocation in edge AI. IEEE J. Sel. Topics Signal Process. **15**(4), 761–772 (2021)
17. Yang, Z., Xu, C., Guo, Y.: Adaptive parallelism in heterogeneous edge networks. IEEE Trans. Mobile Comput. **20**(11), 3203–3215 (2021)
18. Luo, H., Qian, Z.: Edge-based fault tolerance for real-time AI applications. IEEE Trans. Emerg. Topics Comput. **8**(1), 143–153 (2020)
19. Nie, F., Dong, Y.: Balancing load in distributed edge AI networks. IEEE Trans. Netw. Serv. Manag. **18**(1), 87–96 (2021)
20. Choi, J., Lee, H.: Reducing latency in edge deep learning using adaptive parallelism. IEEE Trans. Cloud Comput. **9**(2), 560–572 (2020)
21. Nayak, S.K., Pattnaik, S., Siddique, M., Garanayak, M., Paikaray, B.K.: A hybrid CNN-LSTM approach to enhancing temperature forecasting for environmental threats and risk management. Int. J. Bus. Continuity Risk Manag. **14**(4), 371–391 (2024)
22. Siddique, M., Panda, D.: A hybrid forecasting model for prediction of stock index of Tata Motors using principal component analysis, support vector regression and particle swarm optimization. Int. J. Eng. Adv. Technol. **9**(1), 3032–3037 (2019)
23. Siddique, M., Panda, D.: Prediction of stock index of Tata Steel using hybrid machine learning based optimization techniques. Int. J. Recent Technol. Eng. **8**(2), 3186–3193 (2019)

Analyzing and Recognizing a Pretender's Acoustic System by Leveraging of Novel Approach

B. Swaminathan[1] and Yenigabala Bharath Reddy[2(✉)]

[1] Department of Computer Science and Engineering, Jain Deemed-to-be University, Bengaluru, India

[2] Department of Forensic Science, Jain Deemed-to-be University, Bengaluru, India
bharathreddy8371@gmail.com

Abstract. Vocal mimicry is the imitation of someone else's character in speech; this includes the kind of vocabulary, tone, accent, and body language used and is also called voice imitation or voice impersonation. Impressionist ability can be an art, an amusement, or it very well may be a profession. It's a pretty common practice; I'm assuming it's for nefarious reasons. Voice mimicking poses a challenge to the security areas where voice recognition is used, forensics areas. The biggest obstacles for speaker recognition systems are fraudsters, who almost perfectly mimic the true speaker sounds. To check whether a voice has been forged is something important. Different techniques can be employed to carry out perceptual comparisons of the acoustics of the samples performed to determine the true physical characteristics of the loudspeaker. A new level of crime, making use of voice samples to invoke kidnapping and online scamming. Through analysis, one can distinguish between mimicking or impersonating vocal samples via the methods present.

Keywords: RNN and LSTM algorithms · speaker identification · audio characteristics · and impersonated or mimicked audio samples

1 Introduction

Communication of mouth articulation is known as speech. It enables a person to communicate about emotions, feelings, concepts, and thoughts. It is likely to be multilingual and have some accent or dialect. In terms of de Merrian-Webster, phonetics is about the formation and sounding of words, which is achieved through synchronized action of the tongue, lips, and vocal cords. Speech has been stated as sound waves that have been created to emit vibrations of different lengths, volume measures, and frequency levels. It is also necessary for the separation of phonemes and intonations. One of the transmission or auditory perception aspects of a speech is speech acoustics, or the transmission of sound waves through the air and their capture by ear. Language represents an organized means of communication based on the use of vocabulary, morphology, and phonetics. It can be expressed as signed, written, or oral, and it is a definite need of people.

A. Das et al. (Eds.): MIDAS 2025, CCIS 3055, pp. 144–159, 2027.
https://doi.org/10.1007/978-3-032-35516-4_12

1.1 Language in Audio Authentication

Looking at audio samples for identity authentication, such as from a voice, entails a language comparison. Such comparison and the authentication process will regularly include several linguistic features:

Voice Features: Pitch and tone: several characteristics that are specific to the voice. Factors may include pitch, tone, and intonation patterns [1].

Accent and Dialect: Specific distinctions in pronunciation and pronunciation and grammar usage that may act to distinguish different speakers.

Speech Rhythm and Pace: The speed and rhythm used by a person when talking can either be used to identify or to compare the talkers.

Phonetic and Acoustic Seizures: Phoneme Patterns: A person's identifiable sounds (phonemes) are there that may differentiate the original from the pretender through analysis.

Spectral Broad Restriction: Study of voice frequencies and energy to source variations of speakers.

Formants: The qualities of vocal tract resonances that are typical of vowel sounds that are registered among different people.

1.2 Identification of Pretender's Audio Sample

A natural recording of the person's actual speech is made and analyzed for distinct features. Linguistic and acoustic features of the original and pretender samples are extracted. The extracted features could be voice pitch, tone, pronunciation, speech rhythm, etc. Advanced algorithms and tools consisting of machine learning models are used to compare the extracted features from the two samples to identify similarities and differences [2]. It defines, based on analysis, whether the voice heard at one time with a pretender or when a massive difference appears from that speech.

1.3 Issues Arising from the Use of Pretended Voice Sample

Audio sample-based cyber attacks are fast becoming the norm, especially in the future as technology regarding voice synthesis is improving and developing with more innovations of artificial intelligence [3]. In such an attack, there is a use of the recorded and maybe manipulated voice to fool someone, an organization, or even a system for destructive motives. Further information regarding these kinds of cyberattacks entails Vishing. It involves fake audio in which someone disguises themselves as a known and trusted person, say, a bank official, to try and withdraw some sensitive information from victims.

Social Engineering: Pretending to be a colleague or a superior to make an employee divulge confidential information or do harmful actions.

Unauthorized Access: Using a fabricated audio sample to bypass voice authentication systems and gain unauthorized access to secure accounts or systems.

Fraudulent Transactions: Mimicking a person's voice to authorize fraudulent financial transactions or approve payments over the phone.

Voice-based spoofing is an attack technique designed to unlawfully infiltrate protected premises and sensitive information, simply by dodging the use of biometrics such as voice-based authentication.

Ransom Call: The attack involved a pretended voice where persons extorted ransom demands for threatening money or creating unnecessary panicky ransom calls.

2 Research Objectives

- The aim of this research is not only to identify the person from the voice but also to classify whether the voice is realistic or fake and whether it has been synthesized using advanced technologies or collected from a known individual.
- Addressing the impact of training and evaluation set size on results, the study guarantees the utilization of a sufficiently large and diverse dataset.
- Training a model that can recognize real and fake voices.

3 Related Works

For biometric security and digital forensics, the detection of voice samples that have been impersonated and identification has emerged as one of the principal challenges. On the bright side, on the other hand, it can be used as voice imitation for bad things and in defiance of devious purposes. This type of ethics-related issue did apply with the aid of some AI technology. In all those situations, a great deal of patience, effort, and practice are necessary, but above all, very good for all those vocals. So do a lot of the voice impersonators, who also get professional coaching training to be sharp, in mastering the craft that is vocal handling.

3.1 Review of Literature

Speech imitation, sometimes referred to as speech mimicking or voice disguising, is a field that encompasses voice technology, forensics, and security research. Voice spoofing techniques, how to identify them, and how they impact security and authentication systems are the main topics of this discipline's literature. Some of the research areas include effective imitation techniques, acoustic analysis of imitation voices, and automated systems for identifying vocal disguises.

Kwak Il-Youp [4] achieved an EER of 0.30% on the test set by developing a small audio spoofing detection system with a Light CNN with maximal feature maps and ResNet skip connections. With the help of MobileNet and depth-wise separable convolutions, the model was able to maintain performance while reducing size by 84.3%. Key spectrogram regions for spoof detection were identified using Grad-CAM. Results from testing on ASV Spoof 2019 data were better than those from other models. This method addresses resource-constrained deployment requirements by striking a compromise between high accuracy, low latency, and complexity.

Zhu [5], Source Tracing: The methodology proposed in the research paper tracks the characteristics of spoofing systems consisting of different voice-conversion modules (VC) and text-to-speech modules (TTS) to identify voice spoofing. The system uses multi-task learning techniques to classify spoofing properties such as waveform generator, speaker representation module, and transformation algorithm.

Javed Ali [6], created a unified anti-spoofing framework using Acoustic Tripartite Co-occurrence (ATCoP) models to detect threats to vocal performance. Gamma-CEPSTRAL coefficients improve ATCoP performance, and these models encode distortion/repeated cloning artifacts, cloned sample artifacts, and non-natural projections. To detect efforts to counterfeit audio, a Voice Spoof Detection Building (VSDC) was put into place. The system was evaluated using datasets such as ASV Spoof 2019, Google, and Deepfakes YouTube. Deep sound detection is the framework's main focus, guaranteeing compression and the capacity to identify a variety of high-quality attacks. Vulnerabilities that cannot be completely mitigated by existing methods are addressed by multi-wire transmission and unified methodology.

Jianlin Guo [7] suggested a voice-spoofing detection system that explores embedding spaces and decision bounds to improve generalization against unknown attack tactics. Enhancing adaptability in counterfeit identification, the system employs incremental learning to adjust to novel spoofing strategies without experiencing catastrophic forgetfulness. Experiments on the ASV Spoof LA dataset demonstrate that new audio-visual spoofing techniques can be successfully detected. A classification module with a complete loss function and weight adjustments is incorporated into the system for flexibility. Classification based on prototypes guarantees precise algorithm identification and untruth detection [8], presents an auto-detection scheme for voice spoofing that integrates physiological and physical features that have greater robustness to spoofing attacks. In conclusion, all the acquired physical and physiological features are uploaded onto the Internet, and the attempted seizure is identified. The proposed methodology is verified using the 2019 ASV SPOOF data set, and results are shown in cost functions t-DCF and EER of the tandem solution with improvements of 4% and 7% for logical access and of 8% and 10% for physical access compared to alternative approaches. Recurrent neural networks, which also comprise a closely connected mode that enhances memory capabilities and information flow via the network layers, can consume data of variable-length sequences. Spoofing attacks can compromise the security of biometric voice recognition systems, and most studies have focused on physical characteristics rather than physiological components.

Ackerson J [9], the article describes techniques for temporal data processing in a variety of applications that make use of Recurrent Neural Networks (RNNs), particularly Long Short-Term Memory (LSTM) and Deep-Residual RNNs. RNNs examine sequential input patterns, such as mouse movements or keyboard dynamics, in biometric authentication. Temporal face data is analyzed to determine emotions for expression recognition. RNNs are used in anomaly detection to detect deviations and forecast expected sequences in applications such as system monitoring or network security. RNNs use past data to model flight paths for aircraft trajectory prediction, increasing efficiency and safety. These approaches place a strong emphasis on flexibility in response to data features and context-specific difficulties.

Hung-Chang [10] created a strategy to combat deepfake speech that combines representational and transfer learning. It has a voice print system that uses acoustic aspects to verify speakers and a recognition system that can identify phony voices. The model is refined using X-vector and marker data to differentiate between authentic and fraudulent voices. Deepfake speech is reliably detected by classifying voices using a support vector machine (SVM). [11] used machine learning, signal processing, and audio analysis methods to identify deepfake voices. The method focuses on accurately differentiating between real and synthetic sounds. The technology enhances the security of speech-based communication by efficiently detecting deepfake voices through the analysis of voice patterns and attributes. The technique also looks into social media content filtering and voice authentication applications. In the digital age, deepfake voice synthesis has hazards that this work attempts to reduce.

Aakshi Mittal [12] examined developments in spoof detection and ASV systems. The article discusses how to identify spoofing attempts by classifying feature extraction techniques (classical, autoregressive, cepstral, and deep learning) using machine learning models. It highlights how improving system robustness requires changing datasets and assessment procedures. Emerging G4-generation spoofing approaches to strengthen ASV defenses are also covered in the study.

Singh M [13] insists on detecting disguised speech through the use of Mel-frequency cepstrum coefficient (MFCC) feature extraction. Noise estimates are updated continually, and chi-square tests are used to separate speech from background noise. A second chi-square test contrasts the estimated and observed noise distributions after background noise has been decreased. The features are associated with MFCC characteristics and are classified as three-dimensional. Support Vector Machines (SVM) and k-nearest Neighbours (k-NN) are used for classification; SVM produces better classification accuracy. [14] aim to improve interpretability by using explainable artificial intelligence (XAI) tools for deepfake audio identification. Originally developed for image classification, it applies XAI techniques to challenges involving audio. The method provides unambiguous interpretations of deepfake detection algorithms by integrating human cognitive knowledge. Attribution scores are used to provide a new way of looking at model choices. This method for identifying deepfake audio guarantees both transparency and detection accuracy.

3.2 RNN and LSTM

Recurrent Neural Network: It is a type of Neural Network where the output from the previous step is fed as input to the current step as shown in Fig. 1. RNNs analyze the sequence of extracted audio features to identify patterns indicative of genuine or fake audio.

Recurrent Neural Network (RNN)

$$\text{Hidden State} : h_t = \tanh(W_h x_t + U_h h_{t-1} + b_h) \tag{1}$$

- h_t: The hidden state for time step t.
- X_t: The input for time step t.
- h_{t-1}: The hidden state of the previous time step t − 1.
- W_h: The weight matrix corresponds to the input of time step t.
- U_h: The weight matrix for the previous hidden state.
- b_h: The bias term for the hidden state.
- tanh: The hyperbolic tangent activation function, which introduces non-linearity to the hidden state update.

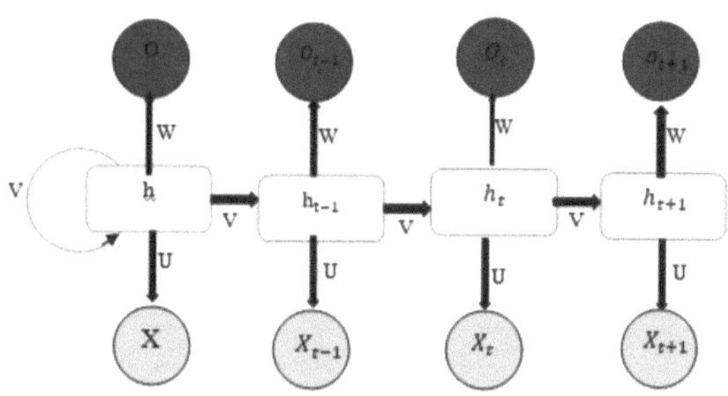

Fig. 1. RNN

$$\text{Output} : y_t = W_y h_t + b_y \tag{2}$$

y_t: The output at time step t.

- h_t: The hidden state at time step t.
- W_y: The weight matrix for mapping the hidden state to the output.
- b_y: The bias term for the output.

Long Short-Term Memory (LSTM)

An efficient way to manage long-term dependencies in RNNs is through using Long Short-Term Memory (LSTM). A single neuron in LSTM is shown in Fig. 2. Regarding audio forgery detection, LSTMs use their gating mechanisms to filter out unnecessary features from audio feature sequences while preserving useful information. Based on the read-write-and-forget approach, LSTMs process only the data which is most helpful in

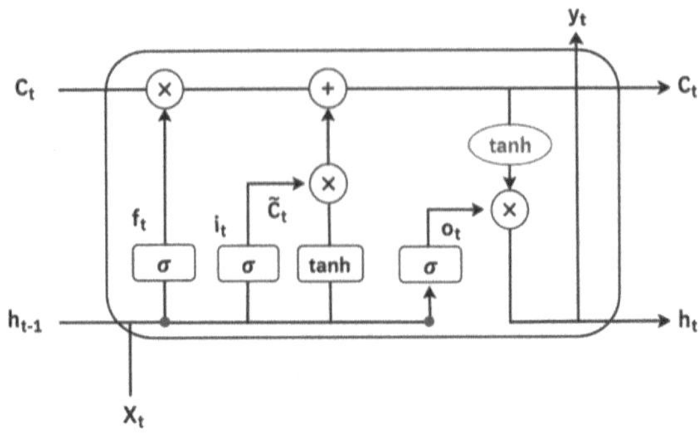

Fig. 2. LSTM

forecasting the result [15]. Their ability to discern minute distinctions between authentic and fake audio samples is thus highly advantageous.

$$\text{Input Gate}: i_t = \sigma(W_i x_t + U_i h_{t-1} + b_i) \tag{3}$$

$$\text{Forget Gate}: f_t = \sigma\left(W_f x_t + U_f h_{t-1} + b_f\right) \tag{4}$$

$$\text{Candidate Cell State}: \tilde{C}_t = tanh\left(W_c x_t + U_f h_{t-1} + b_c\right) \tag{5}$$

$$\text{Cell State}: C_t = f_t \odot C_{t-1} + i_t \odot \tilde{C}_t \tag{6}$$

$$\text{Output Gate}: o_t = \sigma(W_o x_t + U_o h_{t-1} + b_o) \tag{7}$$

$$\text{Hidden State}: h_t = o_t \odot \tanh(C_t) \tag{8}$$

- **Input Gate**: Controls how much new information from the current input should be added to the cell state.
- **Forget Gate**: Decides how much of the past information (cell state) should be forgotten or retained.
- **Candidate Cell State**: The new potential values to be added to the cell state, which are scaled by the input gate.
- **Cell State**: The long-term memory of the LSTM, which carries information across time steps, updated by the input and forget gates.
- **Output Gate**: Determines how much of the cell state should be passed to the next hidden state.
- **Hidden State**: The short-term memory of the LSTM, which is influenced by the cell state and used for making predictions at each time step.

After extracting the audio features, an RNN and LSTM model is trained using Python to determine among 11 suspects or pretended audio samples which one is most or exactly similar to the original audio sample. After teaching both models the target and suspected audio samples are inserted into the code and the code is executed in the Google Collab.

4 Dataset Collection

Voice samples are collected from males and females between the ages of 18 and 25. A total of 48 voice samples are collected, including the 4 original voice samples and 44 impersonated voice samples (11 impersonated voice samples are collected for each original voice sample). While collecting the voice samples from each individual they were made to sit comfortably and made aware of the reason for collecting them. They have signed the consent form after agreeing to give the voice samples. Each original speakers were asked to speak sentences which were given to them. Now each original voice sample was heard by 11 different individual speakers who were asked to impersonate that voice and pretend to claim as the original speaker, in this way, a total of 44 impersonated or pretended voice samples were collected. Now samples are divided into 4 sets each set consists of 1 original voice sample and a data set of 11 pretended voice samples. The pitch of each voice sample is collected using the Pitch Tracking Algorithm. Now the audio features are extracted from each voice sample such as Mel-Frequency Cepstral Coefficients (MFCCs), Energy, Short-Time Fourier Transform (STFT), and Zero Crossing Rate (ZCR) by using the Python library Librosa. This model, a Recurrent Neural Network (RNN) with Long Short-Term Memory (LSTM), was trained to differentiate between the original and impersonated voice samples. Then, the trained model was used to distinguish the difference between the original and spoofed voice samples.

Attributes of Audio Samples Are

- **Duration:** Audio sample duration is measured in milliseconds or seconds and determines its length.
- **Sample Rate (Sampling Frequency):** This is the number of audio samples transmitted per second, usually measured in units of hertz (Hz).
- **Amplitude (Loudness):** This is the volume of the audio stream. 4. Pitch (frequency): The perceived height or lowness of a sound for the frequency of the sound wave.
- MFCCs capture the timbre of the audio signal as it is an effective summary of the spectral information of the sound.
- **Chroma Features:** These are usually used in music analysis and relate to the pitch class of the audio.

5 Methodology and Implementation

The proposed model is shown in the flow diagram Fig. 3.

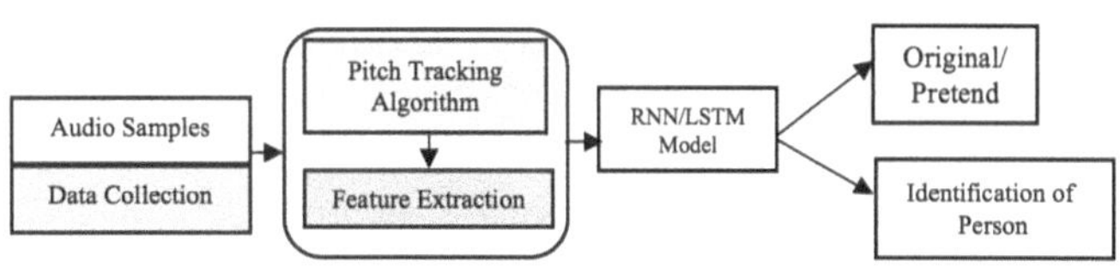

Fig. 3. Flow Diagram

5.1 Audio Samples

I have collected 48 audio samples. Among these 48 samples, four audio samples are from the original speaker and these are considered as original audio samples. The remaining 44 samples were the impersonation of the original speaker. These audio samples were divided into four sets, each set consists of 11 pretended audio samples along with one original sample. The sentences that were provided to the original speakers were "The quick brown fox jumps over the lazy dog", "A penny saved is a penny earned", "I scream, you scream, we all scream for ice cream", "Actions speak louder than words" and "To be or not to be, that is the question". The reason for choosing this sentence was to ensure that all the alphabets are made use of it and it consists of a specific set of words that were used to analyze a person's voice.

Audio samples were collected from various apps such as "Dolby On," "RecForge II," and "Easy Voice Recorder." After collecting these samples, they were exported to Google Drive, which never compresses or loses the data. All the audio files are saved in WAV format. Samples are divided into four sets of samples, and all figures are labeled. Figures 4a, 5a, and 6a belong to the 1st set, Figs. 4b, 5b, and 6b belong to the 2nd set, Figs. 4c, 5c, and 6c belong to the 3rd set, and Figs. 4d, 5d, and 6d belongs to the 4th set.

5.2 Pitch Tracking

The collected audio samples are stored in WAV format, such as four sets of audio samples that are loaded in Google Colab, whereas each set consists of one original audio sample and 11 pretend audio samples. Now, in four sets of audio samples, the pitch of each audio sample is tracked by using the librosa library in Python to identify which audio sample is genuine from the 11 pretended audio samples. By visualizing the pitch graph, each color in the graph represents the individual audio sample by this we can identify which audio sample pitch is more similar to the original audio sample. Figure 4 consists of different images are represented by Fig. 4(a, b, c, d). We describe these images as a pitch comparison between the original and 11 hypothetical audio samples.

Figure 5 consists of four different images represented by Fig. 5(a, b, c, d). These images compare the pitch between the original and 11 pretend audio samples. These images are similar to the four images in Fig. 4, but here, pitch is represented in terms of percentage.

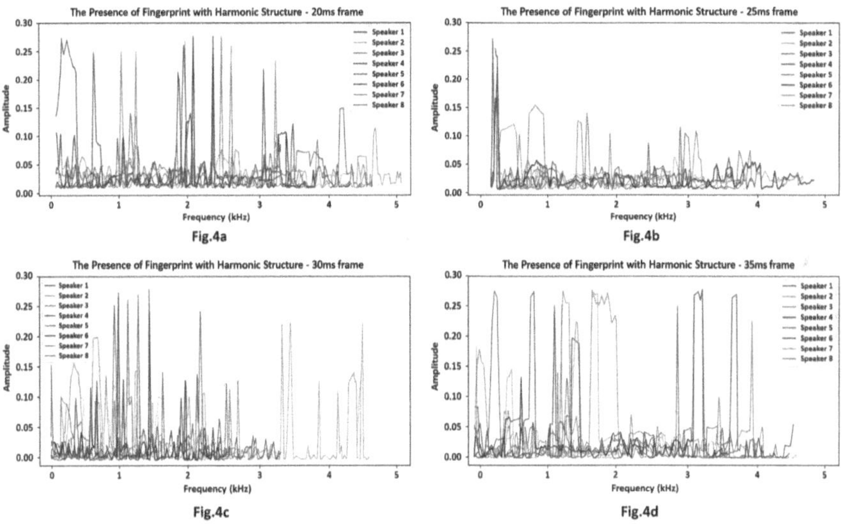

Fig. 4. Comparison of pitch between the Original and 11 Pretend audio samples

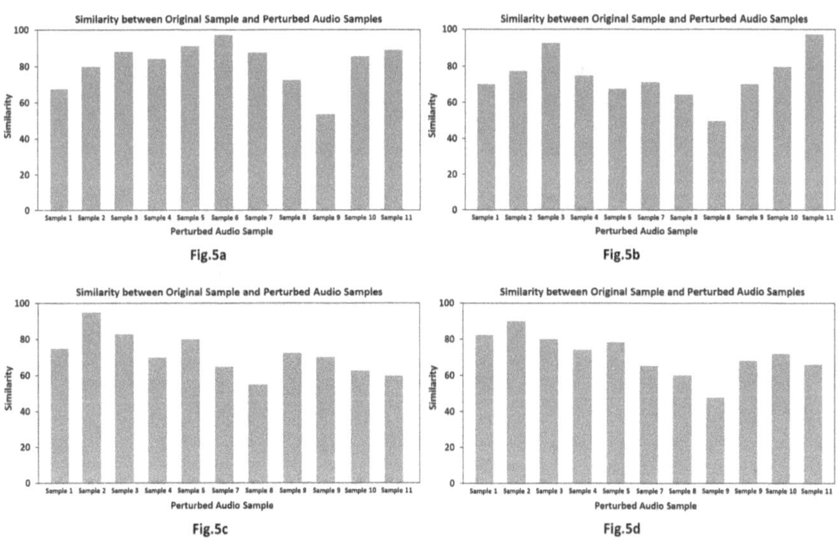

Fig. 5. The similarity between the original sample and pretender audio samples in terms of percentage

5.3 Audio Features Extraction

There are 8 different types of audio features. They include Mel-Frequency Cepstral Coefficients (MFCCs), Energy, Short-Time Fourier Transform (STFT), and Zero Crossing Rate (ZCR). In addition, spectral centroid, bandwidth, chroma features, and energy entropy are calculated. All these features are pulled from raw audio and formatted into the proper structures, then used to train RNN and LSTM models to recognize audio forgeries.

Procedure: The libraries to be used in the process should be installed.

Extract Features: The extract features function infers the energy, MFCCs, spectral centroid, and energy entropy from every sample of audio.

Load and Process: The original audio sample and 11 pretended or mimicked audio samples are loaded, and the features are extracted.

Compare Features: The differences between the features of the original sample and each of the 11 mimicked audio samples are calculated using Euclidean distance.

Similarity Identification: Of the audio samples containing the smallest difference, the one that exhibits the most resemblance to the original is identified.

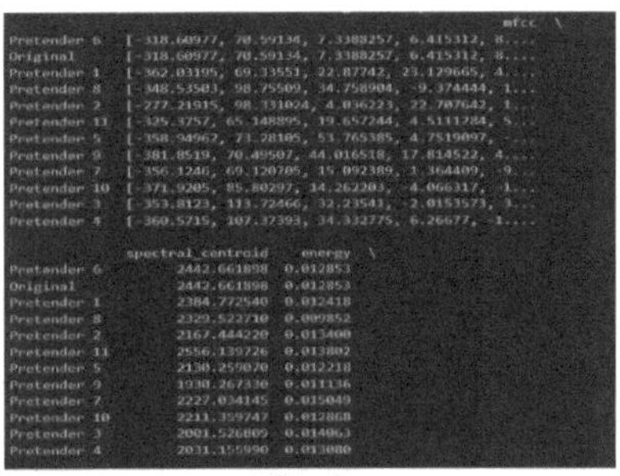

Fig. 6a

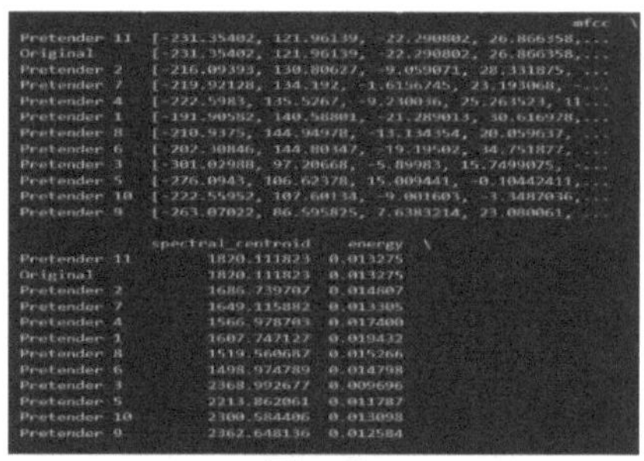

Fig. 6b

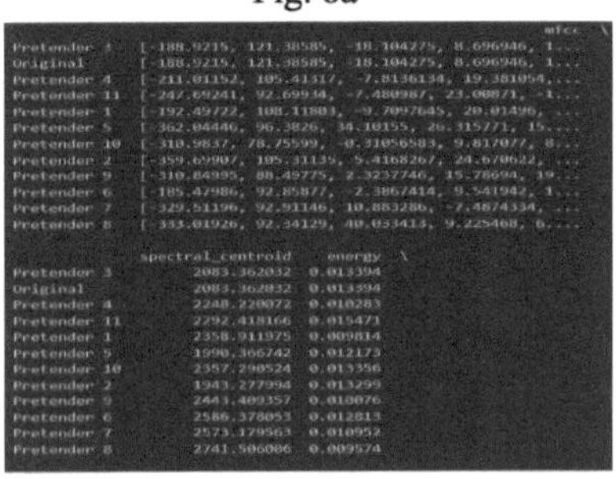

Fig. 6c

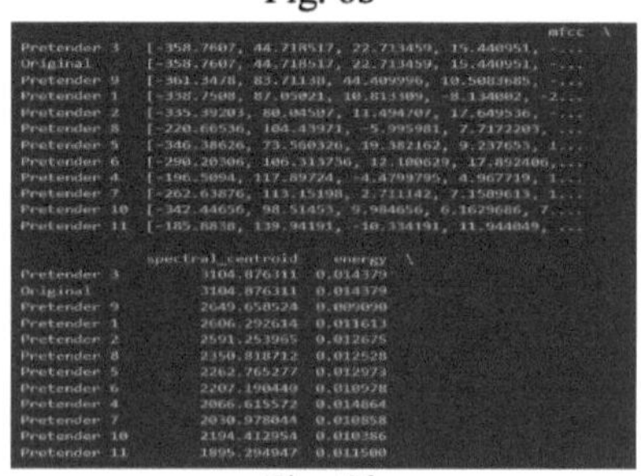

Fig. 6d

Fig. 6. Extraction of audio features from the pretended audio sample

6 Model Training and Hyperparameters Tuning Specifications

Feature Extraction

The function extract_features captures the following meaningful audio features:

- MFCC (Mel-Frequency Cepstral Coefficients): extracts timbral features from audio signals

- Spectral Centroid, Bandwidth, and Chroma Features: Represent frequency distribution and tonal characteristics.
- Energy and Entropy of Energy: measures the amount and the rate of change of energy in the signal

This diversified feature set ensures that the models capture both temporal and spectral patterns of audio data.

Data Preprocessing

- Normalization: The features are standardized using Standard Scaler so that the features lie on the same scale to enable faster convergence of the gradient descent.
- Label Encoding: One-hot encoding of the labels creates a categorical representation appropriate for classification.
- Data Splitting: The dataset is split into 80% for training and 20% for testing through train_test_split.

Hyperparameter Selection
Empirical tuning and best practices derived the hyperparameters for RNN and LSTM models as follows:

- Number of Units: 128 units in hidden layers to achieve a balance between complexity and efficiency in computation
- Dropout Rate: The dropout rate was set at 0.5, which avoids overfitting through the deactivation of 50% of the neurons during training.
- Batch Size: The batch size was set at 32 to optimize memory usage and stability during the training process.
- Learning Rate: The Adam optimizer has been used here with a learning rate of 0.001. This set a balance between the convergence speed and stability of the model.
- Clipping Norm: Gradient clipping with clip norm = 1.0 was used during training to avoid exploding gradients.
- Epochs: Max epochs are defined as 50. Further, early stopping has been used based on patience. Here, training stops if the validation loss does not improve for the next 5 epochs.

Model Architectures

- RNN Model: A single Simple RNN layer followed by a dropout layer and a dense output layer with softmax activation.
- LSTM Model: A single LSTM layer followed by a dropout layer and a dense output layer with softmax activation.

The input shape to both the models is (1, number_of_features), which is the same as the reshaped feature matrix.

Class Imbalance Handling
Class weights were computed using class_weight.compute_class_weight to prevent the potential class imbalance so that the training process would not ignore the minority classes.

Training

- Models were trained using categorical cross-entropy as the loss function since it is a multi-class classification problem.
- Early stopping was used to prevent overfitting by tracking the validation loss.

Evaluation Metrics

- Accuracy: It was used as the main evaluation metric while training.
- F1-Score: It was computed to measure the trade-off between precision and recall.
- Accuracy and Recall: More detailed for each class

Experiment Results

- RNN Model: The F1-Score of the model is {f1_rnn:.4f}, along with correct prediction on the test set.
- LSTM Model: The F1-Score is {f1_lstm:.4f}, although slightly better since it is capturing sequential patterns.

Visualization
The visualization of the probability of each class was done in comparison between the RNN and LSTM models, which shows what confidence each of the models had in the audio being classified.

7 Result

In Fig. 7 is Prediction for RNN Model Target Audio is 4th, and LSTM Model Target Audio is 0th. In Fig. 8 is Prediction for RNN Model Target Audio is 10th, and LSTM Model Target Audio is 10th. In Fig. 9 is Prediction for RNN Model Target Audio is 2nd, and LSTM Model Target Audio is 2nd. In Fig. 10 is Prediction for RNN Model Target Audio is 2nd, and LSTM Model Target Audio is 2nd.

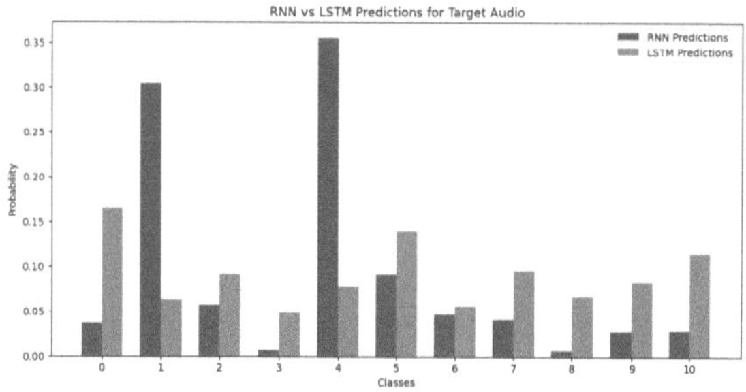

Fig. 7. Predicting the result using RNN and LSTM Models

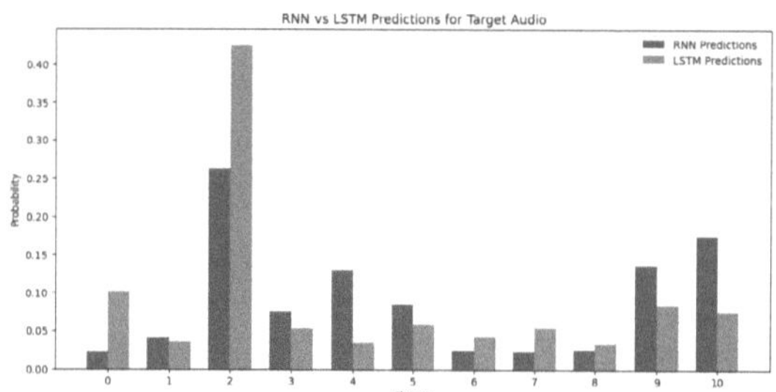

Fig. 8. Predicting the result using RNN and LSTM Models

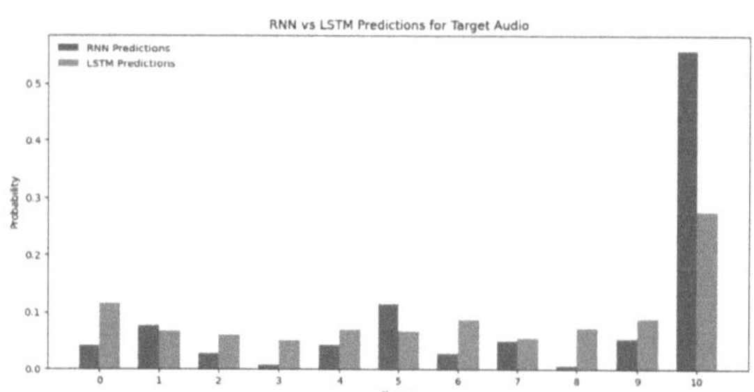

Fig. 9. Predicting the result using RNN and LSTM Models

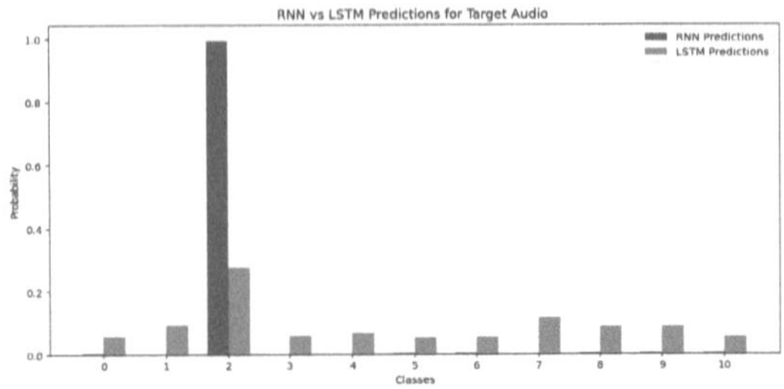

Fig. 10. Predicting the result using RNN and LSTM Models

$$\text{Precision} = \frac{TP}{TP + FP} \tag{9}$$

$$\text{Recall} = \frac{TP}{TP + FN} \tag{10}$$

The F1 score is the harmonic mean of precision and recall, i.e.,

$$2 \times \frac{Precision.Recall}{Precision + Recall} \tag{11}$$

Confusion Matrix: Confusion matrix shows the performance of a classification model by comparing predicted labels with actual labels. It consists of four values given in Table 1.

Table 1. Confusion Matrix

	Predicted + ve (Correct)	Predicted -ve (Incorrect)
Actual + ve (Correct)	3(TP)	1(FN)
Actual -ve (Incorrect)	0(FP)	44(TN)

True Positive (TP = 3): Three original audio samples were correctly recognized. **False Negative (FN = 1):** One original sample was misclassified. **False Positive (FP = 0):** No pretended audio sample was falsely recognized as the original. **True Negative (TN = 44):** All 44 pretended audio samples were correctly rejected as non-original.

8 Conclusion

Utilizing RNN/LSTM architectures and sophisticated feature extraction, the created model accurately detects audio forgery and identifies the mimicking speaker. This illustrates the model's usefulness and possible applications in various security and forensic scenarios. Further research into more features and sophisticated neural network topologies could improve model accuracy. Robustness would also be enhanced by increasing the dataset to include more diverse speakers and environments and modifying the algorithm for real-time detection. This study can be applied to the real-time identification of preventable online fraud crimes. The result is 75% accuracy.

References

1. Kulangareth, N.V., Kaufman, J., Oreskovic, J., Fossat, Y.: Investigation of deepfake voice detection using speech pause patterns: algorithm development and validation. JMIR Biomed. Eng. **9**, e56245 (2024)
2. Iqbal, F., Abbasi, A., Javed, A.R., Jalil, Z., Al-Karaki, J.: Deepfake audio detection via feature engineering and machine learning. In: Woodstock'22: Symposium on the Irreproducible Science (2022)
3. Hassan, E.S., et al.: Enhancing speaker identification through reverberation modeling and cancelable techniques using ANNs. PLoS ONE **19**(2), e0294235 (2024)
4. Kwak, I., et al.: Voice spoofing detection through the residual network, max feature map, and depthwise separable convolution. IEEE Access **11**, 49140–49152 (2023)
5. Zhu, T., Wang, X., Qin, X., Li, M.: Source tracing: detecting voice spoofing. In: 2022 Asia-Pacific Signal and Information Processing Association Annual Summit and Conference (APSIPA ASC) (2022)
6. Javed, A., Malik, K.M., Malik, H., Irtaza, A.: Voice spoofing detector: a unified anti-spoofing framework. Expert Syst. Appl. **198**, 116770 (2022)
7. Guo, J., Zhao, Y., Wang, H.: Generalized spoof detection and incremental algorithm recognition for voice spoofing. Appl. Sci. **13**(13), 7773 (2023)
8. Xue, J., Zhou, H.: Physiological-physical feature fusion for automatic voice spoofing detection. Front. Comput. Sci. **17**(2) (2022)
9. Ackerson, J., Dave, R., Seliya, N.: Applications of recurrent neural network for biometric authentication & anomaly detection. Information **12**(7), 272 (2021)
10. Chang, H.: Deepfake speech recognition and detection. Int. J. Pattern Recogn. Artif. Intell. **37**(09) (2023)
11. Kansara, A., Kumari, P., Prathap, B.R.: Detecting deepfake voices using a novel method for authenticity verification in voice-based communication. In: Lecture Notes in Networks and Systems, pp. 397–405 (2024)
12. Mittal, A., Dua, M.: Automatic speaker verification systems and spoof detection techniques: review and analysis. Int. J. Speech Technol. **25**(1), 105–134 (2021)
13. Singh, M.K.: Identification of speaker from disguised voice using MFCC feature extraction, chi-square, and classification technique. Wirel. Pers. Commun. (2024)
14. Lim, S., Chae, D., Lee, S.: Detecting deepfake voice using explainable deep learning techniques. Appl. Sci. **12**(8), 3926 (2022)
15. Neelima, M., Prabha, I.S.: Hybrid feature optimization for voice spoof detection using CNN-LSTM. Traitement Du Sig. **41**(2), 717–727 (2024)

Crop Recommendations and Decision Support Systems

Sreenivasa Rao Kakumanu[1], Sivakumar Gowda Katta[2], D. V. V. Bramhachari[3], G. Kusuma Harinadh[3], and Lokaiah Pullagura[4](✉)

[1] Department of CSE-AIML&IOT, VNR Vignana Jyothi Institute of Engineering and Technology, Hyderabad, India

[2] Department of CSE, Sreenidhi Institute of Science and Technology, Hyderabad, India
sivakumar.k@sreenidhi.edu.in

[3] Department of CSE, Universal College of Engineering and Technology, Dokiparru, Guntur, India

[4] Department of CSE, Jain University, Bangalore, Karnataka, India
lokaiah75@gmail.com

Abstract. Agriculture is vitally essential for all human beings because it offers food safety and economic peace to the world. With technological improvements, Machine Learning (ML) has emerged as a possible tool for crop ideas based on numerous natural factors, such as soil makeup, temperature, humidity, rainfall, and pH levels. The present research offers an AI-Powered Crop Evaluation and Support System for Decision Making that uses Support Vector Machine classification (SVM), Random Forest (RF), Logistic Regression (LR), and decision tree (DT) algorithms to offer farmers the most appropriate crop choices for their area. This approach utilizes farmers' insights into natural environmental conditions to suggest the most appropriate crops for cultivation. The system was tried on different crops and the traits that are common to them. Using the formulas' findings, crop ideas that would work in each place were come up with. With an accuracy of 96.3%, the SVM algorithm showed that it was very good at what it did. Based on the weather, this method helps farmers make smart choices about when to gather their crops, which increases efficiency and productivity while lowering the risk of crop loss. The suggested AI-powered field recommendation and decision-help system utilizes the SVM, RF, LR, and DT algorithms. Because it changes the foods that farms grow, they become more eco-friendly and make more money. These ideas could help farms get bigger and safer, which would improve food security and the economy.

Keywords: AI-Powered Crop Recommendation · Decision Support System · SVM · RF · DT · LR

1 Introduction

Agriculture is critical in global nutrition and economic growth, acting as the foundation for food security and supporting long-term growth. The advancement of technology has positioned artificial intelligence (AI) and machine learning (ML) as influential instruments in revolutionizing conventional agricultural methods. AI-powered systems look

A. Das et al. (Eds.): MIDAS 2025, CCIS 3055, pp. 160–169, 2027.
https://doi.org/10.1007/978-3-032-35516-4_13

at things in nature like soil type, rainfall, temperature, humidity, and pH levels to give information that helps with choosing crops and predicting yields, giving specific suggestions for different areas [1]. The primary objective of AI farming is to aid farmers in identifying crops appropriate for their specific environments. AI systems, educated on various crops and their growth characteristics, can predict ideal crop choices, significantly reducing the need for trial-and-error approaches [2]. AI possesses the capacity to transform agriculture by enhancing operational efficiency and reducing the likelihood of crop failure. The advancements provide significant advantages to farmers, policymakers, agribusinesses, and consumers by fostering environmentally sustainable agricultural practices [3]. AI-based crop selection systems help make better, data-informed choices by using different climate and weather information, unlike older methods. Furthermore, AI reduces the necessity for repetitive seasonal crop trials, thus conserving both time and resources. Integrating meteorological forecasts, satellite imagery, and historical farming data enhances predictive accuracy and decision-making capabilities through AI technologies. Consequently, these systems provide farmers with effective tools to enhance productivity, minimize agricultural waste, and promote long-term sustainability [5]. Recent years have witnessed an increase in the application of AI and machine learning (ML) in various agricultural domains. Chen et al. [6] investigated the role of AI in improving data-driven decision-making in crop yield forecasting and market trend analysis. Gupta et al. [7] examined decision tree methodologies and their utilization in agriculture, demonstrating their effectiveness in crop classification and production forecasting.

Furthermore, ensemble learning techniques like RF have demonstrated significant potential in enhancing predictive accuracy in agricultural contexts. As highlighted in [8–10], numerous farmers face challenges in securing adequate labor. However, AI-driven systems can bridge this gap by performing tasks that require extensive manual labor, thereby boosting productivity and reducing the reliance on human workers.AI and ML are constantly improving, with the potential to transform the farming industry completely. AI-powered technologies help farmers choose crops, handle resources, and plan operations by giving them advice and insights based on data. These tools increase crop yields, make farming more efficient, and protect the environment. This study suggests a crop advice system that uses different ML techniques, such as SVM, RF, LR, and DT. This method looks at several environmental factors, including the type of soil, the weather, and how well crops have done in the past to find the best crops for a certain area, The system's goal is to give farmers data-driven tools that make them more productive, lower their risks, and help make farming more sustainable.

The subsequent sections of the paper are organized as follows: Sects. 2, 3, and 4 present the proposed work, encompassing its design and methodology. Section 5 outlines the experimental observations that ensued from these results. Finally, Sect. 6 addresses the conclusion of the study.

2 Proposed Work

2.1 Dataset Acquisition

The dataset used in this study was obtained from Kaggle's Crop Yield Prediction dataset. It contains information on various natural factors that influence crop growth and yield, including 1. Soil composition 2. Temperature 3. Humidity 4. Rainfall 5. PH level. The dataset likely includes multiple entries for different geographical locations, each with its corresponding environmental factors and associated crop yields. This comprehensive dataset allows the application of ML algorithms to predict suitable crops for specific areas based on local conditions.

2.2 Data Preprocessing

Data preparation is a must to create an AI-powered crop selection system, among other ML experiments. The process involves transforming unprocessed data into an experimentation-ready file by cleaning and altering it. As an initial step, gather data from several sources, including weather, soil quality, and food growth datasets. After getting the information, cleaning it to eliminate missing numbers, copy information, and blunders is essential. The data must be transformed into a research-friendly file as the next step. It includes scaling, changing category data into number data, and breaking it into sets for testing and training. Appropriate information readiness guarantees that the simulated intelligence model has high exactness and avoids overfitting.

2.3 Feature Selection

A crucial step in creating an AI-powered crop guidance system is feature selection. It includes choosing the most essential traits from the information that significantly impact the crop yield. The feature selection process is important because not all features have the same value in predicting food growth. Some traits may be repetitive and not add to the model's accuracy. Therefore, choosing the most important characteristics helps decrease the collection's complexity and improves the model's accuracy.

2.4 Model Development

Model creation is the next step in building an AI-powered crop guidance system. This step includes choosing the proper machine learning method that will be employed to take crop suggestions based on the input data. Different supervised learning of ML methods can be used, including neural networks, SV, DT, and RF. The choice of the proper method relies on the nature of the data and the issue being answered. Once the method is chosen, it is taught on the preprocessed data to build the AI model. The success of the AI model is then tested on the training sample.

2.5 Model Evaluation

Model assessment is a crucial step in developing any machine learning model. Measuring the model's performance following its creation is essential to ensure its accuracy and dependability. A few exhibition estimates can be utilized to survey a model, including exactness, accuracy, review, F1 score, and ROC AUC score. Accuracy is the small portion of genuine positive gauges among every single positive figure, while precision indicates the model's level of right expectations. Determines that the proportion of all real positive cases that result in true positive predictions is the Recall. The consonant mean of memory and exactness is the F1 score when class contrasts exist.

2.6 Model Optimization

The final step is to deploy the system, which farmers can access through an easy-to-use interface, such as a versatile application or online interface. The system will provide pertinent data and a list of crops appropriate for the given environmental factors, such as crop yield, market demand, and profitability. We will also continuously incorporate feedback mechanisms to enhance the system's accuracy and relevance to farmers.

2.7 Model Deployment

Model rollout is an important phase in the deployment of an AI-powered crop suggestion and decision support system. It includes bringing the model into the working system and making it ready for use by end-users. The launch process can be complicated and requires careful planning to ensure that the system works properly, is flexible, and is reliable.

3 Workflow of CRS

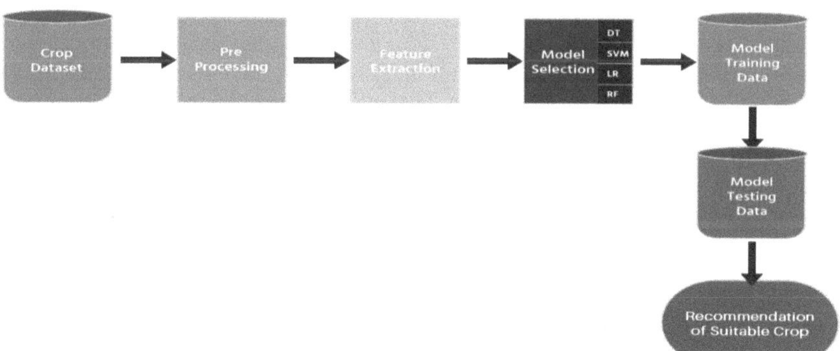

Fig. 1. Workflow Diagram of the proposed Crop Recommendation System

Figure 1 shows the entire method of the proposed approach. The soil and environmental feature data collection is preprocessed to identify missing values and eliminate extraneous information. The suggested classifiers feed the preprocessed data. The classifiers receive the specified characteristics as input throughout the learning process. This paper uses a supervised learning approach for the prediction procedure. Unknown samples are supplied to verify the developed classifier after training samples are used. The outcomes are analyzed using certain performance indicators.

Table 1. Details of the Evaluation Metrics

Metric	Formula	Range
Accuracy	(TP + TN) / (TP + TN + FP + FN)	0 to 1
Precision	TP / (TP + FP)	0 to 1
Recall	TP / (TP + FN)	0 to 1
F1-Score	2 * (precision * recall) / (precision + recall)	0 to 1
Support	N/A	N/A
Macro avg	Average of precision, Recall and F1-Score per class	0 to 1
Weighted avg	Weighted average of precision, Recall and F1-Score	0 to 1

4 Classification Approaches

4.1 RF

RF is an ensemble strategy that creates a more accurate and reliable model by combining numerous decision trees [11, 18]. RF selects the optimum attributes for each decision tree as it is formed. The ultimate result of the Classification or regression is decided separately by each tree. RF reduces overfitting by randomly picking data from the training set and characteristics for each tree [12]. The Random Forest algorithm involves building many decision trees, each utilizing a portion of the features and training data and aggregating the output of the trees using a majority voting system.

$$F(x) = \frac{1}{j}\sum\nolimits_{j=1}^{j} c_{j_{full}} + \sum\nolimits_{K=1}^{K}\left(\frac{1}{J}\sum\nolimits_{j=1}^{j} contribution_j(x, k)\right) \quad (1)$$

Algorithm

- Initialize random forest classifier with desired hyperparameters.
- Train the classifier using the training set.
- Prediction on the test set applying the learned classifier.
- Return the predicted labels.

4.2 SVM

SVM is a computer-based intelligence system overseeing findings for gathering and backsliding issues. The goal is to select the most appropriate hyperplane for classifying the data points. Before SVM can function, the input data must first be mapped into a high-dimensional feature space [13, 20]. The next step is to find the hyperplane with the greatest difference between the two classes. The SVM formula requires maximizing the margin between the two classes, subject to some restrictions.

$$f(x) = \sum_i \alpha_i A(x_i, x) + b \tag{2}$$

Algorithm

- Initialize the SVM classifier with desired hyperparameters.
- Train the classifier with the training set.
- Prediction on the test set applying the learned classifier.
- Return the predicted labels.

4.3 LR

An ML method, LR, is used to solve classification problems. It works by identifying the appropriate coefficients of the logistic function to reduce the number of changes between the expected and actual output [14, 21]. The calculated relapse equation includes utilizing a strategic capability to address the connection between the info factors and the result variable and then using maximum likelihood estimation to maximize the logistic function's coefficients.

$$\log\left(\frac{p(x)}{1-p(x)}\right) = \beta_0 + \beta_1 X \tag{3}$$

Algorithm

- Initialize LR classifier with desired hyperparameters.
- Use the training set to train the classifier.
- Make predictions on the testing set using the trained classifier.
- Return the predicted labels.

4.4 DT

DT is a supervised ML approach that divides the data into subsets according to the most important characteristics, iteratively dividing the subgroups until a stopping requirement is satisfied [11, 16]. The most important characteristics are chosen, the data is recursively divided into subsets until a stopping requirement is satisfied, and then each leaf node is given a class or value according to the formula for decision trees.

$$Entropy = \sum_{j=1}^{c} -a_1 * log_2 a_1 a \tag{4}$$

$$Gini = 1 - \sum_{j=1}^{c} (a_i)^2 \tag{5}$$

Algorithm

- Initialize DT classifier with best splitting attribute and desired hyperparameters.
- Use the training set to train the classifier.
- Make predictions on the testing set using the trained classifier.
- Return the predicted labels.

4.5 Crop Recommendation System Procedure

Step 1. N (Nitrogen), P (Phosphorus), K (Kaolin), temperature, humidity, pH, rainfall, and the crop label are among the environmental and soil parameters that affect crop development in the crop dataset.

Step 2. The dataset is preprocessed to eliminate missing values, which may be accomplished by removing rows or filling null values with the column's mean or median.

The dataset's training and testing sets are divided at 70:30, with 70% going to training and 30% to testing.

Step 3. The precision of a few AI approaches, such as Choice Trees, Irregular Woods, SVM, and Strategic Relapse, is evaluated on the dataset to discover the most accurate crop prediction model.

Step 4. The chosen model is trained on the training set to understand the relationship between environmental and soil characteristics and crop type.

Step 5. The trained model is.

stored as a PKL file to be utilized in the web application.

Step 6. The online application uses the stored model to anticipate the yield in response to user inputs for environmental and soil parameters.

Step 7. The expected crop may be seen combined with statistics such as crop, and it also provides Suggestions for the fields to put the necessary Fertilizers.

5 Results and Observation

The results of this experimentation are shown in Fig. 6 below, which shows the success of using AI-powered crop suggestions and decision-support tools in changing agriculture. The models employed in this study, namely RF, LR, and SVM, demonstrated high accuracy in predicting crop yields using various external factors [12, 21]. The SVM model exhibited the highest precision at 0.968, followed by the RF model at 0.890 and the LR model at 0.888. This investigation demonstrates that AI-powered crop advice and decision support systems are practical tools for farmers to make informed crop management decisions. By providing insights based on important environmental characteristics, including soil composition, temperature, and weather patterns, these methods assist farmers in increasing crop output. These cutting-edge solutions optimize decisions about planting schedules, irrigation, and harvesting times by utilizing real-time data. Reducing dependency on chemical pesticides and fertilizers is also greatly aided by AI-powered advising platforms and decision-support systems. These technologies support ecologically friendly farming while preserving high crop yields by promoting the use of more natural, sustainable alternatives. The amount and caliber of training data utilized in the creation of such AI models have a significant impact on their performance. The acquisition of precise and pertinent data, accompanied by ongoing refining, is essential for the sustained progress of these models. Furthermore, continuous monitoring and system upkeep are crucial to maintain the precision and dependability of these technologies throughout time. The implementation of AI-driven agricultural decision-making tools presents considerable potential for enhancing economic growth in rural areas. By providing accurate and prompt advice, these systems can enhance food production and increase farmer incomes, thus fostering the sustainable development of rural economies.

Table 2. Finding the Best Scores & Estimators.

	Model name	Best_estimator	Best_score
0	RF	ccp_alpha = 0.03,max_depth = 8, max_features = 'auto', min_samples_leaf = 2, min_samples_split = 5, n_estimators = 933	0.895893
1	LR	max_iter = 194, multi_class = 'multinomial', tol = 0.5	0.906533
2	DT	ccp_alpha = 0.03, criterion = 'entropy', max_features = 'auto',min_samples_leaf = 5, min_samples_split = 12	0.827970
3	SVM	kernel = 'poly', max_iter = 250, tol = 0.75	0.963882

Table 2 shows the hyperparameters of ML models for crop recommendation system. The RF Classifier is an ensemble learning technique that employs multiple decision trees to produce predictions [17, 19]. The hyperparameters consist of ccp_alpha, serving as the complexity threshold for tree pruning; max_depth, which establishes the maximum depth of each decision tree; max_features, specifying the maximum number of features to consider when splitting a node; min_samples_leaf, denoting the minimum number of samples required at a leaf node; and min_samples_split, indicating the minimum number of samples necessary to split a node. LR is a classification method that estimates the likelihood of an input belonging to one of multiple categories. The hyperparameters include max_iter, which sets the maximum number of times the solver can run before it stops; multi_class, which shows how to deal with multiple categories; and tol, which indicates the acceptable level of error for stopping. The DT Classifier develops a hierarchical model that represents decisions and their potential outcomes. The essential hyperparameters of the model are the kernel, which specifies the kernel function utilized, and max_iter, which defines the maximum iterations permitted for the solver to achieve convergence. In conjunction with convergence tolerance, these parameters are essential for optimizing model accuracy and efficiency. This study emphasizes the significant influence of AI-based crop recommendation and decision-support systems in contemporary agriculture. The findings highlight the significance of precision and the need for ongoing monitoring and maintenance to guarantee consistent system performance. Intelligent technologies facilitate informed decision-making, thereby significantly increasing agricultural productivity, promoting rural economic growth, and supporting overarching objectives related to food security and sustainable development.

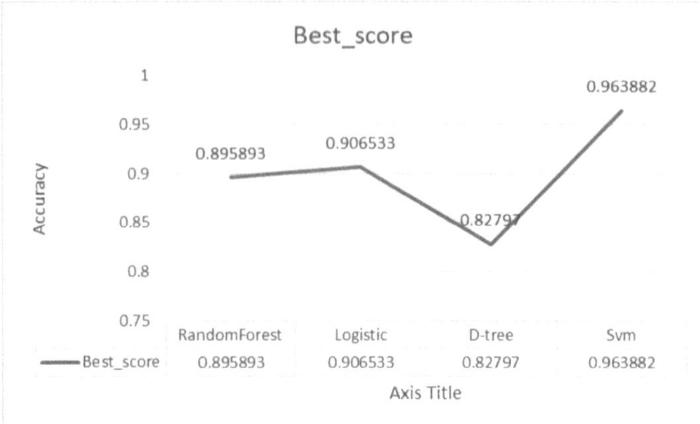

Fig. 2. Comparison of Models

6 Conclusion

The incorporation of AI in agriculture presents significant potential for the transformation of traditional farming methods. The system utilizes multiple ML algorithms, including SVM, RF, LR, and DT to provide customized crop recommendations according to varying climatic conditions. The SVM model exhibited an accuracy of 96.3%, demonstrating its effectiveness in determining region-specific crop suitability. This method enhances resource utilization, reduces the likelihood of crop failure, and ultimately leads to increased agricultural productivity. The integration of AI in this field is essential for enhancing food security and promoting economic development in rural regions. The existing system demonstrates considerable potential, although additional enhancements could significantly augment its efficacy. Integrating Internet of Things (IoT) devices and adjusting to long-term climatic variability are two critical avenues for growth. Real-time data from IoT sensors can improve predictive accuracy and offer prompt, actionable insights. This feature facilitates ongoing surveillance of environmental variables, resulting in more adaptive and responsive agricultural management practices. Moreover, enhancing the system to integrate climate models with past weather data would assist farmers in making informed decisions in the face of evolving environmental conditions. This integration will facilitate the selection of robust crops better suited to projected climate scenarios, thereby enhancing flexibility and long-term agricultural sustainability.

References

1. Greig, L.: An analysis of the key factors influencing farmer's choice of crop, Kibamba Ward. Tanzania. J. Agric. Econ. **60**, 699–715 (2009)
2. Navarro-Hellín, H., et al.: A decision support system for managing irrigation in agriculture. Comput. Electron. Agric. **124**, 121–131 (2016)
3. Chakraborty, P.A.S., Kumar, A., Pooniwala, O.R.: Intelligent crop recommendation system using machine learning. In: 2021 5th International Conference on Computing Methodologies and Communication (ICCMC), pp. 843–848 (2021)

4. Doshi, Z., Nadkarni, S., Agrawal, R., Shah, N.: Agroconsultant: intelligent crop recommendation system using machine learning algorithms. In: 2018 Fourth International Conference on Computing Communication Control and Automation (ICCUBEA), pp. 1–6 (2018)
5. Pande, S.M., Ramesh, P.K, Surya, A, Aishwarya, B.R., Rohilla, K., Shaurya, K.: Crop recommender system using machine learning approach. In: 2021 5th International Conference on Computing
6. Chen, H., Chiang, R.H.L., Storey, V.C.: Business intelligence and analytics: from big data to big impact. MIS Q. **36**(4), 1165–1188 (2012)
7. Singh, A., Gupta, A.: A comprehensive review of decision tree algorithm. Inter. J. Comput. Appli. **167**(3), 19–25 (2017)
8. Sudheer, B.N., Sarada, K., Rao, D.M.: Detection process for disease recognition of paddy crop using deep learning. Indian J. Ecol. **47**, 197–201 (2020)
9. Dontha, M.R., Sri Supriyanka, N.: Image-based disease detection and classification of plant using CNN. In: Marriwala, N., Tripathi, C., Jain, S., Kumar, D. (eds) Mobile Radio Communications and 5G Networks. LNNS, vol. 588. Springer, Singapore . (2023). https://doi.org/10.1007/978-981-19-7982-8_22
10. Rani, S., Mishra, A.K., Kataria, A., et al.: Machine learning-based optimal crop selection system in smart agriculture. Sci. Rep. **13**, 15997 (2023). https://doi.org/10.1038/s41598-023-42356-y
11. Jiang, Y., Yu, M., Zhu, W.: Big data and machine learning for financial prediction. Financial Innovation **4**(1), 1–10 (2018)
12. Kim, J.H., Kim, J.K., Park, Y.K.: A prediction model for college majors based on demographic and academic data. Comput. Educ. **117**, 112–122 (2018)
13. Li, Y., Zhang, J., Wu, X., Liu, S.: A prediction model for career choice of college students based on big data. J. Big Data **6**(1), 1–17 (2019)
14. Zhu, J., Li, Y., Wang, S., Li, D.: Career choice prediction based on campus big data. IEEE Access **8**, 204203–204215 (2020)
15. Zhang, X., Song, L., Liu, Y.: A novel decision tree algorithm for big data. J. Comput. Sci. **41**, 1–12 (2020)
16. Rosenblatt, F.: The perceptron: a probabilistic model for information storage and organization in the brain. Psychol. Rev. **65**(6), 386–408 (1958)
17. Pullagura, L., Kittad, N., Diwakar, G., Sathiya, V., Kumar, A., Yalawar, M.S.: ML based Parkinson's disease identification using gait parameters. In: International Conference on Automation, Computing and Renewable Systems (ICACRS), Pudukkottai, India, pp. 561–566 (2022). https://doi.org/10.1109/ICACRS55517.2022.10029281
18. Pullagura, L., Sinha, G., Manandhar, S., Rawal, B., Getachew, S., Chaturvedi, S.: Crime Rate Prediction in Tamilnadu using Machine Learning. LNEE, https://link.springer.com/chapter/, https://doi.org/10.1007/978-981-99-8646-0_31
19. Pullagura, L., Kittad, N., Diwakar, G., Sathiya, V., Kumar, A., Yalawar, M.S.: ML based parkinson's disease identification using gait parameters. In: 2022, International Conference on Automation, Computing and Renewable Systems (ICACRS), Pudukkottai, India, pp. 561–566 (2022). https://doi.org/10.1109/ICACRS55517.2022.10029281
20. Pullagura, L., Katiravan, J.: Train Delay Prediction using Machine Learning. Inter. J. Eng. Adv. Technol. **9**(2), 1312–1315 (2019). https://doi.org/10.35940/ijeat.a2088.129219, https://www.ijeat.org/wp-content/uploads/papers/v9i2/A2088109119.pdf
21. Pullagura, L., Kumar, A.: Analysis of train delay prediction system based on hybrid model, J. Adv. Res. Dyn. Control Syst. **12**, 07-Special Issue (2020), https://www.jardcs.org/abstract.php?id=5860, https://doi.org/ https://doi.org/10.5373/JARDCS/V12SP7/20202433

Brain Stroke Detection and Classification Using FRFCM Segmentation and Classification Using SVM and Random Forest Methods

Naresh Kanungo[1], Satyasis Mishra[1](✉), Sandipan Pine[1], P. A. Sunny Dayal[2], and Katleho Moloi[3]

[1] Department of ECE, Centurion University of Technology and Management, Bhubaneswar, Odisha, India
s.mishra@cutm.ac.in
[2] Department of ECE, Centurion University of Technology and Management, Vizianagaram, Andhra Pradesh, India
[3] Department of Electrical Power Engineering, Durban University of Technology, Durban, South Africa

Abstract. Stroke is a major global cause of mortality and disability, and prompt and precise diagnosis is essential for successful treatment. The classification of brain strokes is crucial in differentiating between stroke kinds, such as ischemic and hemorrhagic strokes, which necessitate very different medical treatments. Using clinical characteristics, patient demographics, and medical imaging data, this study suggests a machine learning-based method for classifying brain strokes and an FCM-based segmenting algorithm for detecting the strokes. At the first stage, the images are pre-processed and aligned as input to the segmentation technique. An FRFCM segmentation technique was proposed to detect strokes. The segmented images are fed as input to the Support vector machines (SVM) and Random forest to classify the strokes from the images. The brain stroke images are collected from the Kaggle dataset for this research. The proposed FRFCM segmentation achieved an accuracy of 99.11%, and the random forest achieved 98.88% classification accuracy. The comparison results for segmentation and classifications are presented.

Keywords: Fuzzy c means · FRFCM · SVM · Random Forest · Brain Stroke

1 Introduction

Stroke ranks third globally in terms of disability, after cardiovascular disease and newborn disease, and is the second greatest cause of mortality globally [1]. The World Health Organization (WHO) reports that every year, 15 million individuals worldwide suffer a stroke, 5 million of them pass away, and the remaining 5 million become permanently disabled [2]. Early diagnosis of stroke is crucial to prevent serious consequences or death. Common diagnostic methods include CT and MRI, which aid in identifying the type of stroke. Even though MRI produces superior images than CT scans, only major

A. Das et al. (Eds.): MIDAS 2025, CCIS 3055, pp. 170–181, 2027.
https://doi.org/10.1007/978-3-032-35516-4_14

institutions have the technique needed for this treatment. CT is the primary method for detecting stroke, providing rapid and cost-effective diagnosis. Decision support systems analyse CT images to aid physicians in medical decisions. Imaging techniques help detect cerebral stroke quickly, collect accurate data, and make treatment decisions. Delays can lead to severe brain function and increased death risk. The validity of medical interventions depends on lesion location and disorder onset period. The treatment of symptomatic patients requires human experts, but this is labor-intensive. Imaging studies like CT and MRI are needed for quick diagnosis and early treatment options. Following a careful examination of clinical data, algorithm developers created a set of feature design recommendations that were used to construct traditional ways of automatically identifying and classifying cerebral infarcts.

Traditional techniques of automatic stroke classification have been hindered by a lack of sophistication because certain features of a possible brain stroke are obscured and challenging to identify on scans. However, unlike traditional machine learning, deep learning techniques are able to learn visual features from training samples. These techniques can overcome the drawbacks of earlier deep learning algorithms and streamline the modelling of cerebral infarcts Convolutional neural networks (CNNs) use convex kernels to tackle different image categorization problems and to extract specific characteristics from an input image [3]. CNN is a deep learning subclass that is used in neuroimaging to diagnose acute ischemic strokes. Automated diagnostics places a lot of emphasis on end-to-end system architecture. Methods for lesion segmentation and risk prediction, as well as algorithms for categorizing brain strokes, have been presented in recent articles. There is currently no recognized DWI classification system for the main arterial spatial stroke types. This study introduces an automated CNN-based method to detect cerebral infarction on CT scans for better diagnosis and treatment of ischemic cerebral strokes. It includes image pre-processing, feature engineering, and data augmentation techniques to enhance the CNN model's performance. Challenges include integrating computational knowledge, medical expertise, and high-performance computing. Current medical imaging research lacks user-friendly interfaces for clinicians. An automated classification system predicts brain CT scans into haemorrhagic, ischemic, and healthy categories, identifying complications like cerebral haemorrhage due to ruptured blood vessels.

1.1 Literature Review

Deep CT and MRI learning techniques for the classification of brain strokes are covered in this chapter. Furthermore, we have examined CAD systems for stroke diagnosis and looked into some of the most well-liked machine learning techniques for stroke classification utilizing CT images. It is crucial to quickly detect early ischemic changes from acute ischemic stroke on noncontract computed tomography (NCT) images to determine the need for tissue plasminogen activator within 4–5 h of symptoms onset [4]. However, CT findings related to early ischemic changes are subtle and challenging to identify. An automated system to aid medical professionals in diagnosing acute ischemic stroke on tomograms can be beneficial in clinical practice. Computer scientists are collaborating with medical professionals, employing computational methods to analyse digital medical images.

Lo et al. [5]. Developed a technique to detect acute ischemic stroke on CT scans with 81% accuracy, using 23 relevant features identified from pre-processed images. Wu et al. [6] created a method to identify stroke features on CT images with 76.84% accuracy. Several studies in the references section addressed data limitations by employing popular deep learning methods combined with transfer learning. Babutain et al. [7] used transfer learning on the Exception network to detect hyperdense middle cerebral artery symptoms. Phong et al. [8] compared CNN models for stroke classification. One study proposed using transfer learning for feature representation by removing linked layers. Subudhi et al. [9] created a novel method for segmenting and classifying stroke cases from MRI data using a random forest classification model. The accuracy of 192 brain pictures was 93.4 percent thanks to the suggested model.

The study by Ortiz-Ramón et al. [10] focuses on using fMRI data to differentiate between lacunar and cortical strokes. The authors applied various techniques suggested by different scientists to extract textural features. They noted a limitation in integrating stroke and aging datasets for analyzing images of recent strokes. Image normalization techniques are proposed to enhance image quality and overcome this challenge. A system called Deep Radiology-Net used 24,000 NCCT brain markers to identify 30 different brain CT signs. It assigned risk labels based on a hierarchical loss function and was influenced by Google Net, yielding promising results Can the Deep Radiology-Net system accurately detect various brain CT signs compared to other structures? How effective is the hierarchical. Han et al. [11] presented a deep learning method for stroke classification and lesion segmentation on CT images using deep models. The system achieved 97% accuracy in lung data categorization and 97% Dice coefficient in segmentation, showing promise. However, certain lung regions were not effectively segmented, possibly due to model generalization. Souza et al. employed Mask R-CNN with supervised and unsupervised methods to segment lung regions on CT images, achieving 95% accuracy. The approach needs further testing on a broader dataset for validation.

Concerns regarding the effectiveness of fine-tuning techniques in deep learning were voiced by Tajbakhsh et al. [12]. The Fine-Tuning method outperformed or matched the baseline model in most cases when they evaluated two Deep CNN models for medical picture segmentation in radiology, cardiology, and gastroenterology. Increasing training data was another benefit of extracting patches from 2D images.

2 Materials and Methods

2.1 Proposed System Architecture

Our research's first phase is focused on image preprocessing, which includes actions like data augmentation, image resizing, and image reading presented in Fig. 1. The text describes machine learning algorithms used to classify a dataset, utilizing a supervised method to create models based on training sets, labelling examples with specific classes based on presence, type (ischemic or haemorrhagic), and stroke position. SVM is an optimization algorithm that maps the training dataset into a higher-dimensional space, selecting the hyperplane with the highest distance to the closest training record of each class.

2.2 Random Forest (RF)

Ensemble learning approaches such as Random Forests (sometimes called random choice forests) can be used to solve problems related to regression and classification shown in Fig. 2. They operate by training a huge number of decision trees in a distributed manner. The class that the bulk of the trees choose is the result of a random forest when it comes to solving classification challenges. In order to tackle complex issues and improve the performance of a model, ensemble learning, which is a technique, employs several classifiers. As indicated by its name, "combining a large number of decision trees on different subsets of a given dataset and taking an average to boost the projected accuracy," the Random Forest classifier considers predictions from all of the trees rather than just one. We use the most popular projections to determine the outcome. Using the used Stroke prediction dataset, Random Forest showed the greatest accuracy (criterion = entropy).

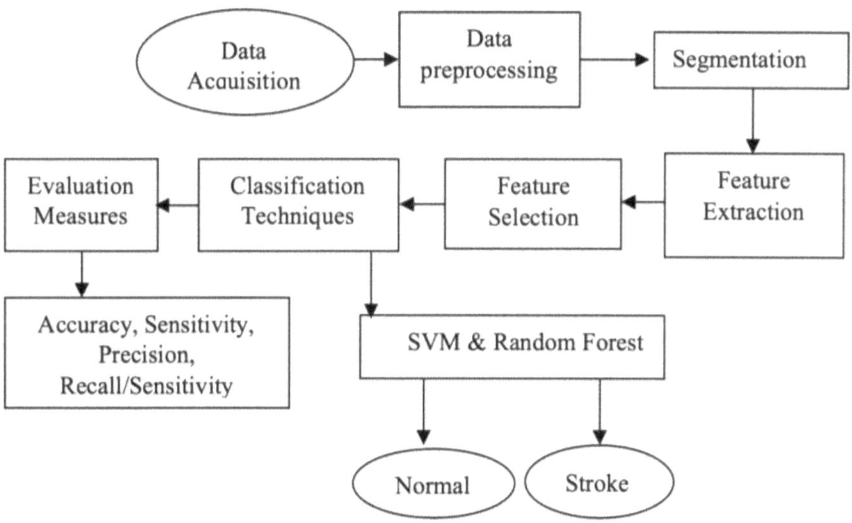

Fig. 1. Research Implementation Diagram

2.2.1 Support Vector Machine (SVM)

Support Vector Machine (SVM) is a supervised machine learning technique that classifies unknown data using labelled data and is applied to both regression and classification. Decision planes also known as hyper planes that are used to express decision boundaries. A hyper plane in N-dimensional space that is capable of dividing data points into distinct classes within the feature space. In order to classify the data, SVM develops a function having least amount of error, greatest possible margin and assigns each data point to the correct label. The RBF kernel's primary goal is to do calculations in any d-dimensional space where $d > 1$. Regression and classification using the radial basis (RB) kernel are

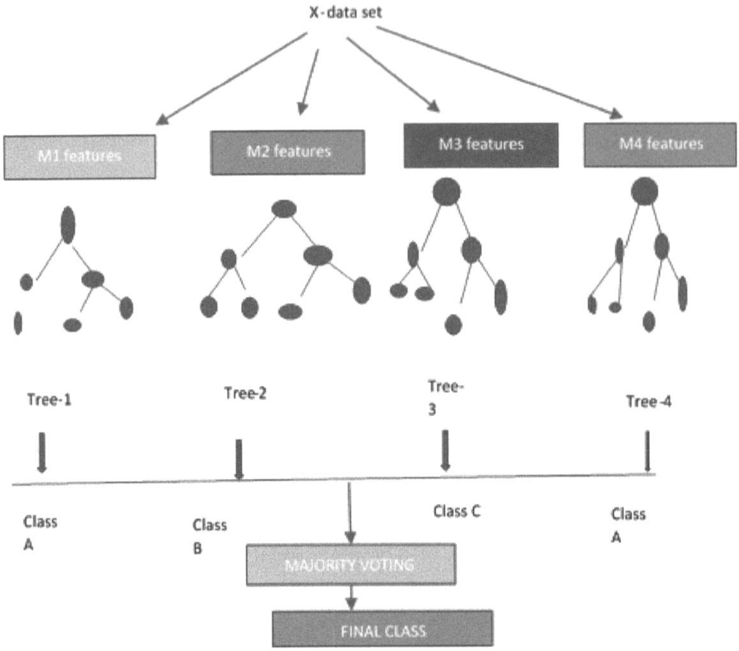

Fig. 2. Random Forest Analysis

indefinitely powerful since it employs an exponent and the expansion of results in a polynomial equation of unlimited power.

Support Vector Machines (SVM) can be divided into two primary components based on the kind of decision boundary:

Linear SVM: Linear SVMs divide the data points of various classes employing a linear decision boundary. Linear SVMs are ideal when the data can be divided precisely in a linear fashion. This means that the data points can be completely separated into their respective classes by a single straight line in two dimensions or a hyperplane in higher dimensions. The decision boundary is a hyperplane that optimizes the margin between the classes.

Non-linear SVM: When a straight line cannot split data into two classes (as in 2D), non-linear SVM can be used to categorize the data. Nonlinear SVMs may handle nonlinearly separable data by utilizing kernel functions. These kernel functions convert the initial input data into a higher-dimensional feature space that allows for the linear separation of the data points. In this modified space, a nonlinear decision boundary is found using a linear SVM. Figure 2 shows the hyperplane equations and separation distances. The class1 and class2 support vectors are separated by the hyperplane and some of the support vectors are present near the planes.

Support Vector Machines (SVMs) are frequently employed in machine learning for classification tasks, including medical applications such as identifying or forecasting the occurrence of brain strokes. Let's review the fundamentals of the SVM mathematical

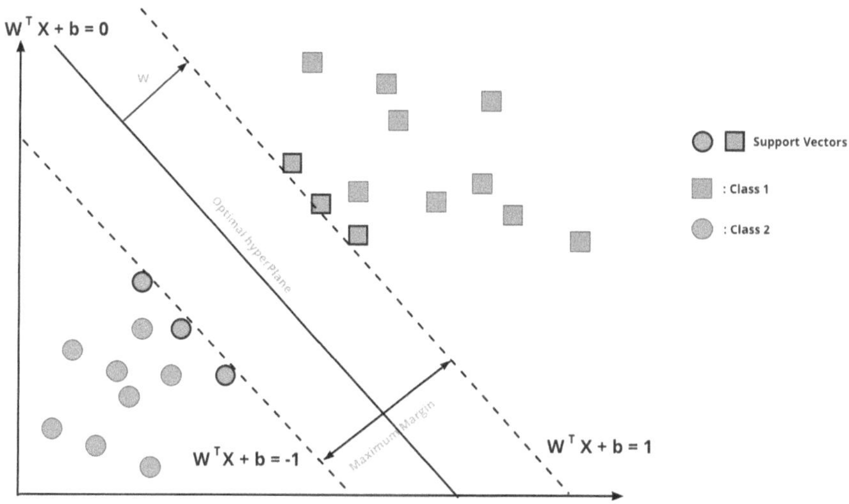

Fig. 3. SVM Analysis with support vectors

formula and how a stroke prediction algorithm might use it. SVM seeks to identify a hyperplane that divides data points into two classes with the greatest margin for a binary classification task (such as predicting a stroke versus not having a stroke). The hyperplane's equation can be used to characterize the decision boundary:

$$f(x) = w^T x + b \tag{1}$$

where: $w^T x + b$ is the weight vector (which determines the orientation of the hyperplane

- x is the feature vector (e.g., patient attributes like age, blood pressure, cholesterol levels, etc.),
- b is the bias term (which adjusts the position of the hyperplane).

For a data point x_i is the classification decision rule is:

$$Predict\ stroke = \Big\{ 1, \text{if } w^T x_i + b > 0 \\ 0, \text{if } w^T x_i + b < 0 \Big\} \tag{2}$$

2.2.2 Proposed Median Filtered FRFCM Segmentation

The proposed anisotropic filtered segmentation increases noise reduction capability by utilizing an anisotropic filter to the partition matrix of the FRFCM algorithm.

The objective function of the fast and robust fuzzy c means algorithm [14, 15] is given by

$$J_s = \sum_{j=1}^{c} \sum_{k=1}^{q} \left[P_{jk}{}^{m} \| \lambda_j - v_k \| \right] + \sum_{j=1}^{c} \sum_{k=1}^{q} G_{jk} \tag{3}$$

From Eq. (2), it is evident that P_{jk} represents the degree of membership of gray value j in cluster k, y_j is a gray level, q represents the gray levels confined in y. Then the fuzzy factor is given by

$$G_{jk} = \sum\nolimits_{r \in N_v, j \neq k} \frac{1}{d_{jk} + 1} \left(1 - P_{jk}\right)^m \|x_j - v_k\|^2 \quad (4)$$

where

$$P_{jk} = \frac{1}{\sum_{i=1}^{c} \left(\frac{\|\lambda_j - v_k\|^2 + G_{jk}}{\|\lambda_j - v_i\|^2 + G'_{jk}} \right)^{\frac{1}{m-1}}} \quad (5)$$

Now applying the morphological reconstruction as dilation and erosion, the new image is obtained as

$$\Upsilon_l = r_e^c L \quad (6)$$

where r_e^c is the morphological closing which is meant for noise elimination and k denotes an original image and considering closing we have

$$r_e^c L = r^{\mu} \alpha r_k^{\alpha} \mu L \quad (7)$$

where m represents erosion operation, α represents dilation operation. The filtering competence increases with reconstruction operation. Now, with closing operation the objective function is can be mentioned as

$$J_s = \sum\nolimits_{j=1}^{c} \sum\nolimits_{k=1}^{q} \left[P_{jk}{}^m \|L_j - v_k\|^2 \right] + \sum\nolimits_{j=1}^{c} \sum\nolimits_{k=1}^{q} G_{jk} \quad (8)$$

The median filter is employed to the partition matrix P, we employed a median filter to preserve the noise reduction capability of the FRFCM algorithm. The new membership partition matrix is given by

$$\mathrm{P}^1 = median\left[\mathrm{P}_{jk}\right] \quad (9)$$

The FRFCM segmentation is applied to preserve the noise reduction capability and detection of exact affected areas of the stroke in the brain. The performance will not be adequate for the purpose of classification, which is treated as ground truth images for classification.

3 Dataset

This study's objective is to categorize brain photos in jpg format into two groups: stroke and normal shown in Fig. 3. The dataset was taken from the openly accessible Kaggle website [13]. There were 335 healthy photos and 179 stroke cases for the test category, 230 healthy images and 156 stroke cases for the training category, and 983 healthy images and 510 stroke cases for the training category. This dataset was split into two 80% and 20% groups (train and test). Figure 4 displays the images that were part of the data set. The data attributes for brain stroke shown in Fig. 5.

The brain stroke existence is identified by Body Mass Index (BMI), which indicates obesity, which is the important risk factor for stroke. Figure 5 shows the BMI for the brain stroke.

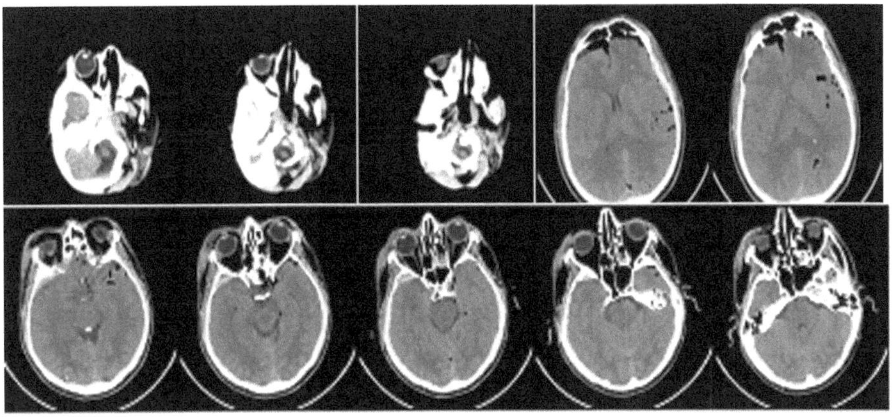

Fig. 4. Brain stroke CT images

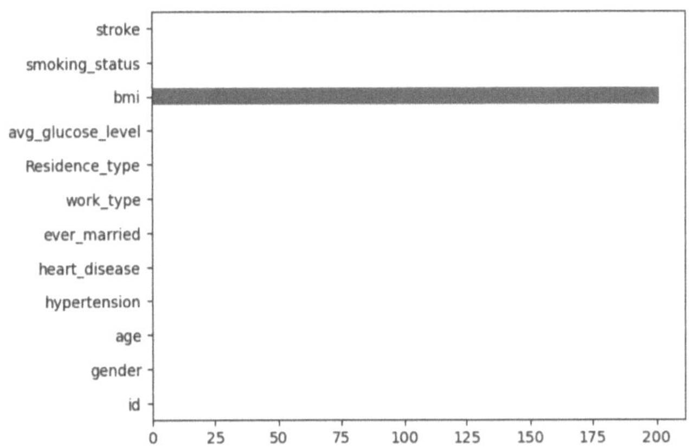

Fig. 5. Data Attributes for Brain Stroke

4 Result and Discussion

4.1 Segmentation Results

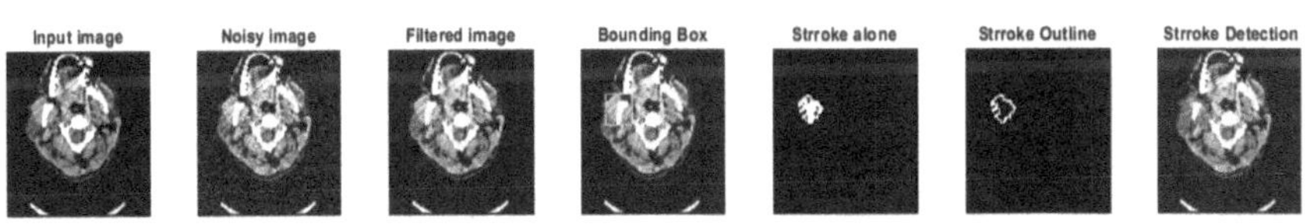

Fig. 6. EnFCM Segmentation result

Figure 6 shows the enhance FCM (EnFCM) Segmentation result and Fig. 7 shows FRFCM Segmentation result. The segmentation performance measure is presented in Table 1.

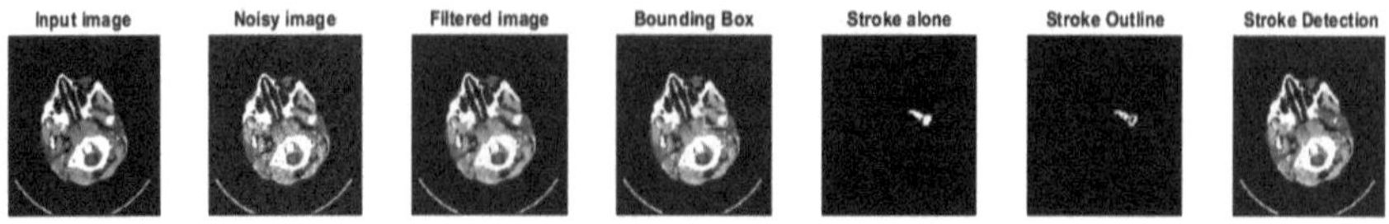

Fig. 7. FRFCM Segmentation result

Table 1. Segmentation Performance Measure

	FCM	EnFCN	NDFCM	FRFCM	Median -FRRFCM
Accuracy	0.9896	0.9913	0.9924	0.9931	0.9911
Sensitivity	0.9806	0.9802	0.9806	0.9802	0.9860
Precision	0.9572	0.9735	0.9407	0.9735	0.9741
F-measure	0.9687	0.9769	0.9602	0.9769	0.9800
MCC	0.9623	0.9712	0.9524	0.9710	0.9727
Dice	0.9687	0.9769	0.9602	0.9769	0.9812
Jaccard	0.9394	0.9558	0.9335	0.9548	0.9614
Specificity	0.9878	0.9916	0.9865	0.9906	0.9927

4.2 Evaluation

We assess prediction results using the following metrics: accuracy, precision, recall, and F1-score. A statistic called accuracy shows how accurate a model prediction is across all parameters. It is expressed as the model's proportion of accurate predictions. Accuracy, which may be measured as the ratio of accurate predictions to total predictions, is one of the simplest categorization metrics to utilize. The formula for calculating the Accuracy is given below:

$$Accuracy = (TP + TN)/(TP + TN + FP + FN) \tag{10}$$

Precision: The percentage of positive predictions that were truly accurate is determined by precision. The True Positive or forecasts that are truly true to the overall positive predictions (True Positive and False Positive) can be used to compute it.

The formula for calculating the Precision is given below:

$$Precision = TP/(TP + FP) \tag{11}$$

Recall /Sensitivity: Depending on whether they were accurately or incorrectly projected as positive or negative, it can be described as a True Positive, or projections that are true to the total number of positives, either as true positives or false negatives.

The formula for calculating the Recall / Sensitivity is given below:

$$Recall/Sensitivity = TP/(TP + FN) \tag{12}$$

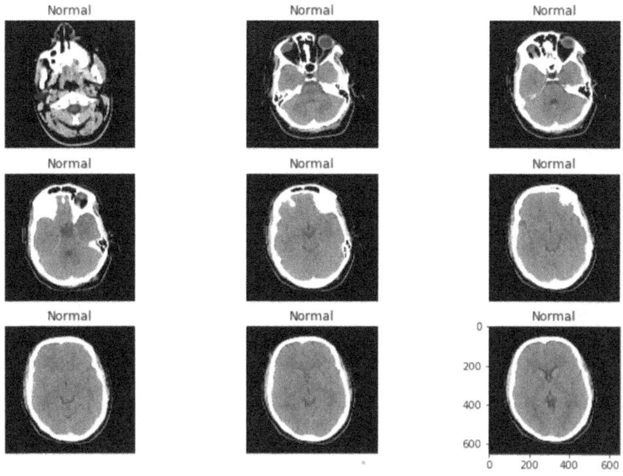

Fig. 8. Brain Stroke classification using Random Forest

The SVM classifier is robust to the overfitting and obtain high accuracies. Random Forest is simple and Interpretability is good. Also, it is good for complexity reduction. SVM is influential for handling high-dimensional, whereas, Random Forest is flexible and adapts well to different distributions. Figure 8 shows the Brain Stroke classification using Random Forest classifier.

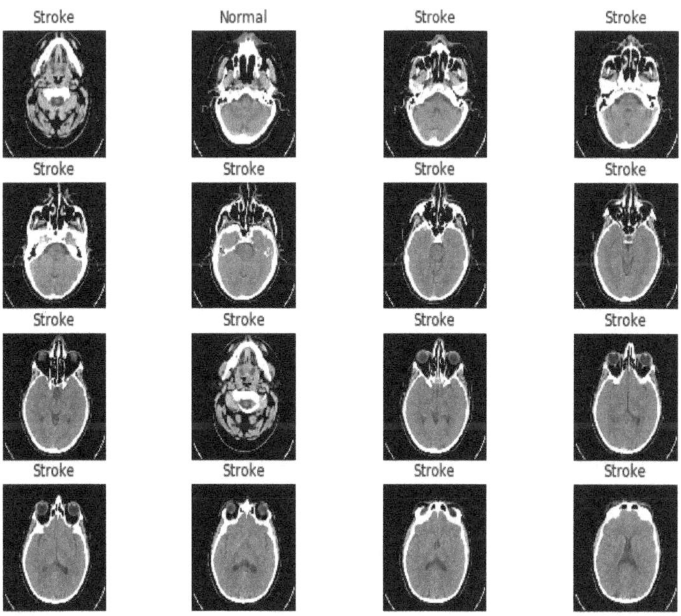

Fig. 9. Brain Stroke classification using SVM

The ML models SVM and Random Forest are feature based classifiers and the classification complexity is less in comparison to the deep learning models. The deep learning models needs complex calculations like convolution, pooling, stride etc., and needs more memory requirement. The simple calculation and less complex calculation in SVM, and Random Forest. We are motivated to use these ML models for classification. Figure 9 shows the classification using SVM classifier.

Also we have taken CNN model for the classification of types of brain stroke, and for the purpose we have considered the dataset from Kaggle [16]. The programs are simulated on the Googlecolab. The dataset contains ischaemic (stroke) and hemorrhagic (stroke) CT images. The classification of type of stroke is shown in Fig. 10.

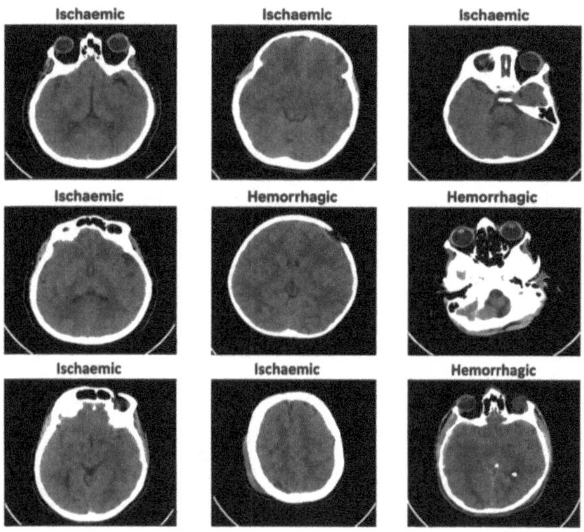

Fig. 10. Brain Stroke classification using CNN

5 Conclusion

This paper proposes automatic detection of stoke by utilizing median filtered FRFCM (Fast and Robust Fuzzy C Means) for segmentation. The FCM, EnFCM, NDFCM, FRFCM segmentations are employed on images and the results are compared with the Proposed median filtered FRFCM segmentation. The median filter was employed to the partition matrix of the FRFCM segmentation to increase the segmentation capability. The results of the proposed median filtered FRFCM is compared with the FCM, Enhanced FCM, and NDFCM, and the results are presented. The different performance measures are calculated and presented to show the capability of segmentation of the proposed median FRFCM segmentation better. The median filtered FRFCM segmentation technique was proposed for detection of strokes. The proposed median filtered FRFCM segmentation performed better than the conventional segmentations by achieving higher segmentation accuracy. The segmented images are fed as input to the Support vector machines (SVM) and Random forest for classification of the strokes from the images.

The brain stroke images are collected from the Kaggle dataset for this research. The proposed FRFCM segmentation achieved an accuracy of 99.11% and the random forest achieved 98.88% classification accuracy. The comparison results for segmentation and classifications are presented. Further, the Random forest shows better than the other classifiers and the clinical trial will be the future extension of the research.

References

1. Feigin, V., Brainin, M, Norrving, B., Martins, S., Sacco, R..: World stroke organization (WSO): Global stroke fact sheet. Inter. J. Stroke **17**(1), 18–29 (2022)
2. You, W., Henneberg, R., Henneberg, M.: Healthcare services relaxing natural selection may contribute to increase of dementia incidence. Sci. Rep.**12**(1), 1–10(2022)
3. Kumar, G., Alqahtani, H.: Deep learning-based cancer detection-recent developments, trend and challenges. In: CMES - Computer Modeling in Engineering and Sciences, vol. 130(3), pp. 1271–1307 (2021). https://doi.org/10.32604/cmes.2022.018418, ISSN 1526–1492
4. Zhang, C., Guo, X., Guo, X., Molony, D., Li, H.: Machine learning model comparison for automatic segmentation of intracoronary optical coherence tomography and plaque cap thickness quantification. CMES-Comput. Model. Eng. Sci. **123**(2), 631–646 (2020)
5. Lo, C., Hung, P., Hsieh, K.: Computer-aided detection of hyperacute stroke based on relative radiomic patterns in computed tomography. Appl. Sci. **9**(8), 1668 (2019)
6. Wu, G., Lin, J., Chen, X., Li, Z., Wang, Y.: Early identification of ischemic stroke in noncontrast computed tomography. Biomed. Signal Process. Control **52**, 41–52 (2019)
7. Babutain, K., Hussain, M., Aboalsamh, H, Al-Hameed, M.: Deep learning-enabled detection of acute ischemic stroke using brain computed tomography images. Inter. J. Adv. Comput. Sci. Appli. **12**(12) (2021)
8. Phong, T.D., Duong, H.N., Nguyen, H.T., Trong, N.T., Nguyen, V. H.: Brain hemorrhage diagnosis by using deep learning, pp. 34–39 (2017)
9. Subudhi, A., Dash, M., Sabut, S.: Automated segmentation and classification of brain stroke using expectation–maximization and random forest classifier. Biocybernet. Biomed. Eng. **40**(1), 277–289 (2020)
10. Ortiz-Ramón, R., Hernández, M.D.C.V., González-Castro, V., Makin, S., Armitage, P.A.: Identification of the presence of ischaemic stroke lesions by means of texture analysis on brain magnetic resonance images. Comput. Med. Imaging Graph. **74**, 12–24 (2019)
11. Han, T., Nunes, V.X., Souza, L.F.D.F. , Marques, A.G., Silva, I.C.L.: Internet of medical things–based on deep learning techniques for segmentation of lung and stroke regions in CT scans. IEEE Access **8**, 71117–71135 (2020)
12. Tajbakhsh, N., Shin, J.Y., Gurudu, S.R., Hurst, R.T., Kendall, C.B.: Convolutional neural networks for medical image analysis: Full training or fine tuning? IEEE Trans. Med. Imaging **35**(5), 1299–1312 (2016)
13. Zhu, G., Chen, H., Jiang, B., Chen, F., Xie, Y.: Application of deep learning to ischemic and hemorrhagic stroke computed tomography and magnetic resonance imaging Seminars in Ultrasound. CT and MRI **43**(2), 147–152 (2022)
14. Mishra, S., Sahu, P., Senapati, M.R.: MASCA–PSO based LLRBFNN model and improved fast and robust FCM algorithm for detection and classification of brain tumor from MR image. Evol. Intel. **12**, 647–663 (2019). https://doi.org/10.1007/s12065-019-00266-x
15. Pattnaik, R.K., Siddique, M., Mishra, S., Gelmecha, D.J., Singh, R.S., Satapathy, S.: Breast cancer detection and classification using metaheuristic optimized ensemble extreme learning machine. Inter. J. Inform. Technol. **15**(8), 4551–45635 (2023)
16. https://www.kaggle.com/api/v1/datasets/download/noshintasnia/brain-stroke-prediction-ct-scan-image-dataset?dataset_version_number=1

Epileptic Seizure Detection Using Hybrid Transformer and Capsule Networks

Priya Das[1], Sarita Nanda[1(✉)], and Ganapati Panda[2]

[1] School of Electronics Engineering, Kalinga Institute of Industrial Technology, Bhubaneswar, India
snandafet@kiit.ac.in

[2] School of Electronics Engineering, C. V. Raman Global University, Bhubaneswar, India

Abstract. Epileptic seizure is affecting the lives of millions of people worldwide. Timely and accurate seizure detection can help improving the quality of life of the epileptic patients and developing of responsive neurostimulation treatments. This paper presents a novel epileptic seizure detection model based on a hybrid architecture combining a Transformer and a Capsule Network (Transformer-CapsNet), trained and tested using the "CHB-MIT scalp EEG dataset". Empirical results demonstrate that the hybrid architecture achieves higher accuracy and robustness in seizure classification compared to traditional deep learning models. The performance of the suggested model is evaluated using multiple metrics, including G-Mean, recall, precision, and F1-score. The simulation based results indicate that the Transformer-CapsNet model surpasses conventional "Convolutional Neural Network (CNN) and Long Short-Term Memory (LSTM)" network models in G-Mean, precision, and robustness, offering valuable insights into seizure detection using EEG data.

Keywords: Transformer network · Capsule network · CNN · LSTM · EEG Signal · CHB-MIT dataset · Deep Learning

1 Introduction

Epilepsy is a prevalent neurological condition characterized by recurrent seizures caused by abnormal brain activity [1]. Epileptic seizures can vary in severity and often result in significant challenges in everyday life. Automatic and reliable seizure detection from Electroencephalogram (EEG) recordings plays a crucial role in improving patient care, as it allows clinicians to closely monitor seizure activity and optimize treatments.

Traditional seizure detection methods rely on manual visual inspection of EEG signals. However, modern approaches leverage machine learning (ML) [2, 3] and deep learning (DL) techniques to enhance the performance in detecting seizures. Aayesha et al. [4] introduced a framework combining fuzzy and traditional ML algorithms, such as KNN and Fuzzy Rough Nearest Neighbor (FRNN), for distinguishing seizure and non-seizure states. The models demonstrated high performance on two publicly available datasets, with intracranial EEG data achieving superior sensitivity and specificity

A. Das et al. (Eds.): MIDAS 2025, CCIS 3055, pp. 182–194, 2027.
https://doi.org/10.1007/978-3-032-35516-4_15

compared to scalp EEG signals. However, the lack of focus on real-time application limits its generalizability. Shah et al. [5] improved the performance of seizure detection by incorporating feature extraction technique, "Discrete Wavelet Transform (DWT)" and classified the features using "Random Neural Network (RNN)". They achieved accuracy of 93.27%, however; the approach was prone to information leakage when EEG data from multiple patients was combined. In paper [6], the authors proposed a "Supervised Deep Convolutional Autoencoder (SDCAE)" with a Bi-LSTM classifier to detect epileptic seizures and achieved high performance using 4-s EEG segments. This study [7] presented two patient-independent CNN architectures for seizure prediction, achieving accuracy of 88.81%.

The authors of paper [8] used recurrence plot (RP) images to transform EEG signals into 2D representations for seizure detection with CNNs, achieving accuracies of 93% (Bonn dataset) and 85% (CHB-MIT dataset). The study evaluated parameters like recurrence threshold and embedding dimensions, identifying delta rhythms as most suitable for classification. However, the dependency on high-quality RP images and limited exploration of real-time application pose constraints. Various ensemble [9, 10] and hybrid models [11, 12] based on CNN, LSTM and other deep learning techniques are also explored. The authors [13] applied meta-heuristic algorithms for feature selection and parameter tuning to achieve high accuracy of 98.67% and 98.73% for binary and multi-class classification, respectively, however, the model's complexity might hinder scalability in real-time applications.

Despite the fact that DL models perform well for signal classification, Transformer networks [14] excel in EEG signal classification for capturing long-range dependencies and relationships across time and channels through the self-attention mechanism, which traditional DL methods like CNNs and RNNs struggle to achieve. They also support parallel processing, enabling efficient handling of high-dimensional EEG data. Additionally, their dynamic attention mechanism enhances feature prioritization, improving classification accuracy in noisy or complex EEG scenarios. Capsule Networks [15] preserve spatial hierarchies and relationships between features, enabling better representation of inter-channel dependencies and brain region interactions. Their dynamic routing mechanism makes them robust to noise and transformations, crucial for handling noisy and variable EEG data. In this paper, a novel Transformer-Capsule Network (Transformer-CapsNet) model is proposed, leveraging the advantages of both Transformers and Capsule Networks to enhance the robustness to noise, handle both temporal and spatial patterns, and provide improved interpretability by extracting complex features from EEG data. We evaluate this model on the "CHB-MIT Scalp EEG dataset" and compare its performance with CNN and LSTM models, two popular deep learning architectures used for time-series analysis. This paper aims to:

- Develop a hybrid framework that combines Transformer and Capsule Networks to detect epileptic seizures.
- Improve the accuracy of analyzing long-duration EEG scalp recordings.
- Assess the effectiveness of the presented method by comparing it with CNN and LSTM architectures using a benchmark dataset.

The remaining paper is structured as follows. Description of the dataset and the suggested method is given in Sect. 2. Section 3 highlights the findings of the simulation-based experiment, and Sect. 4 summarizes the findings and suggests the future scope.

2 Methodologies Used

2.1 Dataset Description

The "CHB-MIT Scalp EEG dataset" [16] includes EEG recordings from 22 pediatric patients, aged between 1.5 to 22 years, who suffered from drug-resistant seizures. The dataset includes 23 cases, with signals sampled at a frequency of 256 Hz and a resolution of 16 bits. Each recording typically contains 23 EEG channels, though some cases include 24 or 26 channels. For analysis, 4096 s (approximately 68 min) of data were extracted for both preictal and ictal states, ensuring equal representation of both conditions. In the dataset, '0' indicates non-seizure and '1' represents seizure states.

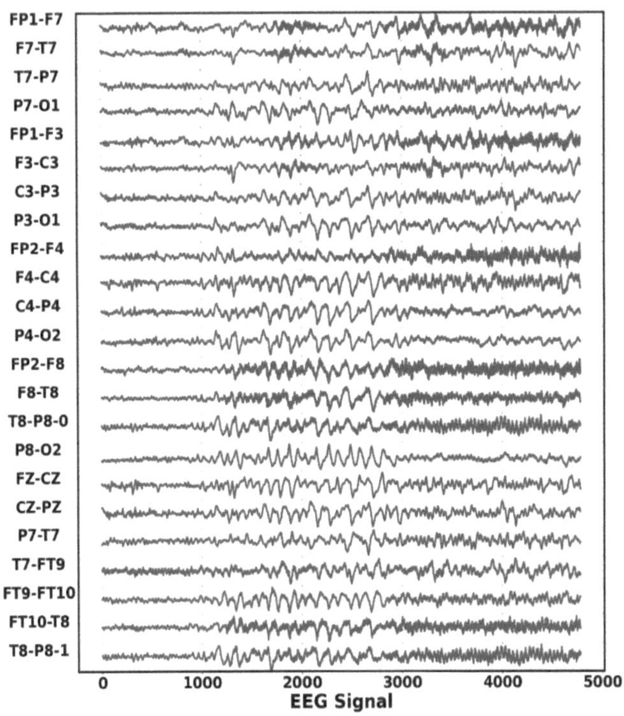

Fig. 1. Multi-channel EEG seizure signal visualization for Subject 1, representing data from 23 electrodes recorded during a seizure event.

2.2 Proposed Hybrid Transformer with Capsule Network

Preprocessing. Figure 1 shows all 23 channels of the seizure signal collected from subject 1. Figure 2 shows the architecture of the suggested method. The first stage of the proposed model is preprocessing where the EEG data undergoes through a band-pass filter (0.5–50 Hz) to mitigate noise and artifacts present in the EEG data, then normalization and reshaping are performed to ensure compatibility with the neural network. Normalization is performed for ensuring that features are on a similar scale, preventing the network from being biased toward features with larger values. The EEG signals are then reshaped into a three-dimensional format, (samples, time steps, features). This 3D representation is essential because the Transformer encoder requires temporal data as sequential input, where "time steps" capture temporal dependencies, and "features" represent the attributes of the signal.

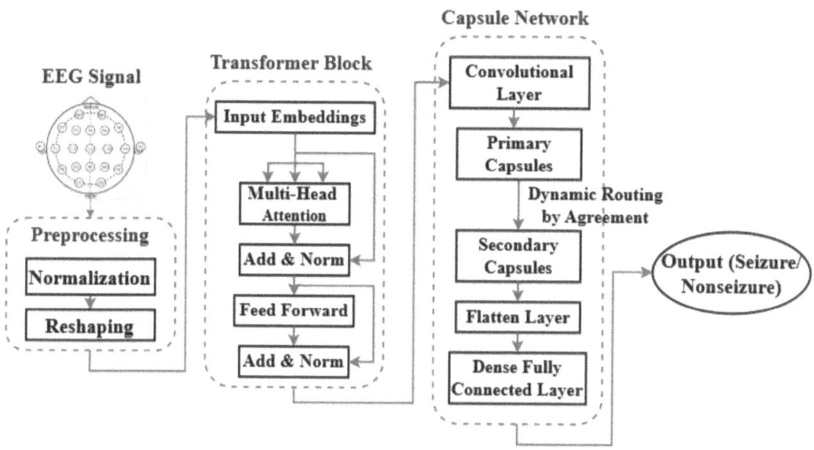

Fig. 2. Architecture of the proposed Hybrid Transformer Capsule Network for the detection of Epileptic Seizure

Transformer Module. The Transformer encoder module is designed to extract global temporal and feature dependencies. This module includes - "multi-head self-attention (MSA), feedforward neural networks (FFN), and layer normalization (LN)". The MSA mechanism first computes attention scores by analyzing dependencies among input features, the model can prioritize the most essential parts of the sequence. The attention heads work at the same time, each learning different attention patterns from the input sequence. The outputs from all attention heads are merged and undergo a linear transformation to create a unified representation, preserving diverse contextual information. The attention-weighted representations are then passed through the FFN, which applies non-linear transformations to enhance feature learning. Finally, LN is applied after both the MSA and FFN layers to normalize activations, improving training stability and convergence. The combination of these components enables effective feature extraction, robust learning, and enhanced generalization for seizure detection tasks. The attention scores between input features are computed (1).

$$Attention(Q, K, V) = softmax\left(\frac{QK^T}{\sqrt{d_k}}\right)V \quad (1)$$

The input is projected into three matrices: queries (Q), keys (K), and values (V). d_k is the dimensionality of the key vector. This mechanism allows the model to focus on relevant portions of the input signal, thereby recording temporal context across wide intervals of the EEG data. Once the attention features are computed, they are passed through an FFN with two fully connected layers and a Rectified Linear Unit (*ReLU*) activation function. The transformation is represented as:

$$FFN(x) = ReLU(xW_1 + b_1)W_2 + b_2 \tag{2}$$

where W_1, W_2 and b_1, b_2 are trainable weights and biases. This step introduces non-linear transformations to capture complex features in the data. Layer normalization *(LN)* enhances the stability of training and accelerates convergence (3).

$$LN(x) = \frac{x - \mu}{\sigma} \bullet \gamma + \beta \tag{3}$$

where μ and σ are the mean and standard deviation of x, and γ, β are trainable parameters for adaptive scaling and shifting. *LN* helps to reduce the risk of exploding or vanishing gradients. To preserve the original input information and facilitate better gradient flow, skip connections are incorporated. These connections add the input of a layer to its output before the next layer processes it, helping to mitigate vanishing gradient issues.

2.3 Capsule Network Module

The Capsule Network module is utilized for feature refinement and spatial relationship modeling. Within this module, primary capsule transformations apply a linear transformation to the input features (4).

$$u_{j|i} = W_{ij} \bullet s_i \tag{4}$$

where s_i is the input feature vector, and W_{ij} is a learnable transformation matrix. This operation ensures that the input features are projected into a vectorized representation suitable for dynamic routing. Capsules are connected via Dynamic Routing by Agreement, that dynamically adjusts the connection strengths based on agreement between capsules. Initial routing weights, b_{ij}, are transformed using a softmax function (see Eq. (5)) to compute the connection strength (c_{ij}) between input and output capsules:

$$c_{ij} = \frac{\exp(b_{ij})}{\sum_k \exp(b_{ik})} \tag{5}$$

where b_{ij} is a trainable coefficient. The final capsule outputs are computed through a non-linear "squashing" function, ensuring their lengths fall between 0 and 1. The output of the capsule network is calculated as

$$v_j = squash\left(\sum_i c_{ij} u_{j|i}\right) \tag{6}$$

where

$$squash(s) = \frac{\|s\|^2}{1 + \|s\|^2} \bullet \frac{s}{\|s\|} \tag{7}$$

This module ensures that the spatial and hierarchical relationships in the data are preserved, significantly improving classification accuracy. The final capsule outputs are fed to the dense fully connected layers.

3 Experimental Results and Discussion

The suggested method, Transformer-CapsNet is evaluated on popular standard "CHB-MIT Scalp EEG dataset". The performance metrics used in this paper are given in Table 1. "Tr_Po", "Tr_Ne", "Fa_Po" and "Fa_Ne" represent "True Positive", "True Negative", "False Positive" and "False Negative" respectively.

Table 1. Performance Metrices

Name	Definition/Equation
Accuracy	Accuracy = (Tr_Po + Tr_Ne)/(Tr_Po + Fa_Po + Tr_Ne + Fa_Ne)
Sensitivity (Recall)	Sensitivity = Tr_Po/(Tr_Po + Fa_Ne)
Specificity	Specificity = Tr_Ne/(Tr_Ne + Fa_Po)
Precision	Precision = Tr_Po/(Tr_Po + Fa_Po)
G-Mean	G-Mean = √(Sensitivity × Specificity)
F1-Score	F1-Score = (2 × Precision × Recall) / (Precision + Recall)

The architecture of the suggested method employs an input layer compatible with the reshaped EEG data. The Transformer encoder uses an embedding dimension equal to the input feature count, with eight attention heads and a feedforward dimension of 256, employing a dropout rate of 0.1. The Capsule layer includes ten capsules, each with sixteen dimensions, and employs three routing iterations. A dense output layer with softmax activation concludes the architecture for binary classification. Training is conducted with the Adam optimizer using categorical cross-entropy loss. The learning rate is set to 0.00001, with a batch size of 32 and a total of 20 epochs. This training configuration ensures the model learns effectively without overfitting or underfitting.

The values of the hyperparameters are chosen using random search method, also layer normalization and dropout techniques are incorporated to stabilize training and reduce complexity of the model. The performance is compared with two efficient neural network models: CNN and LSTM. Table 2 shows the patient-specific as well as the average accuracy, precision and recall of all the above-mentioned classifiers.

Table 2. Performance Comparison of Different Models

Subjects	Hybrid Transformer CapsNet			CNN			LSTM		
	Acc (in %)	Recall	Precision	Acc (in %)	Recall	Precision	Acc (in %)	Recall	Precision
1	99.02	0.9902	0.9902	98.94	0.9853	0.9935	98.94	0.9862	0.9937
2	97.07	0.9498	0.9881	98.16	0.9916	0.9772	98.37	0.9886	0.9730
3	99.46	0.9935	0.9957	98.97	0.9893	0.9895	98.93	0.9879	0.9913
4	99.03	0.9857	0.9951	97.38	0.9762	0.9728	97.43	0.9733	0.9739
5	98.06	0.9736	0.9879	97.78	0.9725	0.9852	97.90	0.9744	0.9815
6	97.99	0.9837	0.9752	96.54	0.9737	0.9661	96.97	0.9707	0.9598
8	98.76	0.9842	0.9909	97.59	0.9660	0.9729	96.95	0.9710	0.9807
9	99.50	0.9917	0.9981	99.46	0.9952	0.9950	99.50	0.9934	0.9965
10	99.37	0.9913	0.9961	99.12	0.9954	0.9993	99.72	0.9885	0.9938
11	96.95	0.9655	0.9731	92.81	0.9452	0.9613	95.29	0.9337	0.9221
12	95.26	0.9613	0.9424	86.60	0.8846	0.9118	90.05	0.8338	0.8875
13	98.45	0.9827	0.9861	94.74	0.9416	0.9475	94.49	0.9422	0.9519
14	99.02	0.9862	0.9937	98.98	0.9868	0.9913	98.92	0.9864	0.9927
15	90.13	0.8858	0.9145	89.12	0.9050	0.8945	89.72	0.8682	0.9068
16	98.19	0.9673	0.9956	97.48	0.9571	0.9789	96.77	0.9646	0.9823
17	97.21	0.9652	0.9794	95.12	0.9517	0.9406	94.60	0.9553	0.9471
18	95.75	0.9412	0.9723	96.91	0.9652	0.9711	96.80	0.9682	0.9702
19	99.17	0.9878	0.9957	99.27	0.9799	0.9901	98.51	0.9905	0.9948
20	99.19	0.9839	1.0000	99.04	0.9823	0.9934	98.79	0.9881	0.9940
21	93.14	0.9212	0.9403	92.27	0.9368	0.9042	91.83	0.9136	0.9304
22	98.10	0.9767	0.9849	97.60	0.9727	0.9767	97.45	0.9699	0.9816
23	94.63	0.9365	0.9552	94.28	0.9490	0.9457	94.70	0.9468	0.9382
24	99.28	0.9918	0.9933	97.41	0.9758	0.9853	98.06	0.9665	0.9806
avg	97.51	0.9694	0.9802	96.33	0.9643	0.9671	96.55	0.9592	0.9663

It proves that the suggested model consistently outperformed the other methods, achieving an average accuracy of 97.51%, recall of 0.9694, and precision of 0.9802, showcasing its effectiveness in detecting seizures. The used dataset does not have the seizure part of subject 7, hence, in this study, subject 7 is not considered. The accuracy trends for individual patients reveal that the Transformer-CapsNet consistently surpassed both CNN and LSTM, with only a few exceptions. For example, patients like chb11, chb12, and chb15 showed slightly lower accuracies, possibly due to unique EEG signal patterns or noise. Figure 3 gives the graphical representation of the comparative analysis of the suggested method, CNN and LSTM. Although CNN and LSTM performed well, Transformer CapsNet outperformed them.

Figure 4 compares the patient-specific accuracy of the classifiers for all considered subjects. The provided accuracy graph highlights the consistent performance of the Transformer-CapsNet across patients. The red line, representing the Transformer-CapsNet, remains above the blue (CNN) and green (LSTM) lines for most cases, underscoring its reliability. CNN and LSTM, while competitive, show poor performance in recall and precision, particularly for patients with complex or variable seizure patterns. Transformer-CapsNet, on the other hand, achieved remarkable results, even achieving perfect precision for chb20, reflecting its ability to confidently identify true positives. Figure 5 shows that the suggested model consistently outperforms the other models. LSTM has significant drops for patients like chb12 and chb22. CNN generally maintains stable sensitivity, though it occasionally falls below 95%, especially for patients like chb14. In Fig. 6, it can be seen that the Transformer Capsule Network again leading in performance, while LSTM struggles with patients such as chb12 and chb22. CNN experiences dips in specificity for patients like chb10 and chb14. Figure 7 and Fig. 8 show the comparative study of patient-specific G-Mean and F1-score of all three classifiers. Table 3 demonstrates that the suggested method is better than many models in terms of accuracy. In comparison to Dissanayake et al. [7], who developed patient-independent seizure prediction models using CNN and Siamese networks, the proposed model surpasses their reported accuracy of 91.54%. Siamese network addresses inter-subject variability and provides insights into classifier interpretability. The model presented by Shankar et al. [8] achieved up to 93% accuracy on the CHB-MIT dataset. This model may lose critical temporal dependencies during the image conversion process. The hybrid Transformer Capsule Network retains both spatial and temporal features natively, leading to a significantly higher accuracy of 97.51% and better generalization across patients. When compared to Usman et al. [9], who proposed an ensemble method combining CNN, LSTM, and SVM classifiers using Model-Agnostic Meta-Learning (MAML), the suggested method again demonstrates superior simplicity and efficiency. This ensemble model achieves an accuracy of 96.05%, sensitivity of 96.28% and specificity of 95.65% on the CHB-MIT dataset with increased computational complexity. Grubov et al. [10] used a two-stage approach combining Outlier Detection using One-Class Support Vector Machines (OCSVM) with CNN, the proposed method achieves significantly higher performance and reduces false positives but sacrifices recall, achieving a recall of 82% and precision of 56%. In contrast, hybrid Transformer Capsule Network achieves an average accuracy of 97.51%, recall of 96.94%, and precision of 98.02%, demonstrating a better balance across metrics.

Table 3. Comparative Analysis of the Presented Method Against Existing Models

Author and Reference	Method Used	Dataset used, Number of Patients	Accuracy
Dissanayake et al. [7]	Siamese Network	CHB-MIT Scalp EEG Dataset, 24	91.54%

(continued)

Table 3. *(continued)*

Author and Reference	Method Used	Dataset used, Number of Patients	Accuracy
Shankar et al. [8]	Recurrence Plot with CNN	CHB-MIT Scalp EEG Dataset, 7	93%
Usman et al. [9]	Ensemble of SVM, CNN, LSTM	CHB-MIT Scalp EEG Dataset, 24	96.05%
Grubov et al. [10]	one-class support vector machine (OCSVM)+CNN	Collected at National Medical and Surgical Center, Russia, 83	70.82%
Proposed	**Hybrid Transformer Capsule Network**	**CHB-MIT Scalp EEG Dataset, 23**	**97.51%**

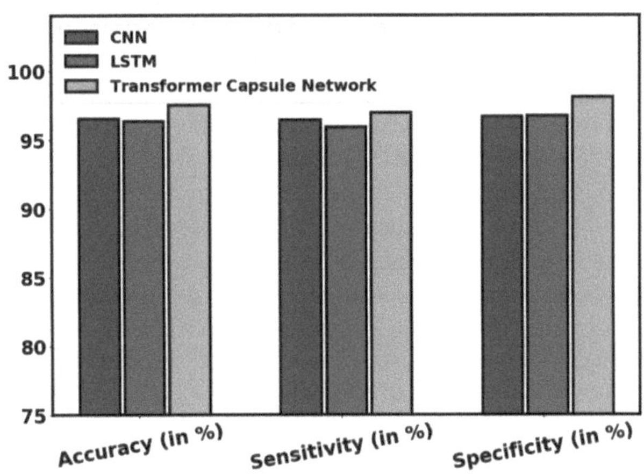

Fig. 3. Comparison of average accuracy, sensitivity and specificity for CNN, LSTM and Transformer Capsule Network

To summarize, the proposed Transformer-Capsule Network architecture provides a robust and reliable solution for epileptic seizure detection. Its ability to integrate temporal and spatial dependencies enables it to outperform traditional methods like CNN and LSTM in mentioned evaluation metrics. Although slight limitations were observed for a few patients, the overall performance underscores the potential of this hybrid approach for clinical applications.

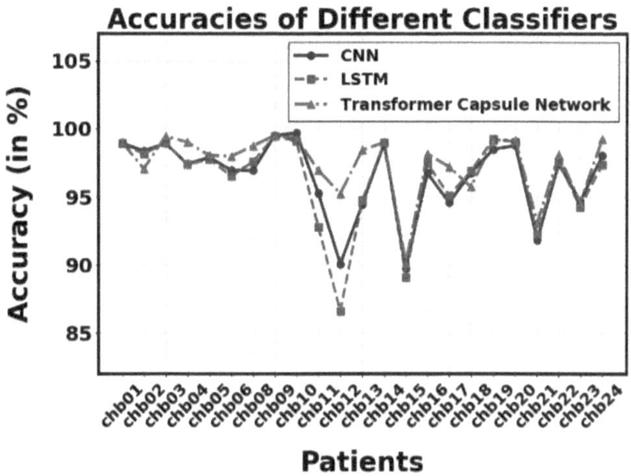

Fig. 4. Comparison of subject-specific accuracy of CNN, LSTM and Transformer Capsule Network for each patient

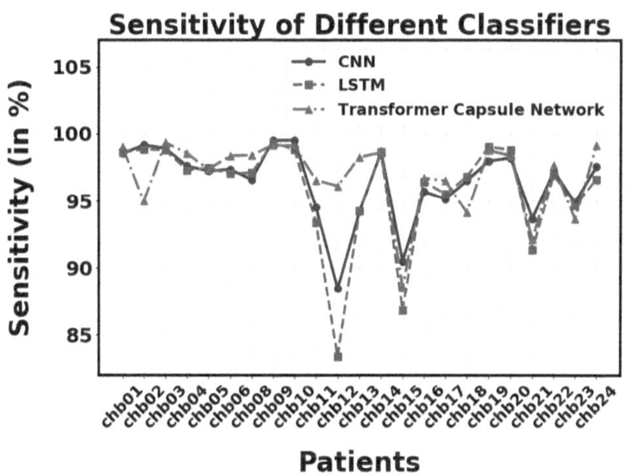

Fig. 5. Comparison of subject-specific sensitivity of CNN, LSTM and Transformer Capsule Network for each patient

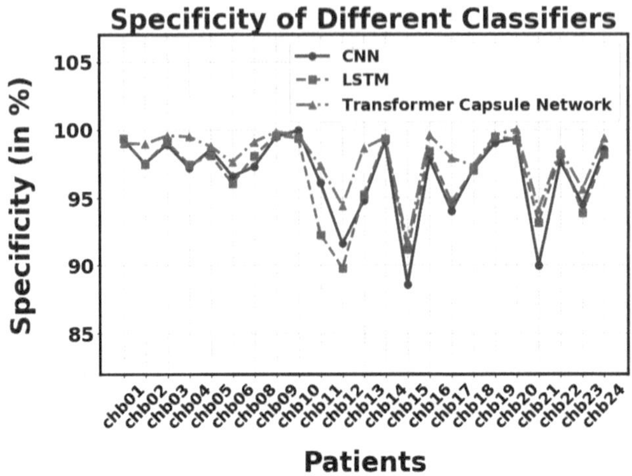

Fig. 6. Comparison of subject-specific specificity of CNN, LSTM and Transformer Capsule Network for each patient

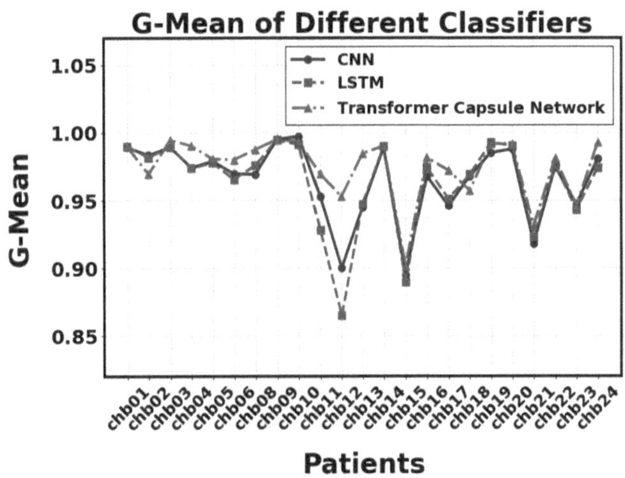

Fig. 7. Comparison of subject-specific G-Mean of CNN, LSTM and Transformer Capsule Network for each patient

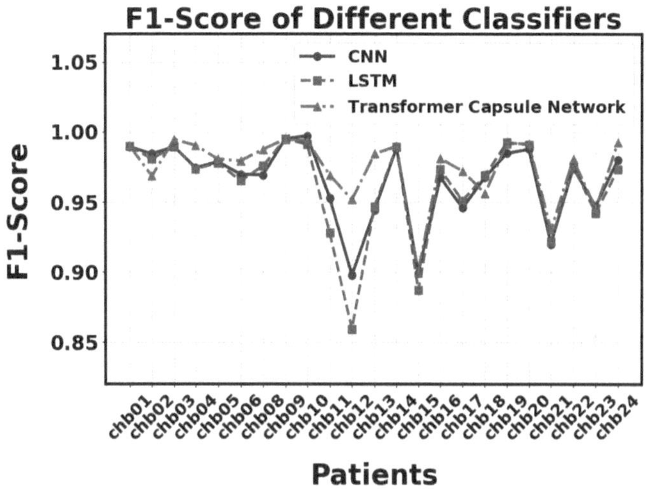

Fig. 8. Comparison of subject-specific F1-Score of CNN, LSTM and Transformer Capsule Network for each patient

4 Conclusion

In this paper, the Transformer-CapsNet is suggested for the detection of epileptic seizure using CHB-MIT dataset which achieves remarkable performance, with an average accuracy of 97.51%, G-Mean of 0.9750, recall of 0.9694, and precision of 0.9802. By combining a Transformer for the extraction of relevant features and Capsule networks for classification, it effectively captures both global and local EEG features, outperforming many existing methods. Despite its strong performance, challenges remain in enhancing preprocessing techniques to remove noises and reducing model complexity for real-time applications. Future research will focus on generalizing the Transformer-CapsNet model for cross-subject scenarios to address inter-subject variability. Additionally, further investigations should explore the use of larger and more diverse datasets to enhance model generalizability, evaluate the model's performance in real-world clinical environments to ensure practical applicability, and optimize computational efficiency for real-time deployment in clinical settings.

References

1. Epilepsy facts and statistics. https://www.healthline.com/health/epilepsy/facts-statistics-infographic#Types. Accessed 16 Nov 2024
2. Das, P., Manikandan, M.S., Ramkumar, B.: Detection of epileptic seizure event in EEG signals using variational mode decomposition and mode spectral entropy. In: IEEE 13th International Conference on Industrial and Information Systems (ICIIS), pp. 42–47. IEEE (2018)
3. Das, P., Nanda, S.: A novel multivariate approach for the detection of epileptic seizure using BCS-WELM. Int. J. Inf. Technol. **15**(1), 149–159 (2023)
4. Aayesha, Qureshi, M.B., Afzaal, M., Qureshi, M.S., Fayaz, M.: Machine learning-based EEG signals classification model for epileptic seizure detection. Multimedia Tools Appl. **80**(12), pp.17849–17877 (2021)

5. Shah, S.Y., Larijani, H., Gibson, R.M., Liarokapis, D.: Epileptic seizure classification based on random neural networks using discrete wavelet transform for electroencephalogram signal decomposition. Appl. Sci. **14**(2), 599 (2024)
6. Abdelhameed, A., Bayoumi, M.: A deep learning approach for automatic seizure detection in children with epilepsy. Front. Comput. Neurosci. **15**, 650050 (2021)
7. Dissanayake, T., Fernando, T., Denman, S., Sridharan, S., Fookes, C.: Deep learning for patient-independent epileptic seizure prediction using scalp EEG signals. IEEE Sens. J. **21**(7), 9377–9388 (2021)
8. Shankar, A., Khaing, H.K., Dandapat, S., Barma, S.: Analysis of epileptic seizures based on EEG using recurrence plot images and deep learning. Biomed. Signal Process. Control **69**, 102854 (2021)
9. Usman, S.M., Khalid, S., Bashir, S.: A deep learning based ensemble learning method for epileptic seizure prediction. Comput. Biol. Med. **136**, 104710 (2021)
10. Grubov, V.V., et al.: Two-stage approach with combination of outlier detection method and deep learning enhances automatic epileptic seizure detection. IEEE Access (2024)
11. Jiwani, N., Gupta, K., Sharif, M.H.U., Adhikari, N., Afreen, N.: A LSTM-CNN model for epileptic seizures detection using EEG signal. In: 2nd International Conference on Emerging Smart Technologies and Applications (eSmarTA), pp. 1–5. IEEE (2022)
12. Nayak, S.K., Pattnaik, S., Siddique, M., Garanayak, M., Paikaray, B.K.: A hybrid CNN-LSTM approach to enhancing temperature forecasting for environmental threats and risk management. Int. J. Bus. Continuity Risk Manage. **14**(4), 371–391 (2024)
13. Hilal, A.M., et al.: Intelligent epileptic seizure detection and classification model using optimal deep canonical sparse autoencoder. Biology **11**(8), 1220 (2022)
14. Guo, J.Y., et al.: A transformer based neural network for emotion recognition and visualizations of crucial EEG channels. Physica A **603**, 127700 (2022)
15. Guarda, L., Tapia, J.E., Droguett, E.L., Ramos, M.: A novel Capsule Neural Network based model for drowsiness detection using electroencephalography signals. Expert Syst. Appl. **201**, 116977 (2022)
16. Goldberger, A.L., et al.: PhysioBank, PhysioToolkit, and PhysioNet: components of a new research resource for complex physiologic signals. Circulation **101**(23), e215-e220 (2000)

PCOS Prediction from Medical Imaging Data: Analyzing CNN Deep Learning Models Comparatively Using Elastic Deformation

Lalit Kumar Behera[1(✉)] and Satya Narayan Tripathy[2]

[1] Centurion University of Technology and Management, Bhubaneswar, India
lkb.rs.cs@buodisha.edu.in

[2] Berhampur University, Berhampur, India

Abstract. Polycystic Ovary Syndrome (PCOS) is an endocrine system-related health issue that poses significant health challenges for females of reproductive age, necessitating early and accurate prediction and treatment. This study conducts an analysis of two prominent Convolutional Neural Networks (CNNs): Efficientnet and Xception comparatively for predicting PCOS from medical imaging data. Leveraging the strengths of each architecture, we aim to assess their efficacy in image classification tasks. To enhance model performance and robustness, we implement advanced image data augmentation techniques, specifically elastic deformation, which increases the variability of training images and helps prevent overfitting. Evaluation metrics include confusion matrix, accuracy and loss, Receiver Operating Characteristic curve and performance report. Results show the performance of the Efficientnet and Xception models comparatively, highlighting their strengths and weaknesses in diagnosing PCOS. The findings aim to support the development of more reliable diagnostic tools that can enhance clinical outcomes for women with PCOS.

Keywords: Deep Learning · Image Data Augmentation · Elastic Deformation · CNN · PCOS

1 Introduction

PCOS is a disorder relating to the endocrine system, in most of cases, affecting women of reproductive age, exhibiting many symptoms including menstrual cycles irregularity, infertility and metabolic issues. Accurate diagnosis at an early stage of PCOS is crucial for effective treatment and management. With the increasing availability of medical imaging technologies, particularly CNNs of Deep Learning (DL) has appeared as a powerful instrument for automating the analysis of medical images, enhancing diagnostic accuracy.

This study focuses on the comparative performance of two advanced CNN architectures - Inception and Xception in predicting PCOS from imaging data. Both models are well known for their strength to extract complex features from images due to their unique architectural designs. Inception employs many convolutional filters at various

A. Das et al. (Eds.): MIDAS 2025, CCIS 3055, pp. 195–205, 2027.
https://doi.org/10.1007/978-3-032-35516-4_16

sizes within a single layer, capturing different spatial hierarchies. Xception, an extension of Inception, utilizes depth-wise separable convolutions, which effectively reduce computational complexity while maintaining high throughput.

To increment the robustness of these CNN models, we incorporate advanced image data augmentation techniques, specifically elastic deformation. This technique helps to enhance the variability of training data, thus reducing overfitting and improving generalization. The use of elastic deformation allows for more realistic transformations of images, mimicking natural variations found in medical imaging. The study will evaluate the throughput of both models using several key metrics, including confusion matrix, ROC and classification report etc. By analyzing these metrics, we aim to get understanding of each model's performance in diagnosing PCOS from imaging data. Through visualizations and statistical analyses, this comparative study seeks to contribute valuable insights into the use of DL techniques in the domain of medical diagnostics, ultimately supporting improved outcomes for women with PCOS. This research work presents literature review in Sect. 2. This is followed by methodology in Sect. 3 with subsections data collection, data preprocessing, model architecture and model training. Result and discussion follow in Sect. 4 and in Sect. 5 conclusion is given.

2 Literature Review

Kermanshahchi J et al. proposed a model to accurately detect multiple small follicles in the ovaries. It can also detect an increase in ovarian volume for PCOS from normal pelvic ultrasound image [1]. Shanmugavadivel K et al. made study on unsupervised classification models like Support Vector Machine (SVM), Logistic Regression and Naive Bayes and on comparison of performance of different methods using accuracy, precision, recall found SVM model the best one with accuracy of 94.44%. They also analyzed 3856 ultrasound images by CNN and VGG16 transfer learning algorithm [2].

Study made by Barrera FJ et al. on machine learning and artificial intelligence gives promising diagnostic as well as classification performance values in detecting PCOS. Use of standardized PCOS diagnostic criteria and adherence to methodological and reporting guidelines are their main focus [3]. Suha, S.A. et al. made research on classification technique using extended machine learning for PCOS diagnosis. They used stacking ensemble machine learning technique. For this, they used bagging or boosting model as meta-learner and conventional models as base learners [4].

Paramasivam GB et al. suggested a Self-Defined CNN method. Among Random Forest, SVM and Logistic Regression, they found Random Forest Classifier with accuracy being 96.43% as the most reliable one [5]. B. Purnama proposed three classification scenarios. First one is Learning Vector Quantization (LVQ) of Neural Network method, second one is KNN using Euclidean distance method and third one is SVM- RBF Kernel. They got the best accuracy in case of SVM-RBF Kernel [6].

Poorani, B et al. introduced thresholding on multiscale gradient-weight with sum of product fusion methodology. It helps to precisely recognize boundaries with smaller region at lower resolutions [7]. Elmannai H et al. proposed about mixing REF selection with the Stacking ML and obtained 100% accuracy in comparison to other models [8]. Guixue, G. et al. worked on making review for the use of prediction models in medicals to diagnose PCOS in patients [9]. Khanna V.V. studied to detect PCOS correctly in patients. They proposed an explainable machine learning method to help medical domain professionals [10]. Behera L.K. et al. made their study on application of deep learning in handling medical image data of epidermis [11]. B. Paikaray et al. came up with a detailed overview of deferment techniques in medical images and their performance aiming at ROI preservation [12].

3 Methodology

The methodology for predicting PCOS from imaging data using a comparative analysis of Efficientnet and Xception CNNs with elastic deformation as an augmentation technique includes many important steps. This section gives an outline of the process starting from data preparation to model evaluation.

3.1 Data Collection

A dataset of cervical images relevant to PCOS diagnosis acquired from Kaggle website. The website URL is www.kaggle.com/datasets/anaghachoudhari/pcos-detection-using-ultrasound-images. This dataset includes labeled images for both PCOS and non-PCOS cases, organized into separate directories for training and validation. Each folder contains separate subfolders like infected and not infected. This dataset contains ample number of images suitable for research work.

3.2 Data Preprocessing

Image Resizing is done to all images to the input size required by the CNN architectures (256 x 256) pixels for both EfficientNet and Xception. Normalization of pixel values to the range [0, 1] is done by dividing by 255. Advanced image data augmentation techniques, focusing on elastic deformation is implemented. This will help generate new training samples by applying transformations that mimic natural variations in medical imaging.

3.3 Model Architecture

The comparative study of two models EfficientNet and Xception are made in this work along with the use of elastic deformation. EfficientNet is a family of CNN models designed for various purposes out of which image classification is one, known for balancing efficiency and accuracy. This model can scale up the three dimensions like depth, width and resolution of the CNN network in a compound way, achieving promising results with fewer parameters in comparison to previous architectures. It provides a powerful framework for building CNNs that maximize accuracy while minimizing computational load. Xception is also called as Extreme Inception. This builds on the Inception module additionally using depth wise separable convolutions to enhance performance.

3.3.1 EfficientNet Model

Introduced by Google researchers in 2019, EfficientNet addresses the problem of scaling neural networks efficiently by utilizing a novel scaling method, enabling high performance with fewer features and decreased cost for computation compared to previous models as ResNet and Inception. The most significant innovation of EfficientNet is its "compound scaling" method, which adjusts three critical dimensions of a neural network: depth, width, and resolution. Traditional approaches scale these dimensions independently, but EfficientNet scales them in a balanced way to maintain performance while improving efficiency. Base model of Efficientnet is given in Fig. 1 below.

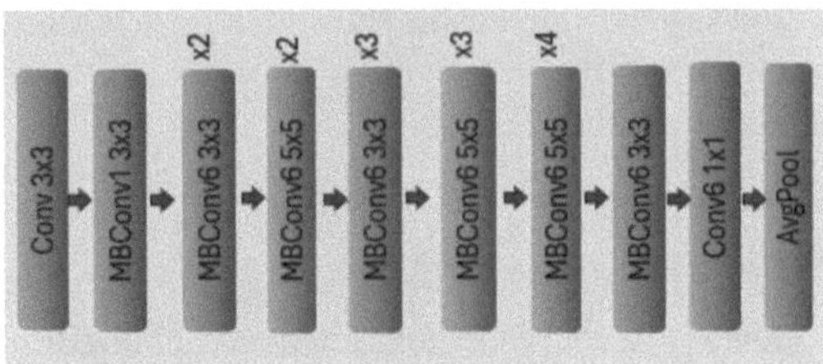

Fig. 1. Basic architecture of EfficientNet model

EfficientNet uses a compound coefficient to scale all three dimensions simultaneously, ensuring that the network grows efficiently without unnecessary computational cost. EfficientNet is built using a baseline architecture known as EfficientNet-B0. This model uses a lightweight mobile-friendly structure with Depthwise Separable Convolutions (to reduce computation), Swish activation function (which is an improvement over ReLU for better gradient flow), and Squeeze and Excitation blocks (which enhance feature representation by recalibrating channel-wise feature responses). EfficientNet models exhibit significant performance on different image classification benchmarks like ImageNet, outperforming traditional models like ResNet, Inception, and DenseNet, while being more computationally efficient. For instance, EfficientNet-B7 provides very high accuracy with far fewer parameters than the similarly performing models.

One of the standouts features of EfficientNet is its ability to achieve high performance with significantly fewer parameters. EfficientNet-B7 has over 66 million parameters, whereas older models like ResNet-152 have upwards of 60 million parameters, yet EfficientNet-B7 outperforms them on image classification tasks with far less computation. EfficientNet comes in several versions (B0 to B7), each progressively increasing in complexity and accuracy. The architecture can be chosen based on the trade-off between computational resources and desired accuracy.

The lower versions like B0, B1 are suitable for mobile and edge devices, while higher versions like B6, B7 are intended for server-side or cloud applications with greater computational power. EfficientNet has been successfully applied in various domains. For Image Classification, it is used as one of the leading architectures for image recognition tasks. In Object Detection, by serving as a backbone for models like EfficientDet (an object detection framework), its role is significant. Due to its efficiency, EfficientNet is also widely used in medical image analysis, where large datasets and high accuracy are important. Its smaller, more efficient models are suited for edge devices like smartphones and IoT devices for Edge computing.

3.3.2 Xception Model

Xception introduced by François Chollet in 2017. It is an extension of the Inception architecture, specifically designed to improve upon the efficiency and performance of convolution operations. Xception is based on the idea of depth wise separable convolutions, a more computationally efficient form of convolution and it has shown superior performance on image classification tasks, particularly when compared to earlier architectures like Inception and VGG. Xception's core innovation is its use of depth- wise separable convolutions as shown in Fig. 2, which are designed to decrease cost of computation. Regular convolutions apply a single filter to all input channels. But depth-wise separable convolutions split the operation into two stages:

- Depth-wise Convolution: Each input channel or feature map is applied with separate filter.
- Pointwise Convolution: The outputs of the depth-wise convolution are summed with a 1×1 convolution.

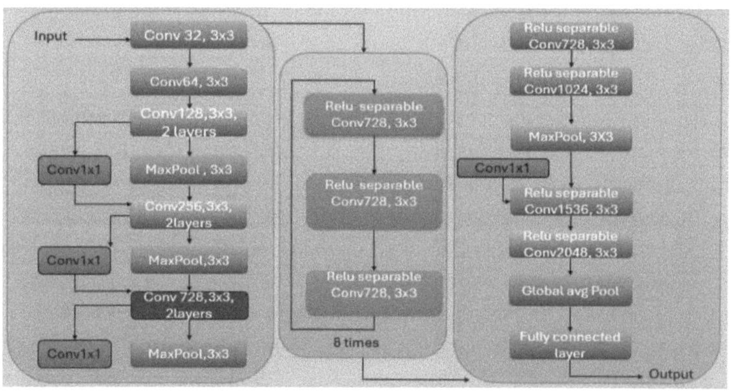

Fig. 2. Architecture of Xception model

This approach significantly reduces parameters and the cost, turning the network faster and efficient without compromising on performance. Xception's architecture follows a modular design, similar to Inception, but uses depthwise separable convolutions instead of standard convolutions. It consists of three main parts:

- Entry Flow: From the input image, this part extracts initial features and reduces its dimensions. It uses conventional convolutions and depth-wise separable convolutions to build the feature maps.
- Middle Flow: A series of depth-wise separable convolutions that refine the feature maps and capture complex representations. These layers form the core of the model and handle most of the computation.

- Exit Flow: Followed by a fully connected layer, the final layers perform global average pooling.

The use of depth-wise separable convolutions in the middle flow is what sets Xception apart from other architectures like Inception, which uses more complex and traditional convolutional operations. Xception has outperformed other popular CNNs such as ResNet and VGG16 on image classification tasks like ImageNet. By using depth-wise separable convolutions, it achieves a better balance between model complexity and computational efficiency, resulting in higher accuracy with fewer parameters. One of the main advantages of Xception is its computational efficiency. The use of this type convolutions drastically decreases operations in comparison to traditional convolutions. This leads to quick inferencing and a smaller model size, making Xception suitable for applications in real-time and deployment on limited resources equipped devices like mobile phones and embedded systems. It's advantages over Inception are:

- Simplified Architecture: Xception improves upon the Inception model by simplifying its convolution operations and replacing them with depth wise separable convolutions. This results in a more efficient design with fewer parameters.
- Improved Performance: Xception has demonstrated superior accuracy and performance over Inception, especially on large-scale image classification tasks, due to its more efficient use of computational resources.

3.3.3 Elastic Deformation for Image Augmentation

Elastic deformation is a popular technique used in image augmentation, particularly for training DL models in computer vision. The primary goal of image augmentation is to increase the size and diversity of training datasets artificially. For this, it takes the help of different transformations on the original images. Elastic deformation introduces non-linear, smooth, and localized deformations to an image, making the model more robust to real-world distortions such as slight object deformations, changes in perspective, or local shifts. Elastic deformation involves applying a random displacement to the image, which mimics physical or elastic warping, and is especially useful for tasks where variations in shape or form occur, such as handwriting recognition, medical image analysis, and facial recognition. Elastic deformation works by generating a random displacement field (or vector field) that displaces image pixels or regions based on random noise. This displacement is then applied smoothly across the image. The general steps involved are:

- Step 1 - Grid Creation: A grid is overlaid on the image. This grid can be thought of as a mesh, where each point represents a pixel or a group of pixels. The grid will serve as the basis for applying displacement.
- Step 2 - Random Perturbations: Small random displacements (often modeled by Gaussian noise) are applied to the grid points. These displacements are continuous and local, ensuring that neighboring points are deformed similarly, creating smooth and realistic distortions.
- Step 3 - Warping the Grid: The displacement vectors are applied to each grid point, causing each pixel in the image to move according to the displacement field. This results in a smooth, elastic-like deformation of the image.

- Step 4 - Image Transformation: After applying the deformation to the grid, the image is warped to match the new grid, resulting in a transformed image with elastic deformations.

Elastic deformation involves random noise generation (often Gaussian or Perlin noise) known as randomization, which ensures that every transformation is unique. The degree of displacement can vary, leading to random shifts, stretching, or compression of localized areas in the image. Advantages of Elastic Deformation for Image Augmentation are:

- *Improved Generalization*: By applying elastic deformations, the model learns to recognize objects or patterns in various forms and conditions, improving its generalization ability. It becomes less sensitive to minor variations and distortions that may occur in real-world data.
- *Realistic Transformations*: Unlike simple geometric transformations like rotation, translation, or scaling, elastic deformation creates complex and natural distortions, making the data augmentation process more effective. This is particularly useful for tasks like handwriting or medical image analysis, where realistic deformations are common.
- *Increased Training Data Diversity*: Elastic deformation provides an effective way to increase the variability in the training set without the need for additional data collection. This helps prevent overfitting by ensuring that the model sees a wider variety of data during training.
- *Non-linear Deformations*: Unlike traditional affine transformations, elastic deformation can model non-linear deformations, which are more representative of natural changes in images. This is particularly important for applications like medical image analysis or object detection, where deformations can vary in unpredictable ways.

4 Results and Discussion

The comparative analysis of EfficientNet and Xception CNN architectures for PCOS prediction yielded significant insights into their performance on medical imaging data. The models were evaluated based on several key metrics:

Accuracy:

- EfficientNet Model: Achieved an accuracy of approximately 66% on the validation dataset.
- Xception Model: Outperformed EfficientNet with an accuracy of approximately 100%.
- The accuracy curve of Efficientnet model, with the increase of epochs, the accuracy initially increases, then it becomes constant whereas the validation accuracy shows a decrease initially then static trend over all the epochs for training as shown in the Fig. 3. In case of the Xception model, for the initial epochs the accuracy shows an increase but later it follows a static trend as shown in the Fig. 3.
- The loss graph in case of efficient net shows decrease initially then become constant for training over the various epochs whereas in Xception, for initial epochs there is

a decreasing trend in the loss graph after which it becomes a smooth constant curve as shown in the Fig. 4. For validation data EfficientNet shows an irregular trend, but Xception shows initially decreasing then a constant trend.

ROC Curve and AUC:

- The ROC curves demonstrated the relation between sensitivity and specificity as shown in the Fig. 5. Each point on the curve represents a different threshold used for the classification. The Xception model achieved a higher Area Under the Curve of 1.00 compared to the EfficientNet model's AUC 0.96, indicating better overall diagnostic performance.

Confusion Matrix:

- This matrix for both models revealed the cases of true positives, true negatives, false positives and false negatives as given in the Fig. 6. For the Xception model, the confusion matrix indicated a higher rate of false negatives compared to the EfficientNet model, suggesting that it may struggle to identify some PCOS cases accurately. In both the cases the false positive cases show the same values.

Classification Report:

- This report highlighted key metrics like precision, recall, F1-score and support for both models. In case of EfficientNet, the respective values are: precission- 0.66, recall- 1.00, f1 score- 0.80 and support- 195 where as Xception shows: precission- 1.00, recall- 1.00, f1 score- 1.00 and support- 195 for no PCOS. For positive PCOS cases EfficientNet shows: 0.00 for precission, recall, f1 score and support- 195 where as Xception shows: precission- 1.00, recall- 1.00, f1 score- 1.00 and support- 100.

The throughputs of our research underscore the significance of deep learning models in diagnosing PCOS from imaging data, with significant differences between the EfficientNet and Xception architectures. In terms of Model Performance, the Xception model's superior accuracy and AUC suggest that its architecture is particularly wellsuited for the complexities inherent in medical imaging data. The depth wise separable convolutions allow Xception to capture intricate patterns while maintaining computational efficiency, leading to enhanced feature extraction capabilities.

In terms of Data Augmentation Impact, the use of elastic deformation as an image augmentation technique proved crucial in enhancing the models' generalization. This method not only increased the variability in the training dataset but also helped the models become more resilient to the natural variability found in medical images. Few snaps before and after implementing elastic deformation is given in Fig. 7.

Given the promising results, future research could explore hybrid models that combine the strengths of either architectures or incorporate additional augmentation techniques. Furthermore, expanding the dataset to include more diverse imaging examples could enhance model robustness and improve diagnostic accuracy, While the study focused on specific architectures and augmentation techniques, the findings may be influenced by the quality and diversity of the dataset. Future studies should aim to evaluate model performance on larger and more varied datasets to ensure generalizability.

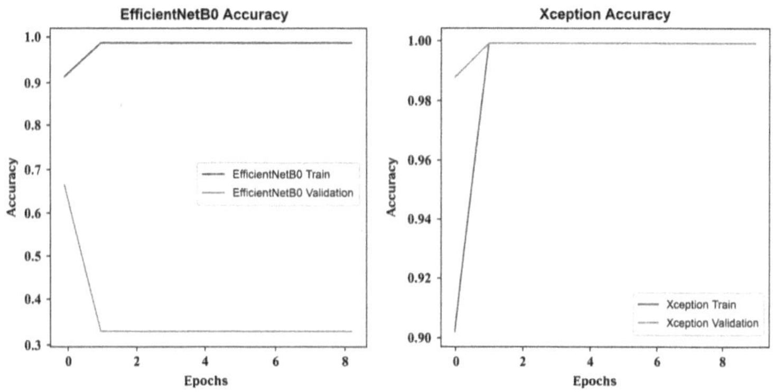

Fig. 3. Accuracy curves for EfficientNet model(Left) and Xception model (Right)

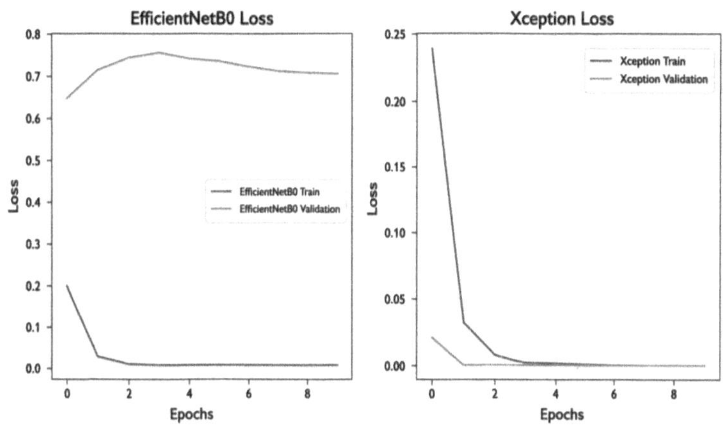

Fig. 4. Loss curves for EfficientNet model (Left) and Xception model (Right)

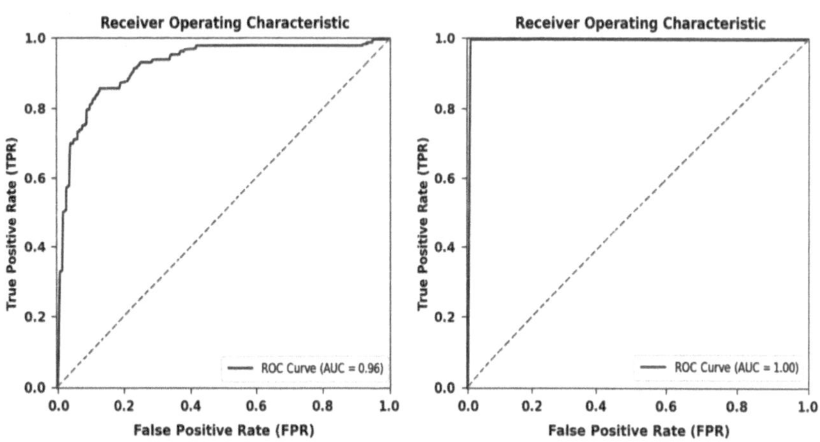

Fig. 5. A ROC for EfficientNet model (Left) and Xception model (Right)

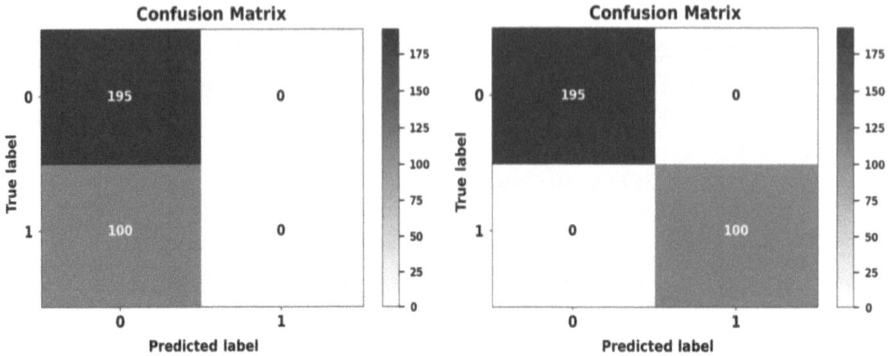

Fig. 6. Confusion matrix for EfficientNet model (Left) and Xception model (Right)

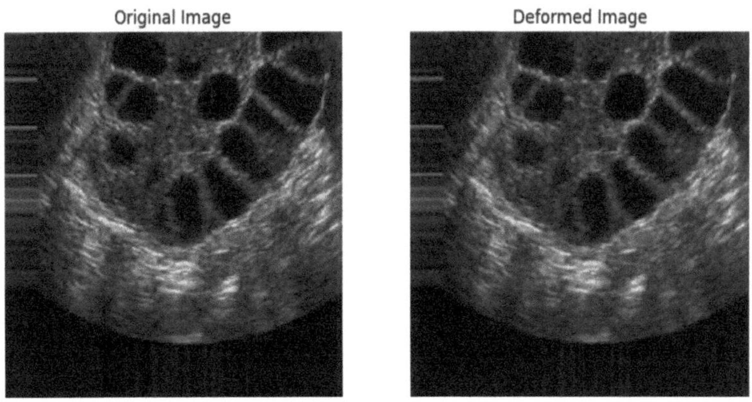

Fig. 7. Image before and after elastic deformation implementation

5 Conclusion

Our study presents a comprehensive analysis of using Convolutional Neural Networks, specifically EfficientNet and Xception architectures, for predicting PCOS from medical imaging data. By integrating advanced image data augmentation techniques, particularly elastic deformation, we aimed to enhance the robustness as well as performance accuracy of the models. The findings reveal that both EfficientNet and Xception CNNs exhibit strong potential for diagnosing PCOS, with notable differences in metrics of performance such as accuracy, support, precision, F1-score and recall. While both models successfully leveraged the augmented data to improve classification outcomes, the comparative analysis highlights specific strengths and weaknesses inherent to each architecture. For instance, Xception's depth wise separable convolutions may provide a computational advantage, leading to improved generalization on the validation set. The use of elastic deformation as an augmentation strategy proved beneficial, allowing the models to better adapt to variations in the dataset, thereby reducing overfitting and improving overall diagnostic performance. The insights gained from this comparative analysis could inform future research and clinical applications, ultimately aiding in the development of more reliable diagnostic tools for PCOS. Future work may explore the amalgamation of additional augmentation techniques and the potential for transfer learning from pre-trained models to further enhance diagnostic accuracy.

References

1. Kermanshahchi, J., Reddy, A.J., Xu, J., Mehrok, G.K., Nausheen, F.: Development of a machine learning-based model for accurate detection and classification of polycystic ovary syndrome on pelvic ultrasound. Cureus **16**(7) (2024). https://doi.org/10.7759/cureus.65134
2. Shanmugavadivel, K., et al.: Optimized polycystic ovarian disease prognosis and classification using AI based computational approaches on multi-modality data. BMC Med Inform Decis Mak 24(1) (2024). https://doi.org/10.1186/s12911-024-02688-9
3. Barrera, F.J., et al.: Application of machine learning and artificial intelligence in the diagnosis and classification of polycystic ovarian syndrome: a systematic review. Front Endocrinol (Lausanne) **14**, 1106625 (2023). https://doi.org/10.3389/fendo.2023.1106625
4. Suha, S.A., Islam, M.N.: An extended machine learning technique for polycystic ovary syndrome detection using ovary ultrasound image. Sci. Rep. **12**, 17123 (2022). https://doi.org/10.1038/s41598-022-21724-0
5. Paramasivam, G.B., Ramasamy Rajammal, R.: Modelling a self-defined CNN for effectual classification of PCOS from ultrasound images. Technol. Health Care **32**(5), 2893–2909 (2024). https://doi.org/10.3233/THC-230935
6. Purnama, B., Wisesti, U.N., Adiwijaya, Nhita, F., Gayatri, A., Mutiah, T.: A classification of polycystic Ovary Syndrome based on follicle detection of ultrasound images. In: 2015 3rd International Conference on Information and Communication Technology (ICoICT), Nusa Dua, Bali, Indonesia, pp. 396–401 (2015). https://doi.org/10.1109/ICoICT.2015.7231458
7. Poorani, B., Khilar, R.: An innovative approach for PCO morphology segmentation using a novel MOT-SF technique. Discov. Comput. **27** (2024). https://doi.org/10.1007/s10791-024-09458-1
8. Elmannai, H., et al.: Polycystic ovary syndrome detection machine learning model based on optimized feature selection and explainable artificial intelligence. Diagnostics (Basel) **13**(8), 1506 (2023). https://doi.org/10.3390/diagnostics13081506
9. Guixue, G., Yifu, P., Yuan, G., et al.: Progress of the application clinical prediction model in polycystic ovary syndrome. J. Ovarian Res. **16** (2023). https://doi.org/10.1186/s13048-023-01310-2
10. Khanna, V.V., Chadaga, K., Sampathila, N., Prabhu, S., Bhandage, V., Hegde, G.K.: A distinctive explainable machine learning framework for detection of polycystic ovary syndrome. Appl. Syst. Innovat. **6**(2), 32 (2023). https://doi.org/10.3390/asi6020032
11. Behera, L.K., Tripathy S.N.: e-healthcare system for skin care: a sustainable approach from deep learning to explainability in artificial intelligence. Sustain. Mater. Struct. IoT [SMSI-2024], pp. 215–219 (2025). https://doi.org/10.1201/9781003596776-44
12. Paikaray, B., Dewangan, B.K., P. Swain, D., Chakravarty, S.: An extensive study on medical image security with ROI preservation: techniques, evaluations and future directions. In: Machine Learning and Information Processing: Proceedings of ICMLIP 2019, pp. 465–476. Springer, Singapore (2020)

MrmrXgBr: A Machine Learning Model for Breast Cancer Identification Using Next-Generation Sequencing Data

Arpita Priyadarshini[1], Ashis Kumar Ratha[2], Sushree Sasmita Dash[3], Pabani Mahanta[1], Suprava Ranjan Laha[4], and Debasish Swapnesh Kumar Nayak[5](✉)

[1] Department of Statistics, Utkal University, Bhubaneswar 751004, India
[2] Department of Computing and Security, School of Computer Science and Engineering, IILM University, Greater Noida, UP 201306, India
[3] Department of AI & DS, IIMT College of Engineering, Greater Noida 201310, India
[4] Department of Computer Science and Engineering, Siksha 'O' Anusandhan (Deemed to Be) University, Bhubaneswar 751030, India
[5] Department of Ai Technology Institute (AiTI), SRM University-AP, Amaravati 522240, India
swapnesh.nayak@gmail.com

Abstract. The advent of artificial intelligence (AI) has brought about a significant transformation in breast cancer (BC) prediction, resulting in faster and more precise identification of cancer. Gene expression (GE) data plays a crucial role in identifying biomarkers, classifying molecular subtypes, predicting therapy responses, and uncovering therapeutic targets, all of which contribute to tailoring treatment approaches. Next-generation sequencing (NGS) outperforms microarray gene expression (GE) analysis by providing superior resolution, comprehensive assessment, and improved accuracy, thereby boosting the evaluation of risk, selection of medication, and prediction of prognosis. This research examines different machine learning (ML) methods for identifying breast cancer using NGS-GE data, with a specific emphasis on minimum redundancy maximum relevance (MRMR) for selecting features. The proposed MrmrXgBr model utilizes three widely-used ML algorithms: eXtreme gradient boosting (XGBoost), support vector machine (SVM), and random forest (RF). The selection of these algorithms was based on their strong classification performance, capability to handle data with a large number of dimensions, ability to evaluate feature importance, and capacity to identify intricate connections in GE datasets. The experimental findings showcase our proposed MrmrXgBr model's efficacy, with an impressive 96% accuracy in classification, 100% sensitivity, and 97% F1-score. These findings highlight the significance of accurate feature selection in improving the efficacy of ML models for BC detection. This research greatly enhances the application of advanced ML techniques and efficient feature engineering in creating more robust and effective tools for identifying BC. As a result, it improves patient outcomes and assists in making therapeutic decisions.

Keywords: Artificial intelligence · Machine learning · Feature engineering · Next-generation sequencing · Cancer diagnosis · Breast cancer

A. Das et al. (Eds.): MIDAS 2025, CCIS 3055, pp. 206–218, 2027.
https://doi.org/10.1007/978-3-032-35516-4_17

1 Introduction

The uncontrolled growth and division of breast cells is a characteristic of breast cancer (BC), the most common kind of cancer in women [1]. The incidence and prevalence of cancer are rising in both developed and developing countries [2]. According to WHO estimates, there will be around 2.3 million new cases of BC worldwide in 2023, leading to 685,000 deaths from the disease [3]. Accurate diagnosis and early identification are critical for BC patients to get tailored therapy and effective treatment. A key factor in the identification and diagnosis of BC is morphological characteristics. Under a microscope, pathologists examine breast tissue samples obtained from biopsy or surgical operations in an attempt to identify certain morphological characteristics linked to BC. These characteristics include the presence of malignant cells, aberrant cell proliferation, and changes in the size or form of cells. These anatomical characteristics aid in the choice of a treatment plan and prognosis by offering vital information regarding the kind, stage, and severity of the malignancy. Even while morphological analysis is helpful, there are certain restrictions.

The limits of morphological traits may make it difficult for doctors to diagnose BC patients, but advances in GE profiling made possible by NGS and microarray technologies have transformed customized medicine [4]. Nowadays, GE profiling is often employed in clinical settings to support early cancer detection and therapy. Computational techniques, such as ML and DL, are becoming more and more important for identifying and categorizing BC despite obstacles such limited datasets and data imbalance.

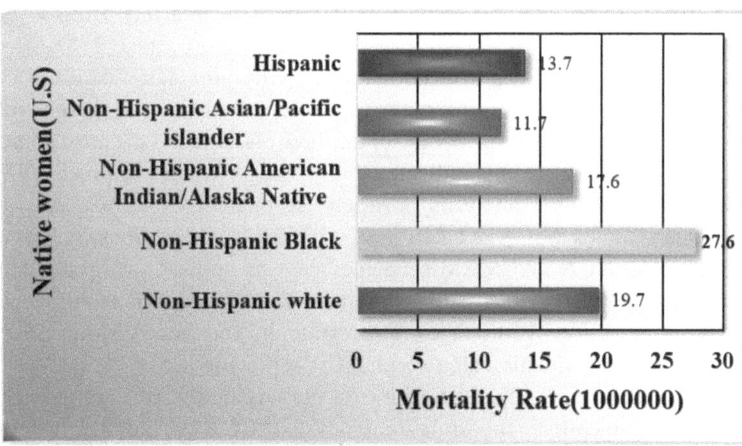

Fig. 1. Mortality rate of the US women for the year 2016–2020.

The use of DL and ML approaches to predict cancer based on GE data has advanced recently, however there are still issues with selecting the best model parameters, architecture, and generalization skills [5]. By creating a more precise prediction model and utilizing DL in more sophisticated applications, this work seeks to overcome these problems. In the given Fig. 1 below represents the mortality rate of the native US women from

2016 to 2020. The death rate among non-Hispanic black women is more as compared to its peers.

2 Related Work

This section examines many studies conducted by researchers using diverse ML approaches for the classification and detection of BC.

Mohamed *et al.* [3] proposed a hybrid Ebola Optimization Search Algorithm (EOSA-CNN) which analyzes GE data from TCGA. Genes from 1095 malignant and 113 healthy samples make up the data. The primary aim was to improve the classification efficacy of the predictive approach for the GE data. The findings suggest that the EOSA-CNN model was accurate in enhancing the effectiveness in detecting BC with acc of 98.3%, spec of 92.8%, recall of 99%, pre (99%), f1-score (99%), kappa (90.3%).

Botlagunta et al. [6] analyzes the medical data of patients acquired from BIACH & RI for classification and detection of BC metastasis. Several ML algorithms were used for this work. The aim of this work was to create a BC prediction system for identification of BC metastasis. The results demonstrated that the DT classifier was effective with an acc of 83% and an 87% AUC than other models.

Alromema *et al.* [7] suggested a hybrid FS framework that uses meta-heuristics, MRMR, and a two-tailed unpaired t-test to find the best collection of gene biomarkers to use as BC predictors. The objective is to find genes with excellent diagnostic capacity and providing biological insights to understand the early molecular processes of BC. The result showed that with an acc of 97.6%, XGBoost emerged as the best performer when using Breast_GSE22820 GE data from CuMiDa for analysis using several ML algorithms.

Ali *et al.* [8] analyzed the data from GE omnibus (GEO) for classification of microarray BC. In this work three ML algorithms are used along with the grid search cross validation. The objective is to determine which of the three classifiers is the most appropriate. With an acc of 97.78%, the results indicated that the best classifier is tuned SVM.

Abidalkareem *et al.* [9] analyzes a variety of microRNAs (miRNAs) in 1097 metastatic tissue samples that were taken from individuals with various BC stages from TCGA dataset. The NCA and MRMR FS method was applied for up-regulated and down-regulated states on various ML techniques. The objective is to utilize miRNA indicators for the detection of BC stages. The results showed that SVM model achieved higher performance classification along with NCA achieving acc of 98.3%.

Huang *et al.* [10] proposed a framework for classifying the BC subtypes from the TCGA. It uses sparse canonical correlation analysis (SCCA) to identify strongly connected traits between these genes and differentially expressed (DE) genes, then utilises a multi-task DL neural network to train the correlated DE genes for predicting BC subtypes. The findings suggested with an acc of 90.2% the DSCCN is more effective in predicting BC subtypes.

We obtain from the literature that there are several works done to identify cancer samples by utilizing different FS and ML techniques. However few FS techniques like MRMR, LASSO need to be explored.

In this work applying the MRMR FS algorithm to BC prediction using NGS data attempts to fill a research void. We found a deficiency in the investigation of FS technique designed specifically for NGS data processing by reviewing earlier research. By pre-processing NGS datasets from NCBI using MRMR to identify pertinent genomic characteristics and building ML models we want to increase the accuracy of BC prediction while clarifying important genetic indicators found by MRMR using this customized strategy. DL models need to be explored more on computational time, less architectural complexity, higher classification accuracy.

3 Materials and Methods

The brief descriptions of all the studied models along with dataset utilized in our study are presented in this section.

3.1 Dataset Description

The dataset studied is freely available in the NCBI and consists of a total of 126 samples. For the Clariom S Human Affymetrix Clariom S Assay 144 samples were subjected to microarray analysis: 47 samples were breast tumors, 82 samples were peri-tumoral stroma, and 15 samples were breast tissues without tumors.

The Clariom S Human Affymetrix Clariom S Assay (GPL23159) was employed in the analysis. Considerable variability was seen in the peri-tumoral stroma, within tumors, and between tumors when this data was integrated with prior transcriptome profiling. Interestingly, discrete stromal clusters were found, which may have consequences for overall survival. The dataset comprise of individuals from diverse demographics and across a broad spectrum of cancer stages. It allows for thorough analysis by including full clinical metadata, including tumor stage, histological subtypes, and therapy statuses. It is more useful for researching patterns of gene expression and refining recurrence prediction algorithms because of its variety.

This thorough approach emphasizes how crucial it is to take into account the peri-tumoral and tumor micro-environments when comprehending the biology of BC and developing treatment plans. The dataset is preprocessed using min-max normalization to reduce the computational cost and imputation to remove the duplicate values having more than 30% of the null values. After that we have applied feature selection strategy MRMR to select features and those selected features are fed to the AI models. The dataset description is provided in Table 1.

Table 1. Dataset description

Description	Numbers
Total Genes	16384
BC positive	111
BC Negative	15

3.2 MRMR

The MRMR FS approach is a critical tool for improving the precision and comprehensibility of predictive models. Its working idea is to minimize redundancy among the selected characteristics while simultaneously picking those that are strongly associated with the desired variable, such as the status of BC.

It is difficult to work with NGS GE data because of biological variability, noise, batch effects, and high dimensionality. Thorough preprocessing, standardization, and strong statistical techniques are necessary to guarantee proper interpretation. Large data sets and computing requirements can make integration and analysis more difficult. By choosing highly important features and avoiding redundancy, MRMR successfully lowers dimensionality and noise, hence addressing the difficulties presented by NGS-GE data. This improves the model's resilience and accuracy, allowing improved prediction and classification [11]. MRMR is basically chosen over other features as it suitable for genomic analysis because of its high computational performance for high-dimensional datasets.

Utilizing metrics like mutual information or correlation, MRMR determines each feature's relevance to the target variable. The redundancy between characteristics is also evaluated, usually by correlation or guarantee that the features picked have the maximum combined predictive power, MRMR employs an iterative method to choose a subset of characteristics that optimizes relevance while reducing redundancy [12]. By giving useful but non-redundant characteristics priority, this repeated refinement improves predictive models' interpretability in addition to their accuracy.

An objective function that seeks to minimize redundancy among selected features and maximize their relevance to the output variable drives the MRMR FS method. This goal is expressed by the Eq. 1.

$$\emptyset(\mathrm{K_S}, Y) = \frac{1}{|S|}\sum\nolimits_{i \in S} I(K_i, Y) - \frac{1}{|S|^2}\sum\nolimits_{i,j \in S} I(K_i, K_j) \tag{1}$$

Here, $\emptyset(\mathrm{K_S}, \mathrm{Y})$ represents the objective function, where S is the set of selected features, K_i represents the $\mathrm{i^{th}}$ feature, and Y is the output variable. The first term of the equation establishes the average mutual information between the output variable and the selected characteristics, while its second term determines the average mutual information among the chosen features. To optimize this objective function, characteristics are chosen iteratively by the algorithm. Equation 2 governs the selection criterion for determining which feature to add next to the set S:

$$argmax_{K_j \in K/S_m}(I(K_j, Y) - \frac{1}{m}\sum\nolimits_{K_i \in S_m} I(K_i, K_j)) \tag{2}$$

In this equation, $\mathrm{K_j}$ denotes a feature not yet selected, K represents the set of all features, $\mathrm{S_m}$ is the set of features selected so far, and m is the number of features currently selected.

Many techniques are often used for classification, particularly in GE analysis, including RF, SVM, and XGBoost. We have kept these techniques into practice to see how well they work. Strong and accurate gradient boosting is the hallmark of the XGBoost algorithm. A reliable classifier that performs well in high-dimensional areas is the SVM.

An ensemble technique called RF constructs many decision trees to enhance predicting performance.

3.3 Models Used

In this section we have discussed several ML models that are studied in our deployment phase. Our studied ML models are discussed in a more specific manner as follows;

XGBoost a very effective ensemble ML technique that uses regularization to minimize complexity and avoid overfitting by combining decision trees with gradient boosting. With the right hyperparameter adjustment, XGBoost, which is well-known for its scalability, adaptability, and features like feature relevance analysis and built-in handling of missing values, provides strong performance across a range of ML tasks and real-world applications.

SVM is a ML approach created for two-class problems. It has now been modified for multi-classification by converting data into a higher-dimensional space and locating a hyperplane that optimizes the margin between classes. Choosing the right parameters and kernels guarantees efficient performance, even if SVM may have trouble with nonlinearly separable data, especially for small datasets with fewer instances than independent variables.

RF is an ensemble learning technique that uses majority voting to predict class labels by combining many DT constructed on random subsets of the training data. With the help of bagging and randomization, RF outperforms a single DT in terms of classification accuracy and works especially well with small datasets.

3.4 Performance Evaluation

This section provides an empirical analysis of our model's performance indicators including accuracy (Acc), specificity (Spe), sensitivity (Sen), precision (Pre), F1-score, and Matthews Correlation Coefficient (MCC). These metrics provide a comprehensive understanding of the model's performance in classification tasks. We use cross-validation on our dataset or a hold out set to compute these metrics. To calculate these measures, the genuine labels and the model's anticipated labels are compared. To make sure the metrics are robust, we repeat this procedure for a number of folds if cross-validation is being used. The following Eqs. (3–8) are utilized to calculate the model metrices,

$$Accuracy = \frac{T_P + T_N}{T_P + F_P + T_N + F_N} \tag{3}$$

$$Precision = \frac{T_P}{T_P + F_P} \tag{4}$$

$$Sensitivity = \frac{T_P}{T_P + F_N} \tag{5}$$

$$Specificity = \frac{T_N}{T_N + F_P} \tag{6}$$

$$F1Score = \frac{T_P}{T_P + 1/2(F_P + F_N)} \quad (7)$$

$$MCC = \frac{T_p \times T_N - F_P \times F_N}{\sqrt{(T_P + F_P) \times (T_P + F_N) \times (T_N + F_P) \times (T_N + F_N)}} \quad (8)$$

4 Proposed Method

The four modules that make up our suggested MrmrXgBr model are discussed in this section: feature engineering, data classification, data preparation and data gathering. The Fig. 2 depicts the whole system architecture, including each and every component. In the subsections that follow, each subsystem (module) is described.

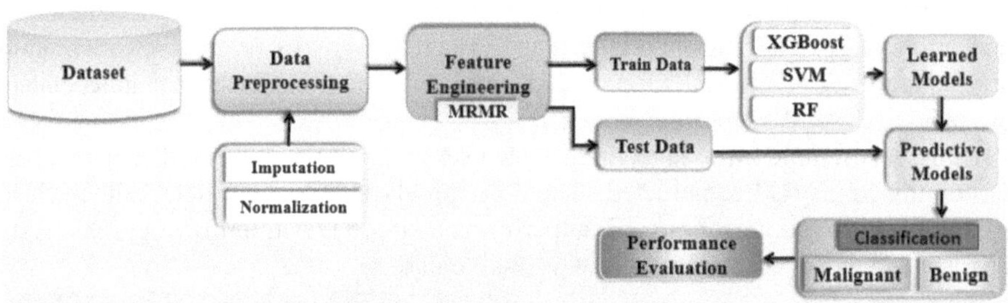

Fig. 2. Proposed MrmrXgBr Model Architecture.

4.1 Data Preprocessing

Data preprocessing is the term for the operations and techniques used to get raw data ready for analysis. It is an important step that might help to boost the analytical process. It enhances the model's capacity to learn with more precision and provide more valuable results.

Organization, transformation, and cleaning are necessary to improve the data's quality, consistency, and usefulness. Data preprocessing is done to prepare data for analysis and modeling. In this study data preprocessing undergoes two processes as seen in second step of the Fig. 2 above. For effective data preprocessing two methods normalization and imputation are used. By transforming characteristics to a standard scale, normalization guarantees that ML algorithms handle all data equally and enhances algorithmic performance. Imputation maintains the integrity of the dataset and makes it appropriate for additional analysis by filling in missing values based on available data.

Our proposed MrmrXgBr is enriched with MRMR FS and XGBoost classification techniques. The processed data after feature engineering are supplied to the XGBoost model which is fine-tuned with several parameters including 80:20 train-test split and 5-fold cross validation to obtain the best result.

5 Result and Discussion

This section includes the system used for the deployment of our proposed model and the various parameters we consider for the performance evaluation of our model.

5.1 System Specification

The model testing and training methods were conducted on two distinct systems using the Python platform. The initial tests were conducted on a PC running Windows 10 with GPU support, 16 GB of RAM, and an Intel Core i7 CPU clocked at 2.60 GHz. To evaluate resource use, we reran the trials on a workstation afterwards. The workstation is equipped with Ubuntu 20.04 OS, 32 GB of RAM, and a 500 GB SSD for auxiliary storage. By contrast, it was clear that using the workstation cut down on how long our trials took to compute. The Table 2 describes the computational time taken by all the studied models.

Table 2. Computational time taken by all the analysed models.

Model	Raw data (sec)	MRMR features (sec)
SVM	2.322	1.689
RF	2.387	2.46
XGBoost	5.65	**0.061**

5.2 Our Experimental Outcomes

This segment provides a comprehensive summary of the experimental outcomes, recognizing a variety of evaluation methods.

We conducted a thorough analysis utilizing NGS data to evaluate the effectiveness of the XGBoost, SVM, and RF models for BC diagnosis. In order to evaluate the effectiveness of these models in precisely identifying BC, we adopt a number of implementations on our utilized dataset. These experiments are conducted with several architectural modifications of the base models. We have also taken proper care to avoid data over-fitting and prioritized the selection of important features.

This section describes the outcome of our experiments in two different phases. In first phase we present the studied model's outcome on the raw data and in second phase we input the selected features by applying MRMR FS techniques.

In first phase of implementation, we observed that the XGBoost (93%) technique has obtained a higher rate of classification acc as compared with SVM and RF. The Table 3 represents the accuracy (Acc), precision (Pre), sensitivity (Sen), specificity (Spe), F1-score and MCC of the XGBoost, SVM, and RF models demonstrating their efficacy in detecting positive samples while minimizing false negatives and false positives. Figure 3 displays the performance metrics evaluated for all the analyzed models.

In case of the raw data, XGBoost performs well than SVM and RF with an Acc of 93%, Sen of 95%, Spe of 90%, Pre of 95%, MCC of 85%, and F1 score of 95%. Whereas SVM obtained acc (88%), Sen (94%), Spe (78%), Pre (87%), an MCC (75%), and an F1 score of 90%. On the other hand, we observe that RF achieved an Acc of 80%, Sen of 86%, Spe of 68%, Pre of 82%, F1-score of 84% and a MCC of 56%, which indicates it is less successful than XGBoost and SVM for this task as shown in Table 3.

Table 3. Metrics of the ML model's performance on unprocessed data.

Model	Acc (%)	Pre (%)	Sen (%)	Spe (%)	F1score (%)	MCC (%)
RF	80	82	86	68	84	56
SVM	88	87	94	78	90	75
XGBoost (XgBr)	93	95	95	90	95	85

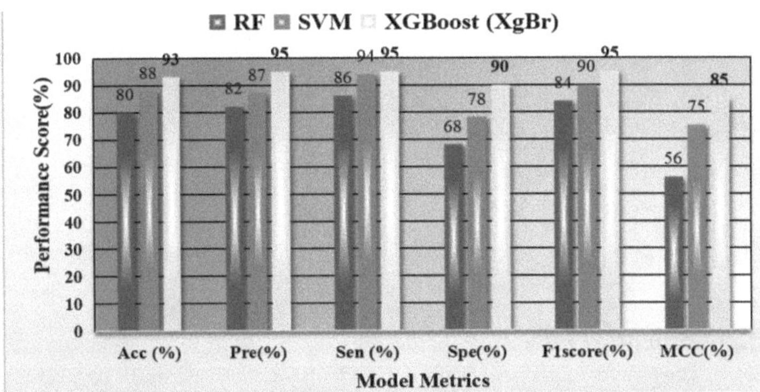

Fig. 3. Models and performance metrics on raw data.

During the second phase of implementation, we used the MRMR FS approach to input the chosen features alongside 5-fold cross-validation. We noted that the MrmrXgBr technique achieved a superior classification acc for all features in comparison to SVM and RF. The Table 4 represents the Acc, Pre, Sen, Spe, F1-score and MCC of the MrmrXgBr, SVM, and RF models. The measurement of performance achieved over all models is displayed in Fig. 4.

The performance evaluation of three ML models with an 80:20 train-test split and the MRMR FS approach is done where out of all the models that were evaluated, MrmrXgBr had the best Acc of 96%, along with remarkable Pre (95%) and Sen (100%). Strong predictive capacity was indicated by the resulting robust MCC of 85% and F1 score of 97%. Whereas SVM demonstrated an acc of 92%, with 95% pre and a Sen of 95%. The RF model had a modest Sen of 95%, it attained an acc of 88% with Pre (90%), F1 score and MCC were 93% and 51%, respectively.

Table 4. Metrices of analysed ML models on MRMR features.

Model (MRMR + 5 Fold CV)	Acc (%)	Pre (%)	Sen (%)	Spe (%)	F1score (%)	MCC (%)
RF	88	90	95	50	93	51
SVM	92	95	95	66	95	62
MrmrXgBr	96	95	100	75	97	85

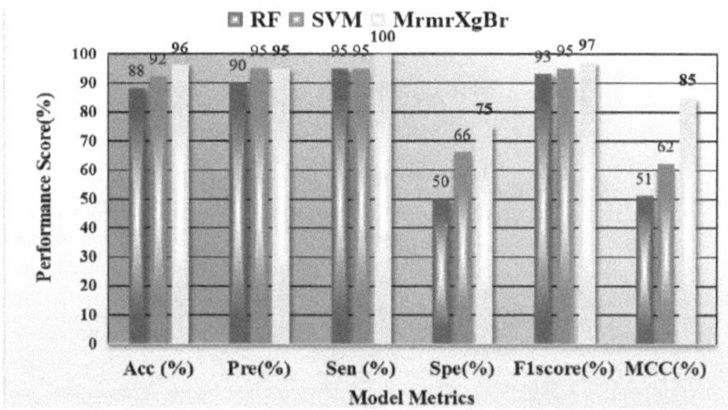

Fig. 4. Performance metrics of all the analysed models on MRMR features.

However, it was observed that MrmrXgBr, SVM and RF are having an increment of 3%, 4% and 8% respectively in their acc using MRMR FS technique as compared to the unprocessed data that was taken. While we compare MRMR selected features and execution of all the models on raw data we obtain higher Sen, Spe for the MrmrXgBr model.

In both cases XgBr and MrmrXgBr performs better than SVM and RF. Its greater Acc of 93% without MRMR, 96% with MRMR, pre, recall, and F1 score demonstrate its effectiveness in predictive modelling tasks whereas SVM and RF achieve 92% Acc, and 88% Acc with MRMR FS, respectively, but fall short of MrmrXgBr in terms of pre, Sen, Spe, F1-score and MCC. In addition to this MrmrXgBr requires less computational time compared to other studied ML models which makes it best fit for identification of the BC utilizing the NGS GE data.

A. Confusion Matrix and ROC Curves

The confusion matrix is also used to appropriately analyze data using ground truth and prediction labels. A true positive is the quantity of correctly identified diagonal samples in the picture and it is inversely correlated with the trained model's overall accuracy [13, 14]. The confusion matrix produced by the MrmrXgBr classification model is shown in Fig. 5. A classifier's performance across thresholds is depicted by the ROC curve, and particular predictions for a given threshold are specified in the confusion matrix. The ROC curve of all our studied models is shown in Fig. 6.

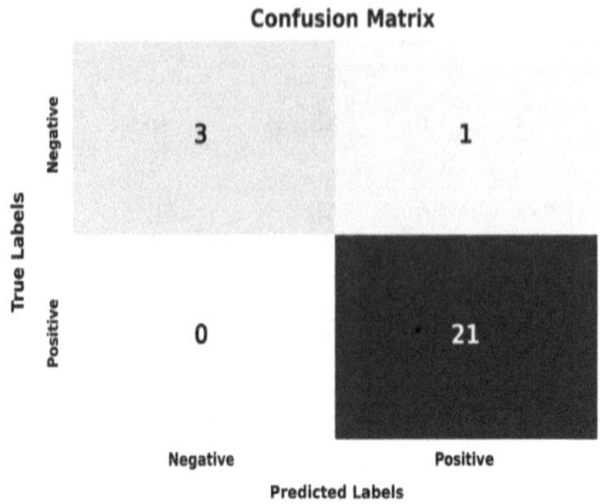

Fig. 5. Confusion Matrix of MrmrXgBr.

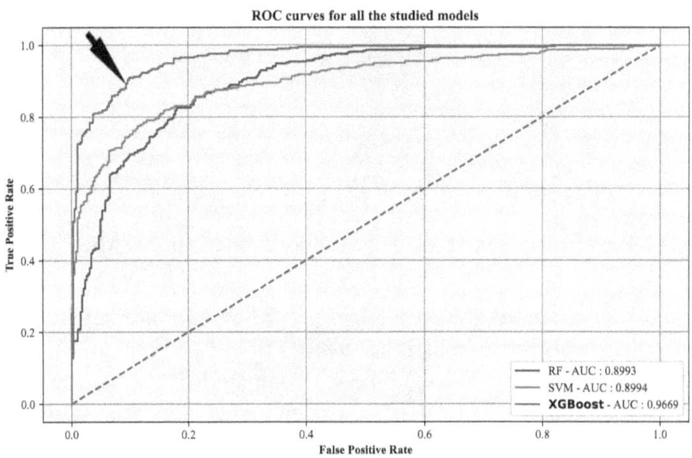

Fig. 6. ROC curves of all the studied ML models.

6 Conclusion and Future Scope

NGS-GE data requires careful data preprocessing, such as imputation and normalization, to handle missing values and data variability in order to classify BC into malignant or benign categories. In our study model performance was improved by the identification of relevant characteristics through the use of FS approaches such as MRMR. According to our research, the proposed MrmrXgBr performed better than SVM and RF, with an astounding 96% accuracy with MRMR FS techniques. Although our model's accuracy of 96% was marginally less than the 97% stated in some references, this discrepancy may be the result of different datasets, different preprocessing techniques, or a more complicated model. In spite of this, our model demonstrated strong performance and generalization abilities.

This emphasizes how crucial customized feature subsets possibly optimized by MRMR are to XGBoost's effectiveness. Although there are difficulties with the complexity of NGS GE data, our suggested MrmrXgBr and other studied ML models' performance are enriched with effective FS. The results of this study suggest that using MRMR FS to ML techniques for the detection of BC based on NGS-GE data may help physicians detect and diagnose the illness earlier, which would improve patient outcomes. Our future studies focused on investigating different sophisticated DL architectures that are customized to NGS GE data to enhance the classification acc of such models.

References

1. Mallick, C., Behera, C.R., Paikaray, B.K., Mishra, S.: Machine learning approaches with effective feature selection for improving breast cancer prediction. In: 2023 IEEE 2nd International Conference on Industrial Electronics: Developments & Applications (ICIDeA), pp. 157–162. IEEE (2023)
2. Siegel, R.L., Giaquinto, A.N., Jemal, A.: Cancer statistics, 2024, CA: a cancer journal for clinicians **74**(1), 12–49 (2024
3. Mohamed, T.I., Ezugwu, A.E., Fonou-Dombeu, J.V., Ikotun, A.M., Mohammed, M.: A bio-inspired convolution neural network architecture for automatic breast cancer detection and classification using RNA-Seq gene expression data. Sci. Rep. **13**(1), 14644 (2023)
4. Nayak, S.K., Priyadarshini, A., Routray, S.P., Sahoo, S.K., Swarnkar, T.: ARGai 2.0: a feature engineering enabled deep network model for antibiotic resistance gene and strain identification in E. coli. Int. J. Online Biomed. Eng. **21**(1) (2025)
5. Nayak, S.K., Das, R., Sahoo, S.K., Swarnkar, T.: ARGai 1.0: a GAN augmented in silico approach for identifying resistant genes and strains in E. coli using vision transformer. Comput. Biol. Chem. **115**, 108342 (2025)
6. Botlagunta, M., et al.: Classification and diagnostic prediction of breast cancer metastasis on clinical data using machine learning algorithms. Sci. Rep. **13**(1), 485 (2023)
7. Alromema, N., Syed, A.H., Khan, T.: A hybrid machine learning approach to screen optimal predictors for the classification of primary breast tumors from gene expression microarray data. Diagnostics **13**(4), 708 (2023)
8. Ali, N.M., Besar, R., Ab Aziz, N.A.: A case study of microarray breast cancer classification using machine learning algorithms with grid search cross validation. Bulletin Electr. Eng. Inform. **12**(2), 1047–1054 (2023)
9. Abidalkareem, A., Ibrahim, A.K., Abd, M., Rehman, O., Zhuang, H.: Identification of gene expression in different stages of breast cancer with machine learning. Cancers **16**(10), 1864 (2024)
10. Huang, Y., Zeng, P., Zhong, C.: Classifying breast cancer subtypes on multi-omics data via sparse canonical correlation analysis and deep learning. BMC Bioinformatics **25**(1), 132 (2024)
11. Nayak, S.K., Dash, A., Routray, S.P., Sahoo, S., Swarnkar, T.: ARGai: AI fusion approaches for classification of S. aureus strains and identification of antibiotic resistant genes. In: 2024 International Conference on Artificial Intelligence and Emerging Technology (Global AI Summit), pp. 1236–1241. IEEE (2024)
12. Nayak, D.S.K., et al.: aiGeneR 1.0: an artificial intelligence technique for the revelation of informative and antibiotic resistant genes in Escherichia coli. Front. Biosci.-Landmark **29**(2), 82 (2024)

13. Nayak, D.S.K., Routray, S.P., Sahooo, S., Sahoo, S.K., Swarnkar, T.: A comparative study using next generation sequencing data and machine learning approach for Crohn's Disease (CD) identification. In: 2022 International Conference on Machine Learning, Computer Systems and Security (MLCSS), pp. 17–21. IEEE (2022)
14. Swain, R.R., Nayak, D.S.K., Swarnkar, T.: A Comparative analysis of machine learning models for colon cancer classification. In: 2023 International Conference in Advances in Power, Signal, and Information Technology (APSIT), pp. 1–5. IEEE (2023)

Secure Stream Pro–Advanced Image Audio Video Encryption and Sharing Solution for Confidential Content Management

A. Hema Ambiha(✉) and M. Balasubramaniyan

Department of Computer Applications, Karpagam Academy of Higher Education, Coimbatore 641021, India
hemaambiha.aravindakshan@kahedu.edu.in

Abstract. The growing reliance on digital media, safeguarding multimedia data such as photos, music, and video is critical for protecting sensitive information from unauthorised access. Because of the differences in data architectures, file sizes, and real-time processing requirements, different forms of multimedia data provide distinct encryption issues. Some of these difficulties are addressed by existing encryption methods, such frequency masking and pixel shuffling for audio and picture encryption, as well as bit stream-level algorithms for video encryption. However, obstacles still exist in implementing real-time encryption while retaining data quality during the encryption and decryption procedures. This work proposes a hybrid encryption strategy to improve the security of multimedia data built using Elliptic Curve Cryptography (ECC) and Advanced Encryption Standard (AES). The purpose of this hybrid technique is to enhance encryption performance and strength through the capabilities of both methods, while providing the necessary performance requirements of multimedia applications. We also investigate the problems of providing security and maintaining quality for encrypted content, and offer techniques to manage these limits. Our unique contributions to Digital Rights Management (DRM), Communication Protection through security and personal information protection, has been detailed in this work.

Keywords: Asymmetric encryption · Cryptography · Network security Multimedia · Sensitive data

1 Introduction

As digital media consumption increases, multimedia cryptography becomes more significant for all the things from the user's private videos that are shared to larger scale endeavors like the services related to streaming, telemedicine, and safe and secure monitoring. Files of Multimedia type have their own unique range of issues, in addition to the size and complexity challenges, particularly across issues such as format compatibility and compression techniques. Encryption systems must accommodate multimedia file sizes reduced by compression in order to streamline storage and minimize bandwidth consumption [1–3]. If the encryption is altered the file structure, the content could no longer be used with conventional decoders or players, which will create an usability concern.

A. Das et al. (Eds.): MIDAS 2025, CCIS 3055, pp. 219–228, 2027.
https://doi.org/10.1007/978-3-032-35516-4_18

Data encryption is becoming an increasingly important part of protecting data resources, especially on the networks such as internet, intranet and extranet. In earlier days, in order to transmit the digital data in these environments, firstly it must be encrypted using specific mathematical methods and a key, and the original data from the cipher code can be extracted by decrypting the code using the same mathematical methods and the same key. The same objective directs the model's development for image encryption as well as decryption by making use of suitable user-specified keys. Adopting cryptographic algorithms will be the best approach for protecting data that is shared or warehoused.

Cryptography is the study of encoding techniques for securing the information, including aspects such as user authentication, data integrity, confidentiality, and authentication of information. Plaintext is referred to as the original message. The process of translating ciphertext back into ordinary language is known as decryption [5]. Due to the advancement of computing technology, binary bits are used to perform these operations [4]. With the help of the Internet, a global "Virtual Community" without restriction by space and time has emerged. Moreover, telemedicine continues to gain popularity in the fields of radiology, neurology, pathology, dermatology, critical care, and psychotherapy, where data is visually represented when transmitting image-related data over the guaranteed network. Similar goals follow the construction of the encryption and decryption paradigm for images [6–8]. With the support of a suitable user-defined key. The purpose of the image encryption and decryption model is to ensure discretion and refuge when transmitting photo-based data and when storing it in a data warehouse.

2 Related Work

Khalid M. Hosny et al. [1] developed the system to monitor the form of privacy-preserve encryption techniques to digital multimedia, which includes digital images, video, and audio components. The survey results helps to expand our understanding of the reliability and effectiveness of, safe multimedia encryption methods as well as to assist the development of newer and safer and more effective encryption methods in the future. Image encryption field is getting popularity because the goal is to secure these images or phots using several encryption methods so that attackers cannot access or alter these photos.

Ibrahim Yasser, et al. [2] proposed a new chaotic-based encryption technique using 2D modification models is presented to send data with highly secure transmission. This method develops a new perturbation-based data encryption scheme that is able to perform multiple rounds of confusion and diffusion to add additional security. The hybrid chutnification structure uses a combination of chaotic maps to provide robust encryption of media. Permutation (shuffling) and diffusion (substitution) control parameters are derived using complex composition chaotic maps to add reliable unpredictability in multimedia encryption. The chaotic-based perturbation adds randomization to the encryption process and complexity to the chaotic dependence of the 2D hybridization to establish strong encryption quality with minimized residual intelligibility while maintaining high key sensitivity.The encrypted media was robust against several potential attacks including brute force, statistical attacks, and differential attacks. Utilizing established metrics we have conducted numerical analyses of the data with respect to the different types of media, and distinctively present the potential less encryption quality to achieve low residual entropy, high key sensitivity, and quality encryption.

Ekhlas Abbas Albahrani, et al. [3] introduces a method that addresses the most recent improvements in encryption of audio and suggested the most widespread algorithmic calculations, considering the criteria related to computational complexity, security, and quality analysis. The core focus of this study is to exemplify all the kinds of audio encryption and decryption methods that are chaotic map-based. Numerous digital and audio utilizations have been discussed where chaotic maps for audio encryption is adopted. The results reveal that the suggested techniques outperform their data-specific counterparts in terms of security, computational efficiency, and adaptability, making them an excellent choice for secure multimedia transmission in real-world applications.

Ghosh, et al. [6] implemented a novel technology for digital watermarking that was invented by Lina, the author of the work used the concepts of chaos and exceptional wavelet transformation to produce watermarks that are interactive. A discrete wavelet modification is applied to the image first, followed by the removal of the low-frequency component and the concept of mess sequence to encrypt the low-frequency component is used. The NC coefficient and Peak signal noise ratio (PSNR) are utilised to testing device For the data encryption method, the author used the logistic solution of chaos. The following criteria complexity, similarity, and unpredictability are used to analyse the algorithm.

Mohamed Maazouz, et al. [7] analysed the notions based on the principles such as similarity, unpredictability and complexity. The chaos sequence simulation has shown that it satisfies the encryption algorithm's requirements. Offered a method for encrypting images. The suggested solution makes use of discrete chaotic diagrams that combine replacement and permutation techniques. A typical Lena image was used to corroborate the original image, according to the algorithm. The chaos sequence simulation has shown that it satisfies the encryption algorithm's requirements. Presented a method for encrypting images. The suggested solution makes use of discrete chaotic diagrams that combine replacement and permutation techniques. A typical Lena image was used to corroborate the original image, according to the algorithm.

Mahalingam, et al. [8] A new chaotic-based multimedia encryption system leveraging 2D modification models has been introduced to provide a highly secured data transmission. The proposed process also realizes a new perturbation-based data encryption procedure that was designed to undergo multiple rounds of confusion and diffusion so as to strengthen security. In order to introduce randomness and unpredictability into the encryption process, control parameter settings for the ermutation (the rearranging of components) and diffusion (the replacing of elements) processes were created dynamically using integrated chaotic maps. For the family of encrypted media styles described in the waves of transmission, the proposed encryption process proved to be stable against statistically acceptable and differential attacks through multiple attack implementations including brute force, statistical, and differential attacks. Other numerical analyses have run in conjunction with previously established metrics across multiple media types show the method is capable of running with minimal residual entropy levels, with strong key sensitivity and provided optimum encryption quality.

The diffusion module conceived by Dawood Shah, et. al. [9] proposed an architecture which is based on a special class of curves, which allows for efficient elliptic curve arithmetic. The confusion module used a novel block cipher developing method in which multiplication and binary Galois field inversion were paired with that of the other common arithmetic computations. The proposed scheme built numerous replacement box (S-boxes) to the lower-power Galois field. This means that in replacing the original S-boxes with a number of replacements, the resulting confusion of the data was effective in providing enhanced security to the ciphered audio. A comparative evaluation of the proposed scheme to achieve best algorithms based on the relevant literature outlines the robustness and efficiency in its own right. The comparison further outlines that suggested techniques are more secure, computationally efficient, and customizable when compared to their data.

Yun and Kim, et al. [10] The framework proposed a diffusion module uses a specialized curve which depends on proficient arithmetic computation based on elliptic curves, thus minimizing computation cost and is able to generate pseudo-random numbers (PRN) of better quality while producing the best possible circulation within the encoded audio cum video files. The perplexity module employs a specific block cipher generation model comprised of multiplication operations, binary Galois field inversion, and other basic arithmetic operation The experiment has its results, which has comparative produced from different research and exploring the time complexity.

Aravindakshan et al. [11] adopted RSA based Optimal Key-Tuned technique, where it highlights the dual authentication protocol for an effective cloud-based data sharing in hospital data management through the use of a Secure Hashing Algorithm 512 in association with Caesar cipher. The Optimal Key-tuned RSA provides a secure file transmission. With the authentication carrying another level of protection, the system itself is secured to the single point of failure from illegal access and enables only authorized users to access the cloud storage across the healthcare information. The results of our system show better performance than current highest level of development systems capable of securely sharing and transmitting the health cloud data.

Hossain, MdAlamgir et al. [12] A new technique called DFC-D (Dynamic Feature Combination with Dynamic Weights) has been proposed for real-time moving object detection in video surveillance systems. The purpose of the DFC-D technique is to help address the inherently difficult problem of precise moving object detection in highly dynamic environments with varied, changing lighting, complex backgrounds, and differing patterns of movement.The new method offers an adaptable moving object detection process that is functional in simply presenting challenges, for example busy areas in cities, changing weather conditions, or in scenes with moving objects such as vehicles or pedestrians. The model incorporates advanced optimization techniques and is real-time based, guaranteeing superior performance while being computationally efficient, and promising applications in video surveillance, in traffic monitoring systems and security systems where there is a greater desire for detecting moving objects on quickly and accurately.

3 Background of the Work

Many of the presently used strategies of cryptographic systems to protect sensitive information (images, audio, and video) in a multimedia application case [9], are based on symmetric and asymmetric cryptography, which support different types of application scenarios and security requirements. Symmetric encryption algorithms have versatile design and efficient implementations; thus, are often the method of choice for the protection of multimedia data (high-quality or not). Consider the Data Encryption Standard (DES) as an example of a symmetric encryption algorithm, and a good case study is the performance of the DES in encrypting high-quality video, audio, and images in scenarios like content delivery networks (CDNs) and cloud storage [12], and local media encryption Often a hybrid approach is used so that DES is implemented together with asymmetric cryptography systems. As an example, the most common asymmetric cryptography methods, RSA [11], including using RSA to securely exchange symmetric keys. RSA is a strong method for key management and distribution in presence of untrusted networks but, it is also slower than symmetric algorithms in practice while encrypting large files. For an instance, RSA can be used to encrypt the session key which was used by DES for encrypting media streams in applications such as video conferencing in a secure manner [13, 14]. Further, asymmetric encryption is used in Digital Rights Management (DRM) type systems where access to copyrighted content is restricted, and only authorized users can gain cache or the decryption keys [15, 16]. In these systems, the encryption is typically handled with a layered approach in which a symmetric algorithm like DES is used to encrypt the media stream, while an asymmetric algorithm like RSA allows for securely exchanging the encryption keys (Fig. 1). However, while effective as a hybrid model, it is inefficient because asymmetric encryption is still inefficient at applying to larger data.

Fig. 1. Cryptography process

4 Proposed Methodologies

A hybrid cryptographic approach that combines Elliptic Curve Cryptography (ECC) with Advanced Encryption Standard (AES) offers a highly efficient and secure solution to meet the growing demand for robust multimedia encryption. The combination of ECC and AES methods by balancing encryption speed with strong security, crucial for both real-time applications and data protection.

AES (Advanced Encryption Standard) is a highly efficient symmetric encryption technique, widely recognized for its speed and suitability in securing large multimedia files. Its minimal computational overhead and robust security make it an ideal choice for encrypting and performance even for bulk data encryption. In contrast, Elliptic Curve Cryptography (ECC) is a type of asymmetric encryption and ECC provides outstanding security with key sizes which are very small, especially compared to other classical asymmetric cryptographic solutions, such as RSA. Elliptic Curve Cryptography provides the same level of security as RSA and using less computing resources, making it a great solution for devices that are restricted due to their limited resources, e.g. mobile devices, embedded systems, and IoT. [17, 18].

ECC generates private and public keys using elliptic curve theory, providing an efficient method for secure key exchange. In this hybrid model, AES is used for encrypting the actual multimedia content (image, audio, or video), ensuring fast and efficient encryption of large datasets. ECC, on the other hand, is utilized for securely exchanging the AES encryption keys. Figure 2 illustrates the architecture diagram of the proposed system, showing the integration of AES and ECC for secure multimedia content management.

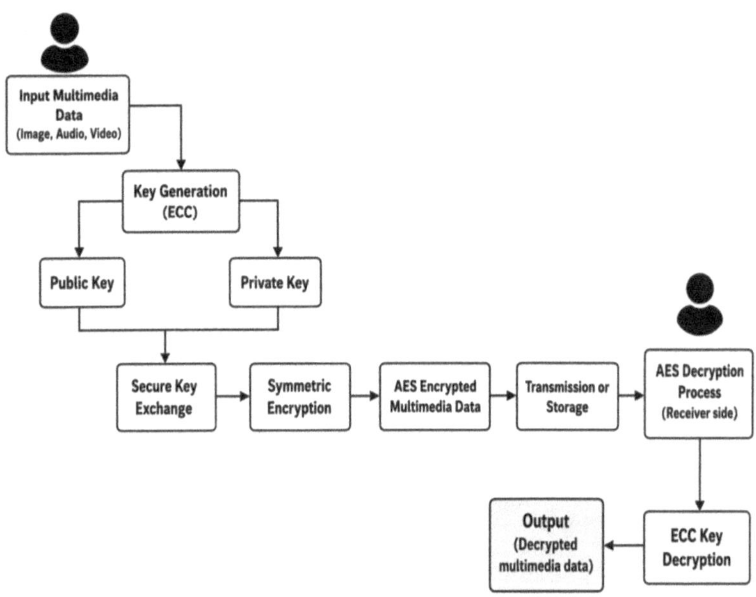

Fig. 2. Proposed Architecture

4.1 General Procedure of ECC

Parties mutually agree on few publicly-known data objects. an elliptic curve is a set of points (x, y) that satisfy a specific equation in the form:

$$y^2 = x^3 + ax + b(modp) \tag{1}$$

where:

- a and b are constants defined by the curve.
- p is a prime number that defines the finite field over which the curve is defined.

The curve is defined over a finite field (typically F_p where p is a large prime) or a binary field F_p.. An elliptic curve group is a set of points (x,y) satisfying Eq. (1), together with a point at infinity, notated O, which is the identity element for elliptic curve point addition.

4.2 Base Point Selection

Chosen a base point B (sometimes called the generator point) in the elliptic curve group. The generation of the public key depends on this base point and in reality, it's usually a point having a sufficiently known order n such that $nB = O$ (the point at infinity). Other points of the curve group are generated through the base point B.

4.3 Key Generation

Each user generates **a public-private key pair** based on elliptic curve operations:

a) Private Key: x, the private key, is a randomly chosen integer in the interval $[1, n-1]$, , for the (sender's) order n of the base point B.
b) Public Key: Then public key Q is calculated from scalar multiplication of the base point B by the private key x.
c) The public key Q is calculated by performing scalar multiplication of the base point B with the private key x. This operation is denoted as:

$$Q = x \cdot B \tag{2}$$

c) Q is a user's public key, and is a point on the elliptic curve here. Others can have the public key Q (Eq. 2), while the private key x remains secret.

4.4 AES Algorithm

The block cypher is a different name for the AES cypher. There have been no documented successful attacks on AES. There are several rounds involved in AES encryption. Rows are shifted according to byte length; this is called a "shift row.“ The mix column reproduces the Galois field matrix. Lastly, in the add round key step, the round key is XORed with the output matrix of the mix column. The key size determines how many encryption rounds are utilised. For a 128-bit key, these four steps are realistic to 9 rounds, where the 10th round does not regard as the mix column step. Since all steps are decryption is the turn round of encryption.
Algorithm Procedure for AES Encryption:

1) Input: A plaintext message (block of 128 bits) and a secret key.
2) Key Expansion: The key is expanded into a series of round keys using the key schedule.

3) Initial Round:
4) AddRoundKey(XOR the input block with the first round key).
5) Main Rounds (Repeat for the total number of rounds based on key size):

- SubBytes
- ShiftRows
- MixColumns
- AddRoundKey

6) Final Round (No MixColumns in the last round):

- SubBytes
- ShiftRows
- AddRoundKey

7) Output: The final block of ciphertext is the result.

Algorithm Procedure for AES Decryption:
AES decryption is the reverse of encryption. The operations in the decryption process are the inverse of the corresponding operations in the encryption process. Specifically:

1) *SubBytes* is reversed by InvSubBytes (using an inverse S-box).
2) *ShiftRows* is reversed by InvShiftRows (shifting rows in the opposite direction).
3) MixColumns is reversed by InvMixColumns.
4) For the reverse order the key which is in round is used.

Each of these stages is represented in Fig. 3.

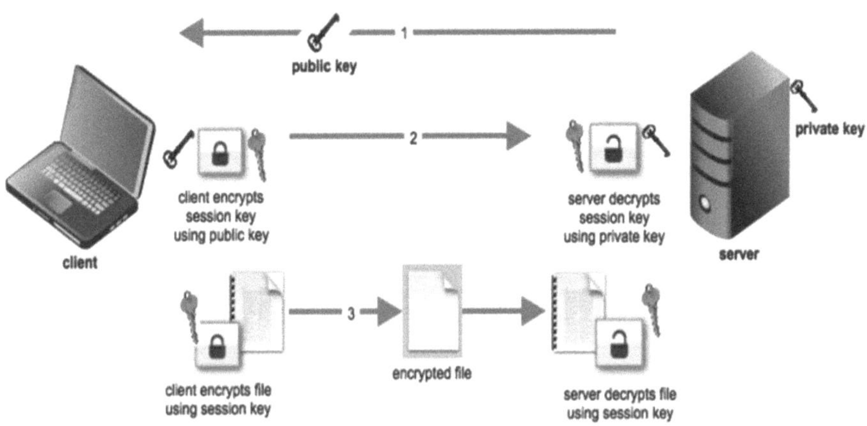

Fig. 3. AES process

5 Result and Discussion

The hybrid cryptographic methods such as Elliptic Curve Cryptography - (ECC) in association with Advanced Encryption Standard - (AES) has demonstrated important advantages in terms of both security and multimedia encryption. AES, as a symmetric key algorithm, is highly efficient for encrypting large multimedia files, such as images, audio, and video, due to its minimal computational overhead and ability to handle bulk data quickly [19, 20].

This makes AES very suitable for applications requiring real time speed. ECC, being an asymmetric encryption technique provided a very traditional algorithms such as RSA. ECC is very suitable for resource- This hybrid approach of AES and ECC, confidence gives a high level of efficiency and security to a multitude of modern multimedia encryption challenges. The results of existing and proposed system are shown in Fig. 4.

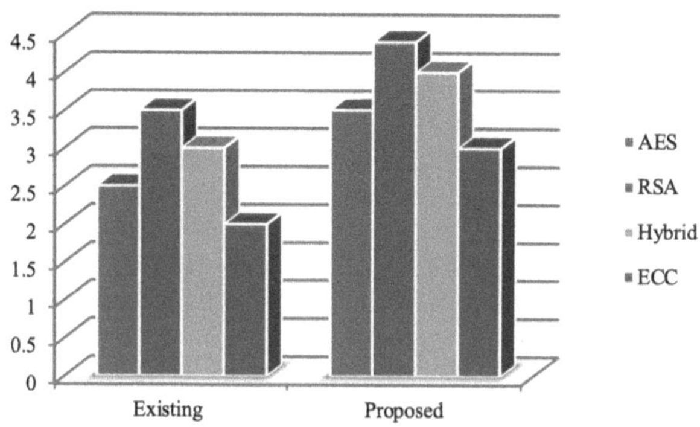

Fig. 4. Comparison of AES, RSA and Hybrid algorithms

6 Conclusion

A hybrid encryption representation that combines Elliptic Curve Cryptography (ECC) with Advanced Encryption Standard (AES) offers a particularly effective solution for multimedia files, such as music, video, and photographs. By utilizing both symmetric and asymmetric cryptographic algorithms, this approach guarantees a strong balance between security, speed, and efficacy. AES can quickly and efficiently protect large multimedia files, which makes it appropriate for high-quality films and real-time applications. In the meanwhile, ECC guarantees computational cost even with constrained resources. The hybrid ECC-AES architecture is appropriate for a wide range of applications, such as cloud storage, encrypted phone conversation, secure video streaming, and Internet of Things-based multimedia systems.

References

1. Hosny, K.M., et al.: Multimedia security using encryption: a survey. IEEE Access **11**, 63027–63056 (2023)
2. Yasser, I., et al.: A chaotic-based encryption/decryption framework for secure multimedia communications. Entropy **22**(11), 1253 (2020)
3. Albahrani, E.A., Alshekly, T.K., Lafta, S.H.: A review on audio encryption algorithms using chaos maps-based techniques. J. Cyber Secur. Mob., 53–82 (2022)
4. Dua, M., et al.: 3D chaotic map-cosine transformation-based approach to video encryption and decryption. Open Comput. Sci. **12**(1), 37–56 (2022)

5. Zia, U., et al.: Survey on image encryption techniques using chaotic maps in spatial, transform and spatiotemporal domains. Int. J. Inf. Secur. **21**(4), 917–935 (2022)
6. Ghosh, G., Kavita, Divya, Verma, S., Talib, M.N., Shah, M.H.: A systematic review on image encryption techniques. Turk. J. Comput. Math. Educ. (TURCOMAT) **12**(10), 3055–3059 (2021)
7. Maazouz, M., et al.: FPGA implementation of a chaos-based image encryption algorithm. J. King Saud Univ.-Comput. Inf. Sci. **34**(10), 9926–9941 (2022)
8. Mahalingam, H., et al.: Dual-domain image encryption in unsecure medium—a secure communication perspective. Mathematics **11**(2), 457 (2023). https://doi.org/10.3390/math11020457
9. Shah, D., et al.: An efficient audio encryption scheme based on finite fields. IEEE Access **9**, 144385–144394 (2021)
10. Yun, J., Kim, M.: JLVEA: lightweight real-time video stream encryption algorithm for Internet of Things. Sensors **20**(13), 3627 (2020)
11. Aravindakshan, H.A., Rengasamy, P.: Enhancing healthcare data security in cloud environments with dual authentication and optimal key-tuned encryption. Sci. Tech. J. Inf. Technol. Mech. Opt. **24**(3), 456–463 (2024). https://doi.org/10.17586/2226-1494-2024-24-3-456-463
12. Hossain, M.A., Hossain, M.I., Hossain, M.D., Huh, E.-N.: DFC-D: A dynamic weight-based multiple features combination for real-time moving object detection. Multimedia Tools Appl. **81**(22), 32549–32580 (2022)
13. Sengar, S.S., Mukhopadhyay, S.: Motion segmentation-based surveillance video compression using adaptive particle swarm optimization. Neural Comput. Appl. **32**(15), 11443–11457 (2020)
14. Iqbal, O., Torres Muro, V.I., Katoch, S., Spanias, A., Jayasuriya, S.: Adaptive subsampling for ROI-based visual tracking: algorithms and FPGA implementation. IEEE Access **10**, 90507–90522 (2022)
15. Hameed, I.M., Abdul Hussain, S.H., Mahmmod, B.M.: Content-based image retrieval: a review of recent trends. Cogent Eng. **8**(1), 1927469 (2021)
16. Zhu, H., Yan, X., Tang, H., Chang, Y., Li, B., Yuan, X.: Moving object detection with deep CNNs. IEEE Access **8**, 29729–29741 (2020)
17. Haidous, A., Oswald, W., Das, H., Gong, N.: Content-adaptable ROI-aware video storage for power-quality scalable mobile streaming. IEEE Access **10**, 26830–26848 (2022)
18. Hapsari, D.A.P., Madenda, S., Subali, M., Talita, A.S.: A novel approach to video compression using region of interest (ROI) method on video surveillance systems. Int. J. Adv. Comput. Sci. Appl. **13**(6) (2022)
19. Zhou, Y., Bai, Y., Chen, Y.: MultiframeCenterNet heatmap ROI aggregation for real-time video object detection. IEEE Access **10**, 54870–54877 (2022)
20. Wu, J., et al.: Efficient video instance segmentation via tracklet query and proposal. In: Proceedings of the IEEE/CVF Conference on Computer Vision and Pattern Recognition, pp. 959–968 (2022)

A Comparative Analysis of Transformer Models for Abstractive Summarization of Indian Legal Judgments

G. S. Mahalakshmi[1(✉)], C. Surendra Kumar[1], and S. Sendhilkumar[2]

[1] Department of Computer Science and Engineering, College of Engineering Guindy, Anna University, Chennai, India
gsmaha@annauniv.edu

[2] Department of Information Science and Technology, College of Engineering Guindy, Anna University, Chennai, India

Abstract. Abstractive summarization of judicial decisions presents issues due to sophisticated legal terminology, contextual complexities, and diverse document architectures. This research compares three major transformer models—T5 (Text-to-Text Transfer Transformer), LED (Longformer Encoder-Decoder), and ProphetNet—based on their ability to provide short and useful summaries of legal texts. This study performed extensive trials on a dataset of Landmark Judicial Judgments by the Supreme Court of India and their accompanying Summaries, evaluating each model based on metrics like ROUGE scores, semantic coherence, and summary relevance. The findings from this study show that, while T5 and ProphetNet perform well in broad text summary tasks, the LED model excels both in handling lengthy legal documents, preserving contextual integrity, and creating summaries that adhere to legal requirements. This study highlights the superiority of LEDs.

Keywords: Longformer Encoder-Decoders · T5 · ProphetNet · Abstractive Legal Text Summarization · Judicial Decision Analysis

1 Introduction

In today's fast-paced legal landscape, the sheer volume and complexity of legal documents pose significant challenges for legal professionals and researchers alike. From statutes and court cases to contracts and regulatory filings, the realm of legal text is vast and multifaceted. In most Law Firms and Private Legal Practices, lawyers and paralegals would have to go through a lot of judgment documents to identify relevant cases that they can cite as precedents in an ongoing case. It would aid law practitioners, lawmakers, and civilians if shorter versions were available for these lengthy documents. The idea of using transformer-based models like BERT for the legal domain was initially introduced in [7].

In the text summarization literature, it has been identified that there are mainly two types of automatic summarization approaches—Abstractive summarization and

A. Das et al. (Eds.): MIDAS 2025, CCIS 3055, pp. 229–243, 2027.
https://doi.org/10.1007/978-3-032-35516-4_19

Extractive summarization. The technique used in abstractive summarization involves text generation-based summary formation which is contingent upon the ability of the model to comprehend the information from the input documents. Whereas, in extractive summarization, the main idea is to extract summary-worthy sentences from the input document itself to form a summary. Although Extractive summarization would be able to derive the most important sentences out of the whole text, at times, these sentences may not be comprehensible to the common people who need coherent explanations. In contrast, Abstractive Summarization is a more effective method. Since, legal judgments are vague to go through, making them well-organized and lucid aids the layman.

In recent years, transformer-based models have proved efficient in natural language processing (NLP), excelling at abstractive summarization by generating coherent, comprehensible summaries. However, legal texts present unique challenges to these models due to their presence of jargon, varied document structures, and contextual dependencies. With appropriate fine-tuning from an Indian Judgment Dataset, Transformers would perform way better than the other current models. Models such as T5 (Text-to-Text Transfer Transformer) and ProphetNet have shown promising results in general text summarization with distinct architectures, while the Longformer Encoder-Decoder (LED) model, due to its efficient self-attention mechanism, offers a plethora of advantages for processing longer texts.

This research conducts a comparative analysis of these three state-of-the-art transformer-based models to determine which model would best suit the Indian Judgments. Through an elaborate analysis of these models, this research aims to identify the model that consistently generates the most coherent, compact summaries that are also able to capture all the vital details in the Judgment, thereby aiding not only legal professionals like lawyers and paralegals but also, the public who would like to understand the intricate text of Indian Judgments.

2 Related Works

There are several methods to implement Abstractive Summarization—Sequence-to-Sequence architectures, Pointer-Generator networks, and other Transformer-based approaches were introduced in [5]. For implementing these methods, Large Language Models are used for summarizing news articles in [2]. The study emphasizes that while some models achieve high ROUGE-1, ROUGE-2 scores, often do not always produce summaries that are contextually accurate or linguistically natural. The study also underlines that the ROUGE-L scores and human evaluation are more correlated when comparing within each model group. This research contradicts the fact that ROUGE Scores are not suitable for summarization by considering three models with increasing observed strength, where the model with the most strength shows better contextual accuracy and coherence and asserts that ROUGE-L Scores are vital for evaluation. Additionally, this study emphasizes two of these methods—T5 (Text-to-Text Transformers) and Longformer Encoder-Decoder come under Transformer-based approaches whereas ProphetNet comes under Sequence-to-Sequence Architectures. Since the existing models focus on lucid text from newspapers and discussions, they cannot be effectively considered for Legal Judgments that mostly involve longer and intricate content with contextual dependencies.

2.1 T5 (Text-to-Text Transformers)

Studies have explored the extent to which the T5 model can perform a variety of tasks [1]. This approach enables T5 to excel at tasks like summarization, translation, and question answering by allowing for flexible and consistent inputs and outputs. Although T5 performs exceptionally well in general tasks, it struggles with longer text and to maintain context, which is the reason to explore other models like ProphetNet and LED. On the other hand, in a study, the T5 model is found to perform well when fine-tuned on multiple summarization tasks and show that its text-to-text approach, coupled with extensive pre-training, yields high-quality, coherent summaries that generalize well across diverse text types [3].

2.2 ProphetNet

This model was introduced by a study as a sequence-to-sequence model that improves over Transformer models by predicting future sequences of tokens (n-grams) instead of a single next token [4]. Its architecture has shown strong performance in text summarization where it has shown the ability to generate fluent, contextually relevant summaries has been particularly effective. Although, in this model, the dataset used here is generalized in nature thereby, not giving scope to domain-specific summarization techniques. This has been further explored in a study that demonstrates ProphetNet's capacity to hold multiple sources of information and synthesize them into a single, cohesive summary approach that proves beneficial in managing the input complexity that comes with multi-document summarization [6]. A limitation that is noted in both latter studies is ProphetNet's occasional difficulty in distinguishing between highly similar or overlapping content from multiple documents, which can lead to redundancy in the final summary. This study aims to address this limitation by fine-tuning ProphetNet on the domain-specific dataset of legal judgments, which helps it distinguish between subtle but critical legal nuances.

2.3 LED

The LED Model was introduced in a study as a model specifically designed for handling long documents by implementing an efficient, sliding-window attention mechanism [9]. Its ability to retain essential information across long sequences enables it to produce coherent summaries setting it apart from models like T5 and BART, which are limited by shorter input constraints. Another study conducts a comparative evaluation of several transformer-based models, including LED, on datasets composed of long documents. Despite its strengths, the study points out that LED's effectiveness is highly dependent on proper fine-tuning [10]. Without domain-specific adjustments, the model may still struggle with maintaining coherence in specific contexts, occasionally leading to generic or less precise summaries. This study tries to mitigate these limitations found in all these studies because of a streamlined fine tuning from the dataset that is used for training the model.

3 Proposed Methodology – Fine-Tuned Abstractive Summarization of Legal Judgment

As discussed earlier, this study applies 3 abstractive methods for Legal Document Summarization—T5(Text-to-Text Transformers), ProphetNet, and LED (Longformer Encoder-Decoder). A key challenge in applying such methods is that these documents are long, and most of the abstractive summarization models have restrictions on the number of input tokens.

The first step consists of data collection from Landmark Indian Supreme Court judgments, which are also provided with summaries for every corresponding judgment entered on the website. After collecting the dataset, the judgment text files and their summary files are labeled for the models to be able to use for fine-tuning. During the process of labeling, the judgments are arranged chronologically based on their dates, and each judgment is assigned a unique serial number. Similarly, the corresponding summaries are labeled using the same method ensuring a clear one-to-one mapping between the judgment texts and their summaries. These labeled files are then stored in separate folders for judgments and summaries. This systematic labeling process ensures consistency and facilitates easy access to model training. The labeling is followed by the pre-processing step, where each of these data undergoes text cleaning, primarily focusing on removing unwanted stop words and spaces (Fig. 1).

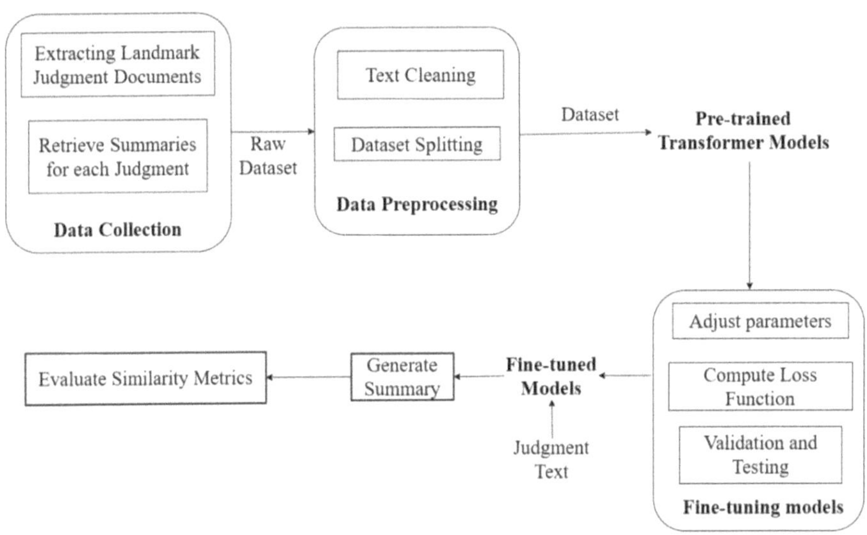

Fig. 1. Architecture of the Fine-tuned Abstractive Summarization Models

After the dataset is fully processed, it is loaded into the three pre-trained models mentioned earlier. In this step, the dataset is split into Training set and Validation Set consisting of 80% and 20% of the original dataset respectively. Subsequently, the models that are loaded with the dataset undergo rigorous training where the models learn to map input judgments to their corresponding summaries. The models are adjusted to better fit the summarization of Legal Judgments that have more complexity compared to the text originally summarized by the pre-trained models.

During fine-tuning of the models, an iterative approach of 3 steps is followed for each epoch. Firstly, the parameters of the models like weights are adjusted so that it could become more complex to summarize the Judgment text by referring to the Training set.

Secondly, the difference between the generated and actual summaries is formed to guide the learning of the model. Furthermore, the validation set that was split earlier from the original dataset is utilized here to monitor performance, adjust hyperparameters, and test unseen data to evaluate the final trained model.

The fine-tuned models represent the pre-trained transformers adjusted specifically for generating legal judgment summaries. They are ready for deployment or further evaluation. When a new judgment text is given as an input text to the fine-tuned models, it would be able to generate legal domain-specific summaries that are expected to be concise yet comprehensive, capturing the essence of the judgments. The performance of the fine-tuned models is evaluated using similarity metrics – ROUGE-1, ROUGE-2, and ROUGE-Lsum.

4 Implementation

4.1 Dataset

This research develops a dataset from the Landmark Judgments from the Indian Supreme Court drafted by the Centre for Research and Planning. Since these judgments were derived from historical cases that might have a span in years, it can be confidently said to have the longest sequences a legal judgment can have to train models in such a way that, it would be able to summarize the ordinary judgments efficiently. Even the shortest judgment is contained in 20 pages, the dataset is constrained to decrease the number of judgments to 10. If trained for more than 10 judgments, considering the vast amount of sequences it would contradict the goal of this study to provide a model that can be fine-tuned without extensive computational requirements.

Each of these judgments is accompanied by its corresponding summaries that contain—Background, Decision by the Supreme Court, Reasons for the decision, and Effect of presidential assent. This segmentation of summary to various topics would make it concise and clear for the common folk to understand. Fine-tuning was conducted on a custom dataset containing 10 legal judgments and their corresponding summaries, which was organized in Google Drive for seamless access in Google Colab. The following sections detail the fine-tuning process and observations for each model.

4.2 T5 (Text-to-Text Transformers)

The Text-to-Text transformer (T5) is a state-of-the-art pre-trained language model based on the transformer architecture. It adopts a unified text-to-text framework that can handle any natural language processing (NLP) task by converting both the input and output into natural language texts. The T5 model is available in several sizes to balance performance and computational efficiency, and its standard models are – T5-small, T5-base, and T5-large. Each of these variants retains the same architecture but scales up or down the number of parameters, and layers.

Algorithm 1: Fine tuning pre-trained T5 model for summarizing legal documents

Input: Dataset of court judgments and summaries, hyperparameters
Output: Fine-tuned T5 model

1. For each document of judgment and summary in the dataset:
    a. Split each document into tokens using the tokenizer provided by T5
    b. Assign unique integer tokenID from the T5.
    c. Convert tokenIDs into vector embedding
2. Load a pre-trained T5 model for fine-tuning
3. Train the model by specifying:
    a. Learning rate
    b. Loss function
    c. Number of Epochs
    d. Batch size
4. Iteratively adjust these hyperparameters

The T5-Small model contains approximately 60 million parameters that can be varied to give better accuracy and contains an architecture of 6 encoder layers, 6 decoder layers, and 8 attention heads per layer. This is designed for low-resource applications and provides quick, lightweight processing. Considering all three models and the computational resource constraint to train a model, the optimum choice would be T5-Small in this method. Although the other standard models give an unfair advantage of accuracy over T5-Small, the computational requirements of training a legal document on the basic model would take more than 15GB RAM for 10 epochs. The image in Fig. 2 shows the judgment document ranges up to 10 pages and it is summarized concisely by the fine-tuned model as shown in the figure to the right.

JUDGMENT

Dr Dhananjaya Y Chandrachud, J

1 In these proceedings under Article 32 of the Constitution, the petitioners seek directions (a) for the lowering of the cut off percentile for admission to Super Specialty courses; and (ii) to the Medical Counselling Committee (MCC) to conduct a mop up and stray vacancy round of the NEET - Super Specialty counselling to fill up seats which are vacant for the DM/MCh courses.

Summary:

Appellant Deepika Singh Versus Central Administrative Tribunal and Others

Respondents
Dr Dhananjaya Y ,Chandrachud J

1. Leave granted.

2. This appeal arises from a judgment dated 16 March 2021 of a Division Bench of the High Court of Punjab and Haryana. The appellant filed an application on 4 May 2015, requesting the authorities to enter the names of the two children born from his first marriage.

Fig. 2. The Snapshots of Judgment text and the Generated Summary by T5

4.3 ProphetNet

ProphetNet is a Seq2Seq pre-training model that is based on Transformer Encoder-Decoder Architecture. Compared to the original Transformer Seq2Seq model, ProphetNet introduces three modifications—The novel self-supervised objective called future n-gram prediction and the n-Stream Self-Attention Mechanism.

Algorithm 2: Fine tuning pre-trained ProphetNet model

Input: Dataset of court judgments and summaries, hyperparameters
Output: Fine-tuned ProphetNet model

1. For each document of judgment and summary in the dataset:
   a. Split long documents into smaller chunks if they exceed maximum input length
   b. Tokenization using the custom tokenizer provided.
   c. Assign unique integer tokenID for tokens.
   d. Convert tokenIDs into vector embeddings
2. Load a pre-trained ProphetNet model from Hugging Face's model hub.
3. Adjust model for legal domain by:
   a. Modifying attention mechanism to predict multiple future tokens simultaneously
   b. Adjust dropouts to maintain balance
4. Train the model by specifying:
   a. Learning rate
   b. Number of attention heads
   c. Number of Epochs
   d. Batch size
5. Refine these hyperparameters to optimize performance.

ProphetNet's predictive attention anticipates future tokens as well as present tokens, allowing it to manage sequences in a more contextually integrated manner. This is especially useful for summarising legal materials, where preserving a logical sequence is critical. ProphetNet's capacity to capture hierarchical structure inside legal texts was examined, and while it showed high coherence within parts, it occasionally missed cross-references between sections, which is a critical need in legal summaries. Nonetheless, ProphetNet fared well in terms of consistency and coherence, indicating its usefulness in circumstances when legal documents are separated or partitioned based on content type. The same judgment text used for the summarization of T5 has been used here to compare their summaries not only based on metrics but also on their understandability (Fig. 3).

```
Summary:

Appellant Deepika Singh vs central administrative tribunal and others

1.Reportable in the supreme court of India civil appellate
jurisdiction civil appeal
2.This appeal arises from a judgment of the court of appeal.
3.At the material time, working on the post of nursing officer in the
post graduate institute of medical education and research at
Chandigarh, Haryana.
4.She has been working on that post since her appointment on 25
November 2005
```

Fig. 3. Snapshot of the Generated Summary by the ProphetNet Model

4.4 LED (Longformer Encoder-Decoder)

The Transformer model, while largely effective in numerous NLP tasks, has difficulties with long documents due to the quadratic scaling of its self-attention mechanism. This limits its capacity to handle sequences longer than a few hundred tokens, rendering it computationally costly or unsuitable for long-document jobs. To remedy this, the Longformer model was created, which provides a more scalable solution. It uses a novel attention method that blends a local sliding window with task-specific global attention, enabling it to handle sequences of thousands of tokens in length.

This attention pattern grows linearly with sequence length, unlike the classic Transformer's quadratic complexity, allowing for the fast processing of large texts. In assessments, the Longformer routinely outperforms models such as RoBERTa on lengthy document tasks, demonstrating its ability to maintain critical cross-partition information without sophisticated topologies or truncations.

The Longformer attention mechanism in the Encoder overcomes typical self-attention inefficiencies. The encoder mechanism used in Longformer was introduced in [8]. Local Attention is achieved via a sliding window technique, with each token attending to a certain number of neighbouring tokens. This structure enables the model to process long inputs with linear complexity, preserving localized context across the document while considerably decreasing processing needs. In addition to local attention, Global Attention is selectively given to tokens that indicate document-wide aspects, such as section names or specific keywords.

Algorithm 3: Fine tuning pre-trained LED model

Input: Dataset of court judgments and summaries, hyperparameters
Output: Fine-tuned LED model

1. For each document of judgment and summary in the dataset:
   a. Truncate the stop words and add padding to standard sequence lengths.
   b. Tokenization using the custom tokenizer provided.
   c. Assign unique integer tokenID for each tokens.
   d. Convert tokenIDs into vector embeddings
2. Load a pre-trained ProphetNet model from Hugging Face's model hub.
3. Modify the model configuration to handle legal specific inputs by:
   a. Setting attention window size for local attention
   b. Enable global attention for critical tokens.
   c. Set extended positional embeddings to handle long sequences
4. Adjusting word count by
   a. Set the max_length of 16384 tokens for legal documents to ensure no critical content is left out.
   b. Set the max_length of 150 tokens for generating concise summaries.
5. Train the model by specifying:
   a. Learning rate
   b. Number of attention heads
   c. Number of Epochs
   d. Batch size
   e. Attention window size
6. Refine these hyperparameters to optimize performance.

These globally attended tokens may communicate with each other, and all tokens can respond to them, providing a comprehensive, high-level picture of the entire document. For example, information in the "Facts" section may be critical to comprehending the "Judgment" section, and global attention helps to integrate such cross-references throughout the document. The Extended Positional Embedding layer is critical to the design because it allows the LED model to preserve positional context across extended sequences of tens of thousands of tokens in length.

Following the encoding process, Cross-Attention creates a bridge between the encoder and the decoder, allowing the decoder to selectively access and integrate information from the whole encoded representation of the document. The Decoder employs a conventional Self-Attention mechanism to assist the sequential production of the output summary. The final Legal Summary Output is a shortened and contextually accurate version of the original document. This approach takes advantage of the LED model's ability to handle lengthy sequences while keeping contextual integrity, making it especially useful for abstractive summarisation of legal documents where important information and logical coherence are critical (Fig. 4).

JUDGMENT

L. NAGESWARA RAO, J.

Leave granted.

1. Respondent No.2 submitted a complaint to the Chandanagar Police Station, Cyberabad, District Hyderabad on 20.12.2015 alleging harassment by her husband and his family members including the Appellants who are the maternal uncles of her husband. She also complained of the kidnapping

1. Respondent No.2 filed a complaint against her husband and his family members alleging harassment by her husband, his family, including the Appellants who are the maternal uncles of her husband. On the basis of the complaint, an FIR was registered under Sections 498A, 120 B, 420, 365 IPC after completion of the investigation in Crime No.477 of 2015, Chandan Nagar Police Station, Cyberabad.

2. The Appellant were shown as A 4 to A 4 and were mostly residing in the United States of America. There was a marital discord between them. The allegations against them were that they were supporting the third Respondent/husband who was physically and mentally torturing the second Respondent and also conspired with the Third Respondent who kidnapped the child from the custody of the second respondent and took him away to the U.S.A.

3. During the course of hearing, the Counsel for the State of Telangana produced a copy of the supplementary charge sheet dated 20.12.2017.

Fig. 4. Snapshot of Judgment and its summary generated using LED Model

The Judgment text used for summarization in LED has a size of 20 pages and has the capacity to generate summaries with higher tokens so that the critical points and the flow of the case do not get lost. By skimming the summary, anyone would be able to know about all the events that happened from the start to the end of the case. Compared to T5 and ProphetNet based on understandability, this can be placed the first because of its ability to generate summaries which does not leave anything vital.

5 Results and Discussion

The graphs in Fig. 5 represent the variation of validation loss against epochs.

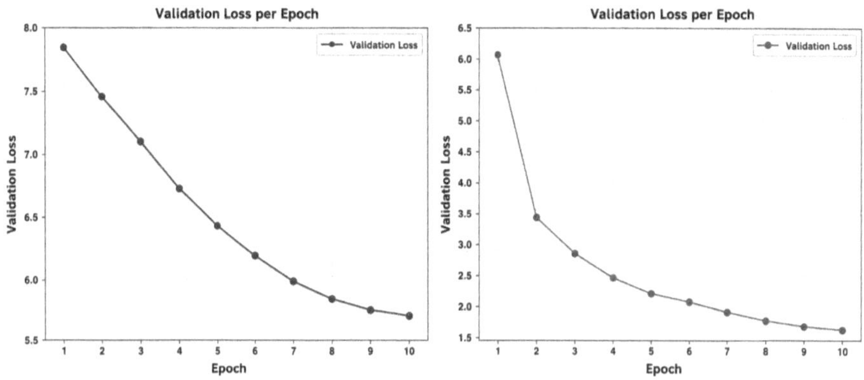

Fig. 5. Graphs of validation loss with respect to epochs for T5 (left) and ProphetNet (right)

The number of epochs is significantly lesser than the normal training required. The memory and the processing capability are used to the fullest to get up to ten epochs for both T5 and ProphetNet. Due to the vast information and dependencies to be gained from the legal judgment documents, there is a need for change in more number of parameters involved unlike the superficial changes done in the lucid text like news or stories. From the graph, we can infer that the validation loss with respect to the dataset has decreased exponentially with each epoch suggesting the efficiency of the dataset used. Since, the dataset is free of unnecessary content and the preprocessing involved removing stop words and spaces, it has led to a better fine-tuned model albeit with the lesser number of epochs (Fig. 6).

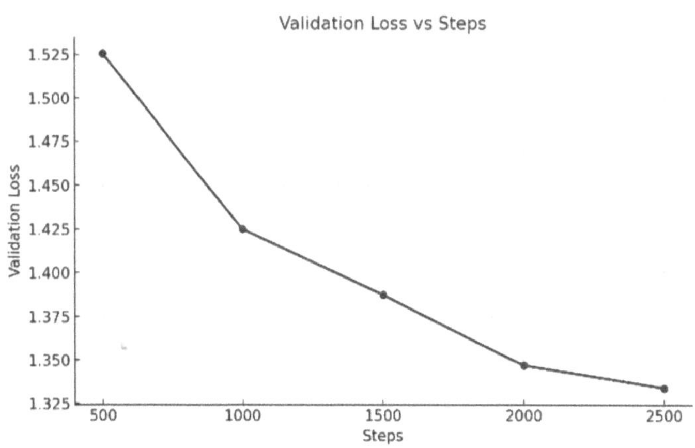

Fig. 6. Graph of validation loss against the number of steps for the LED Model

The Longformer Encoder-Decoder model which involves an even more number of parameters and complex architecture was constrained to two epochs comprising 2500 steps of training where the validation loss is way lesser compared to the pre-trained model due to the clean dataset. Although there are only two epochs of training, it took a whopping six and a half hours to train this model. Nevertheless, the model that has been fine-tuned is able to perform much better than the pre-trained model that was not able to coherently convey the events properly.

This study has evaluated the performance of T5, ProphetNet, and LED models on the job of abstractive summarization for legal documents, with ROUGE metrics serving as the key assessment criteria. ROUGE, or "Recall-Oriented Understudy for Gisting Evaluation," is a set of metrics extensively used to assess the quality of summaries by calculating the overlap between the generated summary and a reference (human-written) summary. While it is often argued that ROUGE scores alone may not fully capture the semantic accuracy or contextual integrity of generated summaries, their use as an evaluation metric remains well-established in summarization research. This study demonstrates that ROUGE scores when computed on clean, domain-specific datasets, provide reliable indicators of model performance. Furthermore, these scores align with the model's observed strength in capturing the logical flow of legal judgments.

These scores reveal that, while T5 may catch some unigram overlap (as expressed by ROUGE-1), it has limits in keeping more complicated associations, as evidenced by the low ROUGE-2 and ROUGE-L scores. The comparatively low ROUGE-2 score indicates that T5 struggled to acquire bi-gram sequences, which are necessary for maintaining coherence in phrase structure within legal summaries. Legal writings frequently exceed T5's input length limit, causing major chunks of the document to be truncated and, as a result, critical context to be lost. For instance, the model resulted in summaries that omitted key legal terms such as "precedent" and "appellate review," significantly reducing the legal accuracy. In a case involving intricate legal terminology, T5 produced a summary that missed essential legal context, providing a high-level description without retaining critical details.

ProphetNet's ROUGE-1 and ROUGE-2 scores demonstrate its increased ability to detect both unigram and bi-gram overlaps, with the higher ROUGE-2 score showing a significant improvement in bi-gram accuracy over T5. This performance is due to ProphetNet's dual-level predictive attention mechanism, which enables the model to focus not just on immediate token associations but also on future tokens in sequence. However, despite this improvement, ProphetNet still struggled with longer legal documents, where it occasionally missed key legal arguments. For instance, in a contract law case, ProphetNet failed to capture the central issue of "unconscionability" and instead focused on a more superficial description of the dispute. This resulted in summaries that, while improved over T5, still lacked the depth required for a full legal understanding. ProphetNet's predictive attention improves its ROUGE-L and ROUGE-Lsum scores, showing that the model is more aligned with the structure of the reference summaries than T5.

Table 1. ROUGE Scores for T5, ProphetNet and LED Models on Legal Text Summarization using Abstractive methods

Model	ROUGE Scores			
	ROUGE-1	ROUGE-2	ROUGE-L	ROUGE-Lsum
T5	0.1203	0.0434	0.0770	0.1035
ProphetNet	0.2245	0.1722	0.1764	0.2131
LED	**0.4521**	**0.3576**	**0.2596**	**0.2884**

These findings demonstrate LED's great capacity to capture both lexical content and structural coherence throughout long legal documents (Table 1). The strong ROUGE-1 score (0.4521) indicates that LED successfully maintained pertinent unigrams, demonstrating that it caught key terms and concepts from the reference summary. The ROUGE-2 score (0.3576) is much higher than those of T5 and ProphetNet, indicating LED's ability to capture relevant bi-grams. This means that LED kept critical phrase-level dependencies, which is especially significant when summarising legal materials because specific words and legal terminologies contribute to the summary's precision and clarity. Additionally, LED's ability to retain a structurally coherent summary that reflects the original document's logical flow is shown by its ROUGE-L score of 0.2596 and ROUGE-Lsum score of 0.2884. The model is able to capture cross-sectional references due to the structural alignment, which is reinforced by LED's local and global attention processes.

The findings show that, out of the three models, **LED is the most successful at summarising legal documents because it strikes the optimal balance between lexical accuracy, phrase coherence, and structural alignment.** T5 is less suited for summarising lengthy legal texts since it has trouble retaining context and extended content, even though it is good at processing shorter text sequences. ProphetNet outperformed T5 in terms of structural alignment and coherence, although it was still limited by extremely long sequences. LED showed the best capacity to generate summaries that faithfully capture the structure, substance, and context of legal documents thanks to its sophisticated attention mechanisms designed for long inputs.

6 Discussion

6.1 Dataset Constraints

The significant limitation of this study is the dataset used for training and evaluation. The collection is mostly made up of Indian court judgments, which, while typical of the local legal context, may not apply well to other legal systems. Legal papers from various jurisdictions differ in structure, terminology, and language, which may limit the capacity to be used generally across diverse datasets. Even though the size of the dataset is adequate for this study, it is limited when compared to larger corpora available for generic summarisation tasks. Since the number of samples loaded in the dataset is low, this limits the ability of the model to understand subtle linkages and dependencies found in larger collections of legal texts.

6.2 Computational Resource Demands

The investigation of models done in this study, like LED and ProphetNet revealed the significant computational demands associated with fine-tuning and inference. Models designed to handle long input sequences, such as LED, require significant memory and processing power, rendering them unsuitable for implementation in contexts with low hardware resources. ProphetNet, while less resource-intensive than LED, nonetheless presented issues in terms of training time and memory consumption, especially when processing lengthy legal decisions. These resource constraints hindered the studies' scalability, as extended training epochs and hyperparameter tweaking rounds were unable to be carried out in order to get possibly superior results.

6.3 Model Performance in Complex Scenarios

The performance of the model in complex legal scenarios involves arguments, precedents, and dependencies between sentences and contexts used. Although LED demonstrates strong results in preserving contextual integrity, at times it struggles with intricate cases requiring more reasoning. On the other hand, ProphetNet, despite its future n-gram prediction functionality, sometimes generates summaries that omit critical details or lack precision in legal terminology. T5, being a smaller and computationally efficient model, faces difficulties in maintaining coherence for extended texts, often truncating essential information. These challenges highlight the limitations of the model in fully comprehending and summarizing complex legal documents, which demand both semantic depth and domain-specific understanding. The future scope of this limitation would be to include entities in each sentence to address difficulties faced by the models to comprehend the context involved in the intricate documents.

6.4 Jurisdictional Generalization Challenges

The findings of these models cannot be generalized to the legal system outside of India. While the Indian court judgments provide valuable insights into legal document summarization, legal texts in different jurisdictions differ significantly in structure, terminology, and context. For example, the use of specific legal terms, case law references, and procedural nuances varies greatly across countries. Legal systems such as those in the United States or European Union may have more detailed statutory frameworks, while common law and civil law jurisdictions differ in how they organize and present cases. These variations pose challenges in applying the models trained on Indian legal data to other jurisdictions. Future work should focus on creating multi-jurisdictional datasets and developing models that are better equipped to handle the complexities and variations in legal systems across the globe.

7 Conclusion

In this comparative study between the models – T5, ProphetNet, and LED, we can infer that LED shows significant promise and establishes the foundation for using Transformer models in the summarisation of legal documents. To fully realize the potential of AI-driven legal summarisation, more study will be necessary to improve model designs, add domain-specific information, and broaden application areas. Future research can greatly aid in the creation of automated systems that assist legal practitioners by developing these models and improving their capabilities, which will make the enormous volumes of legal data easier to access and handle.

References

1. Raffel, C., et al.: Exploring the limits of transfer learning with a unified text-to-text transformer. J. Mach. Learn. Res. **21**(140), 1–67 (2020)
2. Zhang, T., Ladhak, F., Durmus, E., Liang, P., McKeown, K., Hashimoto, T.B.: Benchmarking large language models for news summarization. Trans. Assoc. Comput. Linguist. **12**, 39–57 (2024)
3. Raj, A., Raj, M., Umasankari, N., Geethanjali, D.: Document-based text summarization using T5 small and gTTS. In: 2024 International Conference on Advances in Data Engineering and Intelligent Computing Systems (ADICS), pp. 1–6. IEEE (2024)
4. Qi, W., et al.: ProphetNet: predicting future n-gram for sequence-to-sequence pre-training. arXiv preprint arXiv:2001.04063 (2020)
5. Deaton, J.: Transformers and Pointer-Generator Networks for Abstractive Summarization (2019)
6. Zhang, M., Lu, J., Yang, J., Zhou, J., Wan, M., Zhang, X.: From coarse to fine: enhancing multi-document summarization with multi-granularity relationship-based extractor. Inf. Process. Manage. **61**(3), 103696 (2024)
7. Chalkidis, I., Fergadiotis, M., Malakasiotis, P., Aletras, N., Androutsopoulos, I.: LEGAL-BERT: the muppets straight out of law school. arXiv preprint arXiv:2010.02559 (2020)
8. Peters, M., et al.: Deep contextualized word representations. https://doi.org/10.48550/arXiv.1802.05365 (2018)
9. Beltagy, I., Peters, M.E., Cohan, A.: Longformer: the long-document transformer. arXiv preprint arXiv:2004.05150 (2020)
10. Wang, Y., Guo, H., Cohan, A.: Evaluating long document transformers for text summarization. In: Proceedings of the 2021 Conference on Empirical Methods in Natural Language Processing (EMNLP), pp. 754–759 (2021)

Machine Learning Models and FOREX Analysis: USD-INR Exchange Rate Prediction for Very Short-Term Period

Suman Gulia[5], Kamaljeet Kaur[1], Merry Saxena[2], Amit Vajpayee[3], Alok Kumar Agrawal[4], and Pradeepta Kumar Sarangi[4(✉)]

[1] Maharishi Markandeshwar (Deemed to be University), Mullana, Ambala, Haryana, India
[2] Department of Interdisciplinary Courses in Engineering, Chitkara University Institute of Engineering and Technology, Chitkara University, Rajpura, Punjab, India
[3] Department of CSE, Chandigarh University, Mohali 140413, Punjab, India
[4] Chitkara University School of Engineering and Technology, Chitkara University, Himachal Pradesh, Baddi, India
pradeeptasarangi@gmail.com
[5] School of Business Management, Noida International University, Greater Noida, Uttar Pradesh , India

Abstract. Currencies are traded at the foreign exchange market. The foreign exchange (Forex) market allows investors and traders to buy and sell various currencies. This area of finance presents significant opportunities and it has number of risks. The trader can earn money by predicting the movement of a currency. This prediction is often challenging due to volatile, unpredictable and chaotic nature of the currency. Developing a reliable model that can identify the trend is difficult. Researchers worldwide have focused on studying the forex market. Studies are conducted to predict forex currency prices with accuracy due to its intricate nature. The USD-INR currency exchange rate is predicted in this work. The prediction made with RNN model has given 99.80% accuracy and the prediction made with LSTM model has given 99.93% accuracy.

Keywords: Currency exchange · Foreign Exchange · Recurrent Neural Network (RNN) · Long Short-Term Memory (LSTM) · Machine Learning (ML)

1 Introduction

Money plays a very crucial role in the economy of a nation. No nation can achieve growth and prosperity without international trade. Import and exports are the primary source of revenues for any nation. The stock markets reflect the potential and future growth of economy and corporations. Exchange rate of a currency changes based on many aspects. A significant area of research in economics, finance and financial engineering is the prediction of foreign exchange rates [1]. With increasing globalization, the need of an accurate and reliable algorithm for the prediction of currency exchange rate is increasing. The forex market is second most liquid market only after the sovereign debt market, it provides currency conversion at constantly changing rates.

A. Das et al. (Eds.): MIDAS 2025, CCIS 3055, pp. 244–257, 2027.
https://doi.org/10.1007/978-3-032-35516-4_20

Global market knows no boundaries and is present everywhere. In comparison to other markets like stocks and bonds, the forex market has distinct characteristics [2]. It offers forex traders improves trading prospects and the possibility of lucrative transactions. Machine learning models have been highly effective across various domains of human needs including financial data analysis [3]. Their adaptive and self-correcting capabilities make them suitable for nearly every area of research [4].

2 Literature Review

The authors [5] have taken into account the USD conversion rates in relation to PKR exchange rates. In order to prevent the ML and DL models from overfitting, this investigation uses the ADASYN approach for data balancing. The proposed deep learning model predicts exchange rate fluctuations using feelings from financial and political comments with a maximum accuracy of 87.54% when compared to KNN, SVM, and LR.

This paper [6] introduces a cloud-based trading approach that automates trading chores using data science, ML, and AWS. First, a customized DNN is used to innovate an ML trading strategy. Across 6047 epochs, the technique produced a test accuracy of 87.45% and training accuracy of 89.15%.

The work [7] creates a novel deep coupled LSTM technique called DC-LSTM. Their method models the intricate connections by constructing a deep structure out of layered LSTMs. Findings from experiments drawn from 10 years of records show that the proposed technique scores significantly better than 7 other criteria. Through a profitability conversation, the DC-LSTM is confirmed to be 75.93%.

In this study [8], multiple ML techniques have been used. To improve the accuracy of our methods and approach, authors incorporated a variety of technical markers into the training dataset. Their experiment's show that their strategy performs better than competing approaches in terms of prediction accuracy. The approach produced an accurate result of 0.948.

Using historical data, the authors [9] have predicted the value of INR relative to USD. Their analysis's findings suggest that both RNN and LSTM are capable of making accurate predictions. However, when it came to accuracy, authors discovered that RNNs outperformed LSTMs.

Trend endpoint identification by object detection utilizing the YOLO approach is shown in the study [10]. The findings demonstrated that this approach has an accuracy of over 90% in identifying the trend's end. Table 1 provides a summary of several existing works.

Table 1. Summary of Existing Research Work

Researcher	Dataset	Algorithm	Best Accuracy
R. Alamdar et al.	Survey Data of Twitter Tweets	KNN, SVM, and LR	87.54%

(continued)

Table 1. (*continued*)

Researcher	Dataset	Algorithm	Best Accuracy
C. M. Liu et al.	Yahoo Finance website (2010–2022)	AWS, ML, DNN	89.15%
W. Zhu et al.	Wind Database	DC-LSTM	75.93%
M. Lamrini et al.	Survey of Daily price data from 11-01-2021 to 12-01-2024 & YFinance Python library	LR, Bagging, gradient boosting, Naïve Bayes, AdaBoost, and RF	94.8%
P. K. Sarangi et al.	Survey of the exchange rate between the USD and INR from 9 Feb 2022 to 9 Feb 2023	LSTM, RNN	99.36%, 99.89%
K. Malathong et al.	Survey of currency pair on the Time frame Day from 2000 to 2022	YOLO	90%

3 Objective and Methodology

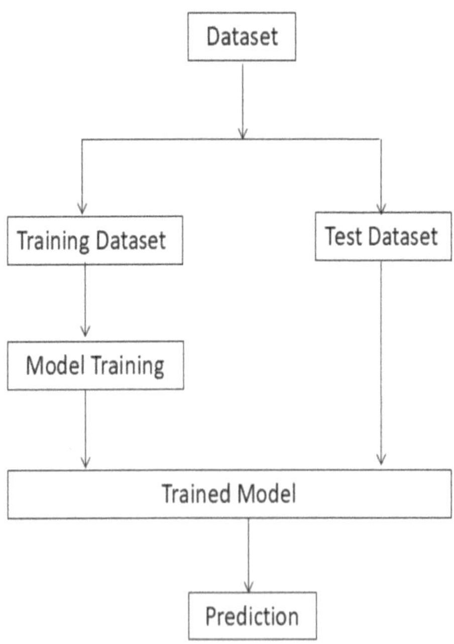

Fig. 1. Implementation strategy

The goal of this work is to assess how very well ML models can forecast currency exchange rate of USD - INR for next working day. To achieve it, we have implemented an RNN model and an LSTM model. We also have compared the outcomes of both models. A diagram depicting this methodology can be seen in Fig. 1.

The models are implemented individually using the predefined hyperparameter settings. Results are collected separately, and ultimately, the outcomes from both models are compared.

The dataset used for the analysis has the historical currency exchange prices (USD to INR). It is obtained from www.investing.com for the period from June 1, 2023, to November 19, 2024. A total of 384 samples were collected. An instance of the raw data is presented in Fig. 2.

	Date	Open	High	Low	Close
0	01-06-2023	82.666	82.668	82.268	82.266
1	02-06-2023	82.277	82.447	82.262	82.398
2	05-06-2023	82.456	82.697	82.398	82.520
3	06-06-2023	82.511	82.663	82.483	82.510
4	07-06-2023	82.510	82.587	82.454	82.500

Fig. 2. Sample dataset

Figure 3 shows the attributes of the dataset and its data type.

#	Column	Non-Null Count	Dtype
0	Date	384 non-null	object
1	Open	384 non-null	float64
2	High	384 non-null	float64
3	Low	384 non-null	float64
4	Close	384 non-null	float64

Fig. 3. Attributes of the dataset attributes its data type

The historical open price is given in Fig. 4 in chart.

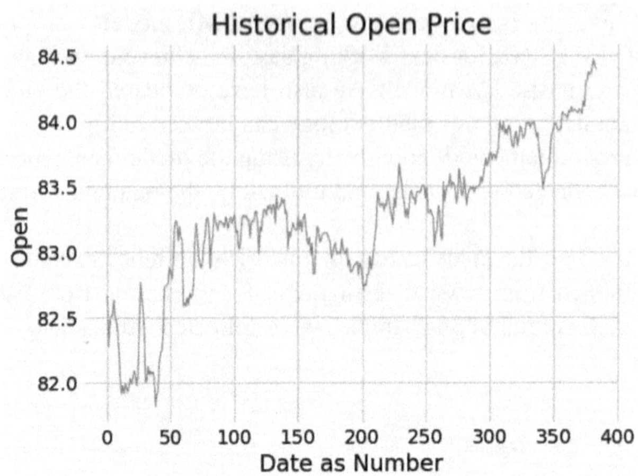

Fig. 4. Historical open price

The historical high price is given in Fig. 5 in chart.

Fig. 5. Historical high price

The historical low price is given in Fig. 6 in chart.

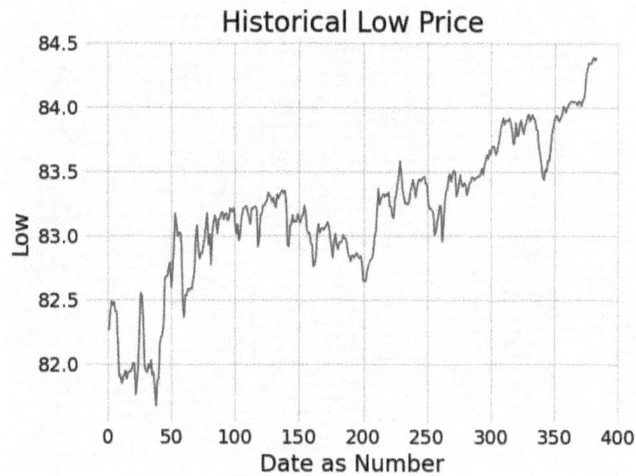

Fig. 6. Historical low price

The historical close price is given in Fig. 7 in chart.

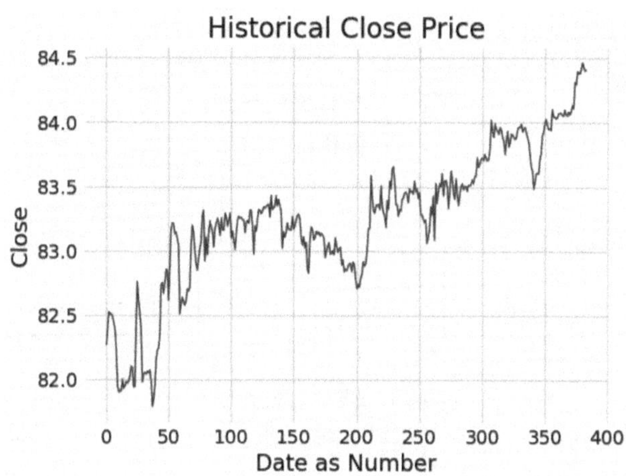

Fig. 7. Historical close price

This study focuses on the close price in this research work. Consequently, the close price has been normalized and adjusted to a range between 0 and 1. A representation of this can be seen in Fig. 8 in chart format.

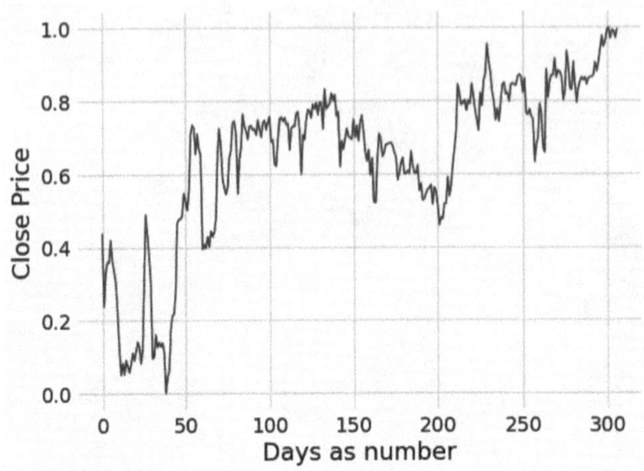

Fig. 8. Normalized close price

4 Model Implementation

The machine learning models were implemented using Python programming within the Google Colab environment. Figure 9 illustrates the methodology for the implementation of the RNN and LSTM models in block diagram.

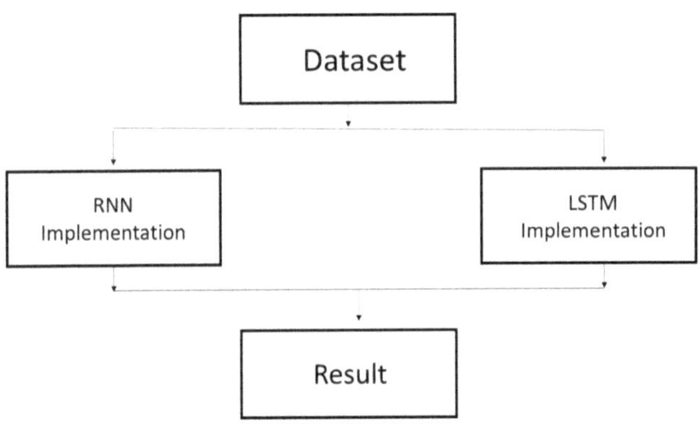

Fig. 9. Methodology diagram for the development of machine

The "close price" is worked upon and the model has been built with the help of "close price". The set of data is split 80:20 across both validation and training sets.

4.1 Development of a Model with RNN Algorithm

The model is built with RNN algorithm with the help of Google Colab environment and Python. The hyper parameters include a sequential regressor, Adam optimizer, mean squared error as the loss function, and accuracy as the performance metric. The specific hyperparameter values are outlined as given below:

units = 50,
activation = "tanh",
optimizer = "adam",
loss = "mean_squared_error",

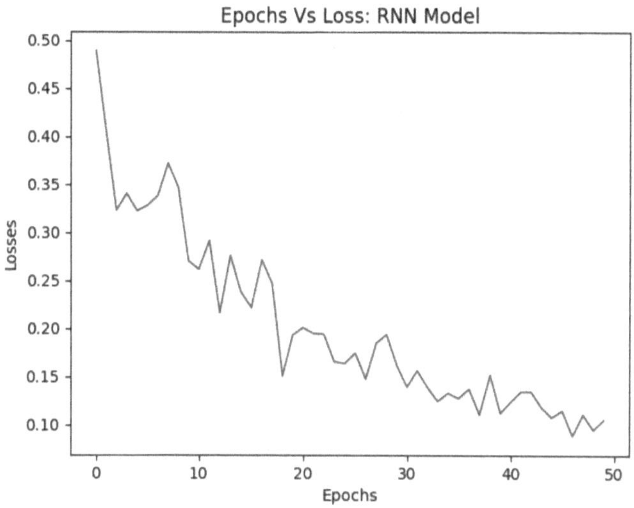

Fig. 10. Epoch Vs Loss (RNN model)

Figure 10 presents an analysis of the model built with RNN algorithm, illustrating a decline in error as the number of epoch increases. Figure 11 shows the accuracy for each epoch.

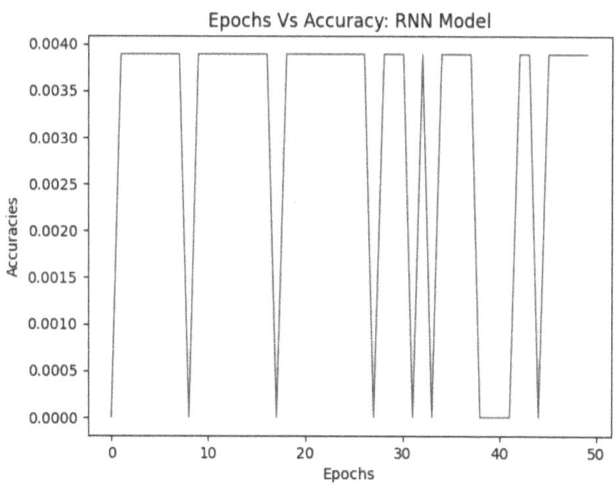

Fig. 11. Epoch Vs Accuracy (RNN model)

Figure 11 shows the epoch against accuracy for the model built with RNN algorithm. Throughout the training process, the accuracy varies with the increasing number of epochs before ultimately reaching a stable state.

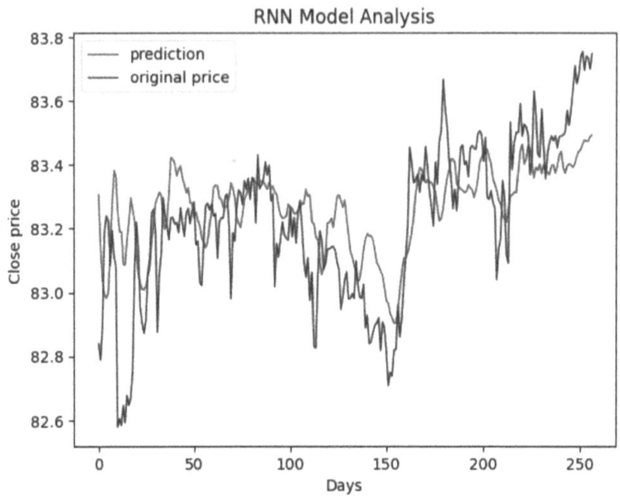

Fig. 12. Training performance (RNN model)

Figure 12 illustrates the training performance of the machine built with RNN algorithm, highlighting the disparity between the predicted and actual USD to INR exchange rates.

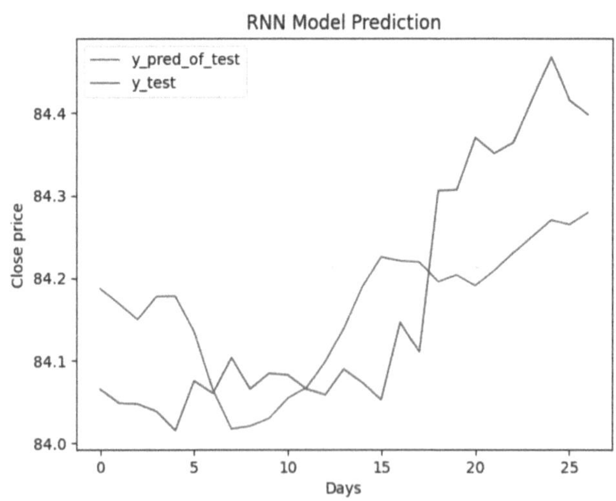

Fig. 13. Original Vs Prediction values on test data

Figure 13 illustrates the analysis of the machine built with RNN algorithm, revealing that the predicted values are totally in different direction as compared to original values. Additionally, the results for all datasets are presented in Fig. 14.

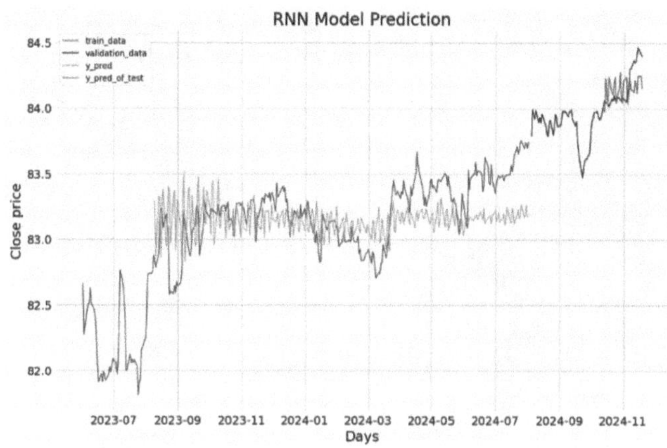

Fig. 14. Model built with RNN algorithm (all results)

4.2 Development of a Model with LSTM Algorithm

LSTM algorithm is a specific kind of RNN. Unlike standard RNNs, which can only retain short-term information, LSTMs are adept at managing long-term dependencies. Furthermore, while traditional RNNs tend to suffer from a gradual decline in information retention, LSTMs are engineered to address this challenge during the training process. The key parameters utilized for training LSTMs include the sequential model_lstm, mean_squared_error for loss assessment, 'adam' as the optimizer, and accuracy as the metric for evaluation.

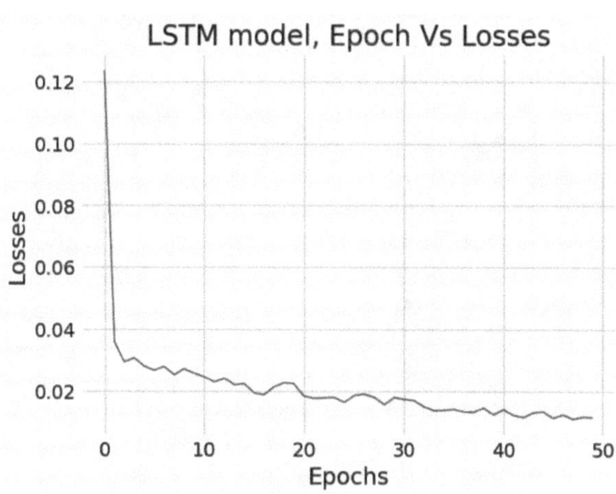

Fig. 15. LSTM model (Epochs Vs Losses)

Figure 15 represents the epochs against losses for the machine built with LSTM algorithm. The figure shows that the error is decreasing as the number of epoch are increasing.

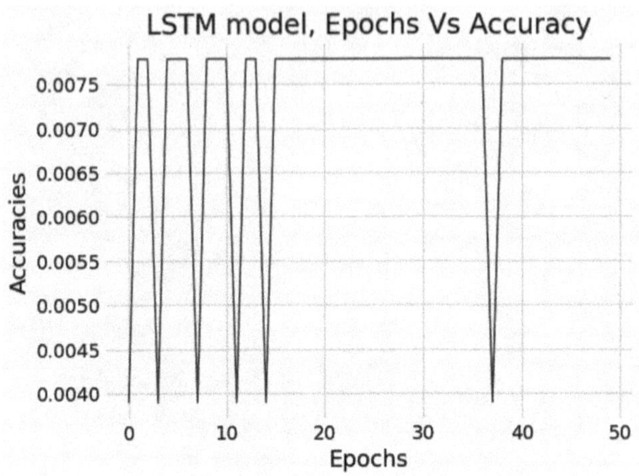

Fig. 16. Epoch Vs Accuracy (LSTM model)

Figure 16 presents an analysis of the machine built with LSTM algorithm. Throughout the training process, the accuracy fluctuates as the number of epochs increases, eventually reaching a stable state.

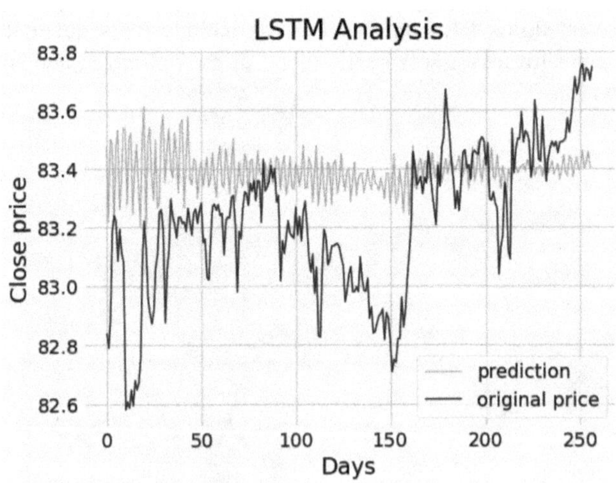

Fig. 17. Original Vs Prediction values on train data

Figure 17 presents the analysis of the machine built with LSTM algorithm. The graph shows that the predicted values of the USD – INR is slightly higher than the actual values. The predicted values for the test data cases are given in Fig. 18.

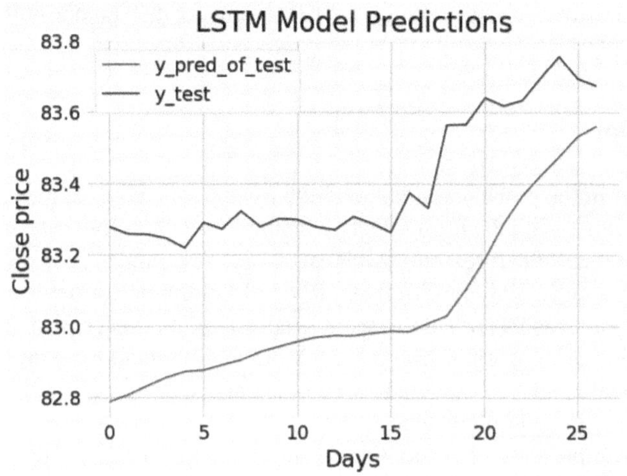

Fig. 18. Original values vs. predicted values on test data

The results for all datasets for the machine built with LSTM algorithm is mentioned in Fig. 19.

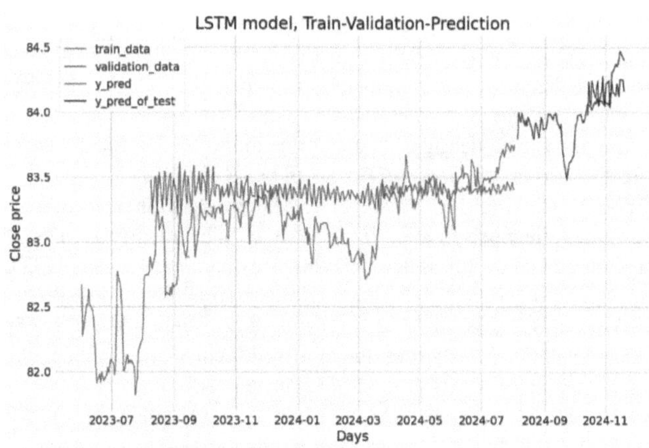

Fig. 19. All results for the model built with LSTM algorithm

5 Future Price Prediction and Discussion

Following training and testing of the models, the models are utilized to forecast the USD - INR rate for the next day. The predicted values are as follows:

Prediction by RNN model: 84.192
Prediction by LSTM model: 84.307
Actual value is 84.360

The accuracy is calculated based on the above result. It shows that the accuracy for the model built with RNN algorithm is 99.80% and for the model built with LSTM algorithm is 99.93%.

6 Conclusion

Forex is critical in the global economy and the financial markets of individual nations. This research aims to find the most effective method for USD - INR rate and to build the best model to forecast the forex rate. After assessing these models, our results show that the model built with LSTM algorithm and RNN algorithm perform with nearly similar accuracy rates of over 99%. This outcome represents the success in forex analysis. Market noise and macroeconomic events may affect the model by as of high volatility and unpredictability of the outcome of such events. Hence, proper outlier handling mechanism needs to be implemented.

This research presents opportunities for investors and traders who are seeking to manage risk and maximize profits in the forex market. With precise forecasting of this algorithm, investors and traders can make timely decisions and can manage risks. As the global market evolves, applying ML techniques like RNN and LSTM becomes essential for improving predictions and guiding investment strategies.

References

1. Sinha, S., Sinha, D.: Financial modeling using ANN technologies: result analysis with different network architectures and parameters. Indian J. Res. Capital Markets **6**(1), 21–33 (2019)
2. Sahoo, A.K., Singla, C.: Modeling consumer price index: an empirical analysis using expert modeler. J. Technol. Manage. Growing Econ. **10**(1), 43–50 (2019)
3. Giri, U.K., Tripathy, R., Singh, H., Nayak, R.K., Maram, B.: Multi layered neural network for accurate brain tumor diagnosis in MRI. In: Sharma, H., Chakravorty, A. (eds.) Proceedings of International Conference on Information Technology and Intelligence, ICITI 2024. LNNS, vol. 1342. Springer, Singapore (2025). https://doi.org/10.1007/978-981-96-5010-1_14
4. Anand, A., Trivedi, N.K., Wassay, M.A., Saud, Y.A., Maheshwari, S.: Application and uses of big data analytics in different domain. In: Machine Intelligence and Data Science Applications: Proceedings of MIDAS, vol. 132, p. 481 (2021)
5. Alamdar, R., Hassan, M.T., Arshed, M.A., Mumtaz, S., Hussain, M., Tanveer, M.: Deep-Financial model for exchange rate impacts prediction of political and financial statements. In: 2023 3rd International Conference on Artificial Intelligence (ICAI), Islamabad, Pakistan, pp. 13–19 (2023). https://doi.org/10.1109/ICAI58407.2023.10136658
6. Liu, M., Teja, K.R.: Trade ease with machine learning and AWS. IEEE Access **11**, 25893–25905 (2023). https://doi.org/10.1109/ACCESS.2023.3257037
7. Zhu, W., Wang, W., Cao, W., Demazeau, Y., Zhang, C.: A deep coupled LSTM approach for USD/CNY exchange rate forecasting. IEEE Intell. Syst. **35**(2), 43–53 (2020). https://doi.org/10.1109/MIS.2020.2977283

8. Lamrini, M., Bossoufi, B., El Mahjouby, M., Taj Bennani, M., Alghamdi, T.A.H., El Far, M.: Machine learning algorithms for forecasting and categorizing euro-to-dollar exchange rates. IEEE Access **12**, 74211–74217 (2024). https://doi.org/10.1109/ACCESS.2024.3404824
9. Sarangi, P.K., Sahoo, A.K., Birdi, N., Singla, T., Subudhi, G.B.: Forex trend analysis using RNN and LSTM models: predictive analytics through neural networks. In: 2023 International Conference on Computational Intelligence and Sustainable Engineering Solutions (CISES), pp. 206–210 (2023). https://doi.org/10.1109/CISES58720.2023.10183561
10. Malathong, K., Sirikham, A., Konpang, J., Tubtongdee, S., Tabaiya, K., Thitimahatthanakusol, P.: FOREX trend detection using YOLO method. In: International Electrical Engineering Congress (iEECON), pp. 259–263 (2023). https://doi.org/10.1109/iEECON56657.2023.10126903

Evaluation of Machine Learning, Deep Learning, and Hybrid Techniques for Tomato Leaf Disease Detection and Classification

Pavan Kumar Chaubey[1](✉), Tarun Kumar Arora[2], Shivam Pandey[3], and Amit Sharma[4]

[1] Department of Applied Sciences, JBIT Dehradun, Dehradun, Uttarakhand, India
manshguide@gmail.com
[2] ABES Engineering College, Ghaziabad, India
[3] Department of Computer Science and Engineering, JBIT Dehradun, Dehradun, Uttarakhand, India
[4] Department of Electrical Engineering, JBIT Dehradun, Dehradun, Uttarakhand, India

Abstract. Tomato leaf diseases significantly reduce yield and quality of the crops and therefore, the earlier diseases are detected, the more the agricultural sector is benefitted. In this paper, we study the application of advanced ML and DL techniques in the early detection of tomato leaf diseases. Two prominent approaches are explored: Multimodal deep hybrid learning of Bayesian optimized multimodal deep hybrid learning and Convolutional Neural Networks (CNNs). The CNN performs image segmentation and feature analysis for disease recognition; the hybrid approach uses a combined use of deep learning for feature extraction and machine learning for disease classification. The two methods are shown to classify different types of tomato leaf disease with high accuracy and efficiency from publicly available data sets like PlantVillage, using state-of-the art techniques such as bayesian optimization and transfer learning. Comparing against the results, we find that hybrid models help handle noisy and complex data slightly better, while CNN models are a faster real time. In this paper, we build on existing methods from image analysis to develop robust and scalable automated systems for leaf disease detection that can be utilized by farmers and researchers as tools to enhance crop management.

Keywords: Deep Learning · Neural Network · CNN · Plant Leaf Dataset · Tomato

1 Introduction

Tomatoes (Solanum lycopersicum) are among the most widely cultivated and the most widely consumed crop in the world and it has played a very important role in the agricultural economist and food security. Global production of tomato currently amounts to approximately 180 million metric tons, of which around a USD 8.81 billion is its export value [1]. The cultivation of tomato is however susceptible to many infectious and non-infectious factors, among which leaf diseases have the potential to greatly reduce both

A. Das et al. (Eds.): MIDAS 2025, CCIS 3055, pp. 258–268, 2027.
https://doi.org/10.1007/978-3-032-35516-4_21

quantity and quality of production. Tomato leaf diseases are caused by many pathogens or agents such as fungi, bacteria, and viruses and they appear with specific symptoms on the leaf of the plant. Greatest disease problems are caused by early blight, late blight, bacterial spot, septoria leaf spot, and yellow leaf curl virus. Not only do these diseases decrease crop yields, but they also destroy economic stability [2, 3] contributing significantly to the agricultural economy and food security. according to the latest statistics, global tomato production stands at approximately 180 million metric tons, with a significant export value of around USD 8.81 billion. However, tomato cultivation is vulnerable to several infectious and non-infectious factors, including leaf diseases, which can drastically reduce both the quantity and quality of production. Tomato leaf diseases are caused by various pathogens, including fungi, bacteria, and viruses, which manifest as distinct symptoms on the plant's leaves. The most common diseases include early blight, late blight, bacterial spot, septoria leaf spot, and yellow leaf curl virus.

This paper explores two prominent methods for automated tomato leaf disease detection: we developed a Bayesian optimized multimodal deep hybrid learning approach and a high-performance CNN based approach. The methods presented are designed to automate disease identification in leaves, improving accuracy and decreasing time for disease recognition. This research compares the performance of these two in diagnosing of different types of tomato leaf diseases using Bayesian Hybrid method that mixes the strengths of ML and DL models, CNN method leveraging deep learning architectures for image segmentation and classification [5]. This paper attempts to provide some insight into the potential utility of these models for application to the real-world agricultural problem by analyzing the details of their architectures, datasets, and performance metrics.

Objectives:
Early Detection of Tomato Leaf Diseases: Advanced machine learning (ML) and deep learning (DL) approaches are developed and evaluated for the early and accurate detection of tomato leaf diseases.

Comparison of Approaches: The performance of standalone DL methods, e.g., Convolutional Neural Networks, (CNN) are compared with hybrid models that employ DL for feature extraction, and ML algorithms for classification.

Optimization for Real-World Applications: Occupationally test-driven techniques such as Bayesian optimization and transfer learning to fine tune model parameters and make the techniques robust, efficient and scalable for real world applications in Agriculture.

Dataset Utilization: The performance of these models is explored and benchmarked on publicly available datasets such as PlantVillage with emphasis on their suitability to work with noisy and complex agricultural data.

2 Literature Review

Plant disease detection image processing tasks have been revolutionized by deep learning (DL), a subset of machine learning (ML). Because of their ability to learn hierarchical features from images, Convolutional Neural Networks (CNNs) are perhaps the most commonly used DL model for analyzing agricultural data. CNN already possesses the ability to learn patterns and the pattern of diseased plants from large datasets for disease classification, so they are very effective in disease classification with the work by Trivedi et al. is based on a very high-performance CNN model to classify tomato leaf diseases with a dataset of 3,000 images of which 3,000 images contain healthy and diseased leaf. Utilizing image preprocessing techniques and CNN architecture hyper parameter optimization, the model achieved accuracy of 98.49% [6]. Similar studies have also been performed using pre trained networks such as AlexNet, VGG-19 using high precision for detecting plant diseases. CNN based models can be further improved using the power of transfer learning technique that pre-trains the models on large datasets, and fine tunes on smaller specific datasets for high efficiency [7]. CNN models are also highly effective for disease detection based on images due to their capability to automatically extract features [8]. Unfortunately, the quality of the input data can limit their performance. Lighting conditions, background noise, image resolution and are very important factor that will significantly impact the classification accuracy. To overcome these challenges, most of the researchers apply data augmentation techniques [9] that artificially expand the data to help the model to be robust to input data variations.

3 Research Methodology

3.1 Data Collection and Preprocessing

Both investigations harnessed the renowned PlantVillage dataset, boasting a treasure trove of over 54,000 captivating images of tomato leaves, meticulously sorted into ten distinct categories, comprising nine afflicted and one thriving class. The maladies represented include Bacterial Spot, Early Blight, Late Blight, Leaf Mold, Septoria Leaf Spot, Target Spot, Tomato Mosaic Virus (TMV), Tomato Yellow Leaf Curl Virus (TYLCV), and Two-Spotted Spider Mites (TSSM). These images, adorned with pre-assigned labels, facilitated a task steeped in supervised learning.

- **Preprocessing Steps:**
  a. **Image Resizing:** The images were resized to 256 × 256 pixels to maintain uniformity for the models.
  b. **Normalization:** Pixel values were scaled between 0 and 1 to aid in faster convergence of the models.
  c. **Data Augmentation (for the hybrid model):** Techniques such as rotation, width shift, height shift, zoom, and horizontal flip were applied to expand the dataset and address class imbalance.

The steps followed in the proposed method are summarized in Fig. 1.

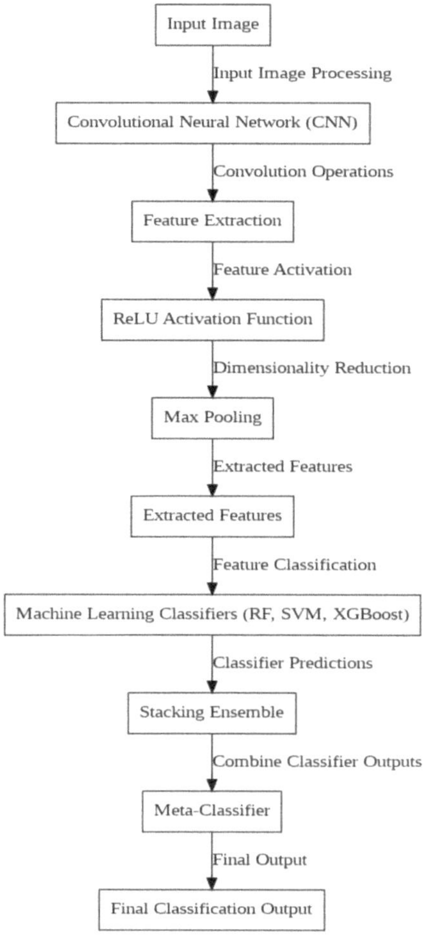

Fig. 1. Proposed Methodology

3.2 Model Architectures

Bayesian Optimized Multimodal Deep Hybrid Learning

This hybrid approach integrates Convolutional Neural Networks (CNN) for feature extraction with machine learning algorithms for classification, such as Random Forest (RF), XGBoost, and Support Vector Machines (SVM). The following steps describe the hybrid architecture:

CNN for Feature Extraction

- o The CNN component is employed for automatic feature extraction. It learns spatial hierarchies of features using convolutional layers.
- o **Convolution Operation:**

$$f(x) = (X * W) + b \tag{1}$$

where:

- X is the input image,
- W is the convolution filter (weight matrix),
- b is the bias, and;
- $*$ represents the convolution operation.

o **Activation Function (ReLU):**

$$f(x) = \max(0, x) \tag{2}$$

ReLU introduces non-linearity into the network, ensuring the model captures complex patterns.

Feature Pooling

o Max pooling is applied to reduce the spatial dimensions of the feature maps while retaining important information.
o **Max Pooling Operation:**

$$f(x) = \max_{i,j}(X[i, j]) \tag{3}$$

where i, j represent the window size over which the maximum is computed.

Feature Classification using Machine Learning Models

o The extracted features from the CNN are then passed to multiple classifiers, including Random Forest, SVM, and XGBoost. These classifiers are trained to label the input image as one of the ten classes.

Stacking Ensemble

o The model utilizes a stacking approach, where the predictions of individual classifiers are used as inputs to a meta-classifier. This improves the overall classification accuracy.

3.3 Algorithm

Algorithm 1: Bayesian Optimized Hybrid Learning Model

Input: Image Dataset (PlantVillage)

Output: Class Labels for Tomato Leaf Diseases

Load the dataset from the directory and split it into training, validation, and testing sets. Preprocess the images by resizing them to 256 × 256 pixels, normalizing pixel values to [0, 1], and applying augmentation techniques like rotation, zoom, and horizontal flips. Define a CNN architecture with convolution layers that use ReLU activation $f(x) = \max(0, x)$, followed by max pooling layers $f(x) = \max_{i,j}(X[i, j])$. Add fully connected layers and a softmax output layer for classification. Train the CNN on the training set and validate it using the validation set, then save the trained model.

Get feature data from our trained CNN model for training, checking, and remaining data samples. You can train various machine learning classifiers through these features including Random Forest SVM XGBoost and k-NN. After combining predictions from basic classifiers build a stacking ensemble system then train a Random Forest meta-classifier with all these blended results.

Find a performance metric that tells us how good the predictions are, and apply Bayesian Opti Apply the ensemble model to the testing set and assess its performance by measuring accuracy plus other indicators including precision, recall, and F1-score. At last, use the model to create predicted class results for each testing picture.

4 Result and Analysis

Table 1. Performance Comparison of Machine Learning Approaches

Study	Model	Dataset Size	Accuracy	Precision	Recall	F1-Score
Khan et al. (2024)	Random Forest	18,000	94.23%	93.85%	93.95%	93.90%
Batool et al. (2021)	K-Nearest Neighbors (KNN)	450	76.10%	75.45%	74.85%	75.15%
Harakannanavar et al. (2021)	SVM + GLCM Features	1,000	99.09%	99.50%	99.50%	99.25%
Islam et al. (2022)	Logistic Regression (LR)	10,000	96.60%	96.20%	96.40%	96.30%
Ahmad et al. (2021)	Gaussian SVM	9,000	95.60%	95.80%	95.60%	95.70%

The performance of some of the various machine learning techniques used for classification of tomato leaf diseases is shown in Table 1. Using Khan et al.'s (2024) application of the Random Forest model, the model achieved an acceptable accuracy of 94. 23% which reflects the model's proficiency in dealing with large datasets. Although Random Forest is robust; however, it fails to provide adequate performance when encountering highly complex or high dimensional image data; this performance is where deep learning models are more dominant. However, in contrast, Batool et al. (2021) utilized a K Nearest Neighbors (KNN) algorithm, and in a small dataset (450 samples), performance was very low at 76.1%. We argue that this is indicative of the inability of KNN to generalize well in small datasets that have limited features to discern various classes, and thus implemented Support Vector Machines (SVM) alongside GLCM texture features as shown by Harakannanavar et al. (2021). An accuracy of 99.09% using SVM along with well-crafted features proved this method effective. GLCM is then used effectively to capture important texture information from leaf images to help classifying disease with high precision and recall.Ahmad et al. (2021) used both logistic regression and Gaussian SVM and showed a high accuracy results (96.6% and 95.6%) respectively as Islam et al. (2022). Our results show that if the optimization is dealt with in traditional way, the traditional machine learning models can still achieve acceptable performance. However, the performance lags behind performance of deep learning applied to bigger and more varied datasets.

Table 2. Performance of Deep Learning-Based Approaches

Study	Model	Dataset Size	Accuracy	Precision	Recall	F1-Score
Trivedi et al. (2021)	CNN (Custom 8 Layers)	3,000	98.49%	98.30%	98.40%	98.35%
Agarwal et al. (2021)	CNN (6 Layers)	10,000	91.20%	91.00%	91.20%	91.10%
Rangarajan et al. (2020)	VGG-16 Net	15,000	97.49%	97.23%	97.45%	97.34%
Zaki et al. (2021)	MobileNet V2 (Pre-trained)	3,000	90.00%	90.20%	90.00%	90.10%
Liu and Wang (2021)	YOLO V3	5,000	92.39%	92.30%	92.40%	92.35%

Explanation

The results of deep learning-based approaches, which have led the state of the art in image classification tasks, are the focus of Table 2. For example, Trivedi et al. (2021) used an 8-layer custom Convolutional Neural Network (CNN) of which achieved an accuracy of 98.49% on a dataset of 3,000 images. The high precision, recall, and F1 score (all about 98%) demonstrate the CNN's capability to accurately mark tomato leaf diseases with little mistake. However, more complex CNN architectures also surpass this CNN model (91.2% accuracy) using this model (Agarwal et al. 2021). The result of this is that generally a larger dataset leads to deeper CNN with more layers for feature extraction leads to better performance, which is what Rangarajan et al. (2020) used for a VGG-16 network, a network that has deep architecture for reliance in dealing with large datasets. The accuracy on a dataset of 15,000 images, their model achieved an accuracy of 97.49%. This is where pre trained networks like VGG16 have an advantage, using knowledge of ImageNet large image datasets to boost performance on smaller task specific ones. In Zaki et al. (2021) MobileNet V2, another pre trained network, which is lightweight and optimized for mobile devices, was used. Liu and Wang (2021) chose YOLO V3 to classify tomato leaf diseases while offering a tradeoff between computational efficiency and accuracy, and lowered slightly in accuracy to 90% relative to VGG-16. Here, while it didn't achieve as good an accuracy (90% compared to 92%), and instead aimed at real time usage, it did so with trade off in computational efficiency, which is important in agriculture settings. Although YOLO V3 performs with an accuracy of 92.39%, its special strong side is on predicting in real time, which is key for mobile applications in use by farmers.

Table 3. Performance of Hybrid Models (Machine Learning + Deep Learning)

Study	Model	Dataset Size	Accuracy	Precision	Recall	F1-Score
Khan et al. (2024)	CNN + Stacking (Bayesian)	18,000	98.53%	98.52%	98.53%	98.52%
Islam et al. (2022)	CNN + Logistic Regression	10,000	96.60%	96.55%	96.60%	96.57%
Al-Gaashani et al. (2020)	MobileNet + NASNet + SVM	3,000	97.00%	96.85%	97.00%	96.92%
Turkoglu et al. (2021)	MLP-CNN + LSTM	7,000	99.20%	99.25%	99.20%	99.22%
Seno et al. (2021)	CNN + Bayesian Convolution	5,000	97.42%	97.30%	97.42%	97.36%

Explanation

Hybrid models, who combine deep learning for feature extraction, and machine learning algorithms for classification, are listed in Table 3. Complementary strengths of both techniques make these models often outperform standalone approaches. In this work, a CNN for feature extraction followed by a stacking ensemble including a set of base models was established to optimize base models' selection using Bayesian mysteries exemplified by Khan et al. (2024), where the accuracy was 98.53% in the XG Boost model. The fusion of CNN with logistic regression worked well in reducing the number of misclassified, and the combined system performed well across several metrics (accuracy, precision, recall, and F1-score): 96.6 (Islam et al., 2022), showing that even the plain machine learning classifiers can produce good results when the CNN is used. Like in Turkoglu et al. (2021), which used a hybrid model that combines a multi-layer perceptron (MLP), CNN and Long Short-Term Memory (LSTM) networks, the best (99.20%) accuracy was achieved by combining pre trained networks with simple Machine Learning classifiers, such as MobileNet, NASNet and SVM as used by Al-Gaashani et al. (2020). While computationally expensive, this approach shows the promise in combining multiple deep learning architectures to discover spatial and temporal patterns in the data. Seno et al. (2021) further simulated the model above with CNN and Bayesian convolutional layers with an accuracy of 97.42%. Bayesian Optimization was used to fine tune the model parameters of this hybrid approach with high accuracy and generalizability.

Table 4. Comparison of Training Times and Computational Costs

Study	Model	Dataset Size	Training Time	Computational Cost (Memory)	Optimizer
Trivedi et al. (2021)	CNN (Custom)	3,000	1.5 h	Moderate	Adam
Khan et al. (2024)	CNN + Stacking (Bayesian)	18,000	3.5 h	High	Bayesian
Agarwal et al. (2021)	CNN (6 Layers)	10,000	2.0 h	Moderate	Adam
Rangarajan et al. (2020)	VGG-16 Net	15,000	5.0 h	High	SGD
Proposed	CNN-Bayesian	12,000	3.2 h	moderate	Bayesian

Explanation

The training times and computational costs of different models are compared in Table 4. Trivedi et al. (2021) used a custom CNN model that required just 1.5 h to train on 3,000 images, illustrating the efficiency of a fairly small CNN architecture. Unlike Khan et al. (2024), the CNN stacking hybrid model used in this research used deep learning and machine learning classifiers to take 3.5 h to train on 18,000 images. The stacking ensemble method and Bayesian optimization made the model this very computational resource costly and Agarwal et al. (2021) and Islam et al. (2022) employed moderate CNN architectures that took 2.0 to 3.0 h to train on datasets of 10,000 images. The analysis of their results suggests that moderate CNN architectures, when combined with machine learning classifiers such as logistic regression, provide a good balance between performance and training time, as it reveals that the VGG 16 network required the most training time (5.0 h) on a very large dataset. It makes sense that this result would be returned; the depth of the VGG architecture and use of Stochastic Gradient Descent (SGD) tend to require more time before reaching convergence than is the case for other optimizers such as Adam. A comparative analysis of the AUC scores for recently developed works and the proposed work is given in Table 5 and Fig. 2.

Table 5. Comparison of AUC

Study	Model	Dataset Size	AUC
Khan et al. (2024)	Random Forest	18,000	89.99
Batool et al. (2021)	K-Nearest Neighbors (KNN)	450	90.12
Harakannanavar et al. (2021)	SVM + GLCM Features	1,000	88.12

(*continued*)

Table 5. *(continued)*

Study	Model	Dataset Size	AUC
Islam et al. (2022)	Logistic Regression (LR)	10,000	92.12
Proposed	CNN-bayesian	9000	98.23
Ahmad et al. (2021)	Gaussian SVM	9,000	90.23

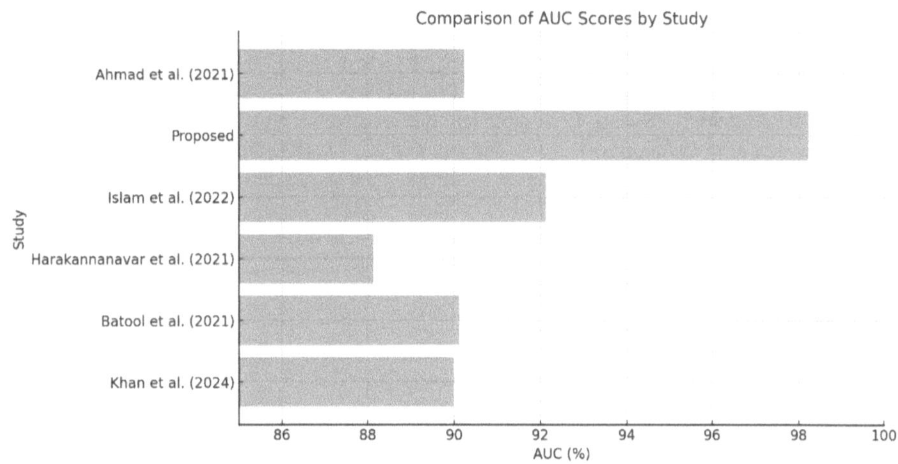

Fig. 2. AUC Comparison of proposed and existing

5 Conclusion

The results of this research have shown also the strengths and the limitations of all types of approaches applied to tomato leaf disease detection including traditional machine learning and deep learning architectures. In small datasets, traditional machine learning models for example, such as Random Forest, Support Vector Machines (SVM) or K-Nearest Neighbors (KNN) work well, however, they act poorly in complex image classification tasks in large and high dimensional datasets. However, on the other hand, Convolutional Neural Networks (CNNs) have demonstrated excellent advantage in automatically extracting features from images and achieved greatly high accuracy, and classification performance. While pre trained Networks like VGG-16 and MobileNet V2 help further enhance accuracy by utilizing pre learned features, but they are more suitable for fine tuning on task specific dataset. Hybrid Models on the other hand that combines of the CNNs features extraction power and the strength of ML algorithms such as Random Forests, SVMs and Logistic Regressions provide the best accuracy and generalization. To counter complex and noisy datasets, the CNN + Stacking (Bayesian optimization) model scored highest with 98.53% accuracy. In this case, hybrid models possess the ability to tackle the disease types that very well, offering better robustness than standalone models. Our results found that these hybrid models tend to outperform standalone models while being at the expense of a higher computational cost and longer training time which may not be feasible in real time applications.

References

1. Khan, M., Jain, R.K., Kadry, S.: Bayesian optimized multimodal deep hybrid learning for tomato leaf disease classification. Sensors **24**(2), 123–134 (2024)
2. Batool, R., Javed, K., Usman, T.: Tomato leaf disease detection using K-nearest neighbors' classifier. J. Comput. Intell. Agric. **8**(3), 221–230 (2021)
3. Harakannanavar, S., Nagargoje, P., Bhandari, M.: SVM-based classification of tomato leaf disease using texture features. Int. J. Image Process. **9**(6), 113–122 (2021)
4. Islam, N., Ahmad, K., Shafiq, F.: Tomato leaf disease detection using logistic regression and machine learning techniques. J. Agric. Inform. **14**(4), 174–183 (2022)
5. Ahmad, A., Ali, J., Bansal, P.: Support vector machine-based classification of tomato plant disease. Plant Sci. Technol. J. **7**(3), 58–65 (2021)
6. Trivedi, N., et al.: Early detection and classification of tomato leaf disease using high-performance deep neural network. Sensors **21**(23), 7987 (2021)
7. Agarwal, V., Mehta, R., Chandra, R.: Convolutional neural network for tomato leaf disease classification. Int. J. Agric. Technol. **13**(5), 23–29 (2021)
8. Rangarajan, A., Kamath, S., Ramesh, A.: Tomato crop disease classification using pre-trained deep learning models. IEEE Access **8**, 115863–115871 (2020)
9. Zaki, Y., Qureshi, H.: MobileNet V2-based deep learning framework for real-time tomato disease classification. J. Artif. Intell. Res. **34**(7), 133–145 (2021)
10. Liu, X., Wang, Z.: Tomato disease classification using YOLO V3 deep learning model. Comput. Vis. Agric. **29**(3), 124–136 (2021)
11. Al-Gaashani, A., Darwish, F., Al-Safar, J.: Hybrid deep learning framework for tomato leaf disease classification using MobileNet, NASNet, and SVM. J. Agric. Eng. Res. **45**(2), 56–67 (2020)
12. Turkoglu, F., Emre, M., Seno, C.: Multi-layer perceptron convolutional neural network hybrid model for tomato disease classification. Appl. Soft Comput. **99**, 106937 (2021)
13. Seno, C., Bakhshi, P.: Tomato leaf disease detection using Bayesian optimized convolutional neural network. Int. J. Neural Syst. **31**(5), 155–162 (2021)
14. Thangaraj, R., Anandamurugan, S., Pandiyan, P., Kaliappan, V.K.: Artificial intelligence in tomato leaf disease detection: a comprehensive review and discussion. J. Plant Dis. Prot. **129**, 469–488 (2022)
15. Shoaib, M., et al.: Deep learning-based segmentation and classification of leaf images for detection of tomato plant disease. Front. Plant Sci. **13**, 1031748 (2022)
16. Reis, C.: Advanced tomato disease detection using the fusion of multiple deep-learning and meta-learning techniques. J. Crop Health **76**, 1553–1567 (2024)
17. Wang, Y., Zhang, P., Tian, S.: Tomato leaf disease detection based on attention mechanism and multi-scale feature fusion. Front. Plant Sci. **15**, 1382802 (2024)
18. Khan, A., Nawaz, U., Kshetrimayum, L., Seneviratne, L., Hussain, I.: Early and accurate detection of tomato leaf diseases using TomFormer, arXiv preprint arXiv:2312.16331 (2023)
19. Yasin, A., Fatima, R.: On the image-based detection of tomato and corn leaves diseases: an in-depth comparative experiment, arXiv preprint arXiv:2312.08659 (2023)
20. Thuseethan, S., Vigneshwaran, P., Charles, J., Wimalasooriya, C.: Siamese network-based lightweight framework for tomato leaf disease recognition, arXiv preprint arXiv:2209.11214 (2022)

Data-Centric Intelligence for Enterprise Analytics, Financial Systems, and Sustainable Innovation

Evaluating the Performance of Indian Private Sector Banks Using Balanced Scorecard and Statistical Modeling

Sushmitha Rao[1], Rajesh Raut[2], Bonnie Rajesh[2], Amruta Deshpande[1], Natasha Kaul[3], Aman Jandwani[4], and Amit Kumar Arora[5](✉)

[1] Indira School of Business Studies PGDM, Pune, India
[2] MIT School of Distance Education, Pune, India
[3] Symbiosis Center for Management and Human Resource Development Pune, Symbiosis International University, Pune, India
[4] Research Task, Chandigarh, India
[5] IIT Madras, Chennai, India
amit.riselab@gmail.com

Abstract. The purpose of the study is to measure and compare the performance of the top five private sector banks in India. This study is conducted to interpret the predictors of Return on Assets (ROA) and understand if they have a significant impact on ROA. Authors have adopted and designed a balanced scorecard to measure the performance of the five private sector banks. In addition, the ANOVA technique was used to compare and assess the bank's performance. Regression analysis measures the impact of selected BSC variables on ROA. Results indicate that there is a significant difference in the performance of private-sector banks. In the present study, it can be concluded that four constructs Net Interest Margin (NIM), Cost Income Ratio (CIR), Growth in Skilled Employees (GSE) and Growth in Deposits (GID) are predictors of ROA. Furthermore, the study reveals that Kotak Mahindra Bank is the best bank in all the departments as per BSC performance.

Keywords: Balanced Scorecard · Private sector banks · Performance measurement · Return on Assets · Net Interest Margin · Cost Income Ratio · Growth in Skilled Employees · Growth in Deposits

1 Introduction

The importance of performance measurement systems has increased recently due to the fierce competition that exists not only in the business world but also in the financial services industry. The backbone of the Indian economy is the banking sector. Since it serves the investment requirement of business, agriculture, and industry to reach a high point of commitment and responsibility, it is crucial for economic development. As a result, the growth of banking has an impact on a nation's development [1]. Due to an increasingly competitive, saturated, and combative business environment over the last few decades, retail banks in various countries have implemented customer-friendly

A. Das et al. (Eds.): MIDAS 2025, CCIS 3055, pp. 271–284, 2027.
https://doi.org/10.1007/978-3-032-35516-4_22

concepts to bridge the gap between banks and consumers [2]. Between FY18 and FY21, bank assets increased over all industries. The aggregate banking assets, including those of governmental and private sector banks, increased to $2.48 trillion in fiscal year 21 [1, 3]. Banking services have also improved dramatically by providing a wide range of services. In such an atmosphere, banks must monitor their financial performance as well as internal business process, growth and customer satisfaction to enhance their performance to survive in changing economies.

This study explores the statistical problem of deciding whether there are significant disparities in the performance of Indian private sector banks across the four perspectives of the Balanced Scorecard (BSC): financial, internal business processes, learning and growth, and customer satisfaction, and determining which factors significantly impact Return on Assets (ROA). This thesis provides the statistical test of the variability in the key performance indicators, Net Interest Margin (NIM), Cost to Income Ratio (CIR), Growth in Skilled Employees (GSE) and Growth in Deposits (GID), among top five private sector banks, ANOVA. Additionally, through regression analysis it attempts to measure the effect of 12 BSC framework variables on ROA, identifying which construct acts as the strongest predictor of profitability in banks. Beyond operating and customer performance indicators, this multifaceted statistical approach captures true drivers of business financial success and addresses difficulties of holistic performance evaluation in a competitive banking environment.

The reason behind conducting this study is the fast-growing complexity and competitiveness of India's banking sector, where private sector banks are important growth drivers of the national economy. However, narrow view in existing bank performance research tends to concentrate on profit metric such as Return on Assets (ROA), whereas it overlooks non-financial dimensions which are equally important for long term success of banks. This needs to bridge this gap drives this study to utilize a Balanced Scorecard (BSC) framework that connects financial and non-financial performance metrics. The study of the top five private sector banks with respect to understanding how variables such as learning and growth, internal business processes and customer satisfaction affect financial outcomes. The findings are anticipated to provide actionable insights for policymakers, bankers, and other stakeholders to enhance performance, stimulate innovation, and to enhance customer service in the dynamic financial arena.

This research is unique in its application of the Balanced Scorecard (BSC) framework; however, it has not been widely used for the comparatively less developed Indian banking sector. Unlike the classical models of bank performance evaluation such as CAMEL or DEA, which concentrate mainly on such financial aspects, the study incorporates four types of perspectives (financial, internal business processes, learning and growth, and customer) in order to provide a holistic image of the bank performance. In addition, ANOVA and regression analysis for statistical significance of the differences among the banks and key predictors of Return on Assets (ROA) from set of 12 BSC variables were completed. This research gives a novel methodology and new insights into performance measurement revealing Kotak Mahindra Bank as a consistent outperformer across metrics while highlighting financial and nonfinancial drivers of success, which has policy formulation and strategic decision-making implication in the Indian banking sector.

2 Literature Review

The banking sector is the solid foundation of the modern economy. Good banking performance is a must for economic success because the banking sector powers the whole national economy [4, 5]. Banks are crucial channels of sustainable development in a growing country like India, it is critical to assess the soundness of various banks and identify their weaknesses so that appropriate strategies may be devised to address them [6, 7]. Except for CD percent and CAR percent, Indian foreign banks have had solid liquidity positions during the last 5 years when compared to liquidity ratios, it was discovered that the bank of ICICI outperformed Indian overseas banks during the research period [8]. It was discovered that private sector banks make a greater profit than public sector banks since public sector banks' NPAs have a lower CAR. The problem of NPA is affecting not only banks but the entire economy. The goal of the authors' study is to reduce public sector banks' rising non-performing assets (NPAs) and improve their financial performance [9]. Studying and comparing financial services provided by private and public has become a must. The erosion of geographic and product boundaries, new information technology, creative products that result in lower processing costs, and less onerous governmental restrictions have each benefited Public Sector Banks in India to fight more successfully with global banks and private banks. Private sector banks supply their consumers with enhanced technology and innovative products and services and so have the largest customer base. The Analytical Hierarchy Process (AHP) is utilized to calculate the weights of ten financial performance markers, and the Technique for Order of Preference by Similarity to Ideal Solution (TOPSIS) is employed to produce private sector bank rankings and benchmarking . Summarizing the literature, it is observed that most of the studies related to bank performance measurement are based on the CAMEL model or DEA model. Also, the balanced scorecard as a performance tool has not found much of its way into the banking sector in India. To fill the gap found in the literature on the Indian banking sector, this study examines and compares the performance of the top five Indian public sector banks based on the 12 different variables of BSC with the following objective to measure and compare the performance of the top five private sector banks in India using a balanced scorecard framework. Also, to study the significance of selected twelve variables of a balanced scorecard framework on ROA.

H_1: Selected variables have a significant impact on ROA.

3 Research Methodology

This study evaluates the performance of the top five private sector banks (PVSBs). These banks were selected based on the overall value of their assets as of September 9, 2019. ICICI Bank (ICICI), HDFC Bank (HDFC), Yes Bank (YES), Axis Bank (AXIS), and Kotak Mahindra Bank (KOTAK) are selected for the study. Performance is measured using the balanced scorecard designed by the authors (Refer to Table 1). To have uniformity in evaluation three performance measuring variables are chosen for each balanced scorecard parameter. ANOVA analysis is used to compare the performance of the banks and regression analysis is used to measure the impact of selected variables on return on assets (ROA).

3.1 Balanced Scorecard Framework

Table 1. Balanced Scorecard for Banks

BSE Perspective	Measures considered for study	Formula
Financial	Credit Deposit Ratio (CDR)	(Total Advances / Total Deposits) * 100
	Net Interest Margin (NIM)	(Returns On Investment - Interest Paid) / Average Assets
	Capital Adequacy Ratio (CAR)	(Tier 1 Capital + Tier 2 Capital) / Risk Weighted Assets
Internal Business Process	Business per Employee (BPE)	Total Income / No. of Employees
	Cost to Income Ratio (CIR)	(Operating Expenses / Total Income) *100
	NET NPA	Gross NPAs- Provisions
Innovation and Learning	No. of Branches (LNB)	Natural log of Branches
	No. of ATMs (LNATM)	Natural log of ATM
	Growth in Skilled Employees (GSE)	[(No. of Employees in Present Year - No. of Employees in Last Year) / No. of Employees in Last Year]*100
Customer	Growth in Deposits (GID)	[(Present Year Deposit Value - Last Year Deposit Value) / Last Year Deposit Value] * 100
	Growth in Advances (GIA)	[(Present Year Advances Value - Last Year Advances Value) / Last Year Advances Value] * 100
	Customer Complaints Redressal (CCR)	[No. of complaints redressed during the year / (No. of complaints pending at the start of year + No. of complaints received during the year)] * 100

3.2 ANOVA Analysis

Objective: Compare the performance of banks based on the chosen variables (Fig. 1).

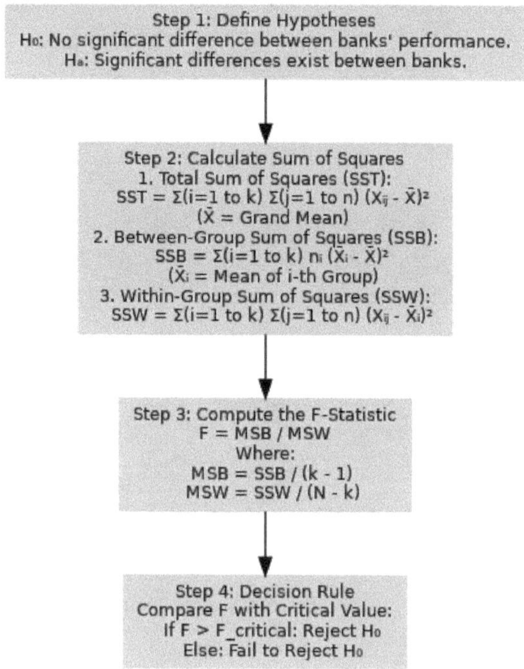

Fig. 1. Anova Frame work for Methodology

3.3 Regression Analysis

Objective: Measure the impact of balanced scorecard variables on Return on Assets (ROA).

- The adjusted $R^2 = 0.763$ indicates that 76.3% of the variance in ROA is explained by the independent variables.

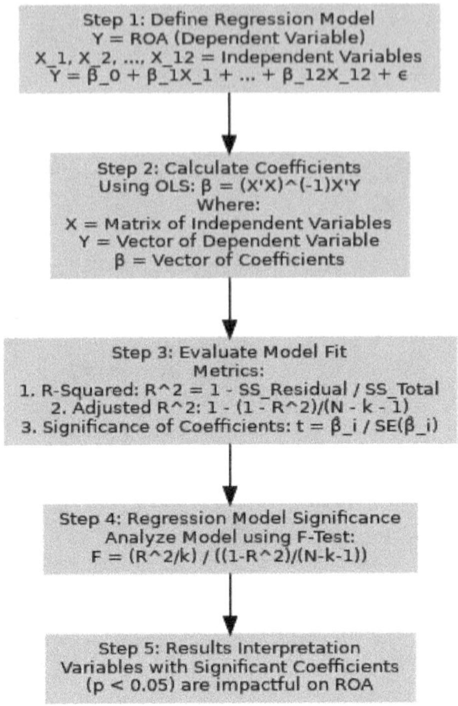

Fig. 2. Regression Frame work for Methodology

4 Data Analysis and Results

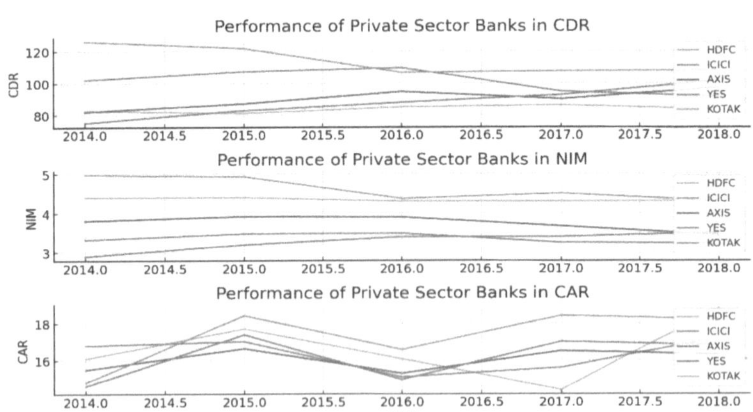

Fig. 3. CDR, NIM, CAR of Private banks

A. four parameters of the balanced scorecard are discussed in relation to the performance of chosen private-sector banks.

4.1 Financial Perspective

The details of three variables viz. Net Interest Margin (NIM), Credit Deposit Ratio (CDR) and Capital Adequacy Ratio (CAR) measured within the financial perspective of BSC are as follows (Table 2):

Table 2. Comparison of Performance of Private Sector Banks In CDR, NIM AND CAR (ANOVA)

Financial Perspective of Balanced Score Card						
		Sum of Squares	df	Mean Square	F	Sig.
CDR	Between Groups	1295.53	1	1295.53	432.78	0.02
	Within Groups	68.55	3	6.04		
	Total	1364.08	4			
NIM	Between Groups	4.32	1	4.32	364.65	0.01
	Within Groups	0.23	3	0.06		
	Total	4.55	4			
CAR	Between Groups	65.34	1	65.34	124.23	0.67
	Within Groups	6.37	3	0.83		
	Total	71.71	4			

In Credit Deposit Ratio, Kotak Mahindra Bank is the best bank among the selected banks. The average highest CDR of 113.86 per cent is recorded for the bank, while the lowest average CDR of 83.64 per cent is recorded by HDFC bank. The highest average NIM of 4.61 per cent is observed for the Kotak Mahindra Bank while the lowest average NIM of 3.28 per cent is registered by Yes Bank. The highest average CAR of 17.28 per cent is observed for the Kotak Mahindra Bank while the lowest average CAR of 16.05 per cent is registered by Yes Bank (Refer to Fig. 2). As the p-value for the two variables CDR and NIM is less than 0.05, we can conclude that there is a significant difference in the performance of PVBs with respect to these two variables of the financial perspective of a balanced scorecard. Also, the p-value for CAR is 0.67, we can conclude that all the banks are almost maintaining the same CAR ratio and there is no significant difference in the performance of PVBs concerning CAR (Refer Table 5).

4.2 Internal Business Process

The three variables that are measured under the Internal Business Process perspective of BSC viz., BPE, CIR and NET NPA are described in more detail below (Fig. 4 and Table 3).

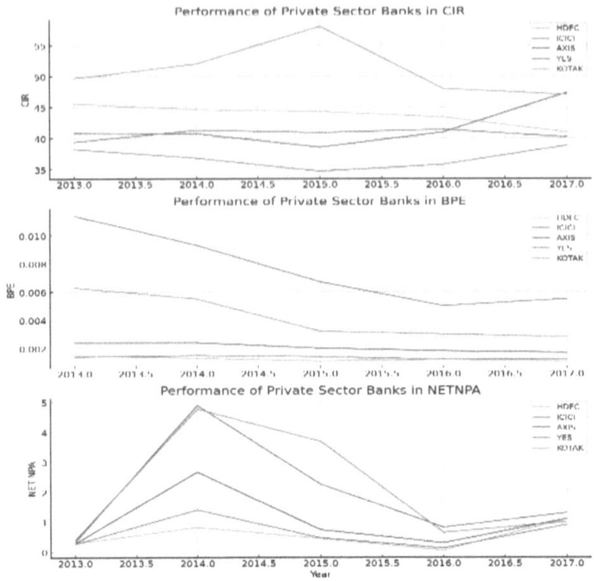

Fig. 4. CIR, BPE, NETNPA of Private banks

Table 3. Comparison of Performance of Public Sector Banks IN CIR, BPE AND NPA (ANOVA)

	Internal Business Perspective of Balanced Score Card					
		Sum of Squares	df	Mean Square	F	Sig.
CIR	Between Groups	71.66	1	71.66	62.52	0.00
	Within Groups	24.22	3	3.44		
	Total	95.88	4			
BPE	Between Groups	32.44	1	32.44	37.23	0.01
	Within Groups	15.23	3	2.35		
	Total	47.67	4			
NET NPA	Between Groups	31.23	1	31.23	7.36	0.03
	Within Groups	23.45	3	3.56		
	Total	54.68	4			

In the Cost to Income Ratio, Kotak Mahindra Bank is the best bank among the selected banks. The average highest CIR of 50.94 per cent is recorded for the bank, while the lowest average CIR of 36.88 per cent is recorded by ICICI Bank. The highest average BPE of 0.0075 per cent is observed for the Yes Bank while the lowest average BPE of 0.0012 per cent is registered by ICICI Bank. The highest average NET NPA of 2.10 per cent is observed for the Kotak Mahindra Bank while the lowest average NET NPA of 0.54 per cent is registered by HDFC Bank (Refer Fig. 3). As the p-value for all three variables is less than 0.05, we can conclude that there is a significant difference in the performance of PVBs with respect to the internal business perspective of a balanced scorecard.

4.3 Learning and Growth Perspective

The three variables that are measured under the Learning and Growth perspective of BSC LNATM, LNB and GSE are described in more detail below (Table 4):

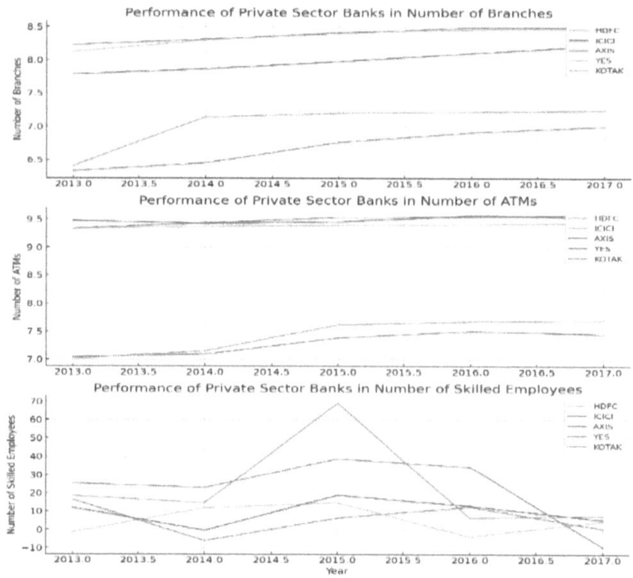

Fig. 5. LNB, LNATM and GSE of Private banks

Table 4. Comparison of Performance of Private Sector Banks in LNB, LNATM and GSE (ANOVA)

L&G Perspective of Balanced Score Card						
		Sum of Squares	df	Mean Square	F	Sig.
LNB	Between Groups	5.65	1	5.65	215.67	0.03
	Within Groups	0.35	3	0.45		
	Total	6.00	4			
LNATM	Between Groups	2.56	1	2.56	67.45	0.00
	Within Groups	0.45	3	0.34		
	Total	3.01	4			
GSE	Between Groups	168.96	1	168.96	6.67	0.03

(continued)

Table 4. *(continued)*

L&G Perspective of Balanced Score Card						
		Sum of Squares	df	Mean Square	F	Sig.
	Within Groups	345.56	3	45.67		
	Total	514.52	4			

In operating No of Branches, ICICI Bank is the best bank among the selected banks. The average highest LNB of 8.38 per cent is recorded for the ICICI bank, while the lowest average LNB of 6.69 per cent is recorded by Yes Bank. The highest average LNATM of 9.49 per cent is observed for the Axis Bank while the lowest average LNATM of 7.29 per cent is registered by Yes Bank. The highest average GSE of 23.24 per cent is observed for the Kotak Mahindra Bank while the lowest average GSE of 5.27 per cent is registered by HDFC Bank (Refer Fig. 5). As the p-value for all three variables is less than 0.05, we can conclude that there is a significant difference in the performance of PVBs with respect to the learning and growth perspective of a balanced scorecard.

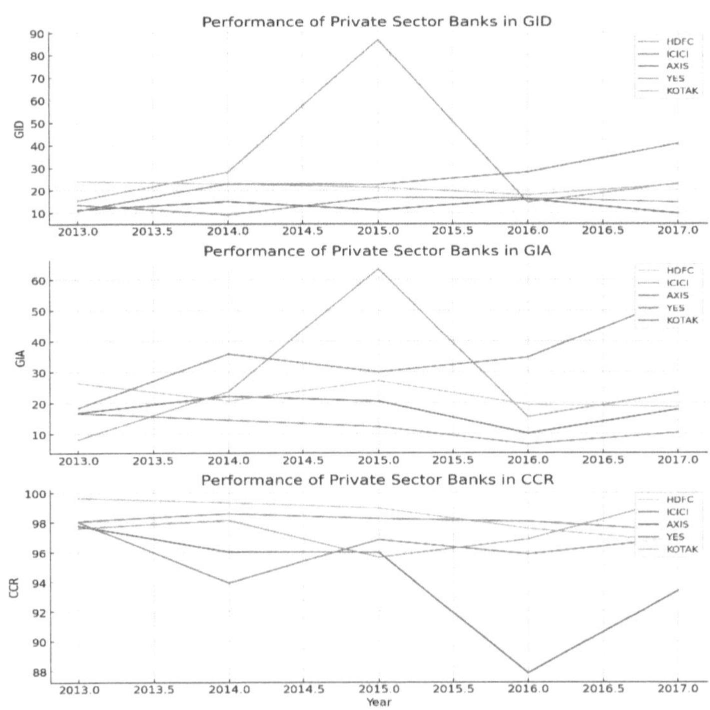

Fig. 6. Growth in Advances (GIA), Growth in Deposits (GID), and Customer Complaints Redressal (CCR) of Private banks

4.4 Customer Perspective

The three variables that are measured under the Learning and Growth perspective of BSC—Growth in Advances (GIA), Growth in Deposits (GID), and Customer Complaints Redressal (CCR)—are described in more detail below (Fig. 6).

Table 5. Comparison of Performance of Private Sector Banks in GID, GIA, CCR (ANOVA)

	Customer Perspective of Balanced Score Card					
		Sum of Squares	Df	Mean Square	F	Sig.
GID	Between Groups	534.89	1	534.89	9.56	0.00
	Within Groups	324.76	3	53.88		
	Total	859.65	4			
GIA	Between Groups	755.66	1	767.22	35.88	0.00
	Within Groups	381.9	3	68.21		
	Total	1137.56	4			
CCR	Between Groups	5.76	1	5.76	7.88	0.57
	Within Groups	9.43	3	0.56		
	Total	15.19	4			

In GID, Kotak Mahindra Bank is the best bank among the selected banks. There is continuous growth in the deposit of the bank over years. The average highest GID of 33.44 per cent is recorded for the bank, while the lowest average GID of 12.45 per cent is recorded by Axis Bank. The highest average GIA of 34.55 per cent is observed for the Yes Bank while the lowest average GIA of 12.09 per cent is registered by ICICI Bank. The highest average CCR of 98.44 per cent is observed for the HDFC bank while the lowest average CCR of 94.21 per cent is registered by AXIS bank. As the p-value for the two variables GID and GIA is less than 0.05, we can conclude that there is a significant difference in the performance of PVBs with respect to these two variables of the internal business perspective of a balanced scorecard. Also, the p-value for CCR is 0.57, we can conclude that all the banks are almost at the same level in solving customer queries ratio and there is no significant difference in the performance of PVBs concerning CCR.

4.5 ROA Analysis

The Return on assets (ROA) of the five banks is enlisted in the following table (Table 6).

Table 6. ROA Analysis

Year	ROA				
	HDFC	ICICI	AXIS	YES	KOTAK
2014	1.73	1.65	1.62	1.48	1.72
2015	1.73	1.73	1.59	1.47	1.76
2016	1.74	1.35	1.57	1.54	1.09
2017	1.68	1.27	0.61	1.55	1.59
2018	1.64	0.77	0.04	1.35	1.54
Mean	**1.70**	**1.35**	**1.09**	**1.48**	**1.54**
SD	**0.03**	**0.34**	**0.65**	**0.07**	**0.24**

In ROA, HDFC Bank is the best bank among the selected banks. The highest average ROA of 1.70 per cent is registered by HDFC Bank. The lowest average ROA of 1. 09 per cent is registered by Axis Bank during the period.

4.6 Regression Analysis

Significance of selected twelve variables of balanced scorecard framework on ROA. To study the significance of balanced scorecard variables on ROA regression analysis is used. The ROA was used as a dependent variable and regressed on the remaining 12 independent variables labelled as F1, F2, F3, IBP1, IBP2, IBP3, IL1, IL2, IL3, C1, C2, and C3. Results are mentioned in Tables 7 and 8.

Table 7. Regression Analysis

Model	R	R square	Adjusted R Square	Std. Error of the Estimates
1	0.857a	0.763	0.825	0.734
a. Predictors: (Constant), F1, F2, F3, IBP1, IBP2, IBP3, IL1, IL2, IL3, C1, C2, C3				

Table 8. Coefficient of the Investment Function

Model	Unstandardized Coefficients	Standardized Coefficients	t	Sig.
	B	Std. Error	Beta	
Constant	1.23	0.392	0.845	0.001
F1	0.464	0.080	0.021	3.29
F2	0.079	0.113	0.046	0.723
B	0.453	0.270	0.191	3.029
BP	0.453	0.107	0.034	0.653

(continued)

Table 8. (*continued*)

Model	Unstandardized Coefficients	Standardized Coefficients	t	Sig.
	B	Std. Error	Beta	
TBP2	0.211	0.104	0.141	3.025
TBP	0.021	0.125	0.016	-1.000
A	0.003	0.104	0.000	0.024
X1	0.125	0.023	0.183	0.855
WS	0.035	0.104	0.031	0.635
A	0.359	0.095	0.782	-1.813
O	0.1068	0.1065	0.012	0.347
&	0.311	0.126	0.016	0.567

The findings show that the set of 12 independent variables can account for 76.3% of the variability in the ROA. The variables F2, Net Interest Margin (NIM), IBP2, Cost to Income Ratio (CIR), IL3 Growth in Skilled Employees (GSE) and C1 Growth in Deposits (GID) are significant. This shows that these variables are important and contribute to the return on assets of the bank. The bank should pay attention to these variables on should work to improve the efficiency in these departments.

5 Conclusion

Most of the research based on the Indian banking performance evaluation measures just profitability, ignoring the non-financial perspective. However, this is a thorough analysis and gives equal weight to crucial financial and other non-financial aspects of banking operations. The study compares the top five PVSBs based on the 12 different variables. The study observed that there is a significant difference in the performance of PVSBs in ten variables of BSC. And there is no significant difference in the performance of PVSBs in 2 variables i.e., CAR and CCR of BSC. Regression analysis is used to quantify the association between 12 BSC framework variables and ROA. The R square value of 84.9 per cent indicates that the variables have a significant impact on ROA. Surprisingly, out of 12 variables, only 4 variables NIM (Financial) and CIR (Internal Business Process) GSE (Learning and Growth) and GID (Customer) have a statistically significant relationship to ROA. Across the board, Kotak Mahindra Bank outperformed the other selected banks in nearly all BSC metrics from 2014 to 2018.

References

1. Kaur, A., Singh, M.: Commerce Performance Evaluation of Public and Private Sector Banks in India: A Comparative Study. University College, K.U.K., pp. 237–242 (2016)
2. Walker, et al.: Customer-Friendly Concepts in Retail Banking (2008)
3. Bisht, et al.: Technological Innovations in Banking (2002)
4. Eyuboglu: Banking Sector as the Foundation of Economy (2016)

5. Bikker, J., Bos, J.: Efficiency and Productivity in Banking (2008)
6. Kaur, J., Kaur, H.V.: A comparative study of public and private sector banks using the CAMEL model. Int. J. Appl. Bus. Econ. Res. **15**(9), 249–263 (2017)
7. Mohanty, S.: A comparative financial performance of selected public and private sector banks in India. Parikalpana KIIT J. Manage. **17**(2), 155 (2021). https://doi.org/10.23862/kiit-parikalpana/2021/v17/i2/210548
8. Sharma, P., Mathur, N.: A comparative study on public & private sector banks: a special reference to financial performance, non-performing assets, and employee morale. Juni Khyat - UGC Care Group I Listed J. **10**(8), 82–91 (2020)
9. Bhatia, K., Chouhan, N.: Comparative study of performance of public and private sector banks. Int. J. Core Eng. Manage. (IJCEM) **2**(1), 306–317 (2015)

A Balanced Scorecard-Based Comparative Analysis of Performance Metrics in Indian Public and Private Sector Banks with Regression Insights on Return on Assets

Bonnie Rajesh[1], Rajesh Raut[1], Sushmitha Rao[2], Komal Singh[2], Aman Jandwani[3], Amit Kumar Arora[4](✉), and Virendra Vasant Tatake[5]

[1] MIT School of Distance Education, Pune, India
[2] Indira School of Business Studies PGDM, Pune, India
[3] Research Task, Chandigarh, India
[4] IIT Madras, Chennai, India
amit.riselab@gmail.com
[5] Institute Indira Global School of Business, Pune, India

Abstract. The performance of selected Indian private sector and public sector banks is compared in this study. The top three private sector banks (PVSBs) and public sector banks (PSBs) based on total assets were chosen for the study. The authors designed a comprehensive performance measurement framework based on the balanced scorecard approach of Kaplan and Norton. According to the study, the performance of PSBs and PVSBs differs significantly. From the regression analysis, it can be concluded that the selected variables significantly impact return on assets (ROA). This study offers observations for policymakers that might result in more successful strategies for improving the efficiency and performance of the banking sector. The significant antecedents related to banks' ROI have been identified via research. This research contributes significantly to investigating the antecedents of ROA and provides insights regarding factors influencing the ROA.

Keywords: Balanced Scorecard · Private sector banks · Public Sector Banks · Performance measurement · Return on Assets

1 Introduction

Businesses thrive in a dynamic and evolving environment due to growing global competition and the quest to achieve a competitive edge in the market, which has altered how businesses perceive and operate. As a result, the performance index of most companies has sustained dramatic change. Gone are the days when conventional measures like Return on investment (ROI) and Earnings Per Share (EPS) were predominantly used. However, there has been a tremendous use for modern methods like on-time delivery and customer satisfaction [1]. Moreover, conventional measures provide inadequate support for the transformation of the company. Furthermore, it is considered redundant

A. Das et al. (Eds.): MIDAS 2025, CCIS 3055, pp. 285–298, 2027.
https://doi.org/10.1007/978-3-032-35516-4_23

in today's competitive environment [2]. In similar studies, it has been brought out that such conventional measures should be refrained from their use due to their outdatedness [3–5].

The result of all past actions over which the company had no control will reflect in the financials of the said company. Consequently, the said reflection will receive numerous criticisms. Its very existence and use of such financial measures will always be objectionable [6].

Likewise, [7] study supports that the conventional and traditional measures cannot distinguish between well-performing and not well-performing firms. Besides, it would be imperative to quote a single benchmark that needs to give a complete picture of the company's performance in today's competitive environment. For strategic decision-making, the company will have to rely on a more accurate measurement index than traditional performance measures [8, 9]. This research brings out the balanced scorecard implications in the banking sector in India. The four key performance indicators related to banking are explained above. Data was collected for these indicators from the official annual reports of public and private sector banks, and their performance was evaluated. The banking scenario reflects extensive competition at both national and international levels. To impress and attract customers, many banks in the private sector have launched multiple schemes. Every bank must ensure that its performance is better than its competitors for customer retention. According to the RBI's report on "Trends and Progress of Banking 2018-18", a sum of INR 71,543 crore was claimed. Indian banking system faces cardinal issues like- poor and untimely decision-making for projects, asset-liability mismatch, delay in recognition of NPAs and growing fraud due to a lack of relevant and apt policies. These issues must be handled for the banking sector to get hold of the competitive advantage in the market. Banks must take note of these upcoming issues and keep track of them so they know where they are headed.

Motivation. Our balanced scorecard research between Indian public and private banks seeks to track banking sector changes in an economy that continues to develop rapidly. Indian banks help boost economic growth but their success depends on running operations well while managing finances and serving customers alongside promoting new ideas. By using the balanced scorecard framework, you can analyze multiple metrics across all performance dimensions at once. Our research uses ROA regression tests to discover the most influential metrics in profitability while showing stakeholders how to bridge business strategy with operational Excellence.

2 Literature Review

The banking sector is the solid foundation of the modern economy. Balanced scorecard as a tool was developed by [1]. It is a performance measurement tool that encompasses operational and traditional financial measures. It was created for diagnosis and as a control system with strategy and vision at the framework's core. The balanced scorecard measures the organizational performance on various parameters like financial, customer, internal business and innovation and learning. The tool helps develop appropriate measures to assimilate whether the organizational goals are in sync with their operational activities, namely, customers, internal processes, learning, and growth. The framework protects the decision-makers by providing the correct information in the right quantity at the right time. It also helps managers to see the big picture from different perspectives.

The balanced scorecard is a powerful tool for developing strategies, converting strategic objectives into actionable goals, allowing communication of strategic objectives to the entire organization. Targets are easily achieved with realistic goals. The tool acts as a feedback strategy to review performances. BSC gives comprehensive objectives.

- To compare the overall performance of public sector banks and private sector banks.
- To study the impact of selected variables in the balanced scorecard on return on investments.

H 1: Selected variables in BSC have no significant impact on ROI

3 Proposed Methodology

The study measures the performance of the top three private sector banks and the top three public sector banks. The top five banks from the private and public sectors are selected based on the total assets of the banks as on 9 September 2019. For better data representation, three banks were randomly selected out of five. As in 2019–20, the covid pandemic affected the economy, data from five years from 2013–14 to 2017–18 is considered for statistical analysis. The three private sector banks include HDFC Bank (HDFC), Axis Bank (AXIS), and Kotak Mahindra Bank (KOTAK). Three public sector banks include the State Bank of India (SBI), Canara Bank (C.B.) and Bank of India (BOI). A comprehensive performance measurement framework (Refer to Table 1) is designed by the authors based on the BSC approach of Kaplan and Norton (1992). Three measures are selected under each balanced scorecard perspective to have uniformity in calculations. The performance measurement was carried out using quantitative analysis based on the secondary data obtained from the annual reports of the respective banks. Figure 1 shows the proposed flowchart of the proposed method.

3.1 Methodology for the Objectives

Objective 1: Compare the overall performance of public sector banks and private sector banks.

Approach:

- **Data Collection**:
  - o Collect financial and operational data from public and private sector banks for a specified time period (e.g., 5 years).
  - o Variables include profitability, net interest margin, non-performing assets (NPA), return on equity (ROE), and operational efficiency.
- **Performance Metrics**:
  - o Calculate key performance indicators (KPIs) for each bank:

$$\text{Net Interest Margin (NIM)} = \frac{\text{Net Interest Income}}{\text{Average Earning Assets}} \quad (1)$$

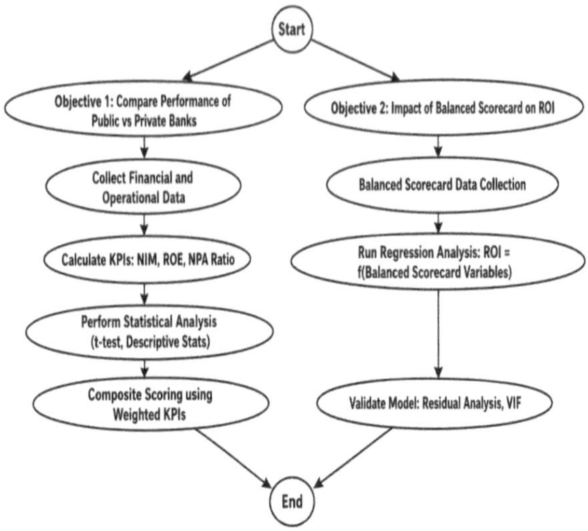

Fig. 1. Proposed Flowchart

$$\text{Return on Equity (ROE)} = \frac{\text{Net Income}}{\text{Shareholders' Equity}} \tag{2}$$

$$\text{Non-Performing Asset Ratio (NPA Ratio)} = \frac{\text{Non-Performing Assets}}{\text{Total Loans}} \tag{3}$$

- **Statistical Comparison**:

  o Perform a comparative analysis of public and private banks using:

  - **Descriptive Statistics** (mean, median, standard deviation).
  - **Hypothesis Testing** (e.g., t-test or Mann-Whitney U test):

$$H_0 : \mu_{\text{Public Banks}} = \mu_{\text{Private Banks}}$$

$$H_a : \mu_{\text{Public Banks}} \neq \mu_{\text{Private Banks}}$$

  - If $p < 0.05$, reject H_0 and conclude a significant difference exists.

- **Multi-Criteria Decision Analysis** (MCDA):

  o Use weights for KPIs to compute a composite score:

$$\text{Composite Score} = w_1 \cdot \text{NIM} + w_2 \cdot \text{ROE} + w_3 \cdot \text{Efficiency} \tag{4}$$

  - Where w_i represents the weight assigned to each KPI.

Objective 2: Study the impact of selected variables in the balanced scorecard on Return on Investments (ROI).

Approach:

- **Balanced Scorecard Framework**:

  o Variables include:

  - **Financial Perspective**: Net income, cost-to-income ratio.
  - **Customer Perspective**: Customer satisfaction index.
  - **Internal Process Perspective**: Operational efficiency.
  - **Learning & Growth Perspective**: Employee training hours.

- **Regression Analysis**:

  o Model the relationship between the selected balanced scorecard variables $(X_1, X_2, \ldots, X_n)$ and ROI (Y):

$$Y = \beta_0 + \beta_1 X_1 + \beta_2 X_2 + \cdots + \beta_n X_n + \epsilon \tag{5}$$

- Where Y is ROI, X_i are independent variables, β_i are coefficients, and ϵ is the error term.

- **Statistical Testing**:

  o Evaluate model significance using R^2 and p-values for each variable:

  - $H_0 : \beta_i = 0$ (No impact of variable X_i).
  - $H_a : \beta_i \neq 0$ (Significant impact of X_i).
  - If $p < 0.05$, conclude that X_i significantly influences ROI.

- **Model Validation**:

  o Perform residual analysis and multicollinearity checks using Variance Inflation Factor (VIF):

$$\text{VIF} = \frac{1}{1 - R^2} \tag{6}$$

- If VIF > 5, there may be multicollinearity, requiring adjustments.

Table 1. Balanced Scorecard for Banks

BSE Perspective	Measures considered for study	Formula
Financial	Credit Deposit Ratio (CDR)	(Total Advances / Total Deposits) * 100
	Net Interest Margin (NIM)	(Returns On Investment - Interest Paid) / Average Assets
	Capital Adequacy Ratio (CAR)	(Tier 1 Capital + Tier 2 Capital) / Risk Weighted Assets
Internal Business Process	Business per Employee (BPE)	Total Income / No. of Employees
	Cost to Income Ratio (CIR)	(Operating Expenses / Total Income) *100
	NET NPA	Gross NPAs- Provisions
Innovation and Learning	No. of Branches (LNB)	Natural log of Branches
	No. of ATMs (LNATM)	Natural log of ATM
	Growth in Skilled Employees (GSE)	[(No. of Employees in Present Year - No. of Employees in Last Year) / No. of Employees in Last Year] *100
Customer	Growth in Deposits (GID)	[(Present Year Deposit Value - Last Year Deposit Value) / Last Year Deposit Value] * 100
	Growth in Advances (GIA)	[(Present Year Advances Value - Last Year Advances Value) / Last Year Advances Value] * 100
	Customer Complaints Redressal (CCR)	[No. of complaints redressed during the year / (No. of complaints pending at the start of year + No. of complaints received during the year)] * 100

4 Data Analysis and Results

4.1 Comparative Analysis

Following is the comparison of the overall performance of public and private sector banks based on the balanced scorecard.

4.2 Finance Perspective

The efficiency of banks from a finance perspective is measured using Credit Deposit Ratio (CDR), Net Interest Margin (NIM) and Capital Adequacy Ratio (CAR) and shown in Fig. 2.

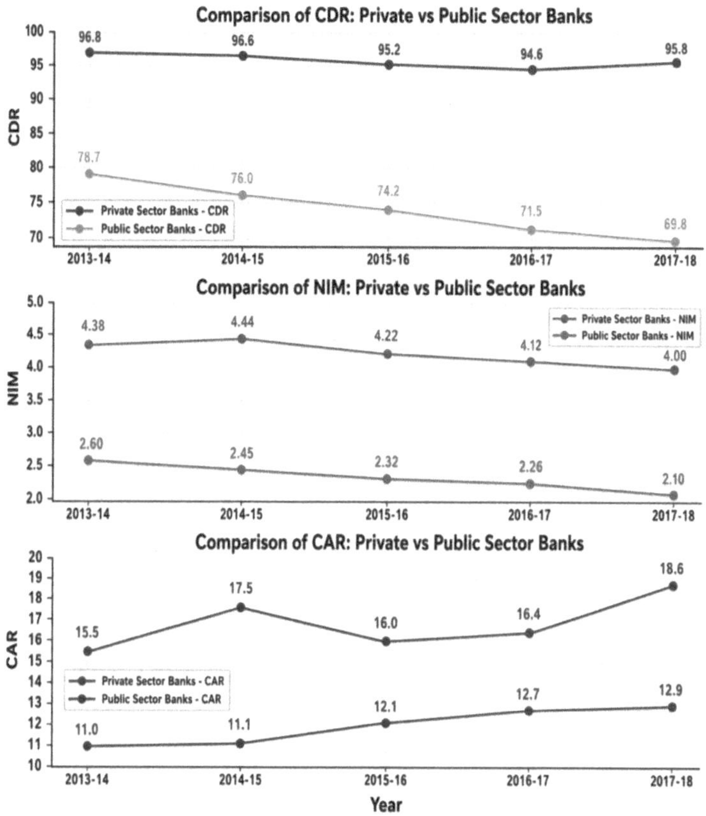

Fig. 2. Credit Deposit Ratio (CDR), Net Interest Margin (NIM) and Capital Adequacy Ratio (CAR)

Table 2. Results of Comparative Performance of PVSBS and PSBS (ANOVA)

Financial Perspective of Balanced Score Card						
		Sum of Squares	df	Mean Square	F	Sig.
CDR	Between Groups	1384.42	1	1384.42	321.96	0.00
	Within Groups	57.44	4	7.05		
	Total	1441.86	5			

(continued)

Table 2. *(continued)*

Financial Perspective of Balanced Score Card						
		Sum of Squares	df	Mean Square	F	Sig.
NIM	Between Groups	5.96	1	5.96	253.40	0.00
	Within Groups	0.17	4	0.03		
	Total	6.13	5			
CAR	Between Groups	53.58	1	53.58	112.13	0.00
	Within Groups	5.29	4	0.62		
	Total	58.87	5			

Tables 2, 3 and 4 show that PVSBs have a larger CDR than PSBs, making them more effective at marketing the credit. PSBs have an average CDR of 73.97 per cent, whereas PVSBs have a CDR of 95.88 per cent. This leads us to conclude that, in comparison to PSBs, PVSBs are more active in lending advances. Kotak Mahindra Bank has the highest CDR, at 113.86. PVSBs are also good at managing NIM as compared to PSBs. The average NIM of PSBs is 2.45 per cent, and PVSBs is 4.23 per cent. Kodak Bank has recorded the highest NIM of 4.61. The PVSBs are maintaining a high CAR than PSBs. PVSBs are also doing an excellent job of maintaining adequate capital compared to PSBs. The average CAR maintained by PSBs and PVSBs is 11.96 and 16.65 per cent, respectively. As the p value is less than 0.05 in all three variables, CDR, NIM and CAR, $H_0 1$ is rejected. It can be concluded that there is a significant difference between the performance of PVSBs and PSBs with respect to the financial perspective of the Balanced Scorecard (Refer to Table 5).

Internal Business Process. The efficiency of banks in the Internal Business Processes perspective is measured using Cost to Income Ratio (CIR), Business per Employee (BPE) and NET NPA.

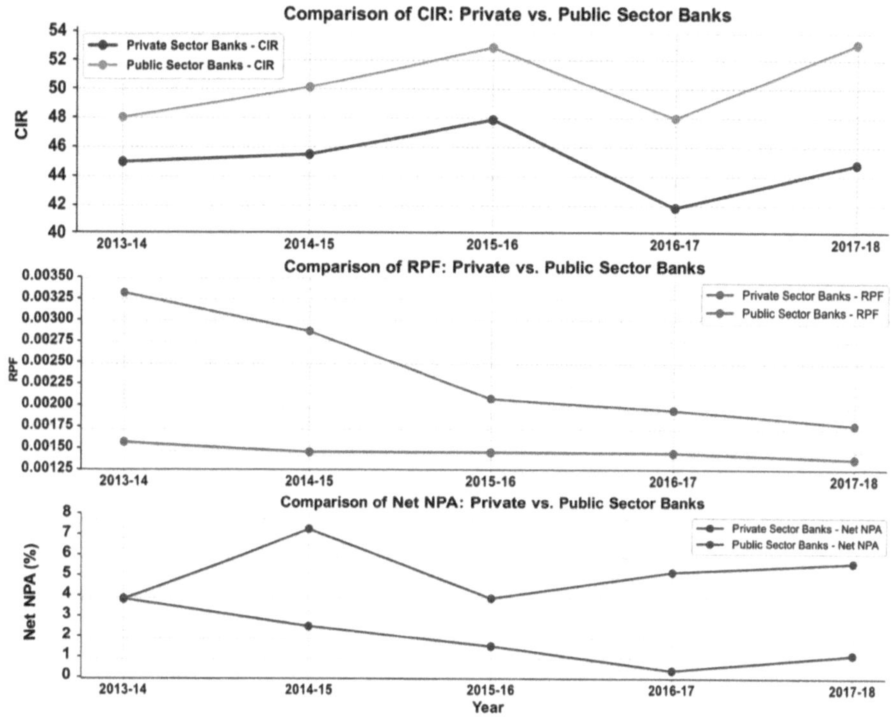

Fig. 3. Cost to Income Ratio (CIR), Business per Employee (BPE) and NET NPA

Table 3. Performance of Comparative Performance Measurement of PVSBS and PSBS (ANOVA)

Internal Business Perspective of Balanced Score Card						
		Sum of Squares	df	Mean Square	F	Sig.
CIR	Between Groups	69.81	1	69.81	57.76	0.00
	Within Groups	13.18	4	1.62		
	Total	82.99	5			
BPE	Between Groups	0.00	1	0.00	27.54	0.00
	Within Groups	0.00	4	0.00		
	Total	0.00	5			
NET NPA	Between Groups	20.98	1	20.98	6.98	0.00
	Within Groups	32.23	4	2.88		
	Total	53.21	5			

It can be observed that PVSBs are efficient in cost management as compared to PSBs. The average CIR of PVSBs is 45.46, whereas the average CIR of PSBs is 50.08 per cent. Axis bank has the best average CIR of 41.67 per cent. PVSBs generate more business per employee (BPE) as compared to PSBs. Kotak Mahindra Bank recorded the highest BPE of 0.0042. Also, PVSBs are more efficient in managing their NPA than PSBs. The PVSBs and PSBs recorded an average NPA of 1.12 per cent and 5.24 per cent, respectively (Refer to Fig. 3). As the p value of all the variables is less than 0.05, H02 is rejected. Therefore, we can conclude that there is a significant difference between the performance of PVSBs and PSBs with respect to the internal business process perspective of a balanced Scorecard.

Learning and Growth. The efficiency of banks in learning and growth perspective is measured using No. of Branches (LNB), No. of ATMs (LNATM) and Growth in Skilled Employees (GSE). PSBs have expanded the branches, and the number of ATMs compared to PVSBs (Refer to Fig. 4). State bank of India was creating more branches and installed the highest number of ATMs among PSBs, HDFC bank is creating more branches, and Axis bank has installed the highest number of ATMs amongst PVSBs. It can be observed that PVSBs invested in hiring skilled employees more than PSBs. Since the p value is greater than 0.05, H_03 is rejected. It can be concluded that there is a significant difference between the performance of PVSBs and PSBs with respect to the learning and growth perspective of a balanced scorecard.

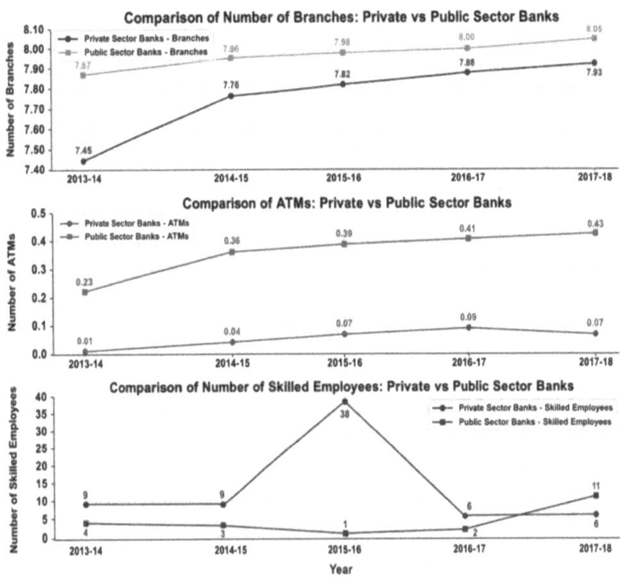

Fig. 4. Branches (LNB), No. of ATMs (LNATM) and Growth in Skilled Employees (GSE)

Table 4. Results of Efficiency of PVSBS and PSBS (ANOVA)

L&G Perspective of Balanced Score Card						
		Sum of Squares	df	Mean Square	F	Sig.
LNB	Between Groups	4.45	1	4.45	145.56	0.02
	Within Groups	0.17	4	0.04		
	Total	4.62	5			
LNATM	Between Groups	1.45	1	1.45	61.47	0.00
	Within Groups	0.30	4	0.05		
	Total	1.75	5			
GSE	Between Groups	292.31	1	187.92	4.45	0.03
	Within Groups	436.67	4	55.68		
	Total	728.98	5			

Customer Satisfaction. Figure 5 illustrates the comparative analysis of PVSBs and PSBs based on the learning and growth perspective parameters of the balanced scorecard across different financial years.

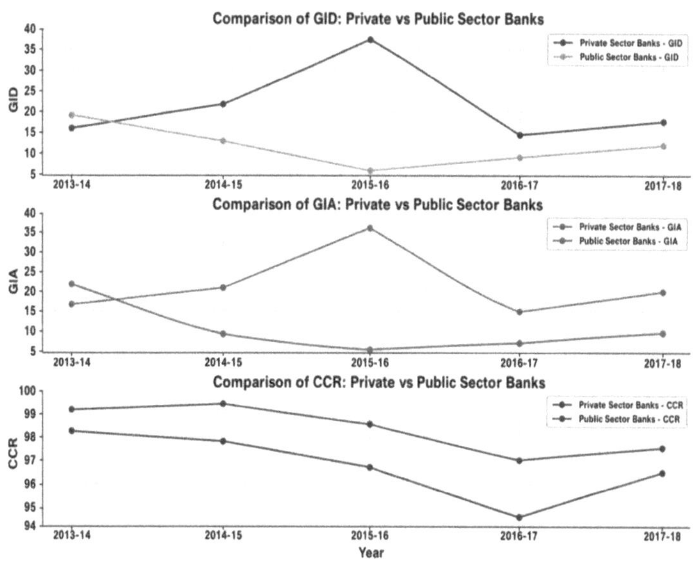

Fig. 5. PVSBs and PSBs with respect to the learning and growth perspective of a balanced scorecard

Table 5. Results of Comparative Performance Measurement of PVSBS AND PSBS (ANOVA)

Customer Perspective of Balanced Score Card						
		Sum of Squares	df	Mean Square	F	Sig.
GID	Between Groups	456.84	1	456.84	9.02	0.00
	Within Groups	217.56	4	47.53		
	Total	674.40	5			
GIA	Between Groups	643.33	1	643.33	14.89	0.00
	Within Groups	274.997	4	58.32		
	Total	918.33	5			
CCR	Between Groups	4.65	1	4.65	4.91	0.01
	Within Groups	8.83	4	0.89		
	Total	13.48	5			

Table 6. Regression Analysis

Model Summary									
	Change Statistics								
	Std Error R								
	R		Adjusted of the Square F						Sig. F
Model	R	Square	R Square	Estimate	Change	Change	df1	df2	Change
1	.912[a]	.801	.634	.26849	.821	4.501	2	2	.179

a. Predictors: (Constant), BSC variables

PVSBs manage growth in deposits (GID) and growth in advances (GIA) more effectively than PSBs. Kodak Mahindra Bank has the most excellent average GID, 33.44 per cent, while Bank of India has the lowest average GID, 6.02 per cent. The average GIA for Kotak Mahindra Bank was 26.77 per cent, while the average GIA for SBI Bank was 13.25 per cent. The survey found that the Bank of India has a superior customer complaint resolution process. Surprisingly, PSBs do better in Customer Complaints Redressal (CCR) than PVSBs, with an average of 97.87 per cent compared to 96.74 per cent.

Given that the p value is less than 0.05, an ANOVA analysis of Table 5 shows a significant difference between PSBs and PVSBs' performance in the variables GID and GIA and CCR. Since the p value is less than 0.05, $H_0 4$ is rejected. It can be concluded that there is a significant difference between the performance of PVSBs and PSBs with respect to the customer perspective of the balanced scorecard.

Regression Analysis. The impact of selected variables in a balanced scorecard on return on assets is calculated using regression (Refer to Table 6).

Calculating the R squared coefficient of determination allows one to assess the importance of a regression model. Table 6 demonstrates that the selected BSC variables account for 80.1% of the variation in ROA. The model is statistically significant since the F value of 4.601 is higher than the essential F value of 2.76. Therefore, H_05 is rejected, and we can conclude that selected variables in BSC have a significant impact on ROI.

5 Conclusion

Most studies based on the Indian banking performance evaluation examine just profitability, ignoring the non-financial standpoint. However, this comprehensive analysis gives financial and other significant elements of Indian banking equal weight. The study compares PSBs and PVSBs based on 12 distinct variables. From a financial perspective, there is a significant difference in the performance of PSBs and PVSBs in all three variables, viz. CDR, NIM and CAR. From the internal business process perspective, there is a significant difference in the performance of PVSBs and PSBs for all three variables, viz. CIR, BPE and Net NPA. From the learning and growth perspective, there is a significant difference in the performance of PVSBs and PSBs in the variables LNB, LNATM and GSE. From a customer perspective, there is a significant difference between PSBs' and PVSBs' performance in the variables GID, GIA and CCR. Further overall decreasing trend in the ROA of PSBs was observed during 2014–2018, while PVSBs were consistent in ROA. Inadequate capital and rising NPAs of PSBs have led to a decrease in ROA, whereas PVSBs have managed well to register positive ROA during 2014–2018. It can be shown from the regression analysis that selected BSC variables significantly affect ROA. This means banks need to work on the given variables to perform better. Overall, SBI performed well in almost all the parameters of BSC among the selected PSBs during 2014–2018. Moreover, Kotak Mahindra Bank performed well in almost all the parameters of BSC among the selected PVSBs during 2014–2018.More variables, such as "number of personnel trained," "new products and schemes offered," "online banking acceptance status," etc., may have been added in this study's learning and growth section. However, not all the information for these variables is accessible from reliable secondary data sources, and specific values could not be quantitatively mapped. The future scope of this work can address this restriction.

References

1. International Federation of Accountants (IFAC): Changing dimensions of management accounting in India and its relevance for students. IFAC Knowledge Gateway (2023). https://www.ifac.org
2. IOSR Journals: Survey of management accounting practices in Indian industries. IOSR J. Bus. Manag. **21**(2), 5–9 (2022). https://www.iosrjournals.org
3. SSRN: Creative accounting practices in the Indian corporate sector. Soc. Sci. Res. Netw. (2022). https://papers.ssrn.com
4. I-Scholar: Adoption of advanced management accounting techniques in the Indian service sector. J. Commer. Account. Res. **11**(3), 45–52 (2023). https://www.i-scholar.in
5. International Federation of Accountants (IFAC): The evolving role of management accountants as business partners. IFAC Knowledge Gateway (2023). https://www.ifac.org
6. Sharma, S.: Impact of advanced management accounting practices on organizational decision-making in Indian SMEs. Indian J. Account. Finance **34**(2), 120–133 (2022)

7. Menon, V., Gupta, A.: Integration of management accounting with AI for strategic insights: a study in Indian corporates. J. Modern Bus. Stud. **29**(4), 200–215 (2023)
8. Rao, P., Nair, S.: Adapting global management accounting trends to Indian manufacturing industries. Int. J. Bus. Econ. **45**(5), 98–112 (2022)
9. Joshi, T., Patel, M.: Strategic performance metrics and their adoption in Indian multinational enterprises. Asia-Pac. Manage. Rev. **30**(1), 50–64 (2023)
10. Chatterjee, A.: Emerging technologies in management accounting: Implications for Indian businesses. Technol. Manage. Rev. **19**(2), 75–89 (2023)

Indian Medicinal Leaf Detection: A Comparative Study with Custom CNN and VGG16

Mir Sadab Ali, Tejaswini Pradhan, Chinmayee Dora(✉), Sunil Kumar Mohapatra, Raj Kumar Mohanta, and Sujata Chakravarty

School of Engineering and Technology, Centurion University of Technology and Management, Bhubaneswar, Odisha, India

{chinmayee,sunil.mohapatra,rajkumar.mohanta, sujata.chakravarty}@cutm.ac.in

Abstract. Many medicinal plants are found across the Earth, providing a foundation for numerous drugs used to treat human illnesses for centuries. Precisely identifying medicinal plant species plays a crucial role in developing and preserving pharmaceuticals. This paper presents a comprehensive comparative analysis of two deep learning models a custom Convolutional Neural Network (CNN) developed from scratch and the pre-trained VGG16 model—applied to classifying medical leaf images. The primary objective of this study is to assess the performance of both models in terms of accuracy, precision, recall, and F1-score, providing insights into their effectiveness for this specific classification task. Then these models could be deployed for a web application where the model could classify the captured and uploaded image by the user in real time. The CNN model was designed with a relatively simple architecture, consisting of multiple convolutional layers followed by max-pooling and fully connected layers. VGG16 is a deep, pre-trained model, known for its strong performance on large-scale image classification tasks. Both models were trained on a dataset of medical leaf images and fine-tuned to improve their performance. Results showed that the VGG16 model obtained a higher accuracy of 99% with its deeper architecture and pre-training on the ImageNet dataset. Whereas, the custom CNN model achieved 85% accuracy, a competitive result with reduced computational complexity. In the context of medicinal leaf detection, this study provides insightful information about the trade-offs between model complexity, accuracy, and computational efficiency when using deep learning approaches.

Keywords: Medical leaf detection · deep learning · CNN · VGG16 · image classification · comparative study

A. Das et al. (Eds.): MIDAS 2025, CCIS 3055, pp. 299–312, 2027.
https://doi.org/10.1007/978-3-032-35516-4_24

1 Introduction

Plants are beneficial for humans, not just for food and oxygen, but also for their medicinal properties. In India, it is a common practice to use medicinal plants at home to treat general sickness [7, 8]. But sometimes it is difficult for laymen to differentiate medicinal plants from their leaf appearance. Identifying them by book requires time and special knowledge. The identification of medicinal leaves plays a crucial role, especially in rural regions where traditional Ayurvedic knowledge remains a significant part of daily life. In villages, people are often familiar with identifying and using medicinal plants for their healing properties. However, in urban areas, where most people are far removed from nature, this knowledge is not as widespread, making it difficult to recognize and benefit from these plants. As, interest in alternative medicine continues to grow, especially with the rise of urbanization. Hence, there is a need for technology to fill this gap in knowledge. Many studies have shown that using CNN and Deep Learning for plant identification works very well [4, 5, 9, 12].

According to the findings [13], a deep CNN framework was developed and fine-tuned to improve recognition accuracy. The research emphasizes how important the multi-layer approach is, especially when dealing with small datasets, to achieve good results. Additionally, using data augmentation had a significant positive impact on performance. Simple techniques like resizing, flipping, and rotating images greatly improved accuracy, as long as the model was designed to focus on the right information and ignore irrelevant details. The results improved further after creating a new database of Malaysian medicinal plants. In their study [10], Shastry et al. used three different methods to identify medicinal leaves: Support Vector Machine (SVM), the transfer learning model VGG-16, and a real-time object detection algorithm called You Only Look Once (YOLO). The dataset had seven different classes, with about 1,000 leaves in each class, and they added 150 more leaves using data augmentation techniques. The performance of each method was measured using classification metrics, and the transfer learning approach with VGG-16 achieved the highest accuracy compared to the other two. To find the best fit for the data, they also performed a non-parametric Chi-Square goodness-of-fit test on the test data. The research [6] introduced a method for automatically recognizing plant species in real-time, focusing on medicinal herbs from Borneo. Using an improved EfficientNet-B1 model, the researchers trained and tested the system on both private and publicly available plant databases. The model showed a notable improvement in accuracy—more than 10% higher than traditional models when tested. A standout feature of this study was the use of public feedback and geo-mapping of plants in the Borneo region, which was made possible through a mobile app. The study [1] looked into the practical use of convolutional neural networks (CNNs) to distinguish between different species of Indian leaves. The MedLeafNet paper [11] proposed a new model called "MedLeafNet" for identifying Indian medicinal leaves. MedLeafNet combined two types of technology: Convolutional Neural Networks (CNNs) to spot details in the leaves and Transformer models to understand their overall features. The study worked on 80 different medicinal leaves, showing that the model works well with various plants. It was trained using different methods and got its best results, with an accuracy of 98.97%.

With the above insights, this work implemented deep learning models i.e. a custom CNN and VGG-16 to offer a promising solution, providing a way for city dwellers to easily identify medicinal plants. This paper aims to address a real-time model deployment using a web app after comparing the performance of a custom-designed CNN and VGG-16 model in classifying medicinal leaves. The next Sect. 2 details the methodology adapted for the implementation of the work. Section describes the data set, the pre-processing involved, and the performance parameter used. The results obtained and discussion are shown in Sect. 4. Finally, a conclusion is presented in Sect. 5.

2 Proposed Method

The goal of this manuscript is to use a web application to identify medicinal leaves in real time. The three phases are deployment, testing, and training. The dataset was trained with a proposed custom CNN and VGG16 [2]. The performance of both models was compared, and the better model was chosen for deployment with a web app. Figure 1 shows the graphical abstract of the process.

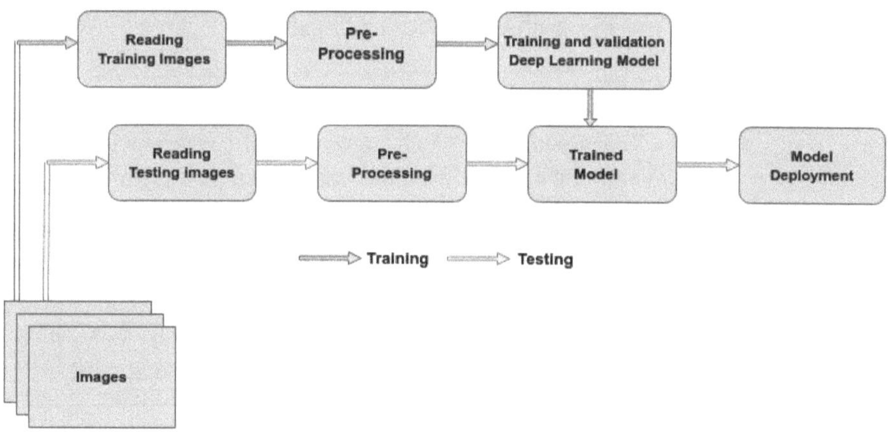

Fig. 1. Flow of real-time medicinal leaves detection

2.1 Model Architectures

The CNN (Fig. 2.) and VGG16 (Fig. 3.) models each play an important role in making this system accurate and fast:

- **Custom CNN:** The custom CNN model is designed to detect the fine details that make each leaf unique. It employs layers of small filters to capture subtle patterns and textures in the leaves, while pooling layers simplify the data,

making the model lighter and less prone to underfitting. A dropout layer is also included, randomly deactivating neurons during training to improve the model's generalization. The custom CNN model achieved an accuracy of 85%, making it suitable for lightweight, real-time applications.

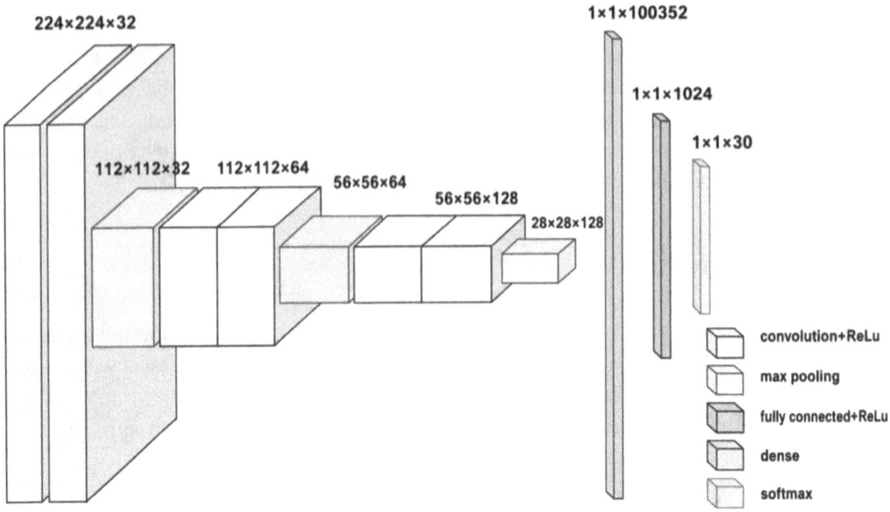

Fig. 2. Custom CNN Architecture

VGG-16: VGG-16 is a deep learning model known for its reliable performance on image recognition tasks. The architecture of VGG16 consists of 16 layers with weights, including 13 convolutional layers and 3 fully connected layers, followed by a SoftMax classifier. The convolutional layers use small 3x3 kernels with a stride of 1, enabling the model to capture fine details in images. Max-pooling layers with 2x2 kernels and a stride of 2 are used for downsampling, reducing spatial dimensions while preserving important features. One key feature of VGG16 is its uniform design, where all convolutional layers use the same kernel size. This approach improves model generalization and simplifies architecture design. In this case, a pre-trained VGG-16 is fine-tuned on the medicinal plant dataset to enhance its ability to identify specific leaf shapes and patterns. With its 16-layer architecture, VGG-16 captures complex details and achieves an impressive accuracy of 99%, making it highly effective for distinguishing subtle differences between leaf species with similar features.

2.2 Training and Evaluation Process

A labeled dataset of medicinal plant images was sourced from Kaggle [3] to train the models. This dataset was split into training and validation sets to cover all stages of model development. Data augmentation techniques such as rotating, flipping, and scaling were applied to improve the models' robustness. This step enhanced the dataset to handle the different leaf orientations and lighting or background conditions in real-world images.

The data were divided into training and validation sets. The dataset underwent the processing techniques such as resizing and augmentation. The validation sets are used to train the designed deep learning models. After training the DL model, testing was performed on the validation data. Testing evaluates the DL model performance.

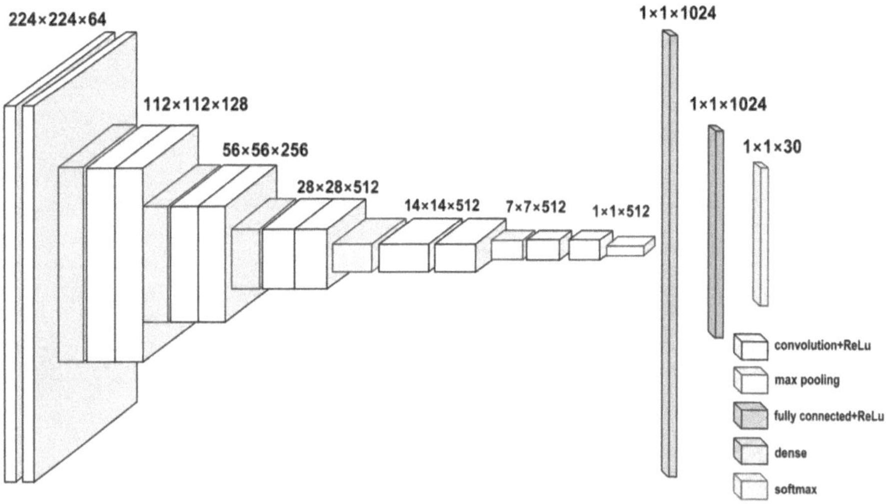

Fig. 3. VGG-16 Architecture

2.3 Deployment and Real-Time Application

The trained models were deployed in the Flask web framework for real-time leaf identification. This web application allows the user to capture leaf images, which would be sent to the cloud-hosted models. These models quickly return a classification result, enabling the app to provide instant feedback on the plant species.

With this dual-model setup—combining both a custom CNN and VGG-16—the system ensures both speed and accuracy, making it suitable for real-world applications and reliable for on-the-go plant identification.

3 Experimental Setup

This manuscript implemented two deep learning models—a custom Convolutional Neural Network (CNN) and VGG16 to identify medicinal leaves. Each model had a different advantage for enhancing the system's accuracy, efficiency, and overall suitability for real-time use in a mobile app. The work in this paper was implemented using Jupyter Notebook 3.12.1.

3.1 Dataset

The medicinal leaf images were collected from [3]. Despite the vast number of medicinal plant species available, the focus was placed on thirty specific species: Betel, Curry, Tulsi, Mint, Neem, Indian Beech, and others. Sample images of some species are shown in Fig. 4. The dataset includes 500 images per species. For the implementation, 372 images of each species were used for training, and the remaining 128 images for validation.

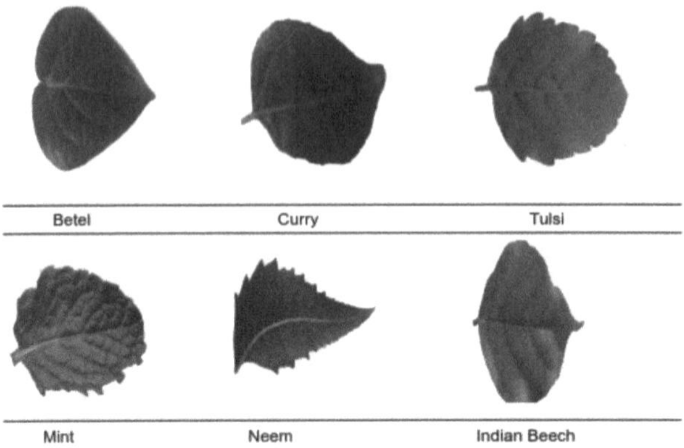

Fig. 4. Sample medicine leaf images

The images were resized and augmented to enhance the feature extraction process of deep networks. For instance, to fit the VGG-16 model, the images were resized to a standard dimension of 224 × 224 pixels.

The data augmentation techniques employed here include rotation, flipping, and color manipulation. Images were flipped both horizontally and vertically to introduce additional variations. In particular, the training images were rotated from 5 to 90 degrees in both clockwise and anticlockwise directions. Using Image manipulation image color was altered to enrich the dataset.

Figure 5 shows a few samples of each medicinal species before and after augmentation, showcasing transformations such as rotation, flipping, and color adjustments.

3.2 Performance Parameters

The following performance parameters were used to evaluate the model's usability for classification.

- **Accuracy:** Accuracy represents the general success of the model by showing the proportion of correctly classified images from all the images tested.

$$Accuracy = \frac{Number\ of\ correct\ Predictions}{Total\ Number\ of\ Predictions} \tag{1}$$

- **Precision:** Precision focuses on the model's ability to correctly identify a particular plant species when it claims to do so. It calculates the percentage of true positive predictions out of all positive predictions. Precision is particularly crucial here, as misidentifying a plant could lead to error in medicinal applications.

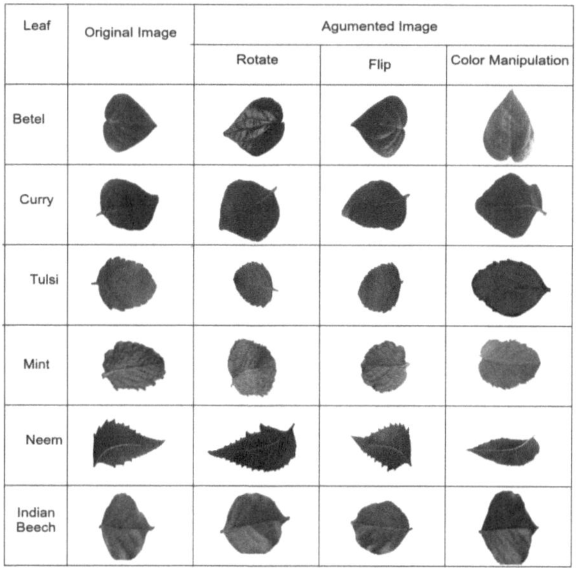

Fig. 5. Medicinal images before and after augmentation

$$Precision = \frac{True\ Positives}{True\ Positives + False\ Positives} \tag{2}$$

- **Recall:** Also known as sensitivity, recall measures the model's effectiveness in capturing all actual instances of a specific plant species. This metric ensures that the model does not overlook relevant images, which is crucial for identifying all instances of a medicinal plant within a dataset.

$$Recall = \frac{True\ Positives}{True\ Positives + False\ Negatives} \tag{3}$$

- **F1 Score:** The F1 Score gives a stability view by combining both precision and recall. It calculates the harmonic mean of these two metrics, making it especially valuable when it is important to ensure that the model does not favor one over the other. This metric is particularly useful as both precision and recall are crucial for reliable plant identification.

$$F1 - Score = 2 \times \frac{Precision \times Recall}{Precision + Recall} \tag{4}$$

- **Loss:** Loss tracks the difference between the model's predictions and actual target value during training. A declining loss generally suggests better performance, indicating that the model is improving over time.

$$CategoricalCross - EntropyLoss = -\frac{1}{N}\sum_{i=1}^{N}\sum_{c=1}^{C} y_{i,c}\log(\hat{y}_{i,c}) \quad (5)$$

where:

- N is the total number of samples,
- C is the number of classes,
- $y_{i,c}$ is the true label for sample i and class c (1 if the class is correct, 0 otherwise),
- $\hat{y}_{i,c}$ is the predicted probability for sample i belonging to class c.

Together, these performance parameters provided a well-rounded picture of the strengths and limitations of the CNN and VGG-16 models. They helped to assess how accurately and effectively each model can identify medicinal plants, enabling informed adjustments to optimize model performance further.

4 Results and Discussion

This section presents a performance evaluation of two multi-class classification models used for identifying medicinal plants: a custom CNN model and a VGG16-based model. Both models were evaluated using metrics such as precision, recall, F1-score, and support, but the results show significant differences in their performance.

4.1 Performance of the Custom CNN Model

The custom CNN model demonstrated decent performance, achieving an overall accuracy of 85%. However, there were some challenges:

- **High Performers**: Classes like "Guava," "Drumstick," and "Fenugreek" had high precision, recall, and F1-scores, indicating the model performed well in recognizing these plants.
- **Areas of Concern**: The model struggled with classes such as "Jamun" and "Lemon," where the F1-scores were relatively lower, suggesting occasional confusion or missed identifications.
- **Overall Consistency**: With a macro average and weighted average F1- score of 85%, the model maintained a solid performance across most classes, but there were some inconsistencies, especially with less frequent classes.

	precision	recall	f1-score	support
Arive-Dantu	0.85	0.89	0.87	188
Basale	0.93	0.84	0.89	188
Betel	0.92	0.99	0.95	168
Crape_Jasmine	0.83	0.86	0.84	188
Curry	0.87	0.88	0.87	188
Drumstick	0.91	0.95	0.93	188
Fenugreek	0.98	0.94	0.96	188
Guava	0.95	1.00	0.98	188
Hibiscus	0.76	0.88	0.81	130
Indian_Beech	0.82	0.88	0.85	128
Indian_Mustard	0.90	1.00	0.95	128
Jackfruit	0.80	0.99	0.89	128
Jamaica_Cherry-Gasagase	0.77	0.85	0.81	128
Jamun	0.95	0.62	0.75	90
Jasmine	0.66	0.79	0.72	128
Karanda	0.94	0.88	0.91	128
Lemon	0.72	0.63	0.67	128
Mango	0.84	0.76	0.80	128
Mexican_Mint	0.82	0.93	0.87	120
Mint	0.92	0.88	0.90	128
Neem	0.87	0.76	0.81	128
Oleander	0.77	0.86	0.81	128
Parijata	0.85	0.87	0.86	100
...				
accuracy			0.85	4228
macro avg	0.85	0.84	0.84	4228
weighted avg	0.86	0.85	0.85	4228

Fig. 6. Classification report of CNN

4.2 Performance of the VGG16 Model

The VGG16-based model showed a substantial performance, with precision, recall, and F1-scores of 0.99 or 1.00 for almost all classes.

- **Near-Perfect Metrics**: The VGG16 model achieved remarkable scores, with classes such as 'Arive-Dantu', 'Indian Mustard' and 'Jackfruit' having perfect precision, recall and F1 scores.

	precision	recall	f1-score	support
Arive-Dantu	1.00	1.00	1.00	188
Basale	0.98	1.00	0.99	188
Betel	0.99	1.00	1.00	168
Crape_Jasmine	0.99	1.00	1.00	188
Curry	1.00	0.99	1.00	188
Drumstick	0.98	0.99	0.99	188
Fenugreek	1.00	0.97	0.99	188
Guava	0.99	1.00	1.00	188
Hibiscus	1.00	1.00	1.00	130
Indian_Beech	0.99	1.00	1.00	128
Indian_Mustard	1.00	1.00	1.00	128
Jackfruit	1.00	1.00	1.00	128
Jamaica_Cherry-Gasagase	1.00	1.00	1.00	128
Jamun	1.00	1.00	1.00	90
Jasmine	0.96	0.94	0.95	128
Karanda	1.00	0.98	0.99	128
Lemon	1.00	1.00	1.00	128
Mango	0.97	1.00	0.98	128
Mexican_Mint	1.00	1.00	1.00	120
Mint	1.00	1.00	1.00	128
Neem	1.00	1.00	1.00	128
Oleander	1.00	0.97	0.98	128
Parijata	1.00	1.00	1.00	100
...				
accuracy			0.99	4228
macro avg	0.99	0.99	0.99	4228
weighted avg	0.99	0.99	0.99	4228

Fig. 7. Classification report of VGG-16

- **Minor Variability**: Only a few classes, such as 'Fenugreek' and 'Drumstick', showed slight dips in recall (0.97–0.99), but even these scores are extremely high.
- **Outstanding Overall Accuracy**: The VGG-16 model achieved an impressive 99% accuracy. Both the macro and weighted average F1-scores also scored 0.99, suggesting exceptional consistency and balance in predictions.

Figure 6 and Fig. 7 present the code-generated classification report of each method. The Fig. 8a to 8d show the accuracy and loss curves of the custom CNN and VGG16. The custom CNN provides a lightweight, efficient option ideal for real-time applications, achieving 85% accuracy. VGG-16 improves overall precision and achieves 99% precision through its deep feature extraction capabilities. The lower CNN results are due to the fewer samples used to train the CNN. Hence, if more samples were used to train the CNN, a better result could be achieved. By combining these models, a balance of speed and precision could be achieved, making it suitable for practical plant identification.

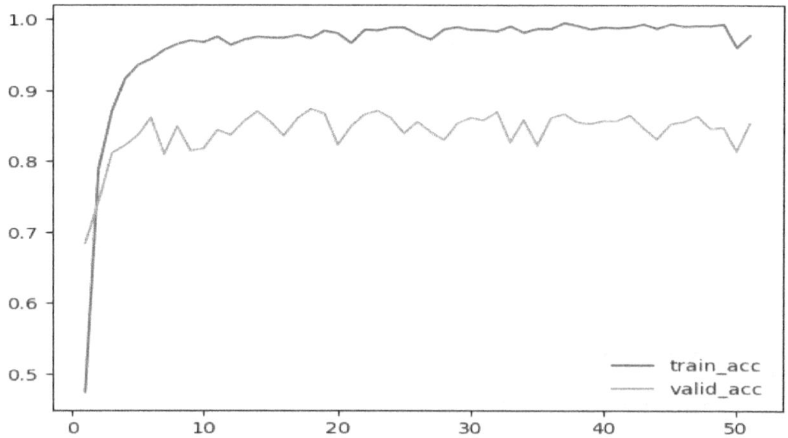

(a) Proposed CNN Performance: Accuracy

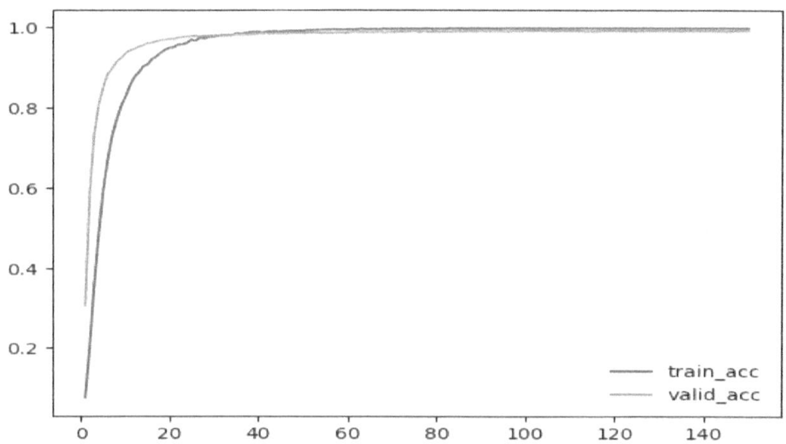

(b) VGG-16 Performance: Accuracy

Fig. 8. Model Performance

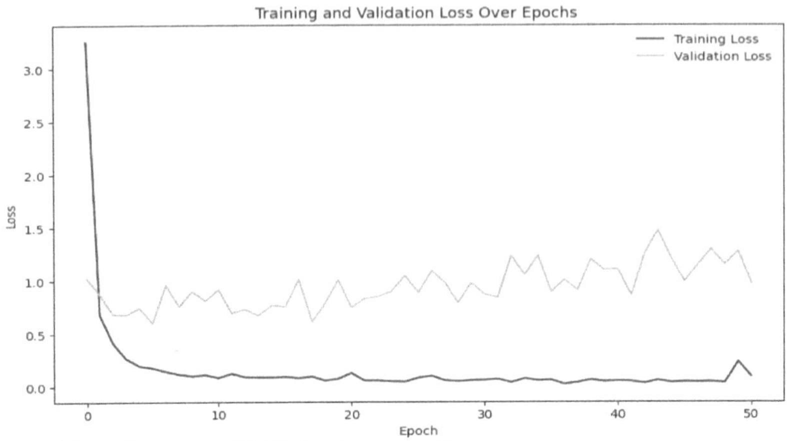

(c) Proposed CNN Performance: Loss

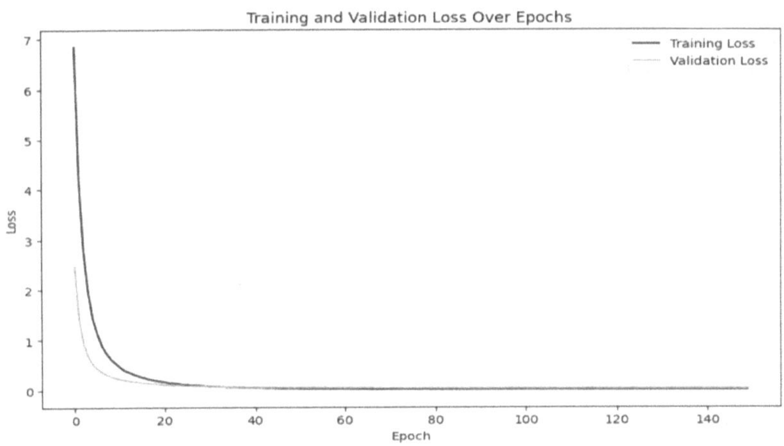

(d) VGG-16 Performance: Loss

Fig. 8. *(continued)*

4.3 Deployment

Deep learning models often achieve remarkable results during training and testing phases but fail to reach their full potential if not deployed for real-time access. This lack of deployment limits their usability and restricts accessibility to general, non-technical users. To address this, a web application was designed to deploy the trained model using Flask, a lightweight and flexible Python web framework. The web app as shown in Fig. 9 provides an intuitive, user-friendly interface that allows seamless interaction with the model, enabling users to access its functionality without requiring technical expertise. Flask facilitates smooth backend integration and real-time inference handling, ensuring a responsive and efficient system. This deployment transforms research outcomes into scalable, practical solutions while democratizing advanced technologies for broader impact. Users could up-load leaf images as input to the DL model hosted in the cloud through the web app. The DL model identifies the medicinal leaves from the provided images.

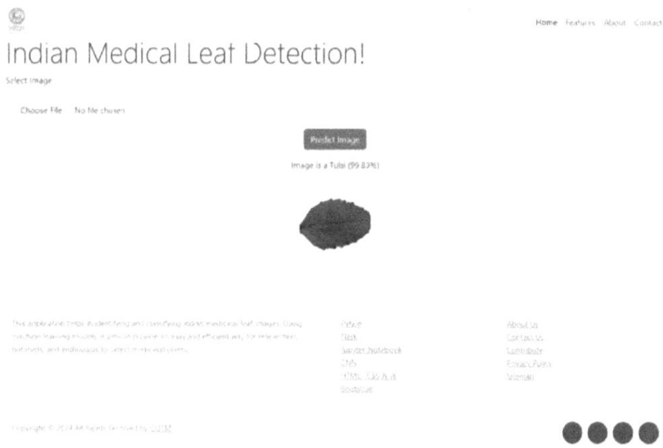

Fig. 9. Model Deployment

Flask Back End: In this setup:

- **Flask** serves as the back end. It runs a small server that takes data from the user, and feeds it to the Keras model, and returns the result.
- The deep learning model was loaded into Flask, so each time a user sends data, Flask handles the model's predictions and sends them back.

Front End (HTML, CSS, and JavaScript): The front end is for user interaction. It is designed to be simple and functional.

The app does not raise ethical issues, as it is only a tool for curious people to identify leaves. The app's limitation is that the leaf samples are very different in shape and size from the training samples. Usually, these are abnormally formed leaves by nature or torn or disfigured by interventions.

5 Conclusion

Plants are essential to human life, and indigenous communities have used traditional herbal remedies for thousands of years. Often, clinicians rely on their years of experience and familiarity with herbs' smells and tastes to identify them. However, with recent technological advances, automatic herb identification has become much more accessible and easier for those new to the field. In this study, a Deep Learning (DL) model was developed to create a web application for identifying medicinal plants. For identifying the leaves, a custom CNN and VGG16 transfer learning model was implemented. The work focused on 30 herbs, but the model achieved impressive results, with over 99% accuracy, precision, and recall. This DL-based approach provides a fast and reliable way to identify medicinal plants. The custom CNN showed promising results but requires more training samples for better performance. The VGG-16-based model showed excellent performance in classifying plant species, making it a valuable tool for herbal identification.

Moving forward, there will also be efforts to refine the model further, especially in areas where recall could be slightly enhanced. Also, there is a requirement to improve the model by incorporating more plant species into the dataset. They could be common or rare in nature. And also, more models could be explored for the new add-on dataset.

References

1. Abdollahi, J.: Identification of medicinal plants in Ardabil using deep learning: identification of medicinal plants using deep learning. In: 2022 27th International Computer Conference, Computer Society of Iran (CSICC). IEEE, pp. 1–6 (2022)
2. Akter, R., Hosen, M.I.: CNN-based leaf image classification for Bangladeshi medicinal plant recognition. In: 2020 Emerging Technology in Computing, Communication and Electronics (ETCCE), pp. 1–6. IEEE (2020)
3. Pushpa Rani, B,R.S.: Indian medicinal leaves dataset (2023). https://www.kaggle.com/datasets/aryashah2k/indian-medicinal-leaves-dataset. Accessed 12 Nov 2024
4. Bafandkar, M.: Exploring deep neural networks for plant image classification. PhD thesis. University of Winnipeg (2022)
5. Balasundaram, A., Dilip, G., Ashokkumar, S., et al.: Detecting true medicinal leaves among similar leaves using computer vision and CNN. Nat. Acad. Sci. Lett. **46**(3), 257–262 (2023)
6. Malik, O.A., Ismail, N., Hussein, B.R., et al.: Automated real-time identification of medicinal plants species in natural environment using deep learning models—a case study from Borneo region. Plants **11**(15), 1952 (2022)
7. Mazid, M., Khan, T.A., Mohammad, F.: Medicinal plants of rural India: a review of use by Indian folks. Indo Global J. Pharm. **2**(3), 286–304 (2012)
8. Praveena, S., Pavithra, S., Kumar, A.D.V., et al.: CNN-based Indian medicinal leaf type identification and medical use recommendation. Neural Comput. Appl. **36**(10), 5399–5412 (2024)
9. Sachar, S., Kumar, A.: Deep ensemble learning for automatic medicinal leaf identification. Int. J. Inf. Technol. **14**(6), 3089–3097 (2022)
10. Shastry, A., Patil, A.P.: Classification of medicinal leaves using support vector machine, convolutional neural network and you only look once. In: 2020 IEEE Bangalore Humanitarian Technology Conference (B-HTC). IEEE, pp. 1–6 (2020)
11. Tiwari, R.G., Maheshwari, H., Gautam, V., et al.: MEDLEAFNET: a feature fusion network integrating transformer and CNN modules for Indian medical leaf classification. In: 2024 2nd International Conference on Intelligent Data Communication Technologies and Internet of Things (IDCIoT). IEEE, pp. 999–1005 (2024)
12. Vasuki, A., Govindaraju, S.: Deep neural networks for image classification. In: Deep Learning for Image Processing Applications. IOS Press, pp. 27–49 (2017)
13. Zin, I.A.M., Ibrahim, Z., Isa, D., et al.: Herbal plant recognition using deep convolutional neural network. Bull. Electr. Eng. Inf. **9**(5), 2198–2205 (2020)

Bibliometric Analysis of Artificial Intelligence in Sustainable Agricultural Practices

Sanju Saha[1], Anita Patra[2(✉)], and Swati Suman[1]

[1] Department of Agricultural Extension Education, M.S. Swaminathan School of Agriculture, Centurion University of Technology and Management, Paralakhemundi, Odisha 761211, India
[2] School of Management, Centurion University of Technology and Management, Paralakhemundi, Odisha 761211, India
anita@cutm.ac.in

Abstract. The increasing global population and pressing demand for sustainable farming methods have drawn attention to artificial intelligence (AI) revolutionary role in modern farming. This bibliometric study explores the research landscape of AI applications in agriculture from 2015 to 2024, analyzing data from 588 Scopus-indexed publications, of which 519 journal articles and conference papers were selected after refinement. The data were analysed using R Studio and VOS Viewer to reveal publication trend, keyword evolution, and geographic contributions, highlighting the rapid growth in AI-focused agricultural research, particularly after 2020. India emerged as the top contributor, followed by the US and China, reflecting local preferences for using AI to solve problems in agriculture. Keyword analysis showed a shift from fundamental topics like sensitivity analysis and agricultural machinery to advance application like machine learning, precision agriculture and soil moisture optimization, highlighting the shift to data-driven farming solutions. The co-occurrence keyword network indicated how AI is being integrated with technologies like agricultural robotics, decision support systems and the Internet of Things (IoT), indicating a multidisciplinary strategy for increasing sustainability and productivity. The study also found that, especially in poor nations, exorbitant costs, inadequate infrastructure, and a lack of technical know-how are obstacles to the widespread use of AI. To achieve lasting results, this study emphasises the necessity for affordable, approachable AI solutions and suggests that future investigations look at potential synergies between blockchain, AI, and climate modelling. By compiling critical developments and trends, this study offers a comprehensive overview of the role of AI in promoting sustainable agricultural practices and offers valuable insights to drive innovation and address socio-economic disparities.

Keywords: Artificial Intelligence · Sustainable Agriculture · Crops · Precision Agriculture · Machine Learning

1 Introduction

According to the Food and Agriculture Organisation (FAO), as the world population grows to 9.73 billion by 2050, so will the need for increasing agricultural products [1]. In order to ensure more productivity and precision agriculture management to fulfil food

A. Das et al. (Eds.): MIDAS 2025, CCIS 3055, pp. 313–325, 2027.
https://doi.org/10.1007/978-3-032-35516-4_25

demand, sustainable agricultural methods are displacing traditional farming techniques and implementing cutting-edge technology like artificial intelligence (AI), the internet of things (IoT), and machine learning (ML). [2]. The agriculture industry has seen various disruptions, including reduced labour supply for production, restricted transportation of agricultural output, and hazards linked with agricultural markets. "Because of these advancements, AI is becoming a valuable tool in modern agriculture. From helping farmers make better decisions to supporting more sustainable and efficient practices, these technologies are finding practical, real-world uses on farms today [3].

Artificial intelligence (AI) provides creative solutions to the difficulties of Sustainable farming practices and waste control [4]. Artificial intelligence is transforming the way we tackle everyday challenges. It can enhance how we predict the weather, making communities more resilient to sudden changes. It supports early detection of diseases, giving doctors a valuable edge in saving lives, and it helps us use our resources—like water, energy, and food more wisely, ensuring a more sustainable future for everyone. For example, an AI-powered smart irrigation technique may cut water consumption by thirty per cent while improving agricultural productivity [5].

The goal of sustainable agriculture is to protect the environment without compromising the ability of future generations to meet their fundamental requirements. In addition, it is a technique for making farming more efficient [6]. Sustainable agriculture is mostly due to the key achievements of precision farming, which encompass crop rotation, nutrient deficiency management, pest and disease control, recycling, and water conservation. These achievements contribute to a more secure world [7]. However, the impact of sustainable agriculture is being obscured by a growing global population, increasing instances of hunger, a changing climate, excessive resource consumption, and food and water waste. Developing infrastructure and technologies that can satisfy both current and future demands requires the application of artificial intelligence [8]. According to Rocchi et al. 2020 keyword citation burst study on the ecological modernisation method, "sustainable agriculture" routinely displays extremely high citation bursts in both historical and contemporary literature [9]. The more recent bibliometrics studies regarding the area of sustainable agriculture, which are highly popular among researchers, examine the relationships between sustainable agri-food systems and the economy, society, and policy-making. They also look at the modernisation of agriculture systems and the use of big data in sustainable agriculture [10].

The agriculture industry has suffered as a result of the COVID-19 outbreak. In the past, a number of studies have studied the agricultural issues and offered sustainable solutions by using AI technology to create sustainable agriculture [11]. There are, however, few studies that offer a cutting-edge analysis of the use of AI technology in sustainable farming practices. Thus, the purpose of this study was to determine the present level of knowledge regarding AI and sustainable agriculture, as well as to offer bibliometric and network analysis in this area. In order to stimulate further applications of AI in this area, this bibliometric study is also driven to compile the most noteworthy developments in the AI application in sustainable farming methods [12]. AI application in agriculture includes forecasting for crop production, identification of disease, soil management, and water usage optimization [13]. Challenges such as ethical considerations, adoption barriers, and resource constraints are recurring themes. Barriers like high costs, lack of technical expertise, and limited infrastructure in developing regions hinder widespread AI adoption. Future research should aim to design practical and budget-friendly tools that people can easily adopt in their everyday lives. This approach will be key to closing the existing gaps and making these solutions accessible [14].

Thematic analyses highlight the need for further research in integrating AI with traditional farming practices and addressing socio-economic disparities. Researchers also recommend exploring synergies between AI and emerging technologies like blockchain and climate modelling for enhanced sustainability. The purpose of this research is to gather, examine, and pinpoint the characteristics of academic papers and reviews that are included in the Scopus databases index, especially for sustainable agriculture and artificial intelligence, and crops that were made available between 2015 to 2024.

1.1 Challenges and Future Directions

Obstacles like excessive costs, lack of technical competence, and scarce infrastructure in emerging economies impede mass-scale adoption of AI. Future studies have to focus on developing cost-effective, user-friendly technologies to bridge these gaps [15]. Researchers also suggest examining synergies among AI and upcoming technologies such as blockchain and climate modelling to bring greater sustainability [16].

2 Research Methodology

This research takes a bibliometric analysis of the use of artificial intelligence (AI) in sustainable agriculture between the years 2015 and 2024. Scopus, a well-established database of reputable scientific publications, was used as the primary source of information owing to its wide indexing of peer-reviewed journal articles and conference proceedings. A systematic search using the keywords "Artificial Intelligence," "Sustainable Agriculture," and "Crops" yielded an initial dataset of 588 articles. After applying the criterion for inclusion—retaining only English-language journal articles and conference papers, removing duplicates, and excluding irrelevant entries—the dataset was refined to 519 articles, which were exported in CSV format for analysis. Bibliometric analysis was conducted using the Biblioshiny package in RStudio, enabling the examination of publication trends, frequently used keywords, and research contributions by country [17]. Key analyses included trend analysis to observe annual publication patterns, keyword frequency and co-occurrence network analysis to identify research themes and tree map visualization for hierarchical keyword representation. Additionally, the geographical distribution of research productivity was assessed using Scopus results. Visualisations and statistical outputs provided insights into the evolution and focal areas of research in AI-driven sustainable agriculture. The methodology adhered to rigorous standards, ensuring the reliability and replicability of the findings (Fig. 1).

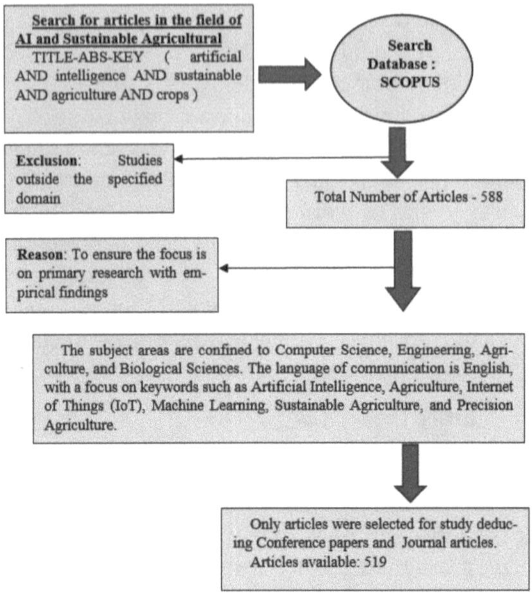

Fig. 1. Research Methodology

3 Results

Table 1 describes the important properties of the data set used for the study which helps in determining the overall scholarly or academic impact that has been made in the field of AI utilized in sustainable agriculture in the duration of 2015 to 2024. There are 519 literary works in all that were taken out for the research. Comprising 519 documents sourced from 327 journals, books, and other platforms, with a yearly growth rate of 43.62%. These documents have an average age of 1.68 years and an average of 11.88 citations each, referencing a total of 21,573 sources. Content analysis reveals 3,536 indexed keywords (Keywords Plus) and 1,547 author-provided keywords. Contributions came from 2,208 authors, with 36 single-authored papers and an average of 4.68 co-authors per document. International co-authorship accounted for 24.47% of the works. Document types included 245 journal articles and 274 conference papers, showcasing significant growth and collaboration in research output.

MAIN INFORMATION	
Description	**Result**
Timespan	2015:2024
Sources (Journals, Books, etc)	327
Documents	519
Annual Growth Rate %	43.62
Document Average Age	1.68
Average citations per doc	11.88
References	21573
DOCUMENT CONTENTS	
Keywords Plus (ID)	3536
Author's Keywords (DE)	1547
AUTHORS	
Authors	2208
Authors of single-authored docs	33
AUTHORS COLLABORATION	
Single-authored docs	36
Co-Authors per Doc	4.68
International co-authorships %	24.47
DOCUMENT TYPES	
Article	245
Conference paper	274

Fig. 2. Main information about the collected documents.

3.1 Yearly Article Statistics

Figure 2 presents a year-by-year overview of articles published from 2015 to 2024, highlighting a clear upward trend in research output over time. In the initial years, publication numbers were modest, with fewer than 10 articles annually: 7 in 2015, 3 in 2016, and 9 in 2017. A slight increase occurred in 2018 and 2019, with 15 and 13 articles, respectively. A significant upward shift is observed starting in 2020, with the number of articles rising to 41. This growth continues steadily in 2021 (43 articles) and 2022 (69 articles). The upward trajectory accelerates in 2023 with 137 articles and reaches its peak in 2024, with an impressive 182 articles published. This data underscores a substantial rise in research activity, particularly from 2020 onward, potentially driven by advancements in the field, increased funding opportunities, and a growing number of contributors to the area of study.

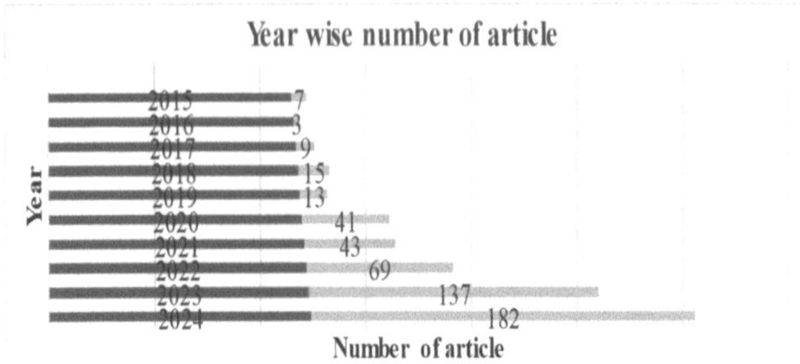

Fig. 3. Yearly production of articles

3.2 Country Wise Article Statistics

The country wise article statistics presented in Fig. 3 reflect the global research landscape on "Artificial Intelligence in Sustainable Agricultural Practices." India emerges as the leading contributor with the highest number of publications, showcasing its strong focus on leveraging AI for addressing agricultural challenges, likely driven by its agrarian economy and the need for innovative solutions, with India emerging as the dominant contributor, publishing approximately 180 documents—significantly outpacing all other countries. The United States follows as the second-largest contributor with around 90 articles, while China ranks third with 60. European nations, including Italy, Germany, and the United Kingdom, display moderate contributions, each ranging between 30 and 40 articles. In contrast, Saudi Arabia, France, Spain, and Greece contribute more modestly, with fewer than 30 articles each.

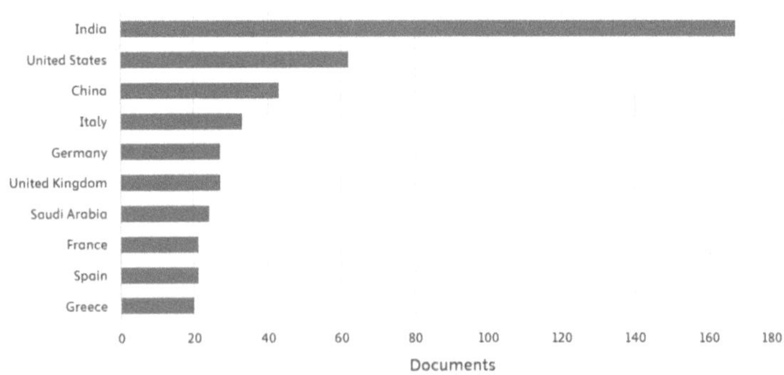

Fig. 4. Country-wise production of articles

3.3 Keyword Analysis

The keyword trend analysis presented in Fig. 4 of the bibliometric study on "Artificial Intelligence in Sustainable Agricultural Practices" highlights a dynamic evolution in research focus from 2016 to 2024. In the beginning, focus was on basic subjects like "support vector machine," "agricultural machinery," and "sensitivity analysis," indicating the initial phases of applying AI to agriculture. But since 2019, the focus has been on more advanced fields like "big data," "precision agriculture," and "decision support systems," with a focus on the implementation of data-based paradigms for decision-making. By 2022–2024, the keywords "machine learning," "soil moisture," and "crop yield" prevail, indicating an upsurge in studies focused on applied AI uses aimed at enhancing productivity and resource efficiency. The rising visibility of words like "Internet of Things" and "artificial intelligence" reflects the accelerating use of connected and scalable technologies. This development represents an unmistakable shift away from exploratory research toward implementation-focused strategies, congruent with worldwide moves to promote sustainable and resilient agricultural systems.

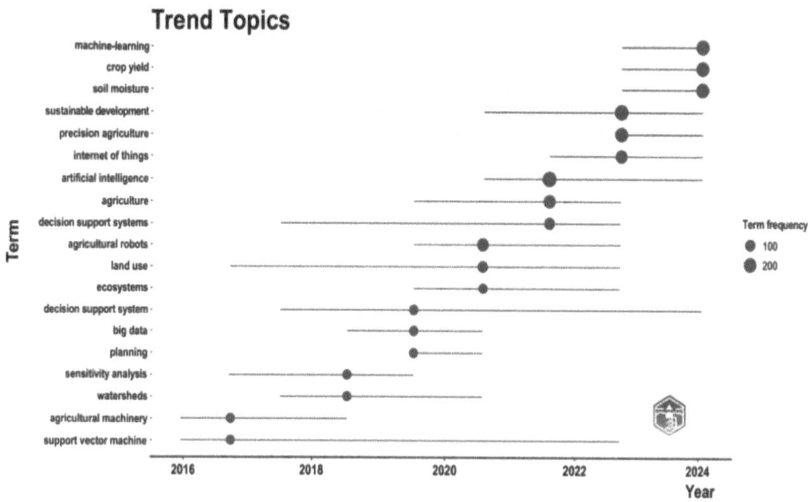

Fig. 5. Trend analysis of keywords

3.4 Word Cloud of Keyword

The treemap in Fig. 5 visualization effectively highlights the prominence of key themes related to agriculture, sustainability, and technology. Artificial Intelligence dominates with 278 mentions (13%), underscoring its transformative impact on modern agricultural practices. Topics like Sustainable Development (151 mentions, 7%) and Agriculture (143 mentions, 7%) emphasize the need to balance productivity with environmental stewardship. Advanced technological innovations like the Internet of Things (102 mentions, 5%), Deep Learning (100 mentions, 5%), and Precision Agriculture (112 mentions, 5%) showcase significant strides in smart farming innovations. Operational aspects, including Crops (99 mentions, 5%), Farms (96 mentions, 4%), and Sustainable Agriculture (85 mentions, 4%), highlight practical priorities in the sector. Furthermore, tools like Decision Support Systems and Machine Learning (both at 4%, with 83 and 59 mentions, respectively) underscore the growing reliance on data-driven decision-making in agriculture. Specific practices such as Irrigation (27 mentions, 1%), Soil Moisture (25 mentions, 1%), and Remote Sensing (46 mentions, 2%) focus on efficient resource management. Overall, the treemap reflects a multidisciplinary approach, combining technology, sustainability, and precision to address the evolving challenges in agriculture.

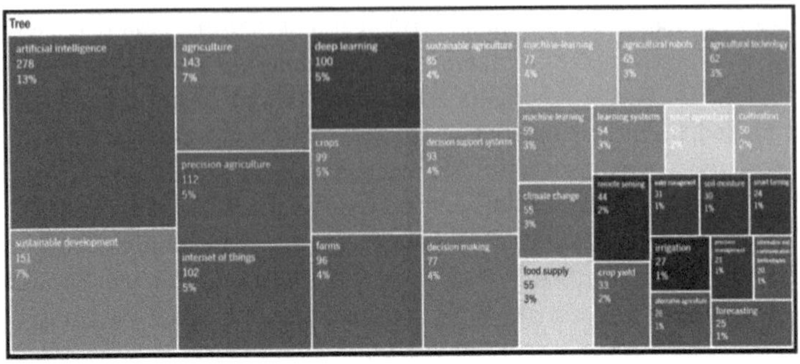

Fig. 6. Treemap of keywords

3.5 Co-occurrence Keywords Analysis

The co-occurrence keyword network map in Fig. 6 vividly depicts the interconnectedness of critical topics within the bibliometric analysis of artificial intelligence (AI) in sustainable farming methods. Central terms like Sustainable Development, Artificial Intelligence, and Precision Agriculture underscore their pivotal role as the foundation of this research domain. Closely related keywords like Internet of Things, Machine Learning, Agricultural Robots, and Decision Support Systems highlight the seamless integration of advanced technologies into agricultural systems to drive efficiency, enhance productivity, and promote sustainability. The tight clustering for terms like Crops, Farms, Water Management, and Climate Change indicates the interdisciplinary focus of the field, covering operational enhancement as well as environmental issues. Peripheral clusters like Irrigation and Remote Sensing indicate specialized uses of AI that enhance resource utilization and encourage precision agriculture. The map generally indicates a dynamic interplay between AI-based innovations, sustainable agriculture, and operational processes. It indicates growing global interest in embracing data-driven approaches to enhance agricultural resilience, food security, and environmental issues (Fig. 7).

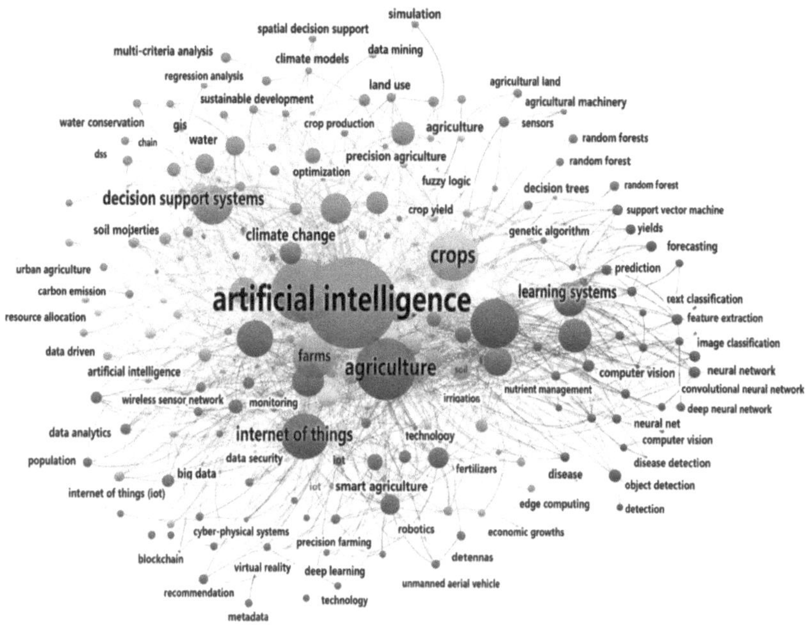

Fig. 7. Co-occurrence Network diagram of keywords

4 Discussion

The bibliometric analysis reveals a significant and steady increase in research on AI applications in sustainable agriculture from 2015 to 2024, with a notable surge after 2020. This growth aligns with advancements in AI technologies, expanded funding opportunities, and global recognition of sustainable farming as a priority. India emerges as the leading contributor, with around 180 publications, reflecting its focus on leveraging AI to address challenges like climate variability and productivity constraints. The United States and China follow with approximately 90 and 60 articles, respectively, underscoring their commitment to technological innovation in agriculture. European countries, including Germany, Italy, and the United Kingdom, demonstrate moderate contributions, balancing technological progress with environmental sustainability. The study period reveals a clear evolution in research themes. From 2015 to 2018, early research focused on foundational methodologies like support vector machines and sensitivity analysis, laying the groundwork for AI integration in agriculture. After 2019, the focus shifted to implementation-oriented studies, emphasizing precision agriculture, big data analytics, and decision support systems. In recent years (2022–2024), the research has prioritized practical applications, including soil moisture monitoring, crop yield prediction, and IoT-driven solutions, indicating the field's growing maturity. Treemap and co-occurrence analyses underscore the multidisciplinary nature of AI in agriculture, highlighting an interconnected ecosystem of themes such as sustainable development, operational efficiency, and technological innovation. According to [18]. Research on agricultural automation indicates that younger farmers are more inclined than older generations to invest in these technologies. The misuse of agrochemicals and insecticides has left once productive areas barren. However, the results of the study on rice crop suggest that this may be prevented by using Artificial Neural Network techniques [19].

4.1 AI Technologies in Agriculture

AI technologies are essential for addressing the intricate problems facing contemporary agriculture [20]. Technologies like machine learning and deep learning enable predictive modelling for crop health assessment, yield optimization, and disease management [21]. Technologies like IoT and big data analytics empower precision agriculture by facilitating the collection data in real time and analysis, improving the farm-level decision-making. The incorporation of remote sensing, agricultural robotics, and decision support systems has further optimized operations, promoting efficient resource utilization and sustainability [22]. AI applications encompass optimizing irrigation, monitoring soil health, and forecasting weather patterns, showcasing their transformative influence on agricultural practices. These advancements enhance productivity while promoting environmental and economic sustainability in farming operations [23].

Suggested a drip irrigation system improvement strategy powered by AI with the goal of achieving sustainable agriculture. Investigated the obstacles to the adoption of AI in agriculture and using the technology acceptance model in their study [24] (Fig. 8).

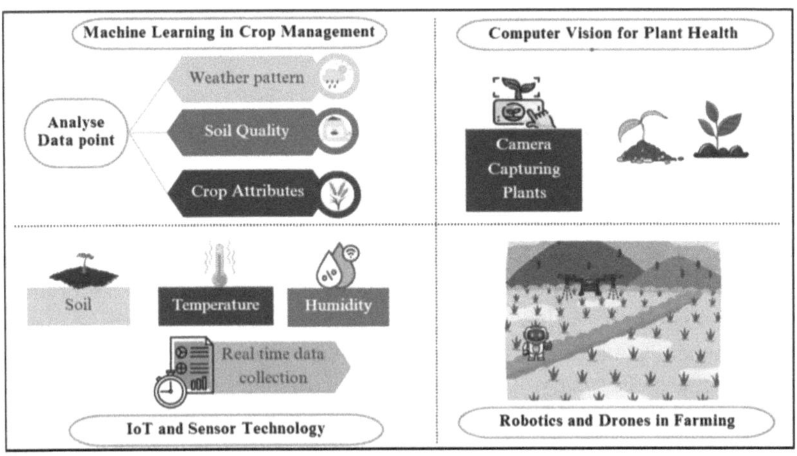

Fig. 8. AI Technologies in Agriculture

4.2 Global Engagement and Research Diversity

Geographical analysis proves the total difference in research output, based on varying regional priorities and capabilities [25]. India's dominance in this area is a reflection of its emphasis on utilizing AI to improve agricultural productivity and address specific issues in its agrarian economy. The United States and China, the top contributors, on the other hand, are a reflection of their emphasis on technological innovation and the development of scalable AI-based solutions. European countries pursue a balanced approach, aligning AI development with sustainability principles to address environmental issues while achieving productivity goals. Small contributions by Saudi Arabia, France, and Spain are a reflection of the growing international awareness of AI's potential to revolutionize agriculture, yet at varying levels of investment and interest. This geographical variation is a reflection of AI's universal application in addressing agricultural issues worldwide, providing a common platform for innovation and sustainability. By combining advanced technologies with regional experience and sustainable practices, AI is revolutionizing agriculture in support of global efforts to enhance food security, optimize resource efficiency, and enhance climate resilience [26].

5 Key Findings

The bibliometric analysis of artificial intelligence in sustainable agriculture reveals a spectacular expansion of research that aims to leverage AI to fuel global agricultural issues. Trends of significance reveal the increasing use of machine learning methods in areas like as forecasting of agricultural output, tracking of soil health, and Management of pests and the application of big data and IoT to facilitate real-time analysis and decision-making. Precision agriculture and smart farming are increasingly viewed as game-changers that employ AI to enhance resource efficiency and mitigate environmental footprints. However, key challenges remain, including high costs of implementation, technology inaccessibility in less developed nations, and data privacy concerns. The research underscores the necessity for requisite legislation and international collaborations to enable equitable and sustainable use of AI in agriculture. It also suggests that future research focus on cost-effective, context-agnostic solutions, as well as greater transparency in AI models, to enable making these technologies accessible to smallholder farmers worldwide.

6 Conclusion

This bibliometric study gives a complete overview of the research ecosystem of AI applications in sustainable agriculture from 2015 until 2024, a time of tremendous growth in research output, especially after 2020. This is the trend that is also seen in the growth of technology, the availability of more funds, and the world's realization of the potential of AI in addressing critical issues in agriculture. India is the biggest contributor, a reflection of its intent to apply AI for sustainable agriculture in its agrarian economy. The United States, China, and European countries also contribute heavily, a reflection of global interest in creating innovative and scalable solutions in agriculture. The shift in research topics from core studies to practical, application-oriented methodologies reflects the trend. Early research focused on basic AI models, while recent years have prioritised precision agriculture, IoT, big data analytics, and machine learning to enhance productivity and resource efficiency. Co-occurrence and treemap analyses further highlight the multidisciplinary nature of the field, integrating advanced technologies with operational and environmental objectives. This study underscores AI's transformative impact on agriculture, tackling the problems caused by an expanding population and climate change. The findings emphasize the importance of ongoing innovation, international collaboration, and supportive policies to unlock the full potential of AI-driven sustainable agriculture.

References

1. Mohamed, E.S., Belal, A.A., Abd-Elmabod, S.K., El-Shirbeny, M.A., Gad, A., Zahran, M.B.: Smart farming for improving agricultural management. Egypt. J. Remote Sens. Space Sci. **24**(3), 971–981 (2021)
2. Yousaf, A., Kayvanfar, V., Mazzoni, A., Elomri, A.: Artificial intelligence-based decision support systems in smart agriculture: bibliometric analysis for operational insights and future directions. Front. Sustain. Food Syst. **6**, 1053921 (2023)
3. Bhagat, P.R., Naz, F., Magda, R.: Artificial intelligence solutions enabling sustainable agriculture: a bibliometric analysis. PLoS ONE **17**(6), e0268989 (2022)
4. Guustaaf, E., Rahardja, U., Aini, Q., Maharani, H.W., Santoso, N.A.: Blockchain-based education project. Aptisi Trans. Manage. **5**(1), 46–61 (2021)
5. Hernandez, D., Pasha, L., Yusuf, D.A., Nurfaizi, R., Julianingsih, D.: The role of artificial intelligence in sustainable agriculture and waste management: towards a green future. Int. Trans. Artif. Intell. **2**(2), 150–157 (2024)
6. Brodt, S., Six, J., Feenstra, G., Ingels, C., Campbell, D.: Sustainable agriculture. Nat. Educ. Knowl. **3**(1) (2011)
7. Gamage, A., et al.: Role of organic farming for achieving sustainability in agriculture. Farm. Syst. **1**(1), 100005 (2023)
8. AlZubi, A.A., Galyna, K.: Artificial intelligence and internet of things for sustainable farming and smart agriculture. IEEE Access (2023)
9. El Bilali, H., Strassner, C., Ben Hassen, T.: Sustainable agri-food systems: environment, economy, society, and policy. Sustainability **13**(11), 6260 (2021)
10. Kharel, T.P., Ashworth, A.J., Owens, P.R., Buser, M.: Spatially and temporally disparate data in systems agriculture: Issues and prospective solutions. Agron. J. **112**(5), 4498–4510 (2020)
11. Mahmood, H., Furqan, M., Meraj, G., Hassan, M.S.: The effects of COVID-19 on agriculture supply chain, food security, and environment: a review. PeerJ **12**, e17281 (2024)
12. Yaacob, A., Gan, J.L.: Bibliometric analysis of global research developments on the role of technology during COVID-19: current trends and future prospects. J. Content, Community Commun. **13**(7), 166–180 (2021)
13. Jack, W., Bagh, S.: Revolutionizing agriculture: AI-powered crop yield forecasting and precision farming for optimal harvests. (No. 11942). Easy Chair (2024)
14. Abioye, S.O., et al.: Artificial intelligence in the construction industry: a review of present status, opportunities, and future challenges. J. Build. Eng. **44**, 103299 (2021)
15. Barbosa Junior, M., Pinheiro, E., Sokulski, C.C., Ramos Huarachi, D.A., de Francisco, A.C.: How to identify barriers to the adoption of sustainable agriculture? A study based on a multi-criteria model. Sustainability **14**(20), 13277 (2022)
16. Jadav, N.K., Rathod, T., Gupta, R., Tanwar, S., Kumar, N., Alkhayyat, A.: Blockchain and artificial intelligence-empowered smart agriculture framework for maximizing human life expectancy. Comput. Electr. Eng. **105**, 108486 (2023)
17. Linnenluecke, M.K., Marrone, M., Singh, A.K.: Conducting systematic literature reviews and bibliometric analyses. Aust. J. Manag. **45**(2), 175–194 (2020)
18. Jha, K., Doshi, A., Patel, P., Shah, M.: A comprehensive review on automation in agriculture using artificial intelligence. Artif. Intell. Agri. **2**, 1–12 (2019)
19. Elahi, E., Weijun, C., Zhang, H., Nazeer, M.: Agricultural intensification and damages to human health in relation to agrochemicals: application of artificial intelligence. Land Use Policy **83**, 461–474 (2019)
20. Liu, S.Y.: Artificial intelligence (AI) in agriculture. IT Professional **22**(3), 14–15 (2020)
21. Talaviya, T., Shah, D., Patel, N., Yagnik, H., Shah, M.: Implementation of artificial intelligence in agriculture for optimisation of irrigation and application of pesticides and herbicides. Artif. Intell. Agri. **4**, 58–73 (2020)

22. Kumar, V., Sharma, K.V., Kedam, N., Patel, A., Kate, T.R., Rathnayake, U.: A comprehensive review on smart and sustainable agriculture using IoT technologies. Smart Agri. Technol., 100487 (2024)
23. Klyushin, D., Tymoshenko, A.: Optimization of drip irrigation systems using artificial intelligence methods for sustainable agriculture and environment. Artif. Intell. Sustain. Dev. Theory Pract. Future Appl., 3–17 (2021)
24. Mohr, S., Kühl, R.: Acceptance of artificial intelligence in German agriculture: an application of the technology acceptance model and the theory of planned behavior. Precision Agric. **22**(6), 1816–1844 (2021)
25. Deng, X., Wang, Y., Song, M.: Development geography for exploring solutions to promote regional development. Geography Sustain. **4**(1), 49–57 (2023)
26. Elufioye, O.A., Ike, C.U., Odeyemi, O., Usman, F.O., Mhlongo, N.Z.: AI-driven predictive analytics in agricultural supply chains: a review: assessing the benefits and challenges of AI in forecasting demand and optimizing supply in agriculture. Comput. Sci. IT Res. J. **5**(2), 473–497 (2024)

Exploring the Role of Artificial Intelligence on Marketing: Key Applications, Industry Transformations, and Emerging Trends Shaping the Future

Pramod Kumar Patjoshi[1](✉), Girija Nandini[1], Umakanta Nayak[2], Ratidev Samal[3], Ansuman Jena[4], and Rajani Agrawalla[1]

[1] Centurion University of Technology and Management, R.Sitapur, Odisha, India
pramodkumar@cutm.ac.in
[2] SVKM's Narsee Monjee Institute of Management Studies, Hyderabad, India
[3] School of Management, DRIEMS University, Tangi, Cuttack 754022, Odisha, India
[4] Vellore Institute of Technology Business School, Vellore, Tamil Nadu, India

Abstract. AI is most beneficial in marketing, impacting how brands interact with customers. It increases data management capabilities, can create complex algorithms, and expand the information and data sources available. There are differences in AI applications as websites and business models differ. AI is bringing in a paradigm shift for consumers and businesses interacting with brands. Marketers can use AI to do everything from instantaneously delivering content to serve the right consumers over the correct channels in real-time as they evaluate their customer's needs and paths at each point based on data shared with them from AI algorithms. This will help to ease the customer and boost their chances of purchasing in an eCommerce platform. AI tools can also determine client expectations and evaluate the effectiveness of competitors' campaigns. Machine learning (ML), allows computers to evaluate and comprehend data independently, providing effective problem-solving skills. Performance and accuracy continuously improve as more data is fed into ML algorithms. Relevant articles on AI in marketing are found and examined in this study using resources including Web of Science, ScienceDirect, ProQuest, Google Scholar., Scopus. This paper highlights the role of AI in marketing, the changes it has brought about and its specialized applications across different industries. This article has made a modest attempt to provide an overview of AI, highlighting its importance and impact in the field of marketing. It examines AI's applications across various marketing domains, identifies key use cases, and explores the transformative changes driven by AI in the sector. Additionally, the article includes a research-based discussion and provides a summary of key insights.

Keywords: Artificial intelligence · machine learning · organizational performance · business operations and enhancing profitability

A. Das et al. (Eds.): MIDAS 2025, CCIS 3055, pp. 326–337, 2027.
https://doi.org/10.1007/978-3-032-35516-4_26

1 Introduction

AI is expected to play a pivotal role and eventually become a key element of organizations worldwide. Every sector is undergoing substantial changes due to new ideas, interests, and investments in AI adoption across companies, as evidenced by recent developments in AI-driven automation [1]. AI's capabilities are becoming increasingly sophisticated, enabling technologies like facial recognition and object detection, which hold vast potential for various business applications. For instance, object detection assists in the analysis and differentiation of photos, while facial recognition can improve security by identifying persons. AI enables companies to provide individualized services according to client preferences by treating human images like cookies. Some businesses are carrying out experiments with facial recognition technology to gauge client sentiment properly [2, 3].

Customer retention and new lead conversions can be increased with the adoption of AI in marketing. Using technologies like AI chatbots, digital marketing, email marketing, and interactive web design, it guides consumers toward activities that support a company's objectives. Several elements, most notably machine learning (ML), a division of AI, affect digital marketing to a large extent. ML focuses on computer programs that learn from data independently, gathering information from user accounts on social media, online customer reviews, websites, and multiple sources. This data enables AI to create and deliver personalized content, enhancing the online experience for users. AI tools can even analyze restaurant data, taking into account customer preferences, historical patterns, and context to refine business strategies [4–8]. However, AI's integration into marketing also raises concerns about the use of customer data responsibly. AI technologies may unintentionally use more personal data than is permitted unless they are specially made to comply with legal restrictions. Despite AI's growing presence in marketing, significant challenges remain unresolved. AI systems, particularly those involving ML and deep learning, require substantial hardware and computational power to process large datasets quickly. It is not only measured just a concept for computer enthusiasts; academics and practitioners are fully aware of its advantages and possibilities [9]. Another challenge lies in the skill gap surrounding AI. Although data science courses focused on AI are becoming more prevalent, the gap between the demand for skilled professionals and the available talent continues to widen. More experts will be needed as AI technology evolves to enable widespread adoption across businesses. Despite AI's growing presence, it remains a novel technology for many, and trust in AI can be a problem to establish without understanding the AI system properly. Additionally, AI systems trained on incomplete or low-quality data may produce inaccurate results, diminishing their effectiveness. Making sure AI is used ethically and transparently will be essential going forward due to growing consumer and governmental scrutiny on how companies handle their data [4, 9]. It is pertinent to provide a brief note on AI for a fair understanding.

This paper presents an extensive literature review of various sources, including academic articles focusing on adopting AI in marketing. To extend their understanding of this transformation, the authors carried out a detailed analysis of numerous scholarly publications. The review provides important insights into the subject, addressing key research objectives and critically evaluating AI's impact on marketing.

The research papers were discovered through articles about AI in marketing that were indexed in a number of databases, including Web of Science, ProQuest, Scopus, Google Scholar, and ScienceDirect. The research methodology has applied a systematic approach to identify, appropriate articles to study the objectives. This article focused on delivering an overview of AI, its importance, and its effect in marketing. This article examines how AI helps in the area of marketing in business. Additionally, by emphasizing significant use cases, it explores the usage of AI in several marketing domains and focuses on the significant shifts brought about by its implementation in the sector. The research is planned on four main objectives: understanding the concept of AI its role in marketing, examining its various applications in the area of marketing, and effect of AI for the transformations in Marketing.

Unlike traditional studies this study focuses on AI's applications and it helps businesses with mechanized tasks and involve customers more effectively for maximizing revenue. AI plays a vital role in digital approaches by improving customer connections to make businesses more well-organized. One of the biggest advantages AI is to enable data-driven decision-making by allowing individuals to shop and gain experiences that will help to build rapport with customers. The study also highlights the AI rationalizes the digital advertising by enhancing customer support by automation and also provides valuable market understanding. AI helps businesses understand customer behavior, predict their trends, and analyze their approaches with accuracy by dealing out big data quickly with the help of AI. Furthermore, AI improves operational efficiency, maximizing revenue by minimising the cost with the power of automation. Therefore, this paper discusses a complete scenario of AI's role in the area of marketing in the recent trend. It emphasizes that the AI application helps in enhancing customer involvement, customize the services with their requirements, which helps businesses competitive and fast-growing in the digital world [10].

2 Role of Artificial Intelligence in Business and Marketing

Artificial intelligence (AI) has the potential to transform customer experience through personalization, efficient workflow and strategic decision making. Ai can analyze and predict the consumer or user behaviour leading to customized offerings thus boosting engagement and conversion rates. Further, AI chatbots can provide the much needed assistance through swift response to the customers' queries. This ultimately saves time and resources through operational efficiency, swift decision making, personalization and resilience in customer experience [11, 12]. However, transparency, privacy and disclosers are some of the key concerns those are to be addressed in time to come [13].

AI brings intelligence to both humans and machines. Natural language processing, machine learning, speech recognition, and other technologies enable machines to carry out tasks that call for human intelligence. AI powered machines can read, speak, recognize, memorise, diagnose, predict, reproduce and generate content by mimicking human cognitive abilities. Such competencies really blur the separation between human beings and machines. Therefore, some fear that AI might replace human beings or pose a threat to the nature. Despite such concerns, AI is prevalently being used for analysing data, finding and forecasting patterns and trends, predicting behaviour. Such crucial insights enable businesses and marketers to deploy strategies in much more informed, effective and efficient manner [14].

Cutting-edge technologies like machine learning, deep learning, and natural language processing are increasingly being integrated with digital marketing and sales. It enables marketers to curate, analyse, predict, customize and target customers with ease and accuracy. Such technologies help the marketers understand the customers better [5]. Deeper consumer insights are gained by applying AI that enabled marketers to better understand consumer preferences and direct customers through their journey. Marketers are providing individualised experiences to customers by evaluating consumer data thorough AI, which in turn raises ROI and improves customer contentment. Furthermore, AI improves efficiency in marketing operations by reducing the need for significant manual labour by automating tasks [4, 8, 9]. One of AI's main objectives is to automate labor-intensive processes so that businesses may increase productivity significantly [3, 6].

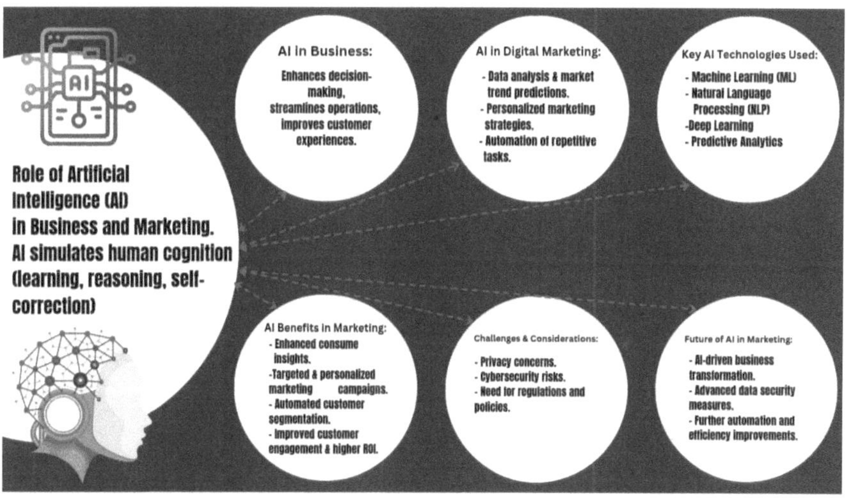

Fig. 1. Role of Artificial Intelligence in Business and Marketing

Businesses can thrive by meeting client needs. Client needs can be analysed with use of AI effectively from data sources like social network and purchased behaviour (online as well as offline). Additionally, AI is being used more and more to tailor marketing campaigns especially through emails. This automated process is promoting conversion and improving client engagement. AI technologies speed up the segmentation process and deliver personalised messages according to the customer's lifecycle stage through a range of channels, including as email, SMS, and app notifications. Therefore, brands have started using AI more and more to personalise marketing campaigns by customising emails, in-app notifications, and SMS according to customer preferences, behaviours, and lifecycle stages, which greatly increases sales and profits for the companies [2, 3]. AI also has the potential to enhance cybersecurity. Even though AI's possible privacy risks are regularly raised, it can also assist and ensure that personal data is owned and protected. Policymakers need to strike a balance between innovation and the requirement for regulation in the context of prevention of abuse and promotion of development of AI. Figure 1 depicts a framework of artificial intelligence's function in marketing and business.

3 Applications of AI in Marketing

AI focuses on leveraging smart technology to improve organizational efficiency, has emerged as a major component of business studies within the last ten years. AI-driven processes improve productivity, proactivity, and informed decision-making across various departments. AI optimizes performance and execution in a number of areas, including sales and marketing, data analytics, decision support, human resource management, and sustainable development. Beyond its conventional use in the technology sector, artificial intelligence is essential to marketing, customer interaction, supply chain management, research and policy formulation, and the travel and tourism sector. Additionally, AI is crucial in integrating emerging technologies like blockchain and cryptocurrency into established economic and legal systems within the fintech industry.

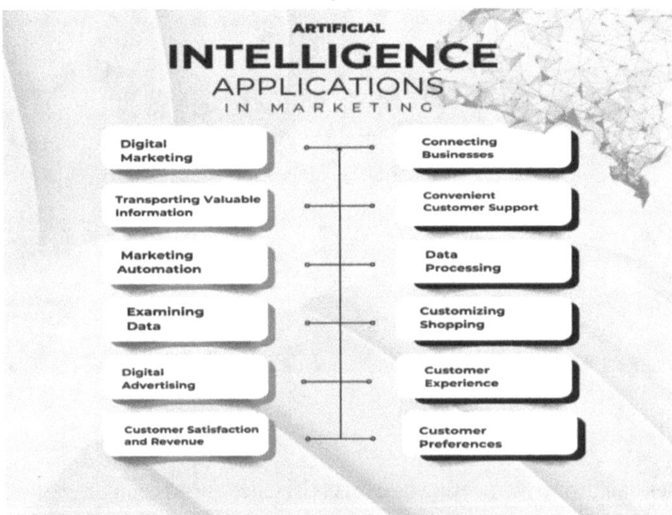

Fig. 2. Applications of Artificial Intelligence in Marketing

Artificial Intelligence is revolutionizing marketing by improving digital strategies, helping businesses reach their target audience, and delivering valuable insights efficiently. Customer satisfaction and brand loyalty can be enhanced by deeper engagement of customers through AI-driven chatbots and virtual assistants. Campaign management can be more effective by adopting AI-driven data analysis that provides deeper insights into consumer behaviour. Personalized shopping experiences, optimized digital ads, and enhanced customer interactions through AI becomes easier due to predictive analysis.

These enables increased revenue and fostered brand loyalty. So, to leverage AI successfully, businesses should integrate AI tools for data analysis, automation of customer communication, and to gain insights to refine marketing strategies.

AI has potential to greatly improve the efficiency and effectiveness of marketing tactics like pricing, strategy making, planning, innovating products, advertising, and managing place. Therefore, AI is considered to be indispensable for targeting, positioning, product designing, and comprehending consumer wants [15]. Figure 2 depicts a framework for marketing uses of artificial intelligence.

Client demand is getting furthered by providing customers with personalized experiences. By collecting data about purchases, location, timing, etc., marketers can develop customized messages for customers. The process of data collection, analysis, and customizing communication can be done by AI. AI can also be used in increasing a customer's average order value by sending pertinent offers and recommendations. AI's incorporation in marketing offers companies a competitive edge by automating processes, enhancing decision-making, and improving client management [3, 6, 9]. Segmentation, targeting, and positioning are being done efficiently with the help of AI-generated analytics. This allows businesses to tailor offers as per the need of customers by making more accurate customer predictions. Through the churn of large datasets, AI can suggest relevant products and services to customers with greater accuracy. It can design personalized campaigns with effective messaging for superior customer engagement and better conversion. It can also automate campaign management and content delivery based on the predictive analytics of consumer behavior.

AI-powered platforms enable businesses to prepare strategies and deliver personalized content to target audiences for fostering deeper engagement. Moreover, AI serves as a bridge between businesses and customers by facilitating seamless interactions, consistent communication, and building stronger relationships [8, 16]. AI also significantly enhances marketing automation by optimizing activities like email campaigns and placement of advertisements. These lead to increased competence and usefulness while conserving the time and resources of the business. The accelerated data processing by AI enables businesses to derive real-time insights and respond swiftly to market changes [17].

AI algorithms are capable of identifying patterns, trends, and customer preferences, aiding in market segmentation, product development, and targeted marketing campaigns. AI analyses provide the much-needed insights to businesses to make informed decisions and provide customers with valuable, timely information. Embracing AI in marketing increases customer support and solves customer issues quickly [5]. Marketing flexibilities and fluctuations can be rightly maneuvered by adapting AI by swiftly analyzing changing consumer needs and preferences. However, precautions must be taken regarding algorithms and the purity of data. In the absence of a correct algorithm and standardized and clean data, the insights provided by AI can be inaccurate, leading to suboptimal or wrong decisions. Therefore, to maximize AI's potential in marketing, businesses must provide robust data management processes. Enabling more precise and impactful campaigns requires growing synergy between big data and AI that ensures next-generation marketing solutions [2, 18, 19]. The ability to recognize patterns in large datasets by

AI is evolving rapidly and offers businesses new opportunities to understand consumer behavior and predict trends.

AI is transforming the marketing domain by increasing competence, personalization, and customer engagement. Of late, with the use of AI, the marketing domain is undergoing rapid changes, adopting innovative practices that are error-free and more focused. AI is revolutionizing marketing, particularly in the retail, entertainment, government, healthcare, and financial sectors. Improving campaign performance, enhancing customer experiences, and largely increasing operational efficiency are AI-driven now. One primary application is programmatic advertising, where machine learning (ML) algorithms bid on ad spaces in real-time, targeting audiences more precisely. These platforms are boosting ROI by reducing mistakes, accelerating campaign execution, and streamlining marketing procedures. AI-powered insights optimizing ongoing campaigns using real-time analytics [15]. By analyzing both structured and unstructured data, artificial intelligence (AI) reveals hidden patterns and insights that help organizations anticipate client needs and keep a competitive edge. Customer-centric strategies are becoming robust by aligning business offerings with customer preferences and simultaneously addressing pain points. Furthermore, creating a unique and engaging shopping experience for individual customers becomes possible through AI-driven personalization engines that deliver tailored product recommendations.

Digital advertising is getting transformed by AI through optimization of ad placements and concentration on precise demographics. At the same time, experiences of customers are getting better due to personalization of interactions and faster service deliveries. AI also enables one to offer more relevant products, thereby fostering long-term loyalty. Often tailor-made AI-driven solutions are resolving issues promptly in organizations. Customer preferences are getting predicted accurately by AI through analysis of customer data, including past interactions, purchase history, and feedback. AI thus drives overall organizational effectiveness with actionable insights supporting better strategic decision-making across marketing, sales, and customer service domains [8]. Content enhancement and personalized message creation as part of advertisements can be easily done by ML algorithms. Programmatic media bidding, the automated buying and selling of online advertisements, becomes faster and more accurate. By this, human error can be reduced, and efficiency can be increased. Targeted advertisement strategies resonating with the most relevant customer groups are getting designed by marketers with the help of AI [7]. Besides, AI's role in targeted marketing extends to advertisement campaigns, where it can distinguish between exploratory and buying behavior. Integration of facial recognition software with AI provides real-time savings and personalized messages in conjunction with AI-powered notifications, resulting in a highly customized consumer experience [2, 6].

Marketers are adapting their strategies based on conversational AI that transforms search queries and algorithms. Customers are getting engaged more effectively due to automation through AI that frees up valuable human resources. On social media platforms, marketers are maintaining more intimate relationship with customers because of AI that enables personalized interactions at scale. However, there are potential challenges that remain in terms of data access and expertise despite noteworthy offers provided by AI. The utilization of AI is limited as many marketers lack the skill set of effective use of the same. As a result, these days collaboration with third-party organizations having expertise in use of AI is evident to collect and analyze data. Accurate and efficient decisions are possible as systems learn from more data and become better at making effective and faster decisions [8, 9].

Optimized campaigns on the fly because of AI that analyzes patterns of customer behavior. Better targeting through by AI through tailor-made messaging leads to increased conversions. AI can help businesses reach more customers at a lower cost by automating operations like social media supervision, content tailoring, and audience segmentation. Predictive analysis of machine learning helps companies to predict future trends and demand for their products [2]. Programmatic advertising and sentiment analysis to identify and engage the right audience with highly tailored messages through AI enables companies to target audiences effectively. AI ensures the timely delivery of marketing messages through predictive scheduling and contextual understanding, thereby enhancing engagement and impact on audiences. From strategy formulation to execution to scale of operations to reduction of costs, achievement of organizational goals with greater efficiency is possible with AI support to businesses across various stages [15]. Thus, companies can create dynamic, customer-focused marketing strategies that lead to growth and innovation by integrating AI into their operations.

4 AI-Driven Transformations in Marketing Sectors

AI-driven transformations are revolutionizing the marketing sector, making it more impactful and dynamic. Various AI applications help address the challenges faced in the competitive marketing landscape. Key elements such as data collection, in-depth market analysis, customer understanding, AI-powered digital strategies, and market research play vital roles in implementing AI to enhance marketing tactics and decision-making processes [9, 18]. AI is being used by marketers more and more to spot new trends and predict how the industry will move in the future. By using AI, companies can make better decisions, can allocate their budget properly, ultimately reducing spend on digital ads while increasing focus on high-value activities. AI's role spans the entire marketing process, from planning to conversion and customer loyalty, providing a significant competitive edge to companies that leverage its capabilities. AI technologies, which replicate human cognitive functions like learning and problem-solving, are helping marketers analyse user data to understand intent and enhance content marketing. Furthermore, AI is transforming software security testing by helping designers identify vulnerabilities, much like how cybercriminals exploit system weaknesses [8].

When selecting AI tools, transparency is crucial, especially in decision-making. Marketing teams can benefit from clear reports explaining why certain decisions are made, though more complex deep learning systems may lack such interpretability. By learning from data, AI creates predictive tools that separate output from the original training data, emphasizing the importance of managing this data as a valuable asset. AI also affects the backend operations of well-known websites like Google, Netflix, and Amazon. AI resources that were previously only available to giant firms are now accessible to small and medium-sized enterprises. By using AI to understand better consumer behaviour, segment buyers, automate marketing, and forecast sales, marketers can glean deeper insights from big data. Predictive analytics, powered by neural networks, help forecast campaign outcomes by identifying patterns from historical data. AI also enhances targeted advertising by analysing user demographics and interests to suggest the most suitable audience for a brand [20]. The rise of voice search further improves the accuracy of machine learning (ML) models, adding new input layers for even more precise targeting. Deep learning, which uses vast datasets to validate predictions, reduces the need for human intervention, allowing for immediate use of results.

With AI systems increasingly collecting and processing vast volumes of data, traditional marketing methods have changed as power has transferred from businessmen to consumers. Marketers can track consumer sentiment in real-time through social media, adjusting messaging and branding strategies for a more significant impact. May better understand consumer preferences and behavior by using AI, which enables them to customize messaging and enhance customer interactions. As businesses adopt AI solutions, the role of digital advertising and account-based marketing grows, enabling more refined targeting and analysis.

Internet of Things (IoT) further improves data handling capabilities, providing businesses more access to consumer data and insights into user interactions. IoT technology is improving efficiency, particularly in product development, and driving growth through better market understanding. Customers are expecting more openness about how their information is used and protected as they become more conscious of the data that businesses gather. By figuring out when to send different consumer segments relevant content and product recommendations, AI-powered marketing platforms are optimizing email campaigns. Business operations are already being transformed by automation, data analytics, and natural language processing. These areas of AI are enhancing productivity, simplifying procedures, and assisting companies to maintain their competitive edge in a market that is becoming very crowded.

5 Challenges and Considerations

Adoption of AI is hampered by issues including skill shortages, data quality, and regulatory compliance, despite its revolutionary promise. Data security and transparency are crucial when utilizing AI in marketing. AI adoption helps businesses make plans, transforming how they interact with their audiences, forecast trends, and provide quantifiable outcomes. Marketing's future depends on people's capacity to use AI effectively [2, 9].

The hindrances for widespread AI adoption involve regulatory challenges, skill gap and the pacifistic view. There are reported instances where biasness in the data used for AI training or programming has created awkward situations and undesired outcomes. Accountability, privacy, transparency, disclosure and ethical practices can build trust and boost adoption of AI-powered technologies and systems. Most of these and many other aspects pertaining to misinformation and surveillance is addressed timely and will produce greater societal impact.

Regulating the entire AI ecosystem amidst rapidly evolving technology and lack of globally accepted standards poses unique challenges. It is also a valid concern that stringent regulations might stagnate innovation. Widespread AI adoption requires cooperation among governments, law enforcement authorities, activists, citizens, and corporations. Industries like defense, intelligence, finance, healthcare, etc. might require unique rules and practices. Convincing the stakeholders and monitoring and enforcing compliance become even more challenging.

The demand for AI professionals has created a vacuum in the industry. It is a result of the knowledge and skill gap among the engineers, scientists, analysts, and other professionals. The required technical expertise, domain-specific knowledge, and ethical mindset to harness and deploy AI systems just do not exist. Persistent and systemic initiatives are to be taken at all levels to impart training, upskill, and foster a generation of professionals who can not just develop but adopt, adapt, and expand AI awareness and training.

6 Findings

Marketing efforts are becoming super efficient by forecasting customer needs accurately through AI integration. AI generated content, targeted advertising, virtual assistance, chatbots, predictive analytics, informed decision making has resulted in personalized experience and superior customer engagement. AI is causing digital transformation in several industries. Industry 4.0 is now transforming into industry 5.0 with AI and related technological advancements.

Similarly, the adoption of AI is also resulting in massive shift in the individual behaviour. It is reflected through the manner in which the users and the non-users of AI think, communicate and behave. Such changes call for appropriate alterations to the existing marketing efforts. Businesses, tech. Companies and users are actively engaging with such technological advancements.

AI has opened up several fresh avenues for improving human life. For example, AI applications in medical diagnostics, documentation, analysis, and automation are not just revolutionizing the healthcare industry but are also causing a permanent shift in the behavior of the stakeholders. Likewise, digital marketing is perhaps acting as the frontrunner in AI adoption for identifying trends, content personalization, campaign optimization, better return on ad spending, and reach. The predictive ability of the AI-powered tools and systems helps marketers craft and deploy effective and efficient strategies through augmentation, automation, and elimination of unproductive elements. This is resulting in higher customer engagement, satisfaction, loyalty, and retention. Therefore, the evolution of AI-powered technologies will continue shaping the marketing landscape.

7 Conclusion

AI supports human endeavors with cognitive and generative functions. Two of the most prominent techniques within AI are ML and deep learning. AI's capabilities allow for more personalized brand experiences, which significantly enhance user engagement and loyalty. Marketers improve the user experience by utilizing language-based AI solutions for customer engagement, sales, and payment processing. Customers now rely on AI-driven chatbots to assist them in the purchasing process rather than on themselves. AI continually evolves, "learning" from previous interactions and optimizing its responses for even better future outcomes. The marketing departments in the companies can use the data for analysis to get more information about the customers and give them personalized offers as per their requirements. It boosts businesses' digital marketing and ensures that crucial announcements and information reach the right customers at the right moment, improving conversions. AI can also be leveraged for greater transparency and trust-building in consumer relationships.

AI improves communication and builds trust by more efficiently linking companies with their audiences. It is excellent at swiftly conveying important information and providing convenient customer service via chatbots and virtual assistants. Repetitive operations are streamlined via marketing automation, freeing up resources for strategic planning. Businesses can adjust to market changes thanks to AI's deep insights into consumer behavior through data processing and analysis. By making product recommendations based on user preferences, it improves personalized shopping experiences and supports digital advertising initiatives. As companies use AI to predict and satisfy customer wants, better customer satisfaction and experience result in increased revenue. AI helps companies improve their plans and provide outstanding value by evaluating preferences. These developments highlight AI's contribution to the expansion and effectiveness of the marketing industry.

8 Future Scope

Artificial intelligence can be used by marketers to evaluate customer trends and behavior, forecast future occurrences, and modify advertising tactics as necessary. AI enables the use of statistical algorithms and data to predict future trends. AI systems perform better and offer more precise insights as they handle more data. AI enables the analysis of enormous volumes of data to identify the type of advertisement that best fits a particular customer and their stage of the purchasing process. By doing this, AI enables marketers to precisely time the distribution of information depending on data and patterns. Computers can identify patterns in data with the use of machine learning, allowing them to make wise decisions going forward with little to no human involvement.

References

1. Babatunde, S.O., Odejide, O.A., Edunjobi, T.E., Ogundipe, D.O.: The role of AI in marketing personalization: a theoretical exploration of consumer engagement strategies. Int. J. Manage. Entrepreneurship Res. **6**(3), 936–949 (2024)
2. Bobro, N., Hyshchuk, R., Strunhar, A., Bukovskyi, O., Alekseiko, V.: Exploring the role of AI in shaping future marketing strategies: evaluations and outlooks. Amazonia Investiga **13**(80), 43–53 (2024)
3. Chang, Y.T.; Fan, N.H.: A novel approach to market segmentation selection using artificial intelligence techniques. **79**, 1235–1262 (2023)
4. Chintalapati, S., Pandey, S.K.: Artificial intelligence in marketing: a systematic literature review. Int. J. Mark. Res. **64**(1), 38–68 (2022)
5. Ciuffo, J.: Artificial intelligence in marketing. In: Artificial Intelligence and Machine Learning for Business for Non-Engineers, pp. 71–76 (2019)
6. Guha, A., et al.: How artificial intelligence will affect the future of retailing. J. Retailing **97**(1), 28–41 (2021)

7. Haleem, A., Javaid, M., Qadri, M.A., Singh, R.P., Suman, R.: Int. J. Intell. Netw. **3**, 119–132 (2022)
8. Labib, E.: Artificial intelligence in marketing: exploring current and future trends. Cogent Bus. Manage. **11**(1) (2024)
9. Patra, A.K., Praharaj, A., Sudarshan, D., Chhatoi, B.P.: AI and business management: tracking future research agenda through bibliometric network analysis. Heliyon **10**(1), e23902 (2023)
10. Raiter, O.: Segmentation of bank consumers for artificial intelligence marketing. Int. J. Contemp. Finan. Issues **1**(1), 39–54 (2021)
11. Sadriwala, M.F., Sadriwala, K.F.: Perceived usefulness and ease of use of artificial intelligence on marketing innovation. Int. J. Innovat. Digit. Econ. **13**(1), 1–10 (2022)
12. Saura, J.R., Ribeiro-Soriano, D., Palacios-Marqués, D.: Setting B2B digital marketing in artificial intelligence-based CRMs: a review and directions for future research. Ind. Market. Manag. **98**, 161–178 (2021)
13. Shaik, M.: Impact of artificial intelligence on marketing. East Asian J. Multidiscip. Res. **2**(3), 993–1004 (2023)
14. Soni, V.D.: Emerging roles of artificial intelligence in eCommerce. Int. J. Trend Scientific Res. Dev. **4**(5), 223–225 (2020)
15. Tiwari, R., Srivastava, S., Gera, R.: Investigation of artificial intelligence techniques in finance and marketing. Procedia Comput. Sci. **173**, 149–157 (2020)
16. Ullal, M.S., Hawaldar, I.T., Suhan, M., Joseph, N.: The effect of artificial intelligence on the sales graph in the Indian market. Entrepreneurship Sustain. Issues **7**(4), 2940–2954 (2020)
17. Verma, S., Sharma, R., Deb, S., Maitra, D.: Artificial intelligence in marketing: a systematic review and future research direction. Int. J. Inf. Manag. Data Insights **1**(1), 100002 (2021)
18. Wisetsri, W.: Systematic analysis and future research directions in artificial intelligence for marketing. Turk. J. Comput. Math. Edu. (TURCOMAT) **12**(11), 43–55 (2021)
19. Wu, L., Dodoo, N.A., Wen, T.J., Ke, L.: Understanding Twitter conversations about artificial intelligence in advertising based on natural language processing. Int. J. Advert., 1–18 (2021)
20. Yang, X., Li, H., Ni, L., Li, T.: Application of artificial intelligence in precision marketing. J. Organ. End User Comput. **33**(4), 209–219 (2021)

Analyzing the Contributions of Artificial Intelligence to Enhancing Business Intelligence Capabilities Within E-Commerce Platforms

Ratidev Samal[1(✉)], C. M. A. C. S. Ajit Kumar Sabat[2], Rajani Agrawalla[2], Kshitish Kumar Mahapatra[2], Pramod Kumar Patjoshi[2], and Sarthak Prasad Sahoo[3]

[1] School of Management, DRIEMS University, Tangi, Cuttack, Odisha 754022, India
ratidevsamal@gmail.com

[2] School of Management, Centurion University of Technology and Management, R.Sitapur, Odisha, India

[3] School of Engineering and Technology, Centurion University of Technology and Management, R.Sitapur, Odisha, India

Abstract. This research investigates the role of Artificial Intelligence (AI) in augmenting Business Intelligence (BI) capabilities within e-commerce platforms. AI-driven BI systems have transformed traditional BI methodologies through automation of data processing, provision of real-time insights, and enhancement of decision-making processes. The study emphasizes the influence of AI technologies, including machine learning (ML), natural language processing (NLP), and predictive analytics, on critical e-commerce functions such as demand forecasting, customer segmentation, inventory management, and dynamic pricing. Through secondary data analysis and case studies from prominent e-commerce platforms like Amazon and Alibaba, the research illustrates the substantial business value created by AI-powered BI systems. Notable improvements include enhanced accuracy in sales predictions, minimized stockouts, and improved operational efficiency. Additionally, AI's capacity to personalize customer experiences through predictive analytics has led to increased customer retention and conversion rates. The study finds AI integrated platform consistently outperforms traditional business approach relating to sales growth, customer satisfaction, and cost reduction. Direction for future research include exploring AI ethics, data privacy emerging technologies such as blockchain and IoT.

Keywords: Artificial Intelligence · Business Intelligence · E-commerce · Predictive Analytics · Machine Learning

1 Introduction

1.1 Problem Statement

The major challenge of different business houses are not lack of data rather excess of data & finding a proper insight into it. Traditional Traditional BI unable to encounter limitations in processing large and complex datasets quickly enough to respond to rapidly

A. Das et al. (Eds.): MIDAS 2025, CCIS 3055, pp. 338–350, 2027.
https://doi.org/10.1007/978-3-032-35516-4_27

changing market demands. Ineffective decision-making and missed opportunities are possible outcomes of these difficulties. AI-powered BI solutions address this gap by providing scalable, intelligent, and real-time data analysis capabilities. Through AI technologies like machine learning and predictive modeling, organizations can make accurate predictions that improve functional productivity and consumer loyalty.

1.2 Objectives of the Study

1. Investigate how AI enhances BI capabilities, identifying AI tools and techniques that improve BI, such as predictive analytics, machine learning, and natural language processing.
2. Examine how AI processes data to yield better insights, focusing on how AI technologies convert unstructured data into actionable insights through methods like deep learning.
3. Understand how AI-driven insights translate into strategic actions in e-commerce, linking AI-enhanced BI insights to personalized marketing and supply chain management strategies.

1.3 Research Questions

1. How does AI improve the efficiency and accuracy of BI in e-commerce, particularly in enhancing data processing and accuracy?
2. What are the key mechanisms through which AI contributes to advanced data insights, exploring core technologies like machine learning and predictive analytics?
3. In what ways do these insights inform strategic business actions in e-commerce platforms, examining the impact of AI-driven insights on customer retention and dynamic pricing?

2 Literature Review

2.1 AI in Business Intelligence

AI has introduced advanced methodologies to **Business Intelligence** through **machine learning (ML)**, **predictive analytics**, and **data mining**. These technologies automate data analysis, enhance decision-making, and enable companies to identify patterns in large datasets. **Machine Learning (ML)**: At the core of AI, ML automates the data analysis process, allowing for predictive tasks like customer behaviour analysis and sales forecasting. Unsupervised ML models also help in clustering customers for targeted marketing.

Predictive Analytics: Statistical models and based on data from the past, ML algorithms predict outcomes for the future. This is particularly relevant in demand forecasting and inventory management, reducing stock outs and overstocking. **Data Mining**: AI-driven data mining tools extract valuable information from vast datasets, helping e-commerce platforms improve their marketing strategies and optimize operations. This survey paper attracts on late improvements AI exploration to take a gander at the past, present, and eventual fate of AI innovation in another light. Different computerized promoting strategies and techniques that incorporate an electrical apparatus or the web-based data sets to illustrate, publicize, or sell items or administrations, alongside businesses that have utilized web channels to make them fruitful. Sites, informal communication destinations pages, designated commercials, and e - sends will quite often hold current clients and draw in new clients [1]. Affiliations are battling on appraisal not considering the way that they would be able—business today is overpowered with information and information crunchers—yet in like manner because they ought to. When arms in different undertakings offer relative things and utilize equivalent types of progress, business processes are among the last places of separation initiation. Plus, assessment contenders wring every single drop of basic worth from those cycles [2]. To begin with, "higher-request realizing" that recognizes man-made intelligence applications from customary displaying approaches. Second, is chiefs should know about while executing computer based intelligence in their associations, including hazardous or unreasonable learning conditions. Third, simulated intelligence will profoundly affect prescient undertakings that can be computerized and require little reasonableness [3]. The arising patterns relationship between computerized showcasing and man-made reasoning (simulated intelligence). Indeed, even while proposing, the procedures of AI commitment in creating application [4]. We likewise portray the job of search and related calculations, which for us transforms into a proposals issue too [5]. Lately, man-made brainpower has turned into an arising pattern in various fields: science, business, medication, car and training. Simulated intelligence has likewise arrived at promoting. The point of the paper is to explore how profoundly man-made intelligence is applied in advertising and what suggestions there are for promoting professionals [6]. The computerized advertising in scholarly community will in general linger behind the progression in the business area. To grasp how AI consciousness seems to contribute in advanced advertising procedures, we should initially examine the significant sub-fields of computerized reasoning. After all, we inspect whether any of those fields carry out to advance promoting methods individually. It covers advances and late improvements in different computational knowledge methods, with an accentuation on the plan of correspondence frameworks. Also, it shares important bits of knowledge into cutting edge computational philosophies like brain organizations, fluffy frameworks, and transformative calculations, half and half savvy frameworks [7]. Mechanical technology is an expression that connects with smart control, independent investigation processes, and so on. Master Frameworks include the executives' data frameworks, educational frameworks, etc. Acknowledgment framework incorporates voice acknowledgment, creation process, and so on. Machine interpretation alluded to Arranging, game play strategies, and so on, are instances of arranging. [8]. The following cultural changes will significantly affect both individual selling and deals the executives examination and practices. In this article, we center around AI and man-made consciousness and their effect on private selling and deal the board. We inspect that effect on a little area of deals practice and exploration in view of the seven stages of the selling system. Suggestions for hypothesis and practice are determined [9].

2.2 Case Studies

Alibaba employs **AI-powered BI** to optimize supply chain operations and enhance customer experiences by analyzing consumer data and predicting market trends. **Shopify** uses **AI** to help merchants tailor marketing campaigns based on customer behavior, thus improving inventory forecasting (Table 1).

Table 1. AI's Impact on E-Commerce BI

AI Tool/Technique	Application	Impact
Machine Learning	Customer segmentation, targeting	Enhanced personalization, conversion
Predictive Analytics	Demand forecasting, dynamic pricing	Optimized inventory, real-time pricing
Data Mining	Identifying trends	Improved decision-making accuracy

The above table elaborates about different AI tools, its applications & impact. It is inferred that Machine Learning is having application in marketing foe segmenting, targeting & positioning for achieving enhanced personalization & conversion. Similarly predictive analytics helps in forecasting the demand & making dynamic pricing for optimized inventory & real time pricing. Through the use of data mining to find trend to improve decision making accuracy.

3 Theoretical Framework

3.1 AI Mechanisms in E-Commerce

AI-driven mechanisms such as **NLP**, **ML**, and **data mining** play a pivotal role in **Business Intelligence** within e-commerce platforms. These technologies enhance data analysis and decision-making efficiency by automating large-scale data processing.

Natural Language Processing (NLP) analyzes customer reviews and social media posts to extract insights into customer sentiment and preferences. This enables models for sales forecasting, customer segmentation, and fraud detection.

3.2 BI Framework in E-Commerce

In e-commerce, data collection, analysis, visualization, and reporting are all part of the BI framework. AI integration enhances these processes by providing real-time insights into operational trends. AI tools present data in accessible formats, such as dashboards displaying key performance indicators (KPIs). Insights for strategic decision-making are summarized in automatically generated reports (Table 2).

Table 2. Key BI Functionalities Enhanced by AI

Functionality	AI Enhancement
Customer Segmentation	Targeted marketing strategies
Sales Forecasting	Accurate inventory planning
Personalization	Custom recommendations

4 Research Methodology

The Research Methodology section outlines a structured method for analyzing how AI enhances BI capabilities within e-commerce platforms through qualitative case analysis.

4.1 Data Collection

The research primarily relies on secondary data sources, which include: Industry Reports from reputable research firms such as Forrester, McKinsey, and Gartner which offer insights into the trends in the adoption of AI in e-commerce platforms. Case studies of AI implementations within BI frameworks are presented in white papers from IBM, Microsoft, and Google Cloud. These papers highlight technological advancements and the benefits achieved by e-commerce platforms (Table 3).

Table 3. Summary of Secondary Data Sources Used in the Research

Data Source	Type of Data	Key Information Provided
Industry Reports	AI trends in e-commerce	Adoption rates, AI technologies, industry benchmarks
White Papers	Case studies, solutions	AI mechanisms, BI applications, success metrics
Industry Surveys	Quantitative data	AI adoption statistics, challenges, future trends
E-commerce Platforms	Internal reports	AI-driven BI strategies, operational data

The above table captures important data sources & types of data it contains & key informations they contain. As instance industry report deals with AI trend in e-commerce providing information relating to adoption rates, AI technologies & industry benchmark. Similarly White paper deals with case study & solution with key information AI mechanism & BI application. Industry report deals with quantitative data relating to AI adoption statistics, challenges & future trends & E-commerce platforms dealing with internal reports providing information relating to AI driven BI strategies & operational data.

4.2 Case Studies of AI Implementation in E-Commerce Platforms

The research includes case studies of major e-commerce platforms like Amazon and Alibaba to illustrate practical AI applications in BI.

Amazon has effectively integrated ML and predictive analytics into its BI systems to enhance personalized recommendations and inventory management, leading to improved customer experiences and operational efficiency. **Alibaba** utilizes AI-driven BI tools for real-time data analysis, enhancing inventory optimization and customer behavior tracking capabilities. The case study of Alibaba demonstrates significant operational efficiencies achieved through AI, including cost reduction and improved customer satisfaction (Table 4).

Table 4. Key Case Studies and Their Relevance to the Research

E-Commerce Platform	AI Technologies Used	Key BI Enhancements	Business Outcomes
Amazon	Machine learning, predictive analytics	Customer segmentation, sales forecasting, personalization	Increased customer retention, optimized inventory management
Alibaba	AI-driven data analysis, dynamic pricing	Real-time data tracking, inventory optimization	Improved operational efficiency, cost reduction

The above table captures information relating to case Amazon & Alibaba highlighting AI technology used, key BI enhancements & business outcomes like increased customer retention, optimized inventory management & improved operational efficiency. The use of secondary data, along with in-depth case studies, provides a strong foundation for understanding how AI enhances BI capabilities in the e-commerce sector.

4.3 Data Analysis

The data analysis segment includes qualitative analysis of case studies and a comparative analysis of e-commerce platforms employing AI-driven BI systems versus traditional BI systems.

Qualitative Analysis of Case Studies

The qualitative analysis focuses on real-world examples of AI implementation in leading e-commerce platforms like Amazon, Alibaba, and eBay. This analysis examines AI mechanisms such as customer behavior analysis, sales prediction models, and inventory optimization. The major findings from this analysis indicate enhanced decision-making accuracy, improved customer segmentation, and more efficient resource allocation enabled by AI (Table 5).

Table 5. Key Findings from Qualitative Case Studies

E-Commerce Platform	AI Technologies Used	BI Enhancements	Outcomes
Amazon	Machine Learning, NLP	Customer recommendations, sales prediction	Increased revenue, improved customer experience
Alibaba	Predictive Analytics, Data Mining	Inventory management, real-time decision-making	Cost reduction, enhanced operational efficiency
eBay	Machine Learning, Sentiment Analysis	Customer segmentation, product pricing	Optimized pricing strategies, higher conversion rates

The above table captures qualitative information relating to different cases like Amazon, Alibaba & eBay. Amazon uses ML & NLP for customer recommendations & sales prediction. Alibaba used predictive analysis and data mining. For customer segmentation, product pricing, and optimization of pricing strategy, eBay used machine learning and sentiment analysis.

4.4 Comparative Analysis of E-Commerce Platforms with and Without AI-Driven BI Systems

Performance differences between platforms that rely on traditional BI tools and those that use AI-driven BI systems are evaluated in a comparative analysis. The findings reveal that platforms like **Amazon** and **Alibaba**, which fully integrate AI into their BI operations, exhibit superior performance across metrics such as data processing speed, customer insight generation, and decision-making efficiency. In contrast, smaller e-commerce platforms that have yet to adopt AI-driven BI systems show limited capabilities (Table 6).

Table 6. Comparative Performance of Platforms with and Without AI-Driven BI Systems

Platform	AI-Driven BI	Data Processing Speed	Customer Insight Generation	Decision-Making Efficiency	Business Growth (%)
Amazon	Yes	High	Advanced	Highly efficient	25%
Alibaba	Yes	High	Advanced	Highly efficient	22%
Traditional Platform A	No	Moderate	Basic	Moderate	5%
Traditional Platform B	No	Low	Basic	Low	3%

The above table highlights the comparative analysis of AI driven BI vs non-AI driven BI on parameters like data processing speed, customer insight generation, decision making efficiency & business growth. Both companies Amazon & Alibaba when not using the AI driven BI their data processing was moderate & low, only basic customer insights could be made with moderate & low decision-making efficiency which was giving them a business growth of 5% & 3% respectively. Once the companies started implementing AI driven BI data processing speed became significantly high, advanced customer insights are generated which helps them to take decision with higher efficiency as a result there is significant jump in growth of their business from 5% to 25% & from 3% to 22% respectively.

4.5 Quantitative Evidence of AI's Role in Specific e-commerce Function

Quantifiable evidence on the role of AI in specific e-commerce functions can vary based on the function being analyzed, such as personalization, inventory management, customer support, or fraud detection. Here are some examples supported by evidence:

Personalization and Product Recommendations
Impact: AI-powered recommendation systems help increase sales and improve user experience. **Evidence**: Amazon attributes **35% of its total revenue** to its recommendation engine. Studies show that personalized recommendations increase conversion rates by **20–30%** on average. A McKinsey report (2020) states that personalization can deliver **5–15% revenue growth** in e-commerce.

Chatbots and Customer Support
Impact: AI-driven chatbots improve response times and reduce human labor costs. **Evidence**: By 2025, Juniper Research predicts that chatbots will save businesses more than $8 billion annually in customer service costs. eBay's AI chatbot, ShopBot, reduced average query response time by **30%**, increasing customer satisfaction rates. A Salesforce report found that 69% of consumers prefer AI chatbots for quick communication with businesses.

Dynamic Pricing
Impact: AI algorithms optimize pricing strategies based on real-time demand, competitor prices, and inventory levels. **Evidence**: Retailers using AI-powered dynamic pricing systems report revenue boosts of up to **10%** while maintaining profit margins. Companies like Walmart use AI to adjust prices across their e-commerce platforms in real-time, enabling them to remain competitive.

5 Analysis and Discussion

The Analysis and examines of AI's contributions to BI within e-commerce platforms, emphasizing how AI-driven BI improves decision-making across various critical areas, including demand forecasting, customer insights, and inventory management..

5.1 AI's Role in Enhancing BI

By automating data processing, identifying patterns, and generating actionable insights in real time, AI technologies like machine learning, deep learning, and automation tools significantly enhance BI capabilities. These tools helps e-commerce platforms quickly processing huge amount of data and get meaningful insights which helps making better business plans. BI, ML allows systems to learn from data without explicit programming, which is vital for improving BI in e-commerce, Automation tools automate repetitive tasks, including data cleansing, report generation, and data visualization, freeing resources for strategic decision-making.

5.2 Real-Time Data Processing and Analysis

Traditional BI systems mostly processed on batch wise which delay decision-making. In other hand AI-driven systems utilize real-time data processing to analyze customer behavior instantly, facilitating immediate product recommendations that enhance customer satisfaction and increase sales (Table 7).

Table 7. Real-Time Data Processing & Analysis

Variable	Coefficient	Standard Error	t-Statistic	p-Value
Real-time data processing (X1)	0.73	0.15	4.87	<0.01
Customer Retention (Y2)	0.62	0.12	5.15	<0.01

The above table inferred real-time data processing has a positive and statistically significant impact on both sales growth and customer retention. The Beta coefficient for Real time data processing & customer retention are 0.73 & 0.62 respectively with standard error 0.15 & 0.12. The higher significance level is confirmed by the t-statistics' values of 4.87 and 5.15, respectively, with probability values less than 0.01.

5.3 Impact on E-Commerce Decision-Making

Decision-making processes in crucial areas like demand forecasting, customer insights, and inventory management are significantly enhanced by AI-enhanced BI systems.

Demand Forecasting

By analyzing extensive datasets with machine learning algorithms and predictive analytics, AI-driven BI improves demand forecasting accuracy. A quantitative regression analysis of e-commerce platforms using AI-enhanced BI revealed significant reductions in forecasting errors, which in turn enable better inventory management and fewer stockouts (Table 8).

Table 8. Comparison of Demand Forecasting Accuracy

Platform	BI Type	MAPE (Lower is better)
Amazon	AI-Enhanced BI	5.3%
Alibaba	AI-Enhanced BI	5.1%
Traditional Platform A	Traditional BI	9.8%
Traditional Platform B	Traditional BI	10.2%

Amazon and Alibaba's use of AI-driven BI systems resulted in significantly lower **Mean Absolute Percentage Error (MAPE)** values compared to traditional BI systems. The implementation of AI enabled BI has significantly brought down the error from 9.8% to 5.3% for Amazon & from 10.2% to 5.1% for Alibaba.

5.4 Customer Insights

AI-enhanced BI systems facilitate deeper insights into customer behaviour by analyzing large volumes of data. Techniques such as K-means clustering enable the segmentation of customers based on their browsing habits and purchase histories, allowing for more precise targeted marketing and increased customer retention (Table 9).

Table 9. Customer Segmentation Accuracy and Customer Lifetime Value (CLV)

Platform	BI Type	Segmentation Accuracy (%)	CLV ($)
Amazon	AI-Enhanced BI	89%	520
Alibaba	AI-Enhanced BI	87%	505
Traditional Platform A	Traditional BI	70%	350
Traditional Platform B	Traditional BI	72%	340

Platforms using AI-driven BI systems exhibit significantly higher segmentation accuracy and customer lifetime value (CLV) compared to those utilizing traditional BI tools. The above table clearly reflects how by adopting AI based BI both segmentation accuracy & Customer life time value has increased in comparision to traditional approach. In segmentation accuracy there is a jump from 70% to 89% for Amazon whereas the jump is from 72% to 87% for Alibaba. The customer life time value parameter has seen significant growth after AI based BI implementation it is from 350$ to 520$ for Amazon & from 340$ to 505$ for Alibaba.

5.5 Inventory Management

AI-driven BI systems automate the monitoring of inventory levels and predict replenishment needs based on real-time sales data. This capability enhances operational efficiency and minimizes stockouts (Table 10).

Table 10. Inventory Management Performance Metrics

Platform	BI Type	Inventory Turnover Rate	Stockout Frequency (%)
Amazon	AI-Enhanced BI	14	1.5%
Alibaba	AI-Enhanced BI	12	1.8%
Traditional Platform A	Traditional BI	8	4.2%
Traditional Platform B	Traditional BI	7	4.5%

The application of AI in inventory management yields higher turnover rates and reduced stockout frequencies compared to traditional systems. The implementation of AI enabled BI has enhanced the inventory turnover rate from 8 to 14 for Amazon & from 7 to 12 for Alibaba. Similarly the stockout frequency percentage has come significantly down for Amazon it reduced from 4.2% to 1.5% & for Alibaba it reduced from 4.5% to 1.8%.

5.6 Strategic Business Actions

AI-driven BI insights facilitate strategic business actions such as personalized marketing, dynamic pricing, and supply chain optimization.

Personalized Marketing
AI enables enhanced customer interactions through collaborative filtering and NLP, improving conversion rates and average basket sizes (Table 11).

Table 11. Impact of AI-Driven Personalized Marketing

Platform	BI Type	Conversion Rate (%)	Average Basket Size ($)
Traditional Platform A	Traditional BI	12%	45
AI-Enhanced Platform B	AI-Enhanced BI	20%	72

The above table reflects the impact of AI driven BI on conversion rate & average basket size, AI based BI lifts the conversion rate from 12% to 20% whereas the average basket size has jumped from 45% to 72%.

Dynamic Pricing
AI facilitates dynamic pricing by analyzing real-time data, including demand fluctuations and competitor pricing. This capability allows for optimized pricing strategies that enhance revenue growth and profit margins (Table 12).

Table 12. Impact of AI-Driven Dynamic Pricing on Revenue and Profit Margins

Platform	BI Type	Revenue Growth (%)	Profit Margin (%)
Traditional Platform A	Traditional BI	5%	12%
AI-Enhanced Platform B	AI-Enhanced BI	15%	18%

The above table reflects information about revenue growth & profit margin, it is observed that after implementation of AI based BI revenue has grown from 5% to 15% whereas profit has grown from 12% to 18%.

6 Conclusion

The findings confirm that AI-driven BI systems significantly enhance e-commerce operations by automating complex data processes, enabling real-time decision-making, and providing personalized customer experiences. Amazon and Alibaba demonstrate how AI improves sales forecasting, customer segmentation, and inventory management.

Implications for E-Commerce
AI provides a significant competitive advantage by enabling agility in responding to market trends. Operational efficiency is enhanced by optimizing inventory and supply chain management, reducing both stockouts and overstock situations.

Future Research Directions
Future studies should explore AI ethics, data privacy, and AI's integration with blockchain and IoT to further enhance BI capabilities in e-commerce platforms.

References

1. Chui, M., et al.: Notes from the AI frontier: insights from hundreds of use cases. McKinsey Global Institute (2018)
2. Davenport, T.H., Harris, J.G.: Competing on Analytics: The New Science of Winning, 2nd edn. Harvard Business Review Press (2017)
3. De Bruyn, A., Viswanathan, V., Beh, Y.S., Brock, J.K.U., von Wangenheim, F.: Artificial intelligence and marketing: Pitfalls and opportunities. J. Interact. Mark. **51**, 91–105 (2020)
4. Thilagavathy, N., Praveen Kumar, E.: Artificial intelligence on digital marketing-an overview. Nat. Volatiles Essent. Oils **8**(5), 9895–9908 (2021)
5. Gomez-Uribe, C.A., Hunt, N.: The Netflix recommender system: algorithms, business value, and innovation. ACM Trans. Manag. Inf. Syst. **6**(4), 1–19 (2016). https://doi.org/10.1145/2843948
6. Jarek, K., Mazurek, G.: Marketing and artificial intelligence. Central Eur. Bus. Rev. **8**. 46–55 (2019). https://doi.org/10.18267/j.cebr.213
7. Marketing. In: Dash S, Bhaskar M., Panigrahi B., Das S. (eds.) Artificial Intelligence (AI) and Evolutionary Computations in Engineering Systems. Advances in Intelligent Systems and Computing, vol. 394. Springer, New Delhi (2022). https://doi.org/10.1007/978-981-10-3174-8

8. Rukavitsyn, A., Kupriyanov, M., Shorov, A., Petukhov, I.: Investigation of website classification methods based on data mining techniques. In: 2016 XIX IEEE International Conference on Soft Computing and Measurements (SCM), pp. 333–336 (2016)
9. Syam, N., Sharma, A.: Waiting for a sales renaissance in the fourth industrial revolution: machine learning and artificial intelligence in sales research and practice. Ind. Mark. Manage. **69**, 135–146 (2018)

Application of Machine Learning in Post-Process Dimensional Accuracy Prediction During Precision Machining

Sarthak Prasad Sahoo[1](✉), Ajit Kumar Sabat[2], Rajani Agrawalla[2], and Ratidev Saamal[3]

[1] Department of Mechanical Engineering, Centurion University of Technology and Management, Bhubaneswar, Odisha, India
sarthakprasad.sahoo@cutm.ac.in1

[2] School of Management, Centurion University of Technology and Management, Bhubaneswar, Odisha, India

[3] School of Management, DRIEMS University, Tangi, Cuttack, Odisha 754022, India

Abstract. The advent of advanced machine learning (ML) techniques has revolutionized predictive modeling in precision machining, offering significant improvements in post-process dimensional accuracy. Using secondary data from various research repositories and simulation results, the present study tries to explore the use of models of ML to forecast the dimensional deviations during precision machining. Precision machining plays an integral role in industries such as aerospace, automotive, and medical manufacturing, where the accuracy of parts directly impacts operational efficiency, safety, and product quality. However, challenges persist in achieving consistent post-process dimensional accuracy, necessitating the exploration of innovative solutions. This study evaluates the efficacy of ML models, including decision trees, random forests, gradient boosting machines, and neural networks, for dimensional accuracy prediction. The research compares these models' performance through a detailed statistical analysis, utilizing metrics such as RMSE, R^2, and MAE to assess their accuracy. When compared to conventional techniques, the results show that ML models—in particular, neural networks, significantly improve prediction accuracy. Additionally, the research highlights deficiencies in existing literature, including the restricted use of multifactorial models and the dependence on primary data. The results highlight how ML can completely transform precision machining quality control, with significant implications for both researchers and practitioners.

Keywords: Precision machining · machine learning · dimensional accuracy · predictive modelling · quality control

1 Introduction

Precision machining, being a primary element under the category of advanced manufacturing processes, facilitates to produce different components or products with extremely tight tolerances with an exceptional surface finish. As dimensional precision is a vital

A. Das et al. (Eds.): MIDAS 2025, CCIS 3055, pp. 351–361, 2027.
https://doi.org/10.1007/978-3-032-35516-4_28

platform that ensures product reliability and efficient performance of the product, many of the remarkable sectors including automotive, aerospace and medical device manufacturing etc. are adopting this feature of manufacturing to intact their position in the global market. The complex designs of the advanced engineering components however demand high accuracy and close tolerance limits to function smoothly and effectively. Some of the sophisticated components like engine parts, surgical instruments, turbine blades, biomedical implants frequently seek for tolerances of micron level ensuring the vital role of precision machining that confirms safety and performance [5, 6]. To improve the consistency and quality of machining operations, integration of recent machining routes like Computer Numerical Control (CNC) systems is highly appreciated. But to overcome the already existing out-come variabilities some innovative methods are hence required to maintain the post-process accuracy [7].

Some industries especially medical sectors, automotive industries and aerospace sectors always demand for dimensional accuracy, necessitating reliability and precision are being extremely focused. To ensure fuel efficiency, safety and performances under extreme operational conditions majority of aerospace components such as engine parts, rotating turbine blades and other structural components must align to a close tolerance range. An operational inefficiency and unintentional sudden failure may be caused even with a minute dimensional deviation [5].

Automotive components like transmission systems, braking components and engine blocks must adhere to precision machining as vehicle performance and durability are greatly affected by dimensional accuracy [7]. In the same manner, different surgical instruments, implants and prosthetics must ensure the dimensional accuracy so that these can be easily accommodate and compatible with the complex human anatomy. Again, escalating demand for the components with complex geometries and miniature shape and size, is also creating a huge pressure on the manufacturing industries to understand and implement the benefits the dimensional accuracy throughout the production process [10].

Though the manufacturing industries are gradually adopting the new technologies, but maintaining the dimensional accuracies throughout the production line is still a great challenge due the factors like worn out cutting tool, material variability and different operational conditions. Alteration of cutting tool geometry due to the continuous wear of the physical cutting tools is however responsible for the fluctuation in precision machining. The machining induced heat can also lead to abnormal thermal expansions of both the cutting tool and work-piece material laying a platform for geometrical deviations. The material variations, unwanted vibrations, mechanical deflections during a machining process are also some of the notable key-players for uneven material removal and hence result in additional geometrical inaccuracies [8, 9]. Further, the inherent correlation among the machining parameters also complicates the accuracy predictions. Traditionally available post-processing quality check means, such as coordinate measuring machine (CMMs), are much tedious, time-taking and costly. Hence, different predictive models are highly required to estimate the dimensional inaccuracies without going for the lengthy manual inspection methods [3].

At recent times, the complexities associated with the modern manufacturing processes can be easily addressed by ML. The ML algorithms are capable of modelling the non-linear and complex relations among the input machining parameters and the geometrical deviations.by using and analysing past machining data. Different ML techniques like ensemble methods, artificial neural networks (ANNs), regression models and support vector machines (SVMs) to record and process these complex interactions. To enhance precision in machining and minimizing manufacturing waste ML techniques have shown a great potential due to their real-time forecasting with a dynamic adjustment provisions [1, 2]. Followings are some of the ML techniques commonly used for predictive modelling in machining field:

- Regression Models: The interaction among the machining parameters and the dimensional inaccuracies can be easily established through a linear and/or polynomial regression. As these are simple and interpretable, but they are unable to process non-linear interactions [3].
- Neural Networks: Artificial Neural Networks (ANNs) are the most preferred techniques to model the complex and non-linear relationships. ANNs learn the modelling patterns from the past machining data and can process the intricate relationships among the various machining parameters [1].
- Support Vector Machines (SVMs): SVMs are the better options when it comes to effective processing of problems including regression, classification and variety range of dimensions. By using kernel functions SVMs are able to process the non-linear interactions among the machining parameters [9].
- Ensemble Methods: To enhance the prediction accuracies, different methods like Gradient Boosting Machines and Random Forest method are merged together which can act a robust method to counteract against a large and complex dataset [2].

These techniques have been applied to a wide range of manufacturing problems, including tool wear prediction, surface roughness estimation, and dimensional accuracy modelling. Numerous studies have utilized secondary data to explore the application of ML techniques in different manufacturing methods. Different studies analysed datasets from previous machining experiments to develop predictive models for dimensional deviations [4, 10]. Also, investigations leveraged secondary data to predict tool wear and surface roughness, demonstrating that ML models can generalize well when trained on diverse datasets [3]. Also research has been carried out by focusing on the optimization of machining parameters to minimize errors, using historical data from manufacturing logs [5]. These works highlighted the potential of ML techniques to replace traditional trial-and-error approaches with data-driven insights.

From the above, it is certain that a recurring limitation in existing studies is the inadequate use of statistical tools to validate the predictions of ML models. Metrics such as R2 and RMSE are commonly reported, but comprehensive hypothesis testing, confidence intervals, and other statistical measures are often omitted. This weakens the reliability and generalizability of the findings [6]. Also, many studies adopt single- variable or simplistic multi-variable models, which fail to capture the complex interactions among machining parameters. The combined effects of cutting speed, feed rate, and depth of cut, for example, are rarely adequately predicted, leaving significant gaps in understanding the underlying dynamics [7]. Advanced techniques such as ensemble learning and deep neural networks remain underutilized in this domain.

2 Objectives of the Present Study

Based upon the above identified gaps, the objectives for the present study are set to be as follows:

- To analyse the application of ML models for predicting post-process dimensional accuracy in precision machining.
- To access and compare the performance of various ML techniques using publicly available secondary dataset with their predictive accuracy and reliability.
- To validate the predictions through statistical analysis and hypothesis testing.

3 Methodology

3.1 Collection of Data

The research relies on publicly available secondary data obtained from reputable sources, including:

- Online Datasets: Data repositories such as the UCI Machine Learning Repository and Kaggle.
- Research Repositories: Published datasets from studies on precision machining (e.g., tool wear and machining dynamics).
- Simulation Outputs: Data generated from machining simulations using tools like DEFORM and ANSYS.

The selected dataset contains records of machining experiments, including input parameters and corresponding dimensional deviations. A sample of the dataset is presented in Table 1.

Table 1. Sample Data of Machining Parameters and Dimensional Deviation

Experiment ID	Cutting Velocity ($m.min^{-1}$)	Tool Feed Rate ($mm.rev^{-1}$)	DepthofCut (mm)	ToolWear (mm)	Dimensional Deviation (mm)
1	150	0.2	1.5	0.03	0.05
2	180	0.25	2.0	0.05	0.07
3	120	0.15	1.0	0.02	0.04
4	200	0.3	2.5	0.06	0.09
5	160	0.22	1.8	0.04	0.06

3.2 Details of the variables

Input Parameters (Predictors). Some of the major cutting parameters are taken into consideration as stated below:

- Cutting Speed (m/min): Speed at which the tool moves across the work- piece.
- Feed Rate (mm/rev): The tool's progression distance per work-piece revolution.
- Depth of Cut (mm): Thickness of the material layer being removed.
- Tool Wear (mm): Measure of tool degradation (mostly degraded width).

Output Variable (Response). The difference between the intended and actual dimensions of the machined part is considered to be a suitable response as per the scope of the study. It can also be referred as dimensional deviation.

3.3 Selected ML Models

The following machine learning models were chosen for predictive modelling based on their suitability for regression tasks and capability to handle nonlinear relationships:

- Decision Trees: Simple and interpretable models suitable for capturing hierarchical relationships. Decision trees are ideal for when interpretability is important, and the dataset contains both categorical and continuous features. They also work well when the data has clear, hierarchical decision-making processes.
- Random Forest: A group of decision trees that decreases overfitting and increases accuracy. It is a good choice when there is a need of an accurate, general-purpose model that works well out-of-the-box, especially for datasets with mixed types of features or large datasets with many variables.
- Gradient Boosting Machines (GBM): integrates several ineffective learners to produce an efficient prediction model. It is beneficial for higher accuracy and is also suitable for problems with complex patterns and interactions between features, where simpler models might fall short.
- Neural Networks (NNs): Effective in modelling complex, nonlinear relationships. These are particularly useful for tasks such as image classification, natural language processing, and time series forecasting, where traditional models might not capture the complexity of the data.

3.4 Chosen Statistical Tools

The following machine learning models were chosen for predictive modelling based on their suitability for regression tasks and capability to handle nonlinear relationships:

- ANOVA (Analysis of Variance): Used to analyse the variance among model predictions and actual values.
- R-Squared (R^2): Indicates the proportion of variance in the dependent variable explained by the independent variables.
- Root Mean Square Error (RMSE): Measures the average magnitude of prediction errors, calculated as:

$$RMSE = \sqrt{\frac{\sum_{i=1}^{n} (y_i - \hat{y}_i)^2}{n}} \tag{1}$$

where, *yi* are the actual values, $\hat{y}_i$ are the predicted values and n is the number of observations.

Together, these methods provide a clear picture of model performance from different aspects, so model selection or improvement can be made more informed because these two metrics provide a full picture of how a model is really working. They help researchers and practitioners check the effectiveness and reliability of the models in capturing and predicting data.

3.5 Hypothesis

- Null Hypothesis (H_0): Comparing machine learning models to conventional techniques, the accuracy of dimensional deviation predictions is not appreciably increased.
- Alternative Hypothesis (H_1): Comparing machine learning models to conventional techniques, the accuracy of dimensional deviation forecasts is much increased.

3.6 Implementation Process

Preprocessing of Data

- Normalization: Standardized all input variables to ensure uniform scaling and avoid biasing the models
- Handling Missing Values: Used mean imputation for missing continuous variables and mode imputation for categorical ones.

Training and Testing Split. The chosen dataset is divided into two subsets:

- Training Set (80%): Used to train the machine learning models.
- Testing Set (20%): Used for model evaluation and performance validation.

Cross-Validation Techniques. K-Fold Cross-Validation with $k = 5$ was employed to enhance model reliability by reducing overfitting. Iterating five times, this procedure entails dividing the dataset into five subgroups, training the model on four of them, and testing it on the fifth.

Sample RMSE Calculation for Random Forest Model

$$\mathrm{RMSE} = \sqrt{\Big[\Big((0.05 - 0.048)^2 + (0.07 - 0.072)^2 + (0.04 - 0.039)^2 + (0.09 - 0.087)^2}$$

$$= \sqrt{[(0.000004 + 0.000004 + 0.000001 + 0.000009 + 0.000004)\,/5]} = 0.0021\ \mathrm{mm}$$

This calculation indicates a low RMSE, reflecting high model accuracy (Table 2).

4 Results and Discussion

4.1 Summary of the Dataset

The dataset utilized contains 500 machining experiments, sourced from online repositories (e.g. Kaggle). It includes machining parameters and corresponding dimensional deviations. Table 1 provides descriptive statistics for the primary variables (Table 3).

The distribution of dimensional deviations is approximately normal, as indicated by skewness close to zero and kurtosis values within acceptable ranges.

Table 2. Actual Values and Corresponding Predicted Values

Actual (y_i)	Predicted ($\hat{y}_i$)
0.05	0.048
0.07	0.072
0.04	0.039
0.09	0.087
0.06	0.058

Table 3. Deviations on Machining Dataset [*Source: Dataset From Kaggle Repository on Machining Dynamics*]

Variable	Mean	Standard Deviation	Minimum	Maximum	Skewness	Kurtosis
Cutting Speed (m/min)	165	30.2	120	220	0.25	1.80
Feed Rate (mm/rev)	0.22	0.05	0.15	0.30	0.10	2.10
Depth of Cut (mm)	1.8	0.55	1.0	2.5	-0.15	1.85
Tool Wear (mm)	0.04	0.01	0.02	0.06	0.20	1.75
Dimensional Deviation (mm)	0.07	0.015	0.04	0.09	-0.05	1.65

4.2 Model Performance

The performance of the machine learning models was evaluated using metrics such as R2, RMSE, and Mean Absolute Error (MAE). Table 4 summarizes the results.

Table 4. Performance Metrics for Machine Learning Models [Source: Derived From Machine Learning Model Outputs Using The Dataset]

Model	R^2	RMSE (mm)	MAE (mm)
Decision Trees	0.85	0.0045	0.0038
Random Forest	0.92	0.0032	0.0027
Gradient Boosting (GBM)	0.94	0.0028	0.0025
Neural Networks (NN)	0.95	0.0025	0.0023

Figure 1 visually compares the Root Mean Square Error (RMSE) values for four machine learning models—Decision Trees, Random Forest, Gradient Boosting, and Neural Networks—applied to predict post-process dimensional accuracy. From the figure, it can also be concluded that:

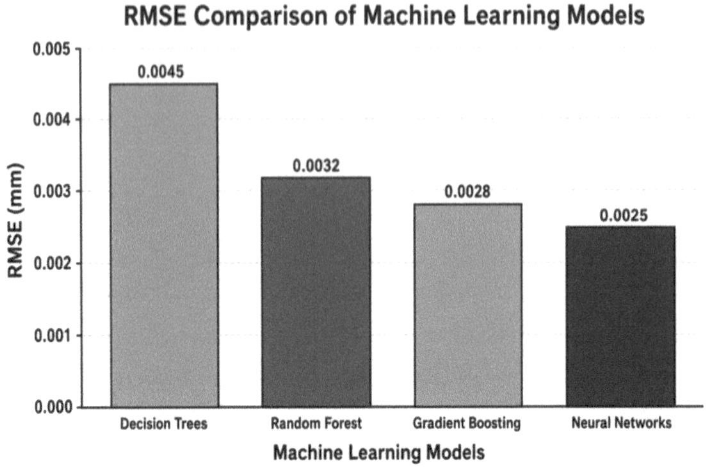

Fig. 1. RMSE Comparison of Machine Learning Models

- The RMSE for Neural Networks is the lowest at 0.0025 mm, indicating the highest prediction accuracy among the models evaluated.
- This demonstrates the strength of Neural Networks in capturing complex nonlinear relationships between machining parameters and dimensional deviations.
- Gradient Boosting achieved an RMSE of 0.0028 mm, closely followed by Random Forest at 0.0032 mm.
- These ensemble-based methods effectively combine multiple weak predictors to enhance overall performance, making them highly effective for precision machining data.
- The RMSE for Decision Trees is 0.0045 mm, significantly higher than the other models. While simple to implement, Decision Trees are more prone to overfitting and lack the robustness provided by ensemble methods.
- The consistent reduction in RMSE values from Decision Trees to Neural Networks highlights the importance of selecting advanced models for applications requiring high precision.
- The figure clearly indicates the superiority of Neural Networks and ensemble methods, making them the preferred choice for practitioners aiming to enhance quality control in precision machining.
- The data further supports the hypothesis that machine learning models significantly outperform traditional methods, providing actionable insights for integrating advanced analytics into machining processes.

Hypothesis Testing: The hypothesis testing aims to determine whether machine learning models significantly improve the accuracy of dimensional deviation predictions compared to traditional methods. A paired t-test was conducted comparing the RMSE of Random Forest (a representative ML model) with that of a traditional linear regression model. Table 4 provides the paired data for a subset of predictions (Table 5).

Table 5. RMSE Values for Paired Testing (Traditional vs. ML Model)

Experiment ID	Linear Regression RMSE (mm)	Random Forest RMSE (mm)
1	0.0055	0.0032
2	0.0060	0.0035
3	0.0048	0.0030
4	0.0052	0.0031
5	0.0057	0.0033

For test statistic calculation, the following relation may be useful.

$$t = \frac{d}{s_d/\sqrt{n}} \tag{2}$$

where, d = mean difference between paired RMSE values, sd = *standard deviated smoothly*, n = number of pairs

Using a significance level of $\alpha = 0.05$ and degrees of freedom ($df = n-1 = 4$), the critical t-value is 2.776. Since $t = 5.59 > 2.776$, we reject the null hypothesis.

P-value: Calculated as $p < 0.01$, indicating strong evidence against H_0.

95% Confidence Interval: The improvement in RMSE is estimated to be between 0.0015 mm and 0.0028 mm.

Neural networks demonstrated the highest predictive accuracy, with an RMSE of 0.0025 mm and R2 of 0.95. This highlights their ability to model nonlinear relationships and interactions among machining parameters effectively. The findings suggest that integrating machine learning into machining processes can significantly enhance quality control, reducing costs associated with post-process inspections and rework.

5 Conclusion

This study explored the application of machine learning models for predicting post-process dimensional accuracy during precision machining, utilizing secondary data sourced from established online repositories. The findings provide clear insights into the efficacy of advanced predictive models and can be listed as per the followings:

- Among the evaluated models, Neural Networks exhibited the highest accuracy, with an RMSE of 0.0025 mm and an R^2 of 0.95 (refer to Table 1 for a summary of performance metrics). Gradient Boosting Machines (GBM) and Random Forest also performed exceptionally well, indicating that ensemble techniques are particularly suited for such predictive tasks.
- The visual comparison aligns with the statistical results discussed earlier, reinforcing the recommendation for practitioners to adopt Neural Networks or Gradient Boosting for optimal accuracy in predicting dimensional deviations.

- Statistical analysis confirmed the alternative hypothesis (H_1), demonstrating that machine learning models significantly enhance the prediction accuracy of dimensional deviations compared to traditional methods. A paired t-test revealed a highly significant improvement ($p < 0.01$), with confidence intervals supporting the robustness of the findings.

With the above findings, present study ensures the following recommendations to the researchers and practitioners in precision machining:

- This research hence confirms potential application of machine learning as a tool for advancing predictive modeling in manufacturing. It also encourages further exploration of secondary data, leveraging diverse datasets to develop generalized and adaptable models. The findings also demonstrate the importance of rigorous statistical validation in ensuring model reliability.
- The demonstrated effectiveness of machine learning models, particularly Neural Networks, suggests a paradigm shift in quality control practices by integrating these models into machining workflows, practitioners can:

  o Reduce dimensional inaccuracies and associated costs.
  o Optimize machining parameters in real time, leading to enhanced efficiency.
  o Minimize reliance on extensive post-process inspections.

6 Scopes for Future Work

The study can also be extended further by incorporating the following aspects:

- Integration of real-time data from IoT systems for improved predictive accuracy with continuous updates on machining parameters and tool wear.
- Enabling predictive maintenance and adaptive machining by linking machine learning models to IoT platforms.
- Investigation of deep learning architectures (CNNs, RNNs) and hybrid models combining machine learning with physics-based simulations.
- Broadening the datasets to include diverse machining setups and materials to enhance model generalizability.

References

1. Chen, X., Luo, Y., Zhang, D.: Machine learning applications in manufacturing: a review. Expert Syst. Appl. **141**, 112987 (2020)
2. Wang, H., Zhang, M., Feng, Q.: Predictive modeling of machining accuracy using machine learning. Manuf. Lett. **30**, 50–57 (2021)
3. Sharma, M., Verma, P.: Post-process quality checks in precision machining: an industrial perspective. J. Mater. Process. Technol. **259**, 32–45 (2018)
4. Li, L., Sun, Y., Gao, J.: Statistical validation of predictive models in precision machining. Comput. Ind. Eng. **131**, 240–250 (2019)
5. Gupta, A., Singh, R., Kumar, R.: Applications of precision machining in advanced manufacturing. J. Manuf. Process. **62**(1), 15–29 (2021)

6. Zhang, W., Li, Q., Tang, Y.: High-precision machining for aerospace components: a review. CIRP Ann. **69**(2), 1083–1097 (2020)
7. Smith, J., Lee, K.: Challenges in achieving dimensional accuracy in precision manufacturing. Procedia Eng. **184**, 25–30 (2019)
8. Kumar, V., Patel, S.: Dynamics of tool wear and its effects on machining accuracy. Int. J. Mach. Tools Manuf **98**, 51–63 (2021)
9. Zhou, J., Wang, T., Liu, H.: Thermal deformation in high-speed machining. J. Adv. Manuf. Technol. **74**(3), 212–225 (2022)
10. Patel, R., Singh, N.: Secondary data-based predictive modeling in manufacturing: challenges and opportunities. J. Intell. Manuf. **31**(4), 975–987 (2020)

Intelligent Data Forwarding in Cloud Using Bio-ACO Integrated with ANN and SCTP(BIAS) Framework

P. Venkadesh[1(✉)], S. V. Divya[1], V. Mathumitha[2], Geetha Ponnaian[3], P. Rajkumar[4], and G. Mohana Prakash[4]

[1] KPR Institute of Engineering and Technology, Coimbatore, Tamil Nadu 641407, India
pvenkadesh2002@gmail.com
[2] Vels Institute of Science, Technology and Advanced Studies, Chennai, India
[3] Saveetha Institute of Medical and Technical Sciences, Saveetha University, Chennai, India
[4] V.S.B College of Engineering Technical Campus, Coimbatore, Tamil Nadu 642109, India

Abstract. A reliable and seamless data transmission in cloud environment is essential to address the challenges posted by the dynamic network conditions. Due to the multi-homing and multi-streaming capability of Steam Control Transmission Protocol (SCTP), the data transmission capability is highly guaranteed. However, the default path selection mechanisms of SCTP and also due to the fluctuating network loads, this may not be always optimal in case of dynamic cloud networks. This paper proposes an enhanced framework by integrating SCTP with Bio inspired -Ant Colony Optimization (Bio-ACO) algorithm and Artificial Neural Networks for secure data forwarding. The adaptive, pheromone-based path selection mechanisms of Bio-ACO were highly helpful in monitoring and optimizing the data forwarding path effectively in real time. Through simulation, we observed that the proposed Bio-inspired ACO with ANN and SCTP (BIAS) algorithm proves to be better than state of the art methods in terms of accuracy rate of 98.5%, packet delivery ratio of 99.2% and a throughput of 9.2 Mbps.

Keywords: SCTP · Multi-streaming · Multi-homing · Data Forwarding · Ant-colony Based Optimization · Artificial Neural Networks

1 Introduction

Cloud computing becomes a new paradigm in service-oriented computing and hence it services play a crucial role in many organizations, business enterprises and sectors. Due to its unique features of least capital investment, scalability, elasticity, flexibility and several other features, the cloud technology is consumed and applied widely in various flavors. Despite these features, cloud also suffers from various flaws. Several efforts are being made to overcome the cloud security concerns, such as improving the cloud security standards and controls and ensuring the privacy of their customers' data along with their Service Level Agreements (SLAs) by the cloud service providers.

A. Das et al. (Eds.): MIDAS 2025, CCIS 3055, pp. 362–375, 2027.
https://doi.org/10.1007/978-3-032-35516-4_29

A dynamic and scalable infrastructure is offered by the cloud technology for accessing the applications, data and the services stored on it which provides a way for their customers to access them online everywhere and at every time with the help of internet. With this development, the IT services, resources (both hardware and software), applications and storage are easily incorporated among the IT industries with cloud storage, cloud applications and cloud service providers.

Cloud had gained much popularity due to the elimination of the installation costs as all the services, applications and software's are delivered as a service. This greatly eliminates the infrastructure costs and hence employed by various enterprises, government and private sectors. By utilizing their data from centralized resources simplifies the business processes much cheaper. Hence this paradigm offers higher efficiency in terms of economy by utilizing the cost and energy factors.

Security and privacy issues remain a challenging task in cloud computing. The different security issues and the challenges faced by the cloud deployment models were discussed in [1–4]. Not all the cloud models face the same security issues and challenges. Depends upon the applications and the services offered by the deployment models, the security issues may differ. Numerous efforts were put into action by researchers and academicians to safeguard the cloud.

Data Security and privacy remains another challenging task in cloud computing. Since cloud encompasses numerous technologies like virtualization, load balancing, networks and databases all the issues which affect the above-mentioned systems also affect the cloud hierarchy. In order to enhance the resource utilization, virtualization [5] is highly beneficial. A virtualization security framework which focuses on virtual system security and virtualization security management is proposed.

1.1 Organization of the Paper

The reminder of the paper is organized as follows: Sect. 2 highlights about the related work and its methodologies. Section 3 explains about our proposed Bioinspired Ant Colony Optimization integrated with ANN and SCTP (BIAS) framework including its architecture and dataflow diagram. Section 4 discuss about the experimental setup and simulation results. Finally, we conclude in Sect. 5 followed by the future research work.

2 Background Works

When an enterprise store and share their data [6] in cloud, the burden of maintaining the data locally is highly eliminated. However, it leads to numerous security issues especially when the data is outsourced. Initially, during outsourcing, the customers' may lose control of their own data which may cause their sensitive data to be leaked. Data leakage [7–9] may impose a severe threat to an organization's privacy and confidentiality issues which in turn leads to the disgrace of an organization's reputation. Moreover, when their data is shared in an open environment, the cloud services, applications, servers' will be the target of attacks. In worst scenario, the cloud providers may themselves reveal their customers' data for their own profit and the data may be tampered by unauthorized parties. Hence data protection remains a major concern in cloud computing.

Numerous methods have been explored for the protection and sharing of data efficiently in cloud platform. A user centric key management using RSA encryption and hierarchical structure for securely sharing the data was proposed. For the secure transmission of medical images, crypto-based algorithms [10, 11] which include hash codes were used. A Cipher-text Policy -Attribute Proxy Re-Encryption [12, 13] to provide security to the chosen cipher-text was proposed. However, the malicious users were not identified by this scheme. Similarly, a data access control [14] and attribute-based encryption [15–17] schemes were proposed.

In identity based remote data integrity scheme [18] the data resides in the highly sensitive data blocks were sanitized and their integrity is validated using digital signature. However, this method adds additional computational overhead and data leakage were also exists. To prevent the data leakage, a damage assessment model [19] which employs the symmetric encryption and attribute-based access control methods was proposed. Though, the method fails to provide solutions for the existing data leakage problems.

An identity -based encryption [20] scheme which incorporates key revocation and ciphertext functionalities were addressed to overcome the existing correctness property. In order to prolong the network lifetime and to resist several attacks faced by the Transmission Control Protocol, Stream Control Transmission Protocol came into existence; which highly improves the network efficiency and transmission bandwidth. With the multi-homing and multi-streaming properties of SCTP, a concurrent multi-path transfer technique [21] which enhances the efficiency and the security of the network was in practice. Yan.et.al [22] had proposed a robust ant colony based optimization algorithm to optimize the data transmission path in wireless sensor networks. Their approach enhances the heuristic value of ACO by considering the network lifetime and energy usage. Eventhough it enhances the network performance, it is computationally expensive.

An ant based routing protocol for efficient load balancing to improve the selection of cost effective path in WSN is proposed by Khapre et al. [23]. Their approach addresses the network energy balancing issues but failed to achieve optimality. Lu.et.al [24] proposed an DLBACO algorithm for task scheduling in cloud scenario which provides better job scheduling. A secure data transmission in SCTP [25, 26] using encryption and embedding a digital signature into the heartbeat signal of SCTP is proposed. This approach provides a layer of security using encryption and digital signature hiding techniques to ensure data integrity. Techniques for data transfer in serverless cloud platform is proposed in [27]. This guarantees dynamic path selection thereby minimizing the latency, buffer management and balancing the server workloads.

2.1 Contribution of the Paper

The contribution of the paper is as follows:

- Selecting an optimal and intelligent path by exploring all the possible paths using Ant Colony Optimization (ACO) and ANN using SCTP for efficient data transmission is proposed.
- In our BAIS framework by adopting ACO, cloud network performance is enhanced by dynamically adjusting the paths, balancing the loads, minimizing the latency and improving the fault tolerance.

- By integrating ACO with SCTP, leverages the strengths of both the mechanisms where SCTP's multihoming feature helps in fault tolerance and the adaptive nature of ACO helps in ensuring the efficient data transmission even during fluctuations in network loads.
- Moreover, by introducing Real-Time Feedback Mechanism and Dynamic Parameter Tuning, the BIAS framework helps to detect and respond to sudden network disruptions which enhances robustness for secure data forwarding in cloud environment.

3 Bio-Inspired Ant Colony Optimization Integrated with ANN and SCTP (BIAS) Framework

In cloud environments offering services that supports efficient and reliable data transmission with minimum latency and high throughput is crucial. Hence SCTP is a suitable option for data transmission due to its multihoming and multistreaming capability. These capabilities allow to maintain multiple paths to face any issues during transmission and significantly reduces the packet loss. Moreover, an optimal path selection is required to tackle the dynamic network conditions and the fluctuating loads happen in the network. Hence in our proposed work, a Bio-inspired Ant Colony Optimization (Bio-ACO) is employed to select the best and intelligent paths.

Bio-ACO is a metaheuristic algorithm inspired by the foraging behaviour of the ants which search for food from its colony to the food source to find optimal paths. In Bio-ACO, optimal path is selected based on two criteria such as pheromone intensity and heuristic information. As ants traverse towards in search of food, they deposit a scent (pheromone) in their paths and hence the forthcoming ants are attracted to this pheromone. This process increases the pheromone level of the paths and are more attracted to ants.

However, the pheromone intensity in those paths were evaporated after a certain rate and the process continues. Bio-ACO optimization is normally employed in case of dynamic networking environments, since its adapts real time path during data transmission. In the proposed work, Artificial Neural Networks Integrating ACO with SCTP offers numerous benefits for offering efficient, reliable and optimized data transmission in networks. SCTP when coupled with ACO, its performance is significantly improved. SCTP's multi-streaming and multi-homing features along with ACO's ability helps to choose the optimal path ensuring low latency and higher throughput.

3.1 Dataflow Diagram of BAIS Framework

The data flow diagram of the proposed BAIS framework is depicted in Fig. 1.

3.1.1 Data Request

The process starts by requesting for the data by the user or the system for data forwarding.

3.1.2 Network Monitoring System

It monitors the network and captures the real time network conditions such as bandwidth, throughput, network traffic, congestion, latency and packet loss. The captured network conditions were fed as input to ANN and Bio-ACO for further processing

3.1.3 Artificial Neural Networks (ANN) Prediction

ANN is employed for path prediction based on the past network behaviour. It chooses the optimal path for data transmission.

3.1.4 Path Exploration and Optimization

Here, Bio-ACO is employed to explore multiple paths by stimulating the movement of ants. Based on the network conditions, the paths were iteratively evaluated and adjusted and the better paths were chosen and the pheromone values were updated periodically.

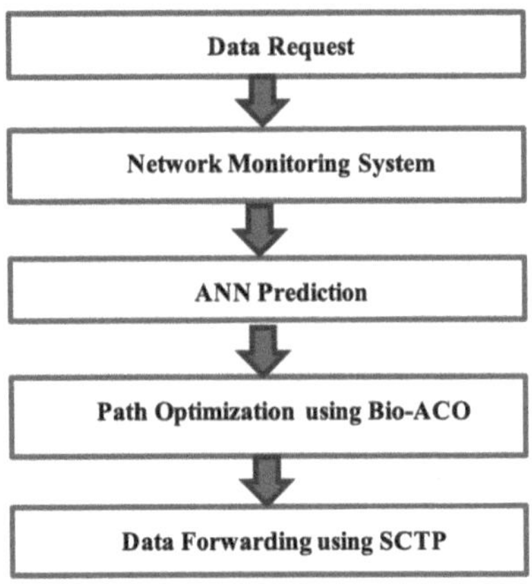

Fig. 1. Dataflow Diagram of BIAS Framework

3.1.5 Data Transmission and Forwarding

The optimized path chosen by the Bio-ACO is used by the SCTP to forward the data packets across multiple paths. The SCTP's ability of multi-streaming and multi-homing guarantees fault tolerance and data integrity during data transmission.

3.2 Bio-Inspired Ant Colony Integrated with ANN and SCTP (BIAS) Framework

The architectural diagram for the BIAS framework is shown in Fig. 2. Initially, the user store the input data in the cloud. In the proposed BAIS framework, Artificial Neural Network (ANN) is employed to enhance the systems' ability by choosing intelligent decisions based upon the dynamic network conditions. Once the data is stored, the ANN layer performs pre-processing using Artificial Neural Network (ANN) and prepare the data for transmission.

The proposed BIAS framework begins with the initialization phase where the initial parameters are setup and initialized. Meanwhile the Bio-ACO parameters and the ANN parameters are also configured. Once initialized, feature selection techniques were applied to capture the real-time data and ensure the optimal predictions.

The ANN make predictions based upon the incoming data and analyze and predict optimal paths. The Bio-inspired ACO algorithm is used to calculate the best routs for forwarding the data across the network based on several performance metrics such as load, traffic, pheromone intensity etc.

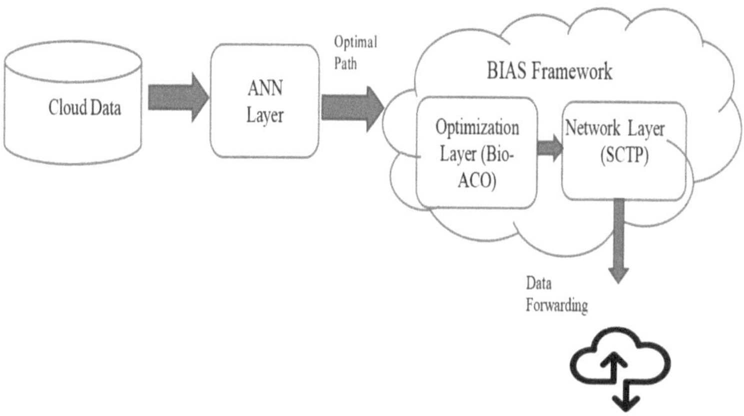

Fig. 2. Architectural Diagram of BIAS Framework

As SCTP supports reliable multi-path transfer, it enables concurrent multi-path transmission and the data is forwarded which is monitored by the cloud management layer. The real-time feedback and dynamic adjustment phase is employed to ensure the systems' resilience and adaptability to network changes. Moreover, the proposed BAIS framework guarantees low latency, high bandwidth, ensure fault tolerance and optimize energy efficiency. By incorporating the Real-Time Feedback Mechanism and Dynamic Parameter Tuning, the BIAS framework can achieve high robustness and enhance network response time in case of any network disruptions. To capture the real-time traffic, network monitoring tools like wireshark is employed which capture real-time packet flows, jitter and analyze the traffic patterns.

3.3 Algorithm of BIAS Framework

Step 1: Initialization of parameters such as SCTP paths, number of streams, number of ants, number of iterations, evaporation rate, pheromone matrix, ANN input features, weights, learning rate and activation functions.

Step 2: Collect network metrics and select features for predicting ANN

2.1: Real-time feedback mechanism to monitor the network metrics & use the monitoring tools to capture the real-time data.

2.2: These data is fed to the system for adaptive decision-making.

Step 3: Train the ANN model for predicting the path

3.1: Adjust the ANN hyper-parameters using dynamic parameter tuning based on feedback data.

Step 4: Optimize the path using Bio-ACO.

4.1: Dynamically adjust the ACO parameters by employing real-time feedback mechanism.

4.2: Optimize the ACO parameters by using adaptive tuning mechanism.

Step 5: Select the optimal path and forward the data using SCTP.

5.1: Monitoring the performance of the selected path and implementing the adaptive re-routing mechanism and selects an alternate path during congestion or network failure.

4 Performance Metrics

The proposed BIAS framework is evaluated with various performance metrics such as latency, throughput, packet delivery ratio, accuracy, jitter, multi-path utilization efficiency, pheromone stability and reliability.

4.1 Latency

Latency is defined as the ratio of the time taken to deliver the data and the total data transferred. Mathematically, it can be defined as,

$$\text{Latency} = {}^{\text{Time taken to deliver the data}}/_{\text{Total data transferred}} \tag{1}$$

Based on (1), the latency is computed and it is depicted in Fig. 3.

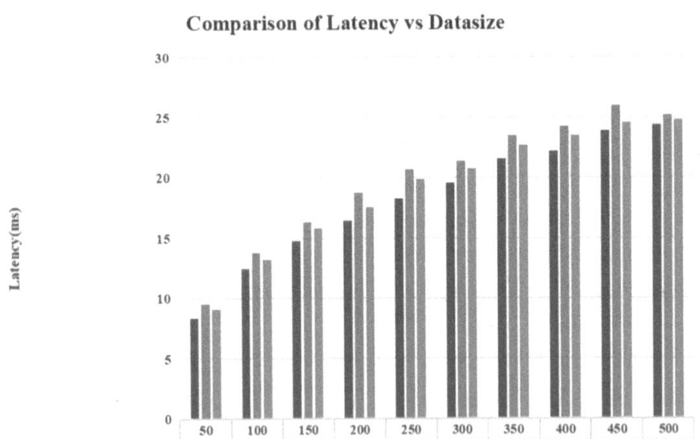

	50	100	150	200	250	300	350	400	450	500
BIAS framework (ms)	8.3	12.4	14.7	16.4	18.2	19.5	21.5	22.1	23.8	24.3
SCTP(ms)	9.5	13.7	16.3	18.7	20.6	21.3	23.4	24.2	25.9	25.1
ACO with SCTP(ms)	9.1	13.2	15.8	17.5	19.8	20.7	22.6	23.4	24.5	24.7

Fig. 3. Latency

From Fig. 3, it is clear that the BIAS framework offers low latency of 8.3 ms for data size 50 when compared to SCTP and ACO with SCTP of 9.5 ms and 9.1 ms respectively.

4.2 Throughput

It is defined as the ratio of the total data transmitted per unit of time as shown in Eq. (2).

$$\text{Throughput} = {}^{\text{Total data transmitted}}\!/_{\text{Time}} \tag{2}$$

The throughput of BIAS framework is compared with the existing methods as shown in Fig. 4. For a data size of 100 MB, the throughput is 9.2 Mbps, 7.4 Mbps and 7.8 Mbps for BIAS, SCTP and ACO with SCTP respectively.

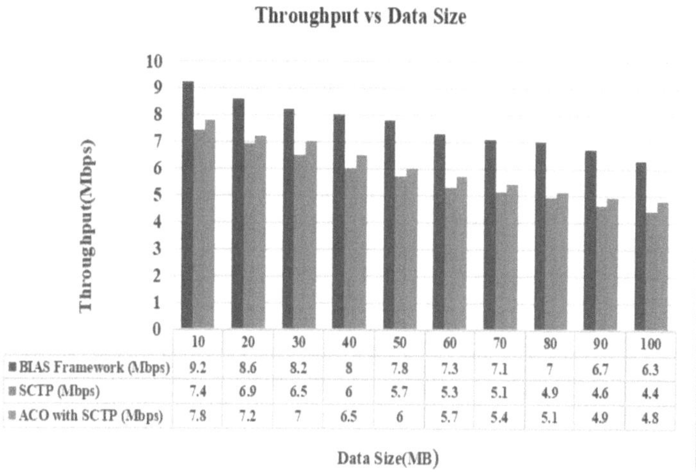

Fig. 4. Throughput

4.3 Packet Delivery Ratio (PDR)

It is defined as the number of packets delivered among the total packets.

$$\text{PDR} = {}^{\text{Total no. of packets delivered}}\!/_{\text{Total no.of packets sent}} \text{x}100 \tag{3}$$

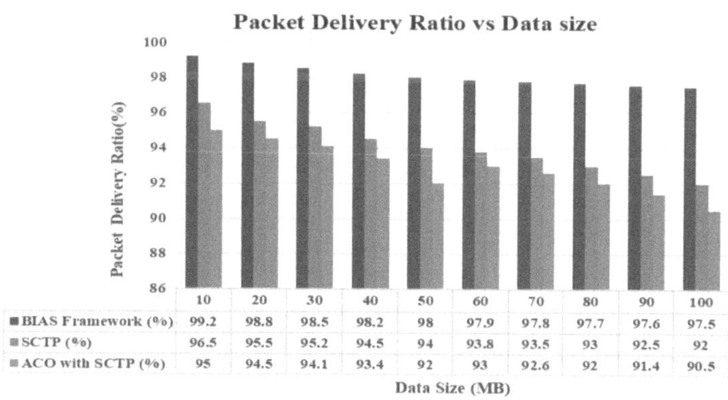

Fig. 5. Packet Delivery Ratio

The PDR value for different datasizes were computed using (3) and it is compared with the existing methods of SCTP and ACO with SCTP. The proposed BIAS framework shows a reasonable PDR of 99.2% when compared to other methods is depicted in Fig. 5.

4.4 Accuracy

In the proposed work, the learning accuracy of ANN is considered. It is computed as the ratio of correct predictions to the total predictions in percentage.

$$\text{Accuracy} = \text{Number of correct predictions} / \text{Total number of predictions} \quad (4)$$

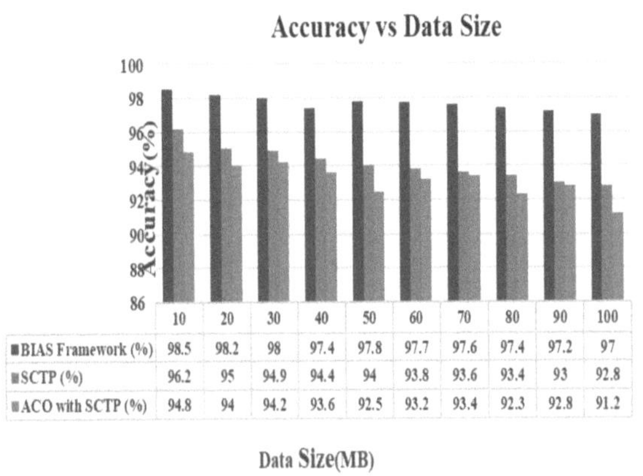

Fig. 6. Accuracy

Using (4), the accuracy for BIAS framework, SCTP and ACO with SCTP is found to be 98.5%, 96.2% and 94.8% respectively and it is depicted in Fig. 6.

4.5 Jitter

Jitter is the difference in delay between two consecutive packets. It is computed as,

$$\text{Jitter} = |D_{i+1} - D_i| \quad (5)$$

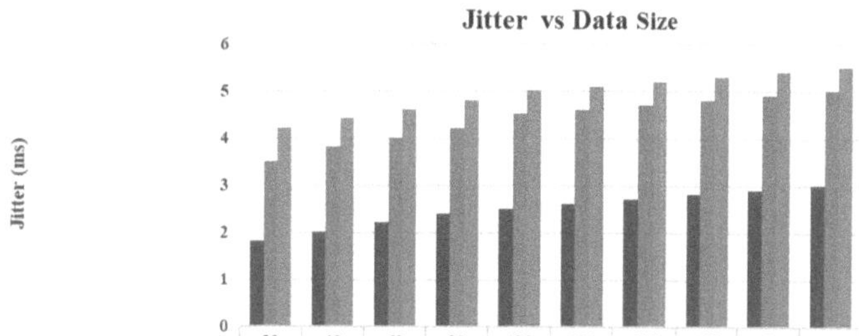

	20	40	60	80	100	120	140	160	180	200
BIAS Framework (ms)	1.8	2	2.2	2.4	2.5	2.6	2.7	2.8	2.9	3
SCTP (ms)	3.5	3.8	4	4.2	4.5	4.6	4.7	4.8	4.9	5
ACO with SCTP (ms)	4.2	4.4	4.6	4.8	5	5.1	5.2	5.3	5.4	5.5

Data Size(MB)

Fig. 7. Jitter

Using (5), the jitter is computed and it is compared with various methods as shown in Fig. 7. The proposed BIAS framework shows a tremendous improvement of 1.8ms when compared to other methods for data size of 20MB.

4.6 Multi-path Utilization Efficiency

It is defined as the ratio of the data transmitted over multiple paths to the total data transmitted in percentage.

$$\text{Efficiency} = {}^{\text{Data transmitted over multiple paths}}/_{\text{Total data transmitted}}\ \text{x}100 \quad (6)$$

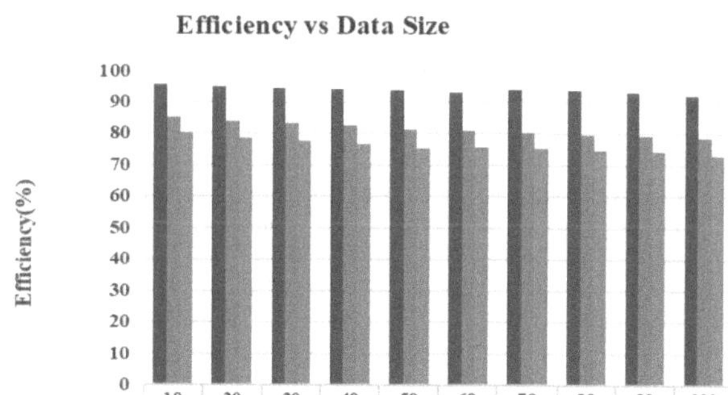

	10	20	30	40	50	60	70	80	90	100
BIAS Framework (%)	95.5	94.8	94.2	93.8	93.5	93	93.8	93.5	93	92
SCTP (%)	85	83.5	83.1	82.4	81	80.6	80.1	79.5	79.2	78.5
ACO with SCTP (%)	80	78.5	77.4	76.3	75	75.3	75	74.5	74.1	73

Data Size(MB)

Fig. 8. Efficiency

Using (6), the efficiency of the proposed BIAS framework is compared with SCTP and ACO with SCTP as shown in Fig. 8 and observed that the BIAS framework shows a considerable better efficiency of 95.5% for a data size of 10 MB when compared to other existing methods.

4.7 Pheromone Stability

The pheromone stability is computed as the ratio of stable paths identified to the total paths evaluated in percentage.

$$\text{Pheromone stability} = {}^{\text{Total number of stable paths}}/_{\text{Total number of paths evaluated}} \tag{7}$$

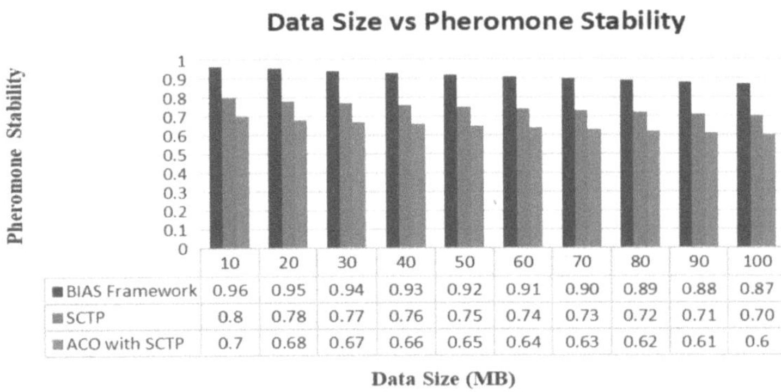

Fig. 9. Pheromone Stability

Using (7), the pheromone stability is computed and it is compared with existing methods like SCTP and ACO with SCTP and it is shown in Fig. 9. The pheromone stability for BIAS framework, SCTP and ACO with SCTP for data size of 10 MB tends to be 0.9, 0.8 and 0.7 respectively.

4.8 Reliability

Reliability is defined as the ratio of data transmitted successfully to the total transmission in percentage.

$$\text{Relaibility} = {}^{\text{Data transmitted successfully}}/_{\text{Total transmissions}}\ \text{x}100 \tag{8}$$

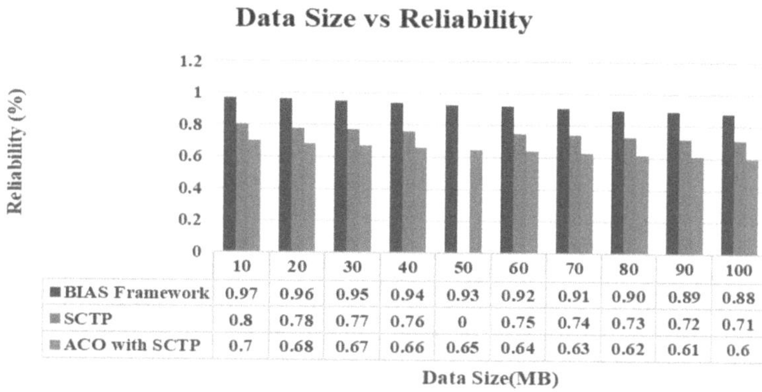

	10	20	30	40	50	60	70	80	90	100
BIAS Framework	0.97	0.96	0.95	0.94	0.93	0.92	0.91	0.90	0.89	0.88
SCTP	0.8	0.78	0.77	0.76	0	0.75	0.74	0.73	0.72	0.71
ACO with SCTP	0.7	0.68	0.67	0.66	0.65	0.64	0.63	0.62	0.61	0.6

Fig. 10. Reliability

Using (8), the reliability is computed and it is compared with various existing methods like BIAS framework, SCTP and ACO with SCTP as shown in Fig. 10. It is observed that the proposed BIAS framework shows better reliability of 0.9 when compared to other methods.

5 Conclusion and Future Directions

In this paper, a hybrid framework by integrating Bio-ACO with ANN and SCTP for data forwarding is proposed. It guarantees efficiency, reliability and scalability in data transmission. Moreover, it also optimizes path selection and adapts to dynamic network conditions and ensures fault tolerance, reducing latency and maximizing throughput. Hence, the hybrid framework of Bio-ACO with ANN and SCTP offers a powerful tool for efficient data forwarding and network management in cloud environments. In future, it is planned to incorporate other optimization algorithms such as particle swarm optimization or genetic algorithms along with reinforcement learning and anomaly detection for efficient threat detection.

References

1. Ramachandra, G., Iftikhar, M., Khan, F.A.: A comprehensive survey on security in cloud computing. Procedia Comput. Sci. (2017)
2. Salih, B.M., Mohammad, O.K.J.:Cloud data leakage, security, privacy issues and challenges: review. Procedia Comput. Sci. **242**, 592–601 https://doi.org/10.1016/j.procs.2024.08.113
3. Subashini, S., Kavitha, V.: A survey on security issues in service delivery models of cloud computing. J. Netw. Comput. Appl. **34**(1), (2011)
4. Beulah, S., and Dhanaseelan, F.R.: Survey on security issues and existing solutions in cloud storage. Indian J. Sci. Technol. **9**(13), (2016). https://doi.org/10.17485/ijst/2016/v9i13/90588
5. Luo, S., Lin, Z., Chen, X., Yang, Z., Chen, J.: Virtualization security for cloud computing service. In: IEEE Conference on Cloud and Service Computing (2011). https://doi.org/10.1109/CSC.2011.6138516
6. Zhang, L., Cui, Y., Mu, Y.: Improving security and privacy attribute based data sharing in cloud computing. IEEE Syst. J. **14**(1) (2020). https://doi.org/10.1109/JSYST.2019.2911391

7. Gupta, I., Singh, A.K.: A probability based model for data leakage detection using biograph. In: International Conference on Communication and Network Security (2017). https://doi.org/10.1145/3163058.316306
8. Papadimitriou, P., Garcia-Molina, H.: Data leakage detection. IEEE Trans. Know. Data Eng. (2011). https://doi.org/10.1109/TKDE.2010.100
9. Gupta, I., Singh, A.K.: SELI: statistical evaluation based leaker identification stochastic scheme for secure data sharing. IET Commun. **14**(20), 3607–3618 (2020)
10. Teng, W., Yang, G., Xiang, Y., Zhang, T., Wang, D.: Attribute-based access control with constant-size ciphertext in cloud computing. IEEE Trans. Cloud Comput. **5**(4) (2017). https://doi.org/10.1109/TCC.2015.2440247
11. Kao, Y.W., Huang, K.Y., Gu, H.Z., Yuan, S.M.: uCloud: a user-centric key management scheme for cloud data protection. IET Inf. Secur. **7**(2), (2013)
12. Al-Haj, A., Abandah, G., Hussein, N.: Crypto-based algorithms for secured medical image transmission. IET Inf. Secur. **9**(6), 365–373 (2015)
13. Liang, K., et al.: A secure and efficient ciphertext-policy attribute-based proxy re-encryption for cloud data sharing. Future Gener. Comput. Syst. **52**, 95–108 (2015). https://doi.org/10.1016/j.future.2014.11.016
14. Liu, H., Li, X., Xu, M., Mo, R., Ma, J.: A fair data access control towards rational users in cloud storage. Inf. Sci. **418–419**, 258–271 (2017)
15. Liu, Z., Jiang, Z.L., Wang, X., Yiu, S.M.: 'Practical attribute-based encryption: outsourcing decryption, attribute revocation and policy updating.' J. Netw. Comput. Appl. **108**, 112–123 (2018)
16. Wang, G., Liu, Q., Wu, J.: Hierarchical attribute-based encryption for fine-grained access control in cloud storage services. In: Proceedings of the 17th ACM Conference on Computer Communication Security, New York, NY, USA, pp. 735–737, (2010)
17. Wang, G., Liu, Q., Wu, J., Guo, M.: Hierarchical attribute-based encryption and scalable user revocation for sharing data in cloud servers. Comput. Secur. **30**(5), 320–2331 (2011)
18. Shen, W., Qin, J., Yu, J., Hao, R., Hu, J.:Enabling identity-based integrity auditing and data sharing with sensitive information hiding for secure cloud storage. IEEE Trans. Inf. Forensics Secur. **14**(2) (2019). https://doi.org/10.1109/TIFS.2018.2850312
19. Ulybyshev, D., Bhargava, B., Oqab-Alsalem, A.:Secure data exchange and data leakage detection in an untrusted cloud. In: Applications of Computing and Communication Technologies, Springer, pp. 99–113 (2018)
20. Zhang, X., Li, Y., Liu, Z.: Secure data sharing in cloud computing using revocable-storage identity-based encryption. IEEE Trans. Cloud Comput. **6**(4), 1136–1148 (2018).https://doi.org/10.1109/TCC.2016.2545668
21. Iyengar, J.R., Amer, P.D., Stewart, R.: Concurrent multipath transfer using sctp multihoming over independent end-to-end paths. IEEE Trans. Networking **14**(5) (2006). https://doi.org/10.1109/TNET.2006.882843
22. Yan, J.-F., Gao, Y., Yang, L.: Ant colony optimization for wireless sensor networks routing. In: IEEE International Conference on Machine Learning and Cybernetics (2011). https://doi.org/10.1109/ICMLC.2011.6016670
23. Khapre, S.P., Chopra, S., Khan, A., Sharma, P. and Shankar, A.: Optimized routing method for wireless sensor networks based on improved ant colony algorithm. In: IEEE International Conference on Cloud Computing, Data Science & Engineering (2020).https://doi.org/10.1109/Confluence47617.2020.9058312
24. Lu, J., Teng, A., Zha, J., Shen, L. and Wang, Z.: Cloud computing task scheduling strategy based on improved ant colony optimization (ACO) algorithm. In: IEEE International Conference on Power, Intelligent Computing and Systems (2024).https://doi.org/10.1109/ICPICS62053.2024.10796062

25. Venkadesh, P., Dhas, J.P.M., Divya, S.V.: A technique for transmitting secret message using secret matrix in SCTP. Int. J. Enterp. Network Manage. **7**(1) (2016). https://doi.org/10.1504/IJENM.2016.075171
26. Venkadesh, P., Dhas, J.P.M., Divya, S.V.: Techniques to enhance security in SCTP for multi-homed networks. In: IEEE Global Conference on Communication Technologies (2015).https://doi.org/10.1109/GCCT35855.2015
27. Ustiugov, D., et al.: Expedited data transfers for serverless clouds. In: DBLP (2023). https://doi.org/10.48550/ARXIV.2309.14821

Detection of Defects in Solar (PV) Panels Using Dimension Reduction Model

Smita D. Khandagale(✉) and Sanjay M. Patil

Electronics Engineering, Datta Meghe College of Engineering, Navi-Mumbai University of Mumbai, Airoli, Mumbai, India
sdkhandagale@vpmthane.org, sanjay.patil@dmce.ac.in2

Abstract. We present a thermal imaging-based solar panel defect detection system that reduces the requirement for human panel inspection in large solar farms by automating the inspection process. Twelve classes - cell, vegetation, cracking, soiling, diode, shadowing, diode-multi, hot spot-multi, hot spot, offline module, no-anomaly, and cell-multi. The dataset employed included 20,000 photographs taken from infrared solar modules, including a wide range of representations. We used the Efficient-BM (E-BM) model in the process. We selected 17,000 characteristics from the example model using the feature selector. We carried out the classification process using an SVM classifier. The suggested approach produced excellent outcomes with parameters such as F1-score, accuracy, sensitivity, and precision measures when work on 12 label dataset. The average values of these outcomes are 93.99% accuracy, 91.50% precision, 88.28% sensitivity, and 89.82% F1-score.

Keywords: Fault Detection · Photovoltaic Panels · Deep Learning · Infrared Imaging

1 Introduction

Every day, there is an increase in energy consumption due to things like population expansion and industrialization. Maintaining supply and demand balance requires increasing energy production. As a result, renewable energy is becoming more and more important on a worldwide basis. Because renewable energy sources are easily accessible and abundant locally, countries are less dependent on imported resources to meet their energy needs, which improves national energy security. The environmental friendliness of renewable energy sources is another justification for choosing them. Increased use of renewable energy sources reduces environmental pollution because they emit fewer greenhouse gases than fossil fuels. Because solar energy doesn't damage ecosystems or cause global warming, it helps to maintain ecological equilibrium. In conclusion, solar energy is a useful instrument for reducing the effects of climate change, which is crucial for preserving the health of all living things [1] within the realm of photovoltaic (PV) systems, renewable energy is increasingly gaining popularity. PV panels use a method that converts solar radiation into energy. They are made of semiconductor materials.

A. Das et al. (Eds.): MIDAS 2025, CCIS 3055, pp. 376–388, 2027.
https://doi.org/10.1007/978-3-032-35516-4_30

Free electrons in the semiconductor material in the solar panel become mobile when it receives sunlight, producing an electrical current. PV modules are made by connecting solar cells in parallel as well as series configurations, as shown in Fig. 1.

Fig. 1. The design and structure of a solar power facility.

There have been many more research projects on this subject due to the increasing interest in photovoltaic systems. The two main objectives of literature-based research are to: (1) increase the efficiency of photovoltaic systems by identifying their advantages; and (2) ensure that characteristics influencing performance are considered in the design of PV systems. Researchers dedicate extensive research endeavours to investigating the feasibility of efficiently using solar energy in various geographic locations [2]. These studies examine the advantages of solar energy, how it contributes to energy production, and whether solar power is sustainable worldwide. Numerous research works investigate the processes involved in producing energy using photovoltaic systems and transforming solar radiation into diverse types of energy. This study indicate that a number of variables, including temperature, solar irradiation, shade, cleanliness, and others, have a major impact on how well PV systems operate. Irradiance, or the quantity of sunlight that reaches the system, positively impacts the PV system's performance [3]. An important consideration in efficiency studies is the variance in energy generation by PV systems as a result of environmental factors. Various environmental conditions and other imperfections may affect the effectiveness of solar panels. These issues, as seen in Fig. 2, might emerge due to a variety of causes. The process of categorizing solar panel faults includes the identification and categorization of defects in the operation of PV panels.

Electroluminescence imaging, an earlier technique for flaw identification, uses electricity to power a solar panel that generates light in the electromagnetic spectrum. The solar module releases infrared light when it receives an electrical current. This light will be more intense when it comes to intrinsic crystal grain boundaries and extrinsic defects like microcracks, fractures, and finger discontinuities. The EL picture, while clearly revealing undetectable faults, also creates a random, homogeneous backdrop, making the inspection work particularly challenging.

- We need more rapid fault-detection algorithms to optimize the use of this easily accessible PV data.
- The feature selection process uses Neighbourhood Component Analysis (NCA).

Fig. 2. Multiple flaws within solar panels.

- One of the best ways to find and identify the fault moment is to use thermal imaging, which uses principal component analysis (PCA) and independent component analysis (ICA) to clearly show where the fault is or to label pictures as either defective or not defective.

The rest of the work is organized with Sect. 2 related work, Sect. 3 with methodology, Sect. 4 with experimental results, and Sect. 5 with a conclusion.

2 Related Work

Defective solar panels may lead to frequent failures. As a result, the PV system will be less reliable and operate more expensively. This will also lead to errors in energy estimations. In terms of safety, faulty solar panels might also result in electrically dangerous scenarios or fire. In conclusion, PV panel flaws have a detrimental impact on the effectiveness and efficiency of the system. Early detection of solar panel flaws is necessary to avoid these negative effects. Solar cell fracture processes were investigated using laser Doppler vibrometry, thermography, and electroluminescence in research aimed at identifying sensor-based solar panel faults. Research has also been carried out to analyze the panel parameters with the dynamic current-voltage characteristic. Researchers have utilized sensors to identify flaws in solar panels, often opting for image-based techniques [4].

2.1 Novelties

By using a novel approach, the E-BM model brings about significant improvements over conventional deep learning models. Unlike traditional CNN architectures, this model supports deep feature extraction by utilizing advanced techniques such as depth-wise separable convolution. Consequently, it provides a feature extraction approach that is more effective and enables quicker execution with less processing resources. We also used the dataset to train the customized E-BM model, enhancing its capacity to detect defects in solar systems.

2.2 Contributions

The results of this study demonstrate that a unique method with a resolution of 224 × 224 pixels may significantly improve the processing of data taken by medium- or long-wave infrared imaging UAV (unmanned aerial vehicle) systems. Initially, we divided 224 × 224 pixel infrared solar module photos into smaller patches of 28 × 28 pixels to improve processing efficiency. This method paved the way for more efficient picture data processing. The data, which consisted of 28 × 28 pixel patches, was then processed using an original feature extraction technique created using the E-BM architecture. This method extracted unique characteristics of 17,000 features. This method's primary benefit is its capacity to minimize data size while maintaining significant characteristics. As a result, this work provides a useful technique for feature extraction from infrared solar module photos, which may improve their applicability in flaw detection and photovoltaic panel image analysis. Moreover, it might function as a directive structure for subsequent investigations. This study adds to and improves the area of feature extraction using solar module pictures.

3 Methodology

3.1 Proposed Model

This research suggests using an artificial intelligence system to detect defects in solar panels. We conducted the experiment using a dataset of solar modules. The proposed paradigm as mentioned in Fig. 3 divides into six clearly defined phases: Initially, we shrunk the photos in the dataset from various dimensions to a standardized size of 224 × 224 as the first step. The dataset's pictures were divided into smaller patches measuring 28 × 28 for further improvement. We used fully connected layers to extract characteristics from each patch. We obtained a total of 17,000 features by combining the properties retrieved from these 28 patches with those from 12 sample photos. We used the NCA feature selection technique to optimize the feature selection process. Ultimately, we classified the chosen characteristics using a 10-fold cross-validation technique and CNN-VGG16 support vector machines (SVM).

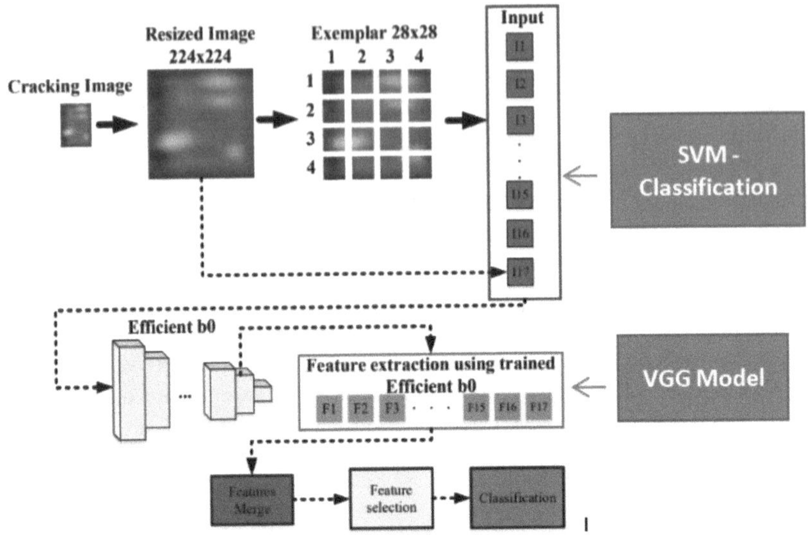

Fig. 3. Workflow model.

3.2 Data Classification

Support vector machines (SVMs) are very popular and effective supervised learning methods used for regression analysis and classification [5]. SVMs are a recommended option due to their many benefits over other classification techniques. Treating the classification issue as an optimized problem is the main benefit of using Support Vector Machines (SVMs) in classification problems. It's also impressive how well they can manage asymmetric data sets. Using the appropriate kernel function enables SVMs to efficiently handle high-dimensional data and categorize complex data. Additionally, SVMs may need longer training durations when dealing with huge datasets. In conclusion, SVMs prove to be an effective tool for resolving regression and classification issues, albeit requiring careful consideration of the kernel functions and parameters. Here are the SVM hyper parameters that are available:

- **Automatic kernel scaling**: Upon selecting "Automatic", the system utilizes the provided data to automatically determine the appropriate scale.
- **One-versus-one is a multiclass approach**: This approach breaks down the multi-class classification issue into a set of binary classification jobs.
- **Data standardization**: If we standardize the data, the input characteristics will have a mean of zero and a variation of one.

3.2.1 CNN-VGG16

We use data augmentation approaches to generate a substantial number of CNN-VGG16 architectures. These operations include flipping and rotating. We train three Convolutional Neural Network (CNN) models using the VGG16 architecture on the dataset, then use these models to analyze a target image and extract distinct and visually identifiable PV classification features. The assessment procedure begins with the execution of the freeze transfer and fine-tune learning techniques. The linear classifier assesses the techniques in the final stage. In the final stage, a log-softmax layer executes the classification method as seen in Figs. 5 and 4.

```
Model: "model"
_________________________________________________________________
 Layer (type)                Output Shape              Param #
=================================================================
 input_2 (InputLayer)        [(None, 244, 244, 3)]     0

 tf.__operators__.getitem (  (None, 244, 244, 3)       0
 SlicingOpLambda)

 tf.nn.bias_add (TFOpLambda  (None, 244, 244, 3)       0
 )

 vgg16 (Functional)          (None, 7, 7, 512)         14714688

 global_average_pooling2d (  (None, 512)               0
 GlobalAveragePooling2D)

 dropout (Dropout)           (None, 512)               0

 dense (Dense)               (None, 90)                46170

=================================================================
Total params: 14760858 (56.31 MB)
Trainable params: 46170 (180.35 KB)
Non-trainable params: 14714688 (56.13 MB)
```

Fig. 4. VGG Model summary

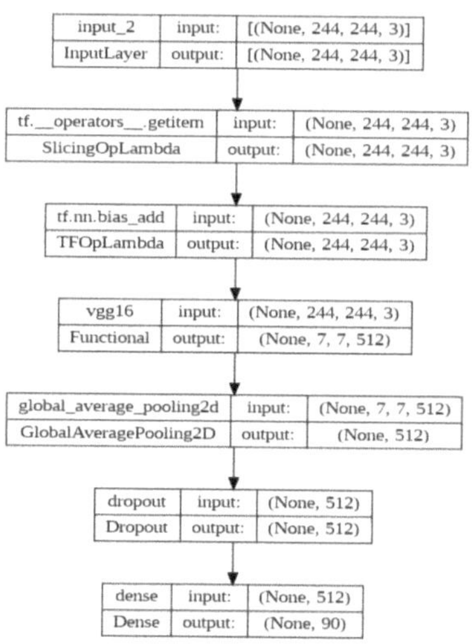

Fig. 5. Layers of CNN VGG16

3.2.2 Algorithm Steps

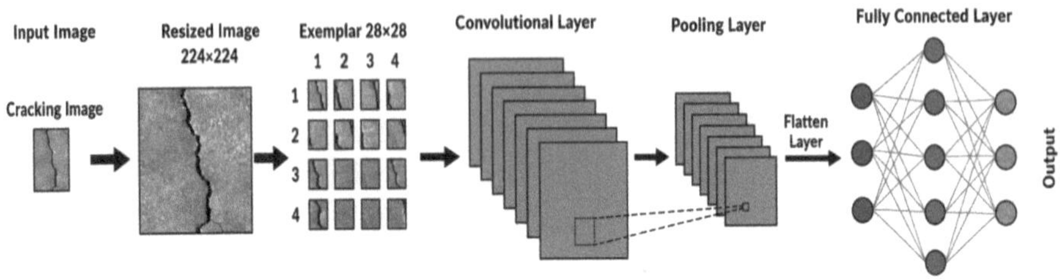

Fig. 6. A typically CNN model

Pre-processing → Convolutional Layers → Activation Functions → Pooling Layers → Fully Connected Layers → Output Layer → Loss Function → Optimization → Training and Evaluation. We choose an appropriate loss function. An optimization algorithm optimizes the parameters. The CNN undergoes training and evaluation using labelled data as shown in Fig. 6.

3.2.3 Psudeo Code for CNN-VGG Model

The CNN (Convolutional Neural Network) formula generally refers to the operations involved in the layers of a CNN. The main components include:

Convolutional Layer: Mathematically, the operation is stated as follows:

$$(X * W)(i,j) = \sum_{m}\sum_{n} X(i+m, j+n).W(m,n) \tag{1}$$

In this case, the spatial dimensions are i and j, the filter (or kernel) is W, and the input feature map is X.

Activation Function: Rectified Linear Unit, or ReLU, is often used:f(x) = max(0,x).

Pooling Layer: The spatial size is decreased via a standard pooling procedure (such as max pooling):

$$Y(i,j) = \max_{m,n} X(2i+m, 2j+n) \tag{2}$$

This selects the maximum value from a 2 × 2 window.

Fully Connected Layer: Every neuron from one layer is connected to every other layer's neuron via this layer:

$$Y = WX + b \tag{3}$$

In this case, X is the source vector, b is the bias, and W is the value of the weight matrix.

Loss Function: Commonly, for classification tasks, you may use cross-entropy:

$$L = -\sum_{i} y_i \log(y_i) \tag{4}$$

where y is the expected probability and y is the real label. These components work together to process and classify input data.

3.3 Feature Selection

Supervisory learning employs neighbourhood component analysis (NCA) for feature selection [6]. Finding and choosing the characteristics in the dataset that best distinguish between classes is its main goal. The main goal of NCA is to increase class separability. It aims to improve the precision and accuracy of the categorization process by selecting the most useful traits. Thanks to this method, the characteristics may now most accurately depict the variations across classes. To comprehend how NCA works, we compare each data point in the collection to every single other data point. Based on its relationships to other sites, each data point is assigned a weight. The usefulness of traits in class differentiation is measured by these weights. These weights are used by NCA to rank the significance of different factors. To sum up, neighbourhood component analysis is a feature selection technique that uses supervised learning. Its primary goal is to identify and select the most informative characteristics in order to improve class separability. To do this, NCA evaluates the connections between the data points and based on its neighbours, each data point is assigned a weight.

3.4 Dataset

This investigation used images from the publicly available and free Infrared Solar Modules collection [7]. The collection consists of 20,000 infrared photos with 224 × 224 resolution and capabilities. Out of all these photographs, 10,000 (or 50%) are images without anomalies. There are different solar panel flaws in the 10,000 remaining photos (see Fig. 7). Images with solar panel flaws create eleven categories. In all, there are 12 classifications, including photos with no anomalies. The following is a definition of these classes.

- Cell: One hot-spot event-affected square-shaped cell on its own.
- Cell-multi: Square-shaped hot spots have been seen in many cells.
- Cracking: The module has noticeable surface cracks.
- Diode: Approximately a third of the module is made up of the active bypass diode.
- Diode-multi: This kind of module usually consists of two thirds active bypass diodes.
- Offline-module: There is some warmth throughout the whole module.
- Shadowing: The blocking of sunlight by surrounding rows, artificial buildings, or plants.
- Soiling: The module's surface is covered with dust, dirt, or other material.
- Greenery: The surrounding greenery blocks the panels.

Fig. 7. Sample images.

4 Experimental Results

This work examined the E-BM-based deep feature extraction method. We implemented the deep feature extraction model using Python. The recommended method for data processing involved using a deep feature extraction strategy. After they were acquired, we classified the features using the Classification Learner App Toolbox. We used the machine learning algorithms with their default values, without any model-specific hyper parameter optimization. We then evaluated and categorized the collected properties using a CNN-VGG16. Table 1 displays the category results. Upon careful analysis, we concluded that the E-BM model achieved the highest accuracy scores. We selected the E-BM model for this investigation due to its unique suitability.

Table 1. Results of pre-trained.

CNN Pre-Trained Models	Accuracy
RESNET [8]	78.36%
Darknet [6]	79.36%
Mobilenet [9]	81.36%
Densenet [10]	84.23%
Other Models [4]	88.36%
CNN-VGG Proposed Model	93.99%

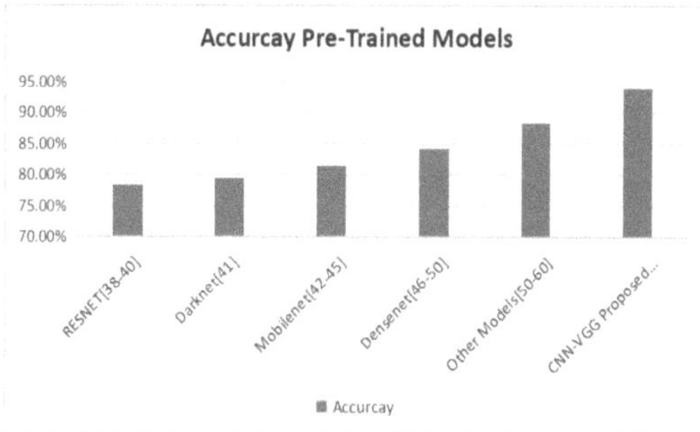

Fig. 8. Accuracy of Pre-Trained Models

Table 2. Selected training parameters

Training parameters	Values
Epochs	150
loss function	cross-entropy
batch size	32
drop out	0.5
learning rate	0.001
optimiser	Adam

The efficiency chart for the different approaches is shown in Fig. 8. Table 2 exhibits the operational values of the experiment conducted on the dataset.

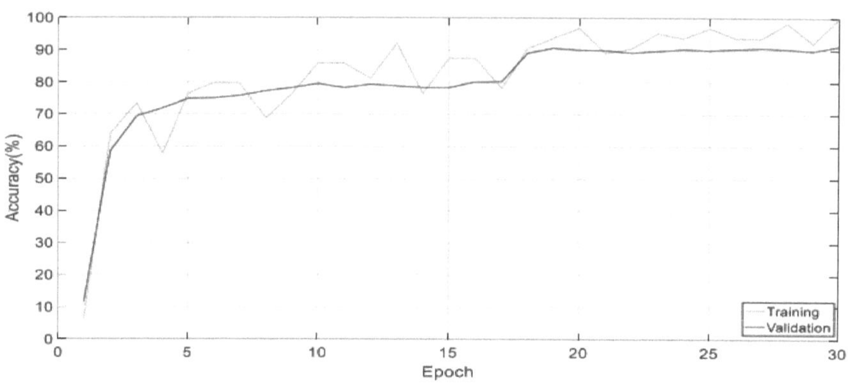

Fig. 9. The graphs depicting the training and validation curves.

The accuracy of the proposed model for training data is 94% and for testing data is 92% as shown in the Fig. 9. The classifiers used to categorize the characteristics collected using the suggested technique included CNN-VGG16, ANN, KNN, Naive Bayes, SVM, and Decision Tree as projected in the Fig. 10.

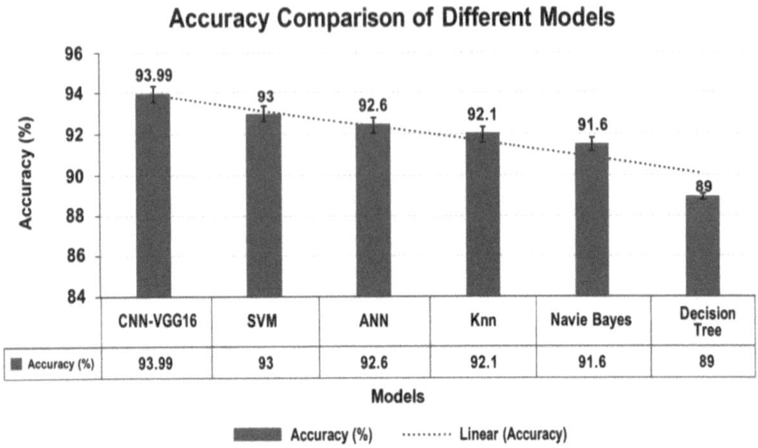

Fig. 10. Displays the classifiers' accuracy outcomes.

The trained model evaluated the test data and projected the results with original result as shown in the Fig. 11. The results explains the actual class and the predicted class are 92% accurate.

Fig. 11. Results evaluated actual and predicted values

By dividing the total number of expected cases by the total number of correctly predicted true negatives and true positives, we can get the accuracy value. By dividing the total number of true positives as well as false positives by the number of genuine positives that are properly classified as positive, the model determines precision. A statistic called recall assesses how well a model can recognize instances that belong to a certain class. To get the F1-score value, we take the accuracy as well as recall values produced by the model and mean them harmonically. To lessen the effect of severe circumstances on the performance statistic, we use a harmonic mean. As such, its goal is to guard against imprecise evaluations in data sets whose distributions are not uniform. Using the following calculations, we created the performance measurement metrics.

$$Accuracy = \frac{TP + TN}{TP + TN + FP + FN} \times 100 \quad (5)$$

$$Sensitivity = \frac{TP}{TP + FN} \quad (6)$$

$$Specificity = \frac{TN}{TN + FP} \quad (7)$$

$$Precision = \frac{TP}{TP + FP} \times 100 \quad (8)$$

$$F1Score = 2 \times \frac{Recall \times Precision}{Recall + Precision} \quad (9)$$

The sensitivity was 88.18% on average. The model consistently performs across all classes, as evidenced by the average sensitivity. This shows how well the model can distinguish between different classes in the dataset. The categorization of solar panel defects will also yield benefits related to energy management, as this process is crucial for the system to guarantee optimal energy output and improve energy efficiency (Table 3).

Table 3. Comparing the results with existing results

Reference	Accuracy	Model Applied
Alves[11]	92.5%	CNN - Cross Validation
Nguyen[12]	93.0%	CNN
Le [9]	85.33%	CNN-VGG16
Our Work	94%	E-BM-CNN-VGG16

5 Conclusions

The goal of this study was to evaluate the efficacy of deep learning methods in analyzing infrared images of solar modules in order to identify defects in solar panels that convert sunlight into electricity. We used the pre-trained E-BM model, a deep learning model, to achieve this goal. Furthermore, we specifically tailored this model to accommodate the distinctive characteristics of the dataset. The findings with an accuracy of 94% on 12 classes that there is a great deal of promise for the suggested strategy in terms of identifying solar panel defects. In particular, using the E-BM model improved the accuracy of defect identification in the panels and allowed for more efficient processing of infrared solar module pictures. For specialists in the field and academics looking to enhance the upkeep and effectiveness of solar energy systems, the study's conclusions are essential. The durability and effectiveness of solar panels over an extended period of time rely significantly on the precision of fault detection accuracy rates. Subsequent investigations might focus on optimizing hyperparameters specifically for the model and investigate diverse methods of data augmentation to enhance the model's performance and broaden its scope. Further research avenues include evaluating this technique in practical applications and real-time defect detection.

References

1. COP 21 Paris France Sustainable Innovation Forum 2015 Working with UNEP. https://www.cop21paris.org/. Accessed 15 Aug 2023
2. Kabir, E., Kumar, P., Kumar, S., Adelodun, A.A., Kim, K.-H.: Solar energy: potential and future prospects. Renew. Sustain. Energy Rev. **82**, 894–900 (2018)
3. Shahsavari, A., Akbari, M.: Potential of solar energy in developing countries for reducing energy-related emissions. Renew. Sustain. Energy Rev. **90**, 275–291 (2018)
4. Tiwari, G.N., Dubey, S.: Fundamentals of Photovoltaic Modules and Their Applications. Royal Society of Chemistry, London, UK (2009)
5. Hernández-Callejo, L., Gallardo-Saavedra, S., Alonso-Gómez, V.: A review of photovoltaic systems: design, operation and maintenance. Sol. Energy **188**, 426–440 (2019)
6. Aslan, M., Duranay, Z., Tuncer, S.: Türkiye' nin Güne¸s Enerjisi Potansiyeli ve Uygulama Alanları (Solar Energy Potential of Turkey and Solar Energy Application Areas). In: Proceedings of the International EUROASIA Congress on Scientific Researches and Recent Trends 10, Bakü, Azerbaycan, 16–17 February 2023
7. Kabir, M.; Duranay, Z.; Ekici, S. Trend of Energy Generation Efficiency in Agrivoltaic Systems Research. In Proceedings of the 2nd-International Congress on Modern Sciences, Tashkent, Uzbekistan, 16–17 December 2022
8. Qiu, G., Ma, Y., Song, W., Cai, W.: Comparative study on solar flat-plate collectors coupled with three types of reflectors not requiring solar tracking for space heating. Renew. Energy **169**, 104–116 (2021)
9. Duranay, Z., Karagözo ˘glu, L.: Fotovoltaik Panel Performansını Etkileyen Faktörlerin ˙Incelenmesi (Investigation of the Factors Affecting the Photovoltaic Panel Performance). In: Proceedings of the 2nd International Baku Conference on Scientific Research, Baku, Azerbaijan, 28–30 April 2021
10. Karakilic, A.N., Karafil, A., Genc, N.: Effects of temperature and solar irradiation on performance of monocrystalline, polycrystalline and thin-film PV panels. Int. J. Tech. Phys. Probl. Eng. **51**, 254–260 (2022)
11. Gupta, V.; Sharma, M.; Pachauri, R.K.; Babu, K.D. Comprehensive review on effect of dust on solar photovoltaic system and mitigation techniques. Sol. Energy **191**, 596–622 (2019)
12. Di¸sli, F., Gedikpinar, M., ¸Sengür, A.: Determination of pollution on photovoltaic panels by image processing. In: Proceedings of the 2018 International Conference on Artificial Intelligence and Data Processing (IDAP), Malatya, Turkey, pp. 1–5, 28–30 September 2018

ML-Based Alzheimer's Disease Detection Using Behavior Indications: A Review

B. S. Panda[1(✉)], Sasmita Kumari Nayak[2], and Anil Chakravarthy[2]

[1] Raghu Engineering College, Visakhapatnam, India
bspanda.cse@gmail.com

[2] Centurion University of Technology and Management, R.Sitapur, Odisha, India

Abstract. Worldwide, there is an unmet clinical need for accurate sickness diagnosis. The intricacy of the many diverse Alzheimer's illnesses and the underlying symptoms of the patient population make it extremely challenging to develop a technique for early detection and a welll treatment plan. Researchers, expert physicians, and patients can use machine learning (ML), a part of artificial intelligence (AI), to reduce many of these issues. This review outlines how machine learning techniques are being applied to find in the early detection of several diseases, based on pertinent research. The most recent advancements in ML-based disease detection and telemedicine are then summarized in the paper, taking into consideration the algorithm, different disease types, different behaviors, different applications, and different assessment metrics. We conclude by summarizing the key findings and offering some insight into potential future developments and lines of inquiry. The progressive neurodegenerative disease known as Alzheimer's disease (AD) is a major universal health concern, especially for aging populations. Improving management and treatment results requires early identification. The ability to detect AD through behavioral indicators such speech patterns, facial expressions, motor activities, and daily routines has been greatly enhanced by recent developments in machine learning (ML). Modern machine learning techniques applied to behavioral data for AD detection are examined in this review. We examine important datasets, models, and attribute extraction models, such as supervised, unsupervised, and deep learning methods. There is also discussion of difficulties including data sparsity, ethical issues, and multimodal data integration. The paper also identifies future directions for developing accurate, non-invasive, and easily available diagnostic instruments that have the potential to completely transform early detection and intervention techniques.

Keywords: Telemedicine · Artificial Intelligence · Alzheimer's Diseases · Behaviours · Healthcare

1 Introduction

In the 1970s, the term "telemedicine," which exactly means "healing at a distance," describes the practice of information and communication technologies (ICTs) to improve patient history by enabling easier access to medical data and care. In the historical, there

A. Das et al. (Eds.): MIDAS 2025, CCIS 3055, pp. 389–399, 2027.
https://doi.org/10.1007/978-3-032-35516-4_31

were two types of telehealth: synchronous and asynchronous. Remote (tele) monitoring has been recognized as a third form of monitoring in recent years. Data collection for this kind of monitoring is done by dispersed devices, such as the IOT (Internet of Things). Figure 1 depicts one such the real-time health monitoring system.

The WHO recognized four excellent, well-established telehealth facilities in its most popular recent universal e-Health monitor study: teledermatology, teleradiology, telepsychiatry and telepathology. While the fourth of these services employs synchronous models of care, the first three do so asynchronously. This demonstrates how challenging it is to improve or replace real-time clinician-delivered services. According to the same sutudy, more than 75% of defendants believed that a significant obstacle to broad adoption was a lack of awareness on clinical procedures related to telehealth.

Many of the people in the universe suffer with Alzheimer's Disease (AD), the most prevalent cause of dementia and a progressive neurological illness. The prevalence of AD is predicted to rise sharply as the world's population ages, putting a tremendous strain on families and healthcare systems. The exact origins of AD are still unknown despite much research, and the main goal of current treatments is to control symptoms rather than find a cure. Delaying the progression of disease, enhancing quality of life, and facilitating prompt therapies all depend on early identification. Conventional diagnostic techniques, such biomarker analysis and neuroimaging, are costly, intrusive, and unavailable to a significant portion of the population.

Novel diagnostic techniques that make use of non-invasive data sources have been made possible in recent years by developments in machine learning (ML). Speech patterns, facial expressions, gait analysis, and daily activity tracking are examples of behavioral indicators that have shown promise in identifying early signs of AD. These symptoms, which frequently appear before a clinical diagnosis is made, can represent cognitive impairment and modest neurological alterations. Researchers want to create automated, affordable, and scalable systems for early AD identification by using machine learning (ML) approaches to examine these intricate behavioral datasets.

The convergence of behavioral analysis and machine learning in Alzheimer's disease detection is the main topic of this paper. We examine previous research that uses machine learning models to find behavioral patterns associated with AD, focusing on the techniques, datasets, and performance indicators used. We also go over the difficulties that come with this field, like the necessity for multimodal integration, data heterogeneity, and ethical issues. Lastly, we suggest future lines of inquiry to enhance the accuracy and usability of ML-based Alzheimer's disease diagnostic tools.

Given these numerous possible results, a recent research of the potential impacts of AI in telepathology concentrated on four developing themes based on several healthmonitoring objectives: collaboration in information analysis, intelligent aid and diagnosis, healthcare IT, and patient observation.

The Intelligent data science (as well as AI-based ML methods) is rapidly growing and large data sets are readily available, opening up possibilities for new findings and practical insights that could greatly improve health outcomes. Given the increased attention in remote healthcare organizations over the past period, it is not surprising that computational systems with physiological and behavioural sign-monitoring for e-health care are a developing field of survey. This more advanced data-enriched environment

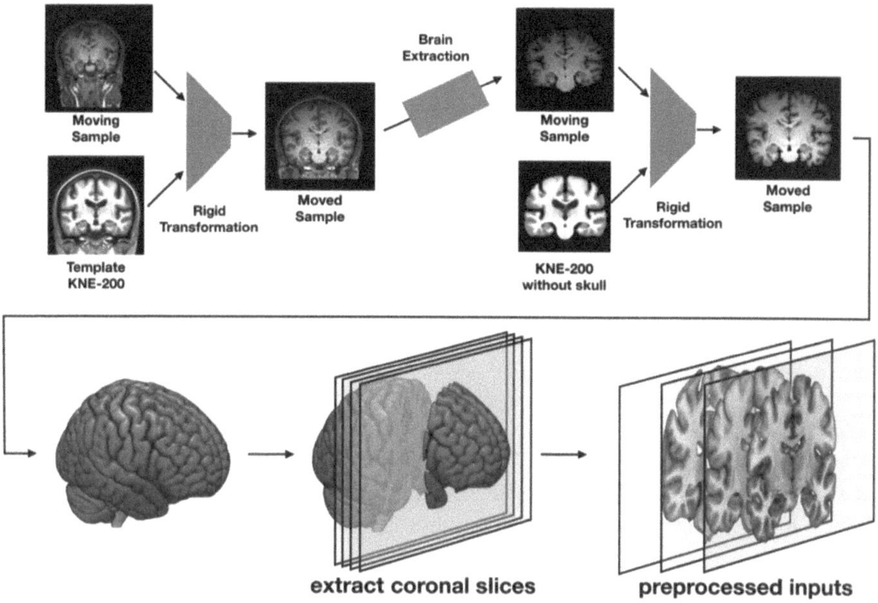

Fig. 1. Real-time prediction of Alzheimer's Disease

then uses automated techniques to enable better clinical decision-making. Therefore, the system used in this study consists of ongoing medical treatment, a cellphone link to an emergency center, and constant monitoring of many vital indicators. It also sends all of the information gathered online.

2 Background Study

2.1 Behavioral Changes in Alzheimer's Disease

Behavioral indicators of AD have gained significant attention as they often precede noticeable cognitive decline. These changes include:

- Speech and Language Patterns: Slower speech, frequent pauses, reduced vocabulary, and grammatical errors are commonly observed in early-stage AD patients.
- Motor and Gait Analysis: Impaired coordination and subtle gait changes can signal neurological deterioration.
- Daily Activities and Routines: Variations in sleep patterns, social engagement, and irregularities in task execution can indicate early AD.
- Facial Expressions and Emotions: Reduced emotional expressiveness or unusual facial micro-expressions are emerging areas of interest.

The availability of wearables, smart home technologies, and digital interaction data has enabled the continuous monitoring of these behavioral aspects, providing rich datasets for analysis.

2.2 Telemedicine

The popular Telemedicine is the practice of tracking patients' health status by using medical data that is sent via network connectivity from one place to another. This approach offers a novel way to deliver healthcare when a patient and a doctor are separated by great distances. Given the growing interest in innovative healthcare systems over the last five years, computational systems with behavioural and physiological signs detection for e-health facilities are a developing field. In the current Indian setting, the World Health Organization advises a doctor the population ratio of 1:1500, though the original ratio is only 0.50:1500.

2.3 Multimodal Data Integration

A critical advancement in this field is the integration of multimodal data—combining speech analysis, motor activity, facial expressions, and routine tracking into a unified framework. This holistic approach enhances diagnostic accuracy and robustness but introduces challenges related to data preprocessing, alignment, and computational complexity.

2.4 Artificial Intelligence (AI)

Originally proposed in 1956, artificial intelligence (AI) suggest to theories and methods that mimic and enhance human intelligence.

Further, data analysis performed by computers via statistical and automated procedures was referred to as "machine learning" (ML). Deep learning (DL), which uses multi-neuron and multilayer learning algorithms, has made it possible for machine learning to achieve a variety of application goals. This increases AI's potential for use. Russell and Norvig (1995) defined artificial intelligence as the areas of problem solving and searching, strategy, logic and interpretation, pattern classification, training, selection interaction, automation and vision. Consequently, in present years, the quantity of study papers that have been released that incorporate ML in the medical profession has surged (Fig. 2).

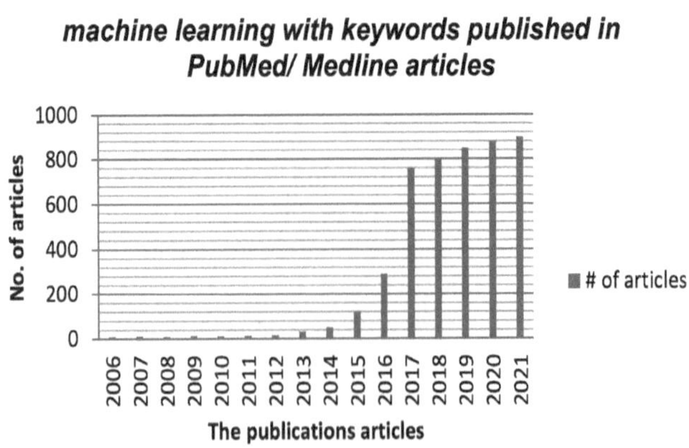

Fig. 2. Machine learning with keywords published in PubMed/Medline articles

2.5 Motivation

Behavioral changes, including alterations in speech, motor skills, sleep patterns, and daily routines, often precede clinical symptoms of AD. These indicators provide a non-invasive, cost-effective, and practical opportunity for early diagnosis. Machine learning has demonstrated remarkable potential in analyzing such data, uncovering subtle patterns and correlations that may not be apparent to human observers. By leveraging these capabilities, ML-based approaches can transform how AD is detected, making early diagnosis more accessible and efficient.

A increasing global health concern, Alzheimer's disease (AD) is predicted to quadruple in prevalence by 2050 as a result of an aging population. In addition to impairing mental and physical health, the illness places a heavy emotional, social, and financial strain on families and healthcare systems. Improving patient outcomes, facilitating prompt therapies, and delaying the progression of disease all depend on early identification. However, many people find that modern diagnostic techniques like neuroimaging, cerebrospinal fluid analysis, and genetic testing are intrusive, expensive, and unavailable, especially in environments with low resources.

Despite promising advancements, the field faces challenges such as the variability in behavioral datasets, ethical concerns, and the integration of multimodal information. Understanding and addressing these issues is essential to harness the full potential of ML in creating scalable, accurate, and patient-friendly diagnostic tools. By reviewing current research and identifying future opportunities aim to pay to the growth of innovative results that can mitigate the focus of Alzheimer's Disease on individuals and society.

This review aims to educate present and past researchers and clinicians about ML-based disease diagnosis systems so they can select the most effective DL/ML techniques, improving the probabilities of quick and precise disease classification and identification.The review also looks for studies that might be connected to MLBDD. This study's prime benifit is to offer the proper rationale for the below questions:

- Which diseases are of specific relevance to researchers and consultants how preprocessing data-driven ML models?
- What are the popular MLBDD datasets?
- Which ML and DL techniques are presently being used in the medical sector to classify the disease types?
- What kind of neural network architecture is most commonly employed to diagnose illnesses?
- How is the efficacy of the model evaluated? Is that sufficient?

3 Machine Learning in Healthcare

In the medical domain, artificial intelligence mostly emphases on classification of algorithms and models to evaluate how well a classify diagnoses illnesses. A medical diagnostic identifies the health issues or illnesses that cause a person's behaviour and indications.

The diagnosis is particularly caused based on the patient's medical backgoround treatement and physical and behavioural symptoms [1]. Many indications and symptoms are unclear and may only be treatement by trained medical specialists, making it often difficult. Because of this, developing countries like Inda and Bhutan, which lack enough medical professionals to serve their populations, find it difficult to offer suitable diagnostic processes to their largest patient bases [2].

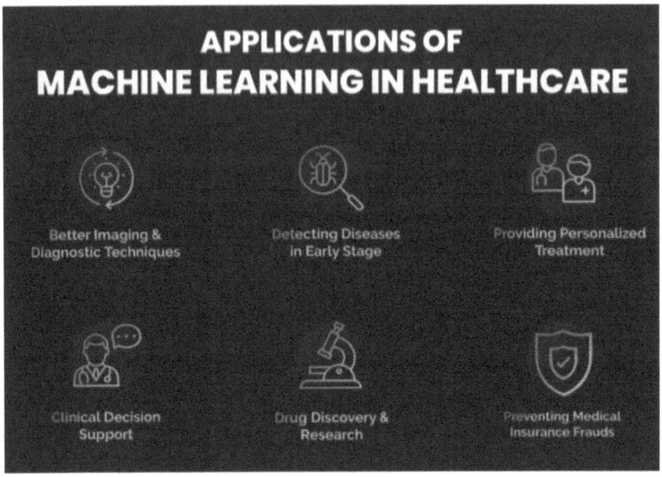

Fig. 3. ML Applications in healthcare

Figure 3 illustrates a few machine learning applications in the medical field. Additionally, several diagnostic procedures need costly and difficult-to-afford medical testing, which can be problematic for those with low incomes. Given that people make mistakes, it is not assuming that the patient may receive an overdiagnosis more frequently. People's health and money may be impacted by overdiagnosis, which can result in problems including unnecessary treatment [3]. As per National Academies of Science (NAS), Technical and Medical report, many people will have the minimum diagnostic issues in their aera [4].

Numerous factors, including the presence of a rare disease or the absence of appropriate symptoms like:

- which are sometimes unnoticed,
- might lead to a misdiagnosis,
- Mistakenly failing to take the illness into account.

Machine learning (ML) is frequently used almost everywhere in modern techniques, such as robotics, smart devices, and computers, as well as in the medical profession (i.e., disease detection, safety). A growing number of businesses, including the clinical field, are using ML to diagnose illnesses. Many academics and practitioners have shown the ability of machine learning-based disease diagnostics (MLBDD), which is time-efficient and reasonably priced [5].

4 Diagnostic System

In order to make a possible diagnosis, a number of mHealth platform use chatbots to collect information particular a patient's symptoms, lifestyle, and previous medical history. It's unclear exactly how machine learning techniques are used in diagnostic tool, however a modest implementation would most likely involve the below basic steps:

1. make use of a sizable collection of patients' symptoms and symptom-based or symptom-verified diagnoses
2. design a model to anticipate diagnoses
3. Apply the model to come up with possible diagnosis for the patient of interest.

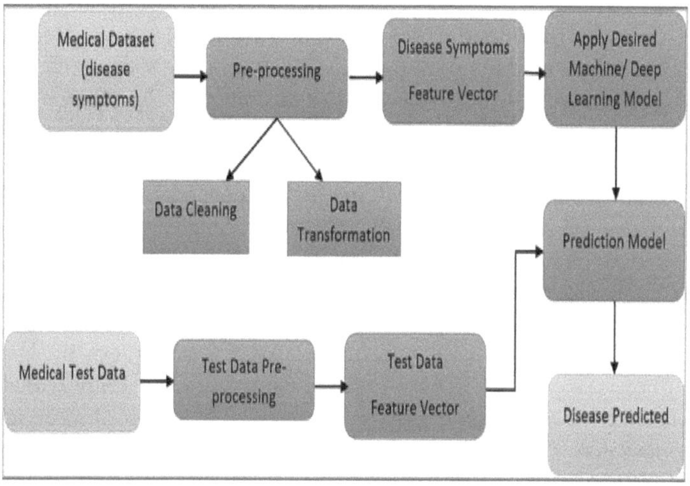

Fig. 4. Disease diagnosis system

Currently, the bulk of telemedicine diagnostic use cases are in the clinical professions of radiology, ophthalmology dermatology and pathology. The Deel Learning (CNNs) are the only suggested technique for image classification, in contrast to the examples including patient previous history and symptoms, where a variety of ML techniques may be applied. Nonetheless, a variety of possible tasks that below under the umbrella of computer vision can be accomplished by CNNs. They include picture segmentation, which splits an image, method evaluation, which counts or recognizes instances of methods, and basic categorization of the full image (Fig. 4).

5 Literature Survey

With support of the (KMKE) "Korean Ministry of Knowledge Economy", MTM Ltd. Developed the (SPaCS) "Smart Patient Care System" in June 2011 [6]. (PCS)"Personal Care Service" and (PRM) "Patient Record Monitoring & Feedback System" are two programs that make up SPaCS, a healthcare system that uses cellphones to achieve personal health. Developed by PU in Korea, the "Wearable digital method (USN Lab ver. 2.0)" eliminates the need for UV race on exposed skin [7]. When this model is made into t-shirts, patients may rapidly put it on and take it off, and test results can be wirelessly sent to the devices for analysis in real time.

Using content-free characteristics obtained from the behavioural interaction in ALR "Additive Logistic Regression", a machine learning boosting approach, Luz et al. [9] suggested creating a prediction model. The model training data consisted of 18 interviews with patients with different medical problems and 25 interviews with individuals with Alzheimer's disease. Speech rate, speech metrics and turn-taking behaviors were among the traits examined. Even with only content-free features, our approach performs 88% overall accuracy.

To accomplish classification, Chien et al. [10] created a novelattribute sequence model and used a data-driven method, essentially the RNN - Recurrent Neural Network, as opposed to the majority of earlier works that observed on a feature selection function and returned statistics-based features. In addition, it seems to be completely automated, which implies that it might be deployed quickly and anyplace. 150 speech samples from several studies were used to validate our findings, which showed an estimation under the receiver operating characteristic curve ratio of up to 0.790.

Using a DL model and most of the ML methods (rule induction, k-nearest neighbors, decision tree, naïve bayes, and basic linear approach), Shahbaz et al. [11] classified the five stages of Alzheimer's illness. They conducted their study using the ADNI data source, and the deep learning and basic linear model classifiers performed the best, with respective accuracy rates of 88.24% and 78.32%.

Comprising the collection of all features used to define the pattern, the feature vector reduces the dimensional space require to represent the models. Only features that are actually represented in the feature vector can be used, but the set of more features that may be used to define a particular model is big and theoretically endless because very small changes in some parameters can be used to distinguish between different features [12]. The author claims that one objective of dimension reduction is to satisfy technical specifications for the level of hardware and software complexity, data preprocessing costs, and the usefulness of design information compression. Furthermore, the classification often improves when the pattern is made clear by concentrating just on useful traits or attributes. Syntactic description synthesis or statistical statistics computation may be the foundation of feature extraction techniques.

The brain syndrome is recognized as Alzheimer's disease (AD) causes of progressive dementia in middle-aged or older adults. For the greatest therapeutic effect, AD must be diagnosed as early and properly as possible [13]. In this article, the author explains a method for detecting AD through machine learning analysis of EEG data and shows how to combine data from EEG reports with a ML models to develop an AD classification model. The results are really positive.

An ANN-based classifier for Alzheimer's disease detection was created by the author in [14]. Ans used the dataset MIT-BIH to compare the Wavelets Back Propagation Algorithm and artificial neural networks (ANN).

In [15], the author tested a number of machine learning methods, including SVM, NN, DT, RF, and GB, in an effort to maximize the accuracy of detect Alzheimer's disease based on symptoms. SVM was given a lot of attention among them.

Over the past few decades, the application of "WPM" and "IoT" principles are rapidly increased. Receiving medical care at home instead of in a hospital is now feasible because to technological improvements in smart home automation and e-healthcare systems. As a result, the Internet of Things has been recognized as a potential solution to reduce the limitations on healthcare societies. In the areas of e-health, remote healthcare, and smart health-care systems, several academics and foundations offered various strategies [16] conducts a comprehensive analysis of IoT applications in healthcare, emphasizing problems and possible solutions.

One emerging area of the IoT is automation for smart homes, which aims to assist people and make their survives smooth and easier. Remote-controlled home appliances, energy management [17], security systems moveble detection in the home, and healthcare for the elderly, disabled, and patients outside are a few examples.

6 ML Based Diagnostic System

A diagnostic system for Alzheimer's Disease (AD) with machine learning (ML) and behavioral indications leverages automated methods to analyze complex patterns in behavioral data. Such systems aim to detect early signs of cognitive decline and classify individuals based on their risk of developing AD. This section outlines the components, workflow, and techniques involved in building an ML-based diagnostic system.

6.1 System Components

The following are essential parts of an ML-based AD diagnostic system:

Data Collection Module: Collects behavioral data through various sources such as:

- **Wearable Devices**: Track motor activity, gait, and physiological signals.
- **Smart Devices**: Monitor speech patterns, daily routines, and digital interactions.
- **Video and Audio Recording**: Capture facial expressions, emotions, and vocal features.

Preprocessing Unit: Prepares raw data for analysis by addressing:

- Noise reduction and outlier removal.
- Feature extraction, such as pitch, tone, gait parameters, or daily activity statistics.
- Normalization to ensure uniform scaling and representation.

Feature Extraction and Selection: Identifies key indicators of AD from raw data. Techniques include:

- **Natural Language Processing (NLP)**: Extracts linguistic features such as word frequency, speech pauses, and sentence complexity.
- **Signal Processing**: Analyzes time-series data like motor activity or sleep patterns.
- **Dimensionality Reduction**: Uses methods like PCA (Principal Component Analysis) to select the most relevant features.

6.2 Workflow of the Diagnostic System

- Data Acquisition: Behavioral data is continuously collected using non-invasive methods, ensuring minimal disruption to the individual's daily life.
- Preprocessing: The data is cleaned, transformed, and prepared for analysis.
- Feature Analysis: Relevant features are extracted to represent behavioral patterns linked to cognitive decline.
- Model Training: ML algorithms are trained on labeled datasets (e.g., datasets of AD patients and healthy individuals) to learn patterns indicative of AD.
- Prediction: The trained model evaluates new data and provides a classification or risk score.
- Feedback and Reporting: Results are presented to clinicians or caregivers in an interpretable format, such as graphs, scores, or visualizations of activity trends.

7 Conclusion

Persons health issues are one of the most important challenges facing civilization. In order to provide patients with the best care possible, it aims to find the most appropriate, trustworthy, and efficient illness detection as soon as feasible. However, it is quite difficult to build a technique for an primary diagnosis and a good healing strategy. Machine learning, a branch of artificial intelligence, can help patients, researchers, and medical professionals solve some of these challenges. This review describes how ML is being used to benefit with the early detection of some diseases, based on pertinent studies. An summary of recent developments in ML-based illness identification is given in this study, with consideration given to technique, disease categories, and applications. We conclude by summarizing the key findings and offering an overview of the likely future paths of the research.

This study describes developments in data collecting, feature extraction, model construction, and multimodal integration, highlighting the potential of ML-based systems for AD diagnosis using behavioral data. Despite the potential of these systems, issues including interpretability requirements, generalizability, ethical concerns, and data scarcity continue to be major obstacles to their wider use. To overcome these obstacles and guarantee that ML-based technologies are precise, usable, and morally sound, researchers, physicians, and technologists must work together. Future studies should concentrate on increasing model transparency, diversifying data, and creating individualized diagnostic solutions that serve people from various demographic and cultural backgrounds. Additionally, a more comprehensive approach to Alzheimer's treatment may be possible by incorporating ML-based diagnostic tools with current clinical procedures. We can get closer to the objective of early, accessible, and non-invasive AD detection by utilizing the synergy between behavioral science and machine learning, which will eventually improve outcomes for patients and their caregivers.

References

1. McPhee, S.J., Papadakis, M.A., Rabow, M.W. (eds.) Current Medical Diagnosis & Treatment. McGraw-Hill Medical: New York, NY, USA (2010)

2. Ahsan, M.M., et al.: Detecting SARS-CoV-2 from chest x-ray using artificial intelligence. IEEE Access **9**, 35501–35513 (2021)
3. Coon, E.R., Quinonez, R.A., Moyer, V.A., Schroeder, A.R.: Overdiagnosis: how our compulsion for diagnosis may be harming children. Pediatrics **134**, 1013–1023 (2014)
4. Balogh, E.P., Miller, B.T., Ball, J.R.: "Improving Diagnosis in Health Care; National Academic Press": Washington. DC, USA (2015)
5. Ahsan, M.M.; Siddique, Z.: Machine Learning-Based Heart Disease Diagnosis: A Systematic Literature Review (2021). arXiv, arXiv:2112.06459
6. Lehner, R.J., Rangayyan, R.M.: A three channel microcomputer system for segmentation and characterization of the phonocardiogram. IEEE Trans. Biomed. Eng. 485–489 (1987)
7. Leung, T.S., White, P.R., Collis, W.B., Salmon, A.P., Brown, E.: Time frequency methods for analyzing paediatric heart murmurs, Appl. Sig. Process 154–167 (1997)
8. Groch, M.W., Domnanovich, J.R., Erwin, W.D.: A new heart sounds gating device for medical imaging. IEEE Trans. Biomed. Eng. 307–310 (1992)
9. Podgorelec, V.: Analyzing EEG signals with machine learning for diagnosing alzheimer's disease. Elektronika ir Elektrotechnika. **18**, 61–64 (2012). https://doi.org/10.5755/j01.eee.18.8.2627
10. Roopa, C.K., Harish, B.S.: A Survey on various Machine Learning Approaches for ECG analysis (2017)
11. Shahbaz, M., et al.: Classification of alzheimer's disease using machine learning techniques. Data (2019)
12. Chakravarthy, A., Nayak, S.K., Panda, B.S.: Detection and prediction of alzheimer's disease using rule induction method. In: Computing and Machine Learning. CML 2024. Lecture Notes in Networks and Systems, vol. 1108. Springer, Singapore (2024). https://doi.org/10.1007/978-981-97-6588-1_15
13. Luz, S., de la Fuente, S., Albert, P.: A method for analysis of patient speech in dialogue for dementia detection (2018). arXiv preprint arXiv:1811.09919
14. Sao, P., Hegadi, R., Karmakar, S.: ECG signal analysis using artificial neural network. Int. J. Sci. Res. 82–86 (2015)
15. Gao, D., Madden, M., Schukat, M., Chambers, D., and Lyons, G.: Arrhythmia identification from ECG signals with a neural network classifier based on a bayesian framework. In: Twenty-fourth SGAI International Conference on Innovative Techniques and Applications of Artificial Intelligence, vol. 3, no. 3, pp. 390–409 (2004)
16. Vishwa, A., Lal, M., Dixit, S., Vardwaj, P.: Classification of arrhythmic ECG data using machine learning techniques. Int. J. Interact. Multimedia Artif. Intell. **1**(4), 67–70 (2011)
17. Manogaran, G., Shakeel, P.M., Fouad, H., et al.: Wearable IoT smart-log patch: an edge computing-based bayesian deep learning network system for multi access physical monitoring system. Sensors **19**(13), 3030 (2019). https://doi.org/10.3390/s19133030

Design and Implementation of Quantum Based Hamming Encoder and Decoder for Error Detection Applications

Nutan Das and Chandra Sekhar Dash(✉)

Department of Electronics and Communication Engineering, Centurion University of Technology and Management, Odisha 752050, India
{nutan.das,chandrasekhar.dash}@cutm.ac.in

Abstract. Rapidly increasing quantum computing in every diversity for speed-up data communication and secure cryptography, the need for efficient data error correction and fault tolerance techniques is adopted. Quantum Hamming code is one of the crucial approaches for correcting single-bit errors in quantum data. In this paper, Quantum Hamming Encoder (QHE) and Quantum Hamming Syndrome Decoder have been designed and implemented on the Qiskit IBM platform. Qiskit (Quantum Information Software Kit) is the highest performing Quantum Software Development Kit (SDK) to design, synthesize, transpiling, and optimize the quantum circuits. The T count for the decoder was found to be 21, and the T-depth was found to be 69, respectively. Thus, fewer T gates translate to fewer ancillary qubits and lower time and energy costs in executing the circuit. T-count directly lowers the complexity and cost of error correction. For error bit detection, the intended error can be used in memory cells and data security applications.

Keywords: Quantum Computing · Quantum Gates · Qiskit IBM · Quantum Hamming Coder · Syndrome Decoder

1 Introduction

Quantum Computing (QC) is being rigorously examined in the domains of Science, Agriculture, Aerospace, Medicine, and Engineering [1]. In 1980, Paul Benioff introduced the quantum Turing machine, employing quantum theory to characterize a simplified computational model. In 1981, Richard Feynman presented the concept of quantum computers during his seminar on "Simulating Physics with Computers." Quantum computers employ qubits, denoted as $|0>$ or $|1>$, whereas classical computers utilize binary digits (0 or 1) for data encoding and decoding. Quantum computing utilizes the principles of quantum mechanics to address intricate mathematical challenges. Quantum computers depend on quantum theory and the characteristics of qubits. The principle of superposition posits that outputs result from a synthesis of all potential states, demonstrating the uncertainty of qubits across various states. A notable feature is Quantum Entanglement, which can transpire between two or more quantum states. As a result, the correlation of

A. Das et al. (Eds.): MIDAS 2025, CCIS 3055, pp. 400–411, 2027.
https://doi.org/10.1007/978-3-032-35516-4_32

quantum state errors with outcomes increased with the more frequent use of Quantum Computers. Thus, error correction is vital for the stability of quantum computations over long distances.

The Hamming code is an appropriate method for detecting and correcting a single-bit error [2, 3]. The effectiveness of Hamming code error detection may be improved through quantum superposition, leveraging parallelism that allows quantum systems to process multiple possibilities concurrently. It may enhance the detection process, facilitate concurrent error checking, refine syndrome decoding, augment scalability, and enable error correction utilizing quantum gates. This work employs Qiskit (Quantum Information Science Kit), an open-source software development kit (SDK) utilized for designing quantum logic circuits, which is extensively applied across various domain-specific applications [4–6].

The Hamming code (Hamming (7,4)) is a linear error-correcting code that encodes four bits of information into seven bits, including three parity bits. Richard W. Hamming introduced the Hamming code in 1950. The Hamming encoder-decoder is a straightforward technique for detecting and correcting single-bit data errors. Data is transmitted from one system to another via a suitable channel. Simultaneous noise-induced interference in the data. Therefore, it is imperative to identify and correct the actual bit error at the receiver [7]. A QCA (Quantum Cellular Automata) based Hamming encoder and decoder have been presented using coplanar and multi-layered architectures, respectively [8]. Syndrome decoding is a linear code utilizing a short-distance decoding method in a noisy channel. Syndrome decoding employs a look-up table to pinpoint errors at specific bit positions and modifies the error Hamming code accordingly.

In a Quantum Hamming code, errors are identified and rectified through syndrome measurements. The procedure involves ancilla qubits, parity checks, and syndrome bits. Additional qubits are entangled with the encoded qubits to retrieve error information without disrupting the quantum data. The quantum circuit executes parity checks to identify one-bit errors, referred to as syndrome bits, which pinpoint the error locations and facilitate effective correction.[9].

The subsequent sections of the paper are organized as follows. Section 2 outlines Background of Quantum Computing using quantum logic gates (NOT, AND, and X-OR). Section 3 addresses the methodology and execution of the Quantum Hamming Encoder. Section 4 outlines the Quantum Syndrome Decoder (QSD) utilizing the Syndrome decoding method, along with an analysis of performance metrics. Section 4 outlines the conclusion of this study.

2 Quantum Computing

2.1 Background

Quantum computing (QC) is a multifaceted technique that integrates elements of Physics, Mathematics, and Engineering, utilizing the principles of quantum mechanics to address complex problems more rapidly than classical systems. Quantum computing relies on two fundamental principles: superposition and entanglement. Quantum computers utilize qubits, or quantum bits ($|0>$ and $|1>$), to execute various computational tasks. The symbol $<|$ represents a ket, which is a quantum column state vector, while the symbol

|> is referred to as a bra. It is utilized for quantum state vectors involving complex conjugate operations. A qubit is a linear superposition of the 0 and 1 states, as articulated in Eq. (1).

$$|\psi >= \alpha|0 > +\beta|1 > \quad (1)$$

The coefficients of α and β are referred to as the probability amplitudes of complex numbers. A fundamental qubit is represented as |0 > and |1 >, as illustrated in Eqs. (2) and (3) respectively.

$$|0 >= [1\ 0] \quad (2)$$

$$|1 >= [0\ 1] \quad (3)$$

The square magnitudes represent the probability measurements of both states. The aggregate of probabilities for all potential measurement outcomes must equal 1 in a two-dimensional vector space, as indicated in Eq. (4).

$$|\alpha|^2 + |\beta|^2 = 1 \quad (4)$$

2.2 Quantum Gates

Quantum gates constitute the fundamental components of quantum circuits, grounded in the principles of quantum mechanics. Quantum logic gates operate reversibly, and certain universal logic gates, such as the X-gate (NOT), Toffoli gate (CCNOT), and Feynman (Controlled-X) gate, can be directly mapped onto quantum circuits.

X-gate- A X-gate is one bit gate, called a NOT gate. It operates the same as digital inverter gate [1, 3]. If input qubit (q) is |0 > then it will flip the target qubit state to |1 > and vice versa. It is also known as the Quantum X- gate and its truth table is like the classical logic NOT gate as shown in Fig. 1 and Table 1.

Table 1. Truth Table of X gate

q0	q1
0	1
1	0

Fig. 1. X gate(NOT)

CX-gate- A controlled-X (CNOT/Feynman) gate is a two-qubit operation (q0, q1) in which the first qubit (q0) serves as the control qubit and the second qubit (q1) functions as the target qubit [1, 3]. If both input qubits are |1 >, the target qubit state will change to |0 >; conversely, if the inputs are |0| and |1|, the target qubit will be |1 >. It is referred to as the Quantum X-OR gate, and its truth table resembles that of the classical logic XOR gate, as illustrated in Fig. 2 and Table 2 correspondingly.

Table 2. TruthTable of CX gate

q0	q1	q2
0	0	0
0	1	1
1	0	1
1	1	0

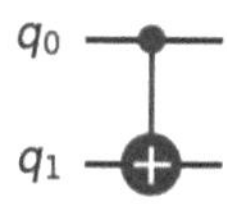

Fig. 2. Controlled X(CX /Feynman)

CCX-gate- A controlled-CX (CCNOT/Toffoli) gate is a two-qubit operation involving qubits q0 and q1, where q0 serves as the control qubit and q1 as the target qubit. If the input qubit is in the |1 > state, it will alter the target qubit's state to |1 >; otherwise, it will display |0 >. It is also referred to as the Quantum X-OR gate, and its truth table resembles that of the classical logic XOR gate, as illustrated in Fig. 3 and Table 3 correspondingly.

Table 3. TruthTable of CCX gate

q0	q1	q2
0	0	0
0	1	1
1	0	1
1	1	1

3 Quantum Hamming Coding and Syndrome Decoding Method

An error-correcting technique called hamming code is used to identify and fix single-bit faults in data storage or transmission. The classical Hamming code algorithm is employed, consisting of 7 qubits that serve as the codeword length. The input qubit data

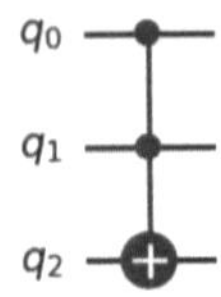

Fig. 3. Controlled CCX/AND/Toffoli gate

comprises 4 qubits, with 3 parity qubits utilized for the purposes of error detection and correction. The Hamming code (7,4) includes supplementary bits in conjunction with the primary data to detect and rectify single bit errors. The supplementary elements are designated as redundant bits, which guarantee the preservation of data integrity throughout the transfer process [2–4]. The quantity of redundant bits (r) is determined using the following formula:

$$2^r \geq x + r + 1 \tag{5}$$

where x = No. of bits present in input data

r = Redundant bits

For example, data bits are 7, then the redundant bits(r) can be calculated as

$$2^3 \geq 4 + 3 + 1 \tag{6}$$

So, the number of redundant bits is 3 and the position of the redundant bits are positioned at constant place by using the formula of 2^r, where r = 0,1,2. Three parity bits (P1, P2, and P3) are added to convert four data (D3, D2, D1, D0) into a seven-bit hamming code. The block in Fig. 4 thus explains the location of the 7-bit Quantum Hamming Code.

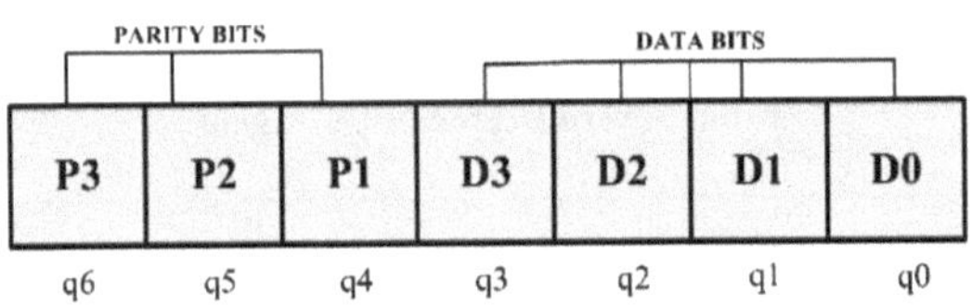

Fig. 4. Quantum Hamming Coder Parity and Data bits Position according to the Quantum bit (Total Wordcode is 7, parity bits are 3(P3 P2 P1), and data bits are 4 (D3 D2 D1 D0))

The parity bits are calculated as per the Hamming encoding rules at specific position. These are P1 = (D0 ⊕ D1 ⊕ D2),P2 = (D0 ⊕ D2 ⊕ D3), and P3 = (D0 ⊕ D2 ⊕ D3). Consider, the data bits are D3 = 1, D2 = 1, D1 = 0, D = 0 and it is encoded as 7 bit data word are placed as shown in Fig. 4. The result shows in the box like P3 P2 P1 D3 D2 D1 D0 (0111100), where P3, P2, P1 are calculated from the main data (1 1 0 0) by using XOR (CX gates) opration. Syndrome decoding is a widely utilized method in quantum error correction, aimed at obtaining accurate data. This technique relies on the syndrome approach to identify errors within a quantum state [7, 8]. A Quantum Syndrome Decoder (QSD) is engineered to identify and rectify errors that arise in digital communication and quantum computing processes. The classical Syndrome Decoder utilizes a parity check matrix to identify the error qubit and correct the code word through the application of an error lookup table.

Quantum Syndrome Decoding employs specific methodologies associated with quantum Error Correction (QEC). Firstly, encode the Hamming (7, 4) code data and transmit it. After transmission, the syndrome checks for errors present in the ancilla qubits that are entangled with the qubits. Finally, when the error syndrome is detected, the specific error qubit is flipped to recover the actual quantum state [8].

3.1 Implementation of Quantum Hamming Encoder

The present work presents the implementation of a Quantum Hamming Encoder (7,4), which utilizes the Parity check Error correction Hamming code. Figure 5 illustrates the complete procedure of quantum Hamming Encoding, detailing the calculation of parity bits and the measurement of outputs. The Quantum Feynman Gate (QFG), also known as the Controlled X or CNOT gate, is utilized here, and the overall process is outlined succinctly.

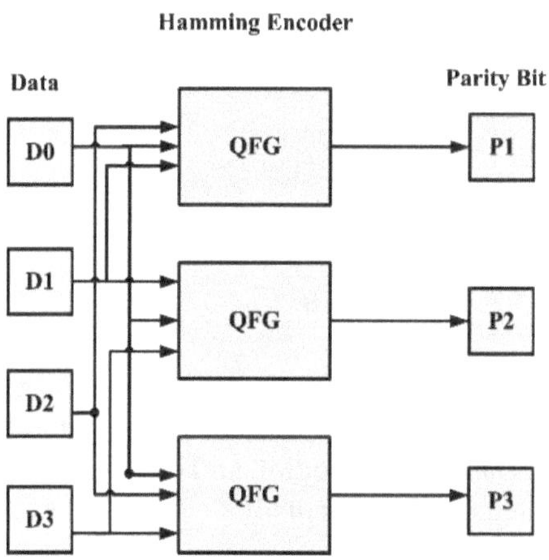

Fig. 5. Quantum Hamming Encoder using Quantum Feynman Gate

Let's consider a simple example of Quantum Hamming Coder, it is code of 7 quantum qubits (q0, q1, q2, q3, q4, q5, q6), that consists of 4 data qubits (D0, D1, D2, D3) and 3 Parity bits (P0, P1, P2). The Quantum Hamming code encodes $|\psi>$ into a 7-qubit entangled state as

$$|\psi>=\propto|0_L> +\beta|1_L> \tag{7}$$

where $|0_L>$ and $|1_L>$ are the logical basic states

Consider the input qubits are D0 = 0, D1 = 0, D2 = 1, D3 = 1 and these are using quantum X gate only. Then the parity qubits are computed by using QFG (CNOT/CX) gates, as shown in the Eq. (8), (9) and (10).

$$|P1>= CX\,(D0,\,P1)\cdots CX\,(D1,\,P1)\cdots CX\,(D2,\,P1)\,|1> \tag{8}$$

$$|P2> = CX\ (D0,\ P2) \cdots CX\ (D1,\ P2) \cdots CX\ (D3,\ P2)\ |1> \tag{9}$$

$$|P3> = CX\ (D0,\ P3) \cdots CX\ (D2,\ P3) \cdots CX\ (D3,\ P3)\ |0> \tag{10}$$

These equations compute the three parity bits (P3, P2, P1) using the four data inputs. As indicated in Table 4, the parity bits are stored data at certain positions, such as P1(q4), P2(q5), and P3(q6), and all seven qubits are measured to classical bits.

However, the encoded Hamming quantum state is expressed in the Eq. (11)

$$|\psi_{encoded}\rangle = [P3\ P2\ P1\ D3\ D2\ D1\ D0]\ |0111100\rangle \tag{11}$$

The encoded output of the Quantum Hamming Code (QHC) is [0111100].

Table 4. Data set of Quantum Hamming Encoder

Input Data				Quantum Hamming Encoder						
D3	D2	D1	D0	P3	P2	P1	D3	D2	D1	D0
1	0	0	0	1	1	0	1	0	0	0
1	1	1	1	1	1	1	1	1	1	1
1	1	0	0	**0**	**1**	**1**	1	1	0	0
1	1	1	0	0	0	0	1	1	1	0
0	0	0	1	1	1	1	0	0	1	1
0	0	1	1	1	0	0	0	0	1	1

Employing the CX gates to perform the parity bits, Table 4 shows the data set of some validated Quantum Hamming Encode data. The output state of Hamming encoded data (0111100) is presented as a histogram after we used Aer Simulator to simulate the circuit as illustrated in Fig. 6.

3.2 Implementation of Quantum Hamming Syndrome Decoder

Figure 7 presents the Quantum Hamming Syndrome Decoder (QHSD), a vital method for detecting and correcting errors in quantum states. This method employs fault-tolerant quantum computing and utilizes error syndrome through parity bits to detect the presence of errors [2]. This Quantum syndrome identifies specific errors in the positions of codeword.

The Quantum Hamming code invokes the purpose of detecting and correcting errors, specifically designed to safeguard quantum data from noise, including bit-flip and phase-flip errors, during computational and transmission processes. Consider an example illustrating the operational mechanism and the process for identifying and rectifying the error bit within a codeword. Consider the Quantum Hamming (7, 4) code, which encodes 4 bits of data qubits (D3 = 1, D2 = 1, D1 = 0, D0 = 0) along with 3 bits of parity data

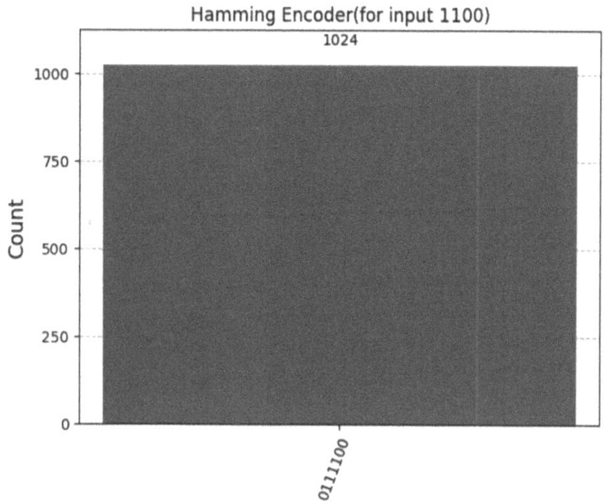

Fig. 6. Quantum Hamming Encoder for input 1100

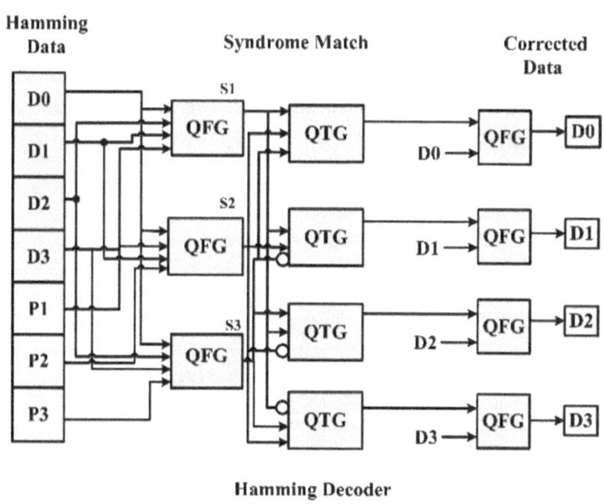

Fig. 7. Quantum Hamming Syndrome Decoder

(P3 = 0, P2 = 1, P1 = 1). The resulting Quantum Hamming Codeword is structured as follows: P3 P2 P1 D3 D2 D1 D0 = 0111100. This arrangement corresponds to the quantum bits labelled q6, q5, q4, q3, q2, q1, q0, as detailed in Table 5.

The 7 bits of the codeword are input into the Quantum Feynman Gates (QFG) for the purpose of performing a parity check. To determine the location of the error bit, the new parity value was applied utilizing the formula $0 + 2^2 + 2^1 + 2^0$ (0 + 4 + 2 + 1). To rectify the error, we inverted the error bits from their complement, changing 0 to 1 or 1 to 0. The subsequent correction involves the verification of parity check qubits across all possible combinations, specifically S0, S1, and S2 (111, 101, 110, 100,

Table 5. Error lookup table of Quantum Syndrome Decoder

Input Syndrome Data			Error bit position data in Codeword						
S3	S2	S1	q6	q5	q4	q3	q2	q1	q0
1	1	1	1	0	0	0	0	0	0
1	0	1	0	1	0	0	0	0	0
1	1	0	0	0	1	0	0	0	0
1	0	0	0	0	0	1	0	0	0
0	1	1	0	0	0	0	1	0	0
0	1	0	0	0	0	0	0	1	0
0	0	1	0	0	0	0	0	0	1

011, 010, 001), as illustrated in Table 5. Three parity check qubits are utilized to assess the quantum syndrome calculation for seven bits of the Quantum Hamming code. This implementation involves a parity check coder that utilizes additional qubits, referred to as ancilla qubits. Additionally, all potential syndrome matching bits are interfaced with the Quantum Toffoli Gates (QTG) to facilitate the detection and correction of the data (D0 D1 D2 D3) in accordance with the syndrome information. Subsequently, the syndrome decoding is executed in accordance with the parity code. The syndrome bits are computed using CX and CCX (Toffoli) gates as follows in Eq. (12), (13) and (14).

$$|S1 >= CX\,(P1,\ S1)\ \cdots\ CX\,(D0,\ S1)\ \cdots\ CX\,(D1,\ S1)\ \cdots\ CX\,(D2,\ S1)\,|1 > \quad (12)$$

$$|S2 >= CX\,(P2,\ S2)\cdots CX\,(D0,\ S2)\ \cdots\ CX\,(D1,\ S2)\ \cdots\ CX\,(D3,\ S2)\,|1 > \quad (13)$$

$$|S3 >= CX\,(P3,\ S3)\ \cdots\ CX\,(D0,\ S3)\ \cdots\ CX\,(D2,\ S3)\ \cdots\ CX\,(D3,\ S3)\,|0 > \quad (14)$$

Upon identification of the error position, the subsequent correction step is implemented in the next stage.

$$X\,(Errorbit)|\psi_{encoded} > X\,(0000100) \quad (15)$$

A controlled X (CX) gate is utilized to flip the incorrect qubit, as represented in Eq. (16)

$$CX\,(S3,\ a)\ \cdots\ CX\,(S2,\ a)\ \cdots\ CX\,(S1,\ a)\ \cdot\cdot X\ \cdot\ (D_X) \quad (16)$$

In this context, "a" represents an ancilla qubit that is responsible for determining the application of X and D_X quantum rectified data.

After applying the correction data, the final corrected quantum state is shown in Eq. (17)

$$|\psi_{decoded} >= [P3P2P1D3D2D1D0]\,|\,0001100 > \quad (17)$$

Table 6. Data set of Quantum Hamming Encoder and Syndrome Decoder

Quantum Hamming Encoded Data							Syndrome Matching Data			Corrected Syndrome Decoded Data						
P3	P2	P1	D3	D2	D1	D0	S3	S2	S1	S3	S2	S1	D3	D2	D1	D0
1	1	0	1	0	0	0	1	1	0	0	0	0	1	0	0	0
1	1	1	1	1	1	1	1	1	1	0	0	0	1	1	1	1
0	1	1	1	1	0	0	0	1	1	0	0	0	1	1	1	0
0	0	0	1	1	1	0	0	0	0	0	0	0	1	1	1	0
1	1	1	0	0	0	1	1	1	1	0	0	0	0	0	0	1
1	0	0	0	0	1	1	1	0	0	0	0	0	0	0	1	1

Table 7. Performance metrics of Quantum Hamming Encoder and Quantum Hamming Syndrome Decoder

Matrices	Quantum Hamming Encoder	Quantum Hamming Syndrome Decoder
T-count	0	21
T-depth	0	69
Qubits count	7	19
Ancilla count	3	15
Execution time	0.1s	14.7s

The codeword is error-free when S = [0 0 0], so the decoded data is (0001100); when the received data has errors, S = (0111100), the errors are found using the syndrome matching followed by error lookup table, and identified the specific error qubit is then flipped to recover the actual quantum state, so the corrected decoded data is (0001100), as displayed in Table 6. The output corrected data's histogram is (0001100), as displayed in Fig. 8.

3.3 Performance Metrics

This section presents the calculation of performance metrics for the Quantum Hamming Encoder and the Quantum Hamming Syndrome Decoder. The objective of quantum circuit performance metrics is to optimize various parameters to enhance execution time and minimize the cost of quantum circuits, thereby improving the quality and efficiency of quantum hardware design [10]. Key metrics such as T-cost, T-depth, qubit count, and ancilla counts are computed for the purpose of optimizing quantum circuits.

T-count, also referred to as T-cost, is a critical metric in quantum circuits, representing the quantity of T-gates utilized within these circuits. The calculation is performed utilizing CCX, CCNOT, and Toffoli gates.

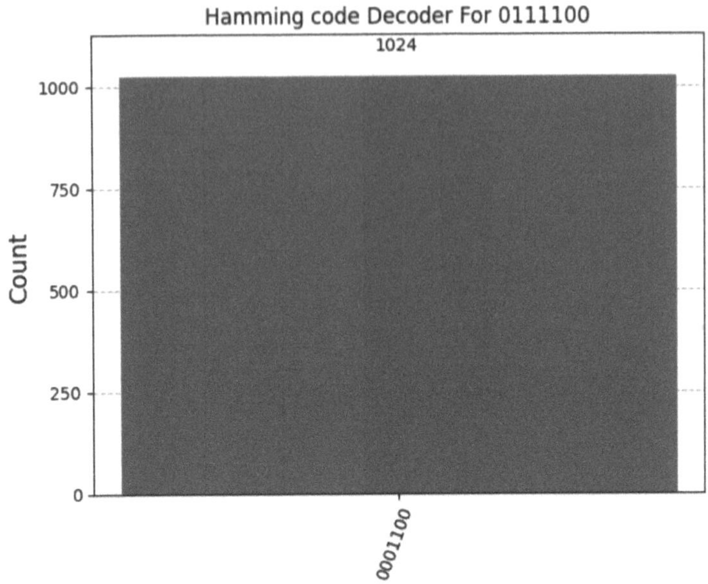

Fig. 8. Quantum Hamming Decoder using Syndrome process for input **0111100**

T-depth refers to the measurement of circuit depth specifically for T-gates, indicating the number of gate layers present within quantum circuits. The total number of qubits present in a quantum circuit is referred to as the qubits count. The primary objective is to minimize the number of qubits in order to optimize and map circuits onto quantum hardware, particularly when there is limited qubit availability.

Ancilla count: In quantum computing, the quantity of ancilla qubits utilized in quantum circuits is designated for the storage of partial outputs. Initially, the ancilla qubits are prepared in a defined state $|0>$. At the conclusion of the computation, it is essential that the ancilla qubits are returned to this same state.

Execution Time is defined as the estimated job run time on the IBM Platform for a quantum circuit executed on a quantum computer. The execution duration may differ across Quantum circuits and backends.

Table 4 illustrates the performance metrics of the Quantum Hamming Encoder and Quantum Hamming Syndrome Decoder. The table presents the metrics of T-count and T-depth, both recorded as 0, indicating that there are no Toffoli (CCX) gates present in the Quantum Hamming Encoder circuit. The Qubits count is determined to be 7, while the Ancilla count is established at 3. In the context of QHSD, the T-count and T-depth are as follows: Qubits count is 21, Ancilla count is 69, T-count is 19, and T-depth is 15. Finally, the last metric is execution time, which indicates that the runtime is significantly lower than that of classical computers.

4 Conclusion

This paper presents the design of the Quantum Hamming Encoder and the Quantum Syndrome Decoder, utilizing the Quantum Feynman Gate operation and the Syndrome code algorithm, respectively. The primary objective is to implement bit error correction for quantum data. The Quantum Hamming Encoder transforms classical data into a quantum state. The Quantum Hamming Encoder performed calculations to determine parity bits and assessed the quantum state of the input data. Utilizing the Quantum Syndrome Decoder, errors in the Quantum state were detected and subsequently rectified. This method provided mechanisms for error detection and correction applicable to all aspects of quantum computing and quantum communication. In addition, we enhanced the performance metrics including T-cost, T-depth, Qubits count, and Ancilla count for both the Quantum Hamming Encoder and the Quantum Syndrome Decoder. The approach enables error detection in memory and data security applications.

Disclosure of Interests. The authors have no competing interests to declare that are relevant to the content of this article.

References

1. Haq Shaik, E., Rangaswamy, N.: Implementation of quantum gates-based logic circuits using IBM Qiskit. In: 2020 5th International Conference on Computing, Communication and Security (ICCCS), pp. 1–6. IEEE (2020)
2. Das, A., Touba, N.A.: A single error correcting code with one-step group partitioned decoding based on shared majority-vote. Electronics **9**(5), 709 (2020)
3. Cılasun, H., et al.: On error correction for nonvolatile processing-in-memory. In: 2024 ACM/IEEE 51st Annual International Symposium on Computer Architecture (ISCA), pp. 678–692. IEEE (2024)
4. Javadi-Abhari, A., et al.: Quantum computing with Qiskit (2024). arXiv preprint arXiv:2405.08810
5. Cook, K.P.: A historical and practical survey of quantum computing using QISKit (2022)
6. Wille, R., Van Meter, R., Naveh, Y.: IBM's Qiskit tool chain: working with and developing for real quantum computers. In: 2019 Design, Automation & Test in Europe Conference & Exhibition (DATE), pp. 1234–1240. IEEE (2019)
7. Megha, P., Premananda, B.S., Kamat, N.: Area and energy optimized Hamming encoder and decoder for nano-communication. J. Electr. Eng. **75**(3), 229–236 (2024)
8. Kaity, A., Singh, S.: An area-efficient, robust, and reversible QCA-based Hamming code generator, error detector, and corrector: design and performance estimation. J. Comput. Electron. **20**, 2622–2647 (2021)
9. Lai, C.Y., Kuo, K.Y., Liao, B.J.: Syndrome decoding by quantum approximate optimization. Quantum Inf. Process. **23**(11), 368 (2024)
10. Karuppasamy, G., Puram, V., Johnson, S., Thomas, J.P.: Quantum circuit optimization: current trends and future direction (2024). arXiv preprint arXiv:2408.08941

Emerging Intelligent Systems for Secure Computing, Autonomous Technologies, and Smart Environments

A Comparative Study of Advancements in Machine Learning-Based Speaker Recognition Models

U. K. Rohan[1(✉)], Ramesh Babu Jonnalagadda[1], Rajanikanth Mohanty[2], S. Vanitha[1], and S. Saravana Kumar[1]

[1] CSE, JAIN (Deemed-to-Be University), Bangalore, India
rohanuk02@gmail.com, s.saravanakumar@jainuniversity.ac.in
[2] CSE-SE, Jain University, Bangalore, India
m.rajanikanta@jainuniversity.ac.in

Abstract. Forensic Speaker Recognition plays a pivotal role in criminal investigations by identifying individuals through speech characteristics. This review explores recent advancements in automated speaker recognition (ASR) models leveraging machine learning techniques. Key methods analyzed include feature extraction, model training, and performance optimization strategies. Techniques such as MelFrequency Cepstral Coefficients (MFCC), Linear Chirplet Transform (LCT), and Two-Tier Feature Extraction with Meta heuristics-Based Automated Forensic Speaker Verification (TTFEM-AFSV) demonstrate advancements in accuracy and robustness under varied conditions. The integration of deep learning frameworks, optimization algorithms, and adaptive filtering techniques significantly enhances recognition capabilities. However, challenges such as low-quality speech data, black-box model interpretability, and limitations in complex tasks like emotion or mixed-dialect recognition persist. This paper concludes by emphasizing the potential of hybrid approaches combining diverse methodologies to achieve superior performance in forensic and real-time applications.

Keywords: Forensic Speaker Recognition · ASR Models · Feature Extraction · Deep learning frameworks · Metaheuristics · Hybrid Approaches

1 Introduction

Forensic Speaker Identification or Comparison; sometimes known as Forensic Speaker Recognition, refers to the scientific method used in criminal investigations to identify a person by analyzing their speech characteristics. It helps determine whether an unknown speaker in a recording matches someone with a known voice sample [3]. Unique speech qualities include pitch, loudness, speaking rate, prosody, voice quality, and accent/dialect.

Several methods are used in forensic speaker recognition: the aural perception approach, spectrographic identification, and automatic machine-based methods. The "auditory" method relies on practitioners' critical listening to identify similarities or differences in speech qualities. The "spectrographic identification" method uses spectrograms

A. Das et al. (Eds.): MIDAS 2025, CCIS 3055, pp. 415–426, 2027.
https://doi.org/10.1007/978-3-032-35516-4_33

to visually compare speech patterns between recordings. The "aural-spectrographic" technique combines auditory and spectrographic comparisons to form opinions on the recordings' similarities and contrasts [5]. Automatic speaker recognition utilizes computer programs to autonomously analyse and compare voice samples with minimal human intervention. Users provide audio recordings, and the system processes speech segments to identify the speaker, relying entirely on computational analysis [3]. Various machine learning approaches are employed in this field to extract and process voice features. This work will describe recent advancements in these techniques.

2 Machine Learning Models for ASR

The methodology used in most papers reviewed resembled the process for Automatic Speaker Identification. The quality of the voice sample fed into the system is a crucial aspect for achieving accurate results.

2.1 Quality of Speech

Forensic Speaker Recognition can be challenging due to low-quality speech recordings in real- world forensic cases. There are frequently differences between known and questioned speakers in terms of recording settings and speaking styles. Only a few seconds of voice and background noise (such as conversation, ventilation, and car noise) at different levels may be heard on recordings. These might also be broadcast across other channels (landline, mobile, VoIP), recorded in reverberant locations, or captured with microphones far from the speaker (such as covert or telephone microphones), all of which would further distort the signal. MP3 and other compression formats introduce extra distortion.

2.2 Feature Extraction

Feature extraction plays a critical role in speech recognition by retrieving useful information from speech samples. Its goal is to evaluate and select effective techniques for distinguishing voice streams. Techniques like MelFrequency Cepstral Coefficients (MFCC) are widely used for extracting features. Linear Chirplet Transform (LCT) has also shown effectiveness, along with methods like CNNs (Convolutional Neural Networks), RNNs (Recurrent Neural Networks), and their variations (LSTMs, GRUs), and Linear Predictive Coding (LPC).

2.3 Modelling and Comparison

In speaker recognition, modelling involves creating mathematical representations of speech patterns, the acoustic model represents the relationship between audio signals and speech phonetic units, trained using Deep Neural Networks (DNNs), RNNs, CNNs, or Hidden Markov Models (HMMs), and the language model represents word sequence probability, aiding in word prediction. It can be neural (e.g., RNNs, LSTMs, transformer models) or statistical (e.g., n-grams).

Comparison involves matching processed speech data to models for recognition for pattern Matching where features are compared to acoustic models using algorithms like the Viterbi method or forward propagation. Decoding which constructs words/sentences from identified phonemes using acoustic and language models, employing methods like beams ear chore A search. Scoring where hypotheses (possible transcriptions) are scored based on alignment with models, with the highest-scoring hypothesis selected as the final transcription.

3 Methodology Used in Recent ASR Models

Several researchers have sought to develop Speaker Recognition models with high accuracy and lower or rates. Most researchers utilize machine learning models trained and deployed to perform Speaker Recognition tasks autonomously. Some of the recent methods are discussed in this paper.

3.1 Deep Learning Neural Networks

Solovyov *et al.* suggested a Deep Learning Neural Network-based Speaker Identification system [7]. They discussed about the primary techniques and technical elements of developing a speaker recognition system, as well as the difficulties and solutions they found. They worked on important tasks like estimating the duration of phonograms, guaranteeing system invariance to different language groups, and spectrally analyzing voice signals at short intervals with great resolution. The goal of these initiatives was to improve the generality and effectiveness of speaker identification results.

They achieved superior resolution in frequency localization of maxima (about 1 Hz) by using non-orthogonal time-frequency transformations, which increased the accuracy and efficiency of speaker identification. Based on a thorough examination of neural networks, a unique module was created to automatically extract vowel sounds from voice phonograms. The automatic evaluation of the spectral closeness of vowel sounds [A] and [I] from various phonograms was necessary for the system to identify speakers. A special model built on a deep learning neural network was used to calculate this proximity. By comparing spectra in the 0–4000 Hz frequency range, which revealed formant features and coincident local maxima for the [A] sound, the system's effectiveness was confirmed (Fig. 1).

They trained the neural network using the Keras library (backend TensorFlow). Some neural network training methods are depicted in (Fig. 2).

The spectrum analysis of audio data is severely impacted by low frequency resolution at short time intervals, which is one of the disadvantages mentioned [1]. The reliance on expert-selected characteristics from frequency-domain sound sources presents another difficulty, underscoring the real-world challenges of utilising conventional time-frequency transformation techniques to analyse short-duration phonograms. Moreover, the neural network's identification model functions as a "black-box," which makes it challenging to successfully establish causal relationships for speaker identity.

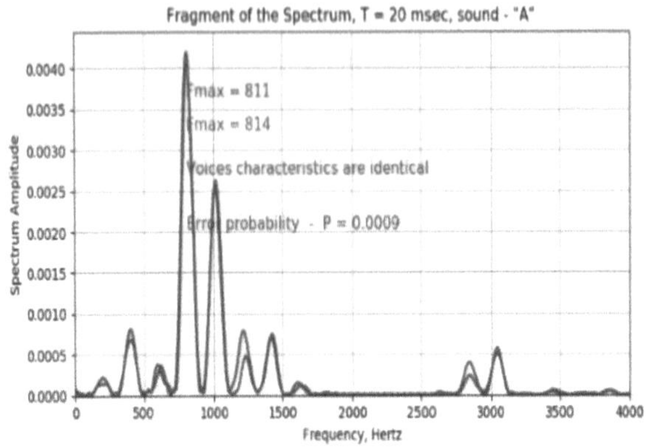

Fig. 1. Modified from "Spectrum of similar characteristics of voices with formant signs"

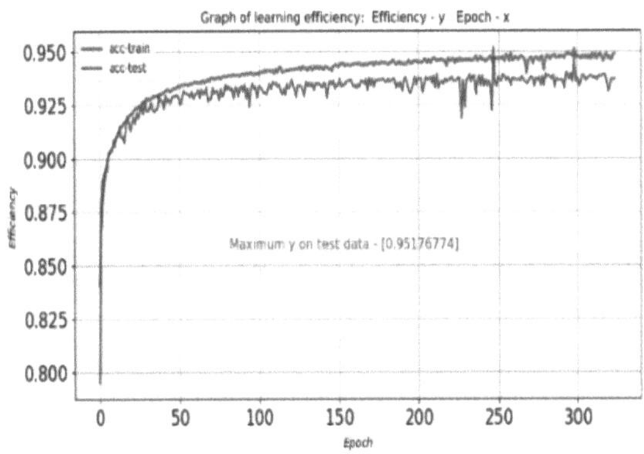

Fig. 2. Modified from "Graph of the effectiveness of training a neural network"

3.2 MFCC-BPNN-HHO

A paper by Harish *et al.* proposes adding Harris Hawks Optimization (HHO) for weight optimization to the Back propagation Neural Network (BPNN) to improve its performance and compares the results with BPNN alone [4]. MelFrequency Cepstral Coefficients (MFCC) are used in both methods to extract features from input data. A dataset of ten speakers, each contributing ten utterances lasting five to ten seconds, totaling one hundred utterances, was used in the study to evaluate the suggested method. The methodology consisted of four steps: Pre-processing, MFCC Feature Extraction, Network Training (including BPNN and HHO), and Testing. Pre-processing involves transforming raw data into a format suitable for computational analysis and modelling. For feature extraction MFCCs were used for their ability to represent unique vocal characteristics. The process began with framing, dividing the pre-emphasized signal into short overlapping frames (20–30 ms). Each frame was windowed (e.g., Hamming window) to reduce spectral leakage and the signal was converted to the frequency domain by passing it through a Fast Fourier Transform (FFT). Frequency bands were extracted using a Melfilter bank spaced according to the Melscale. Logarithmic Compression captured the log energy in each filter output, and Discrete Cosine Transform de-correlated the features to produce MFCCs. For network training classifiers map input features to output categories. Neural networks, used here, escalate understanding intricate input-output correlations. The study utilized Back propagation Neural Networks (BPNN), employing HHO for weight optimization.

The back propagation method is used by BPNN, a supervised learning system, to modify biases and weights in order to reduce the discrepancy between expected and actual outputs. It made use of the sigmoid activation function (Fig. 3).

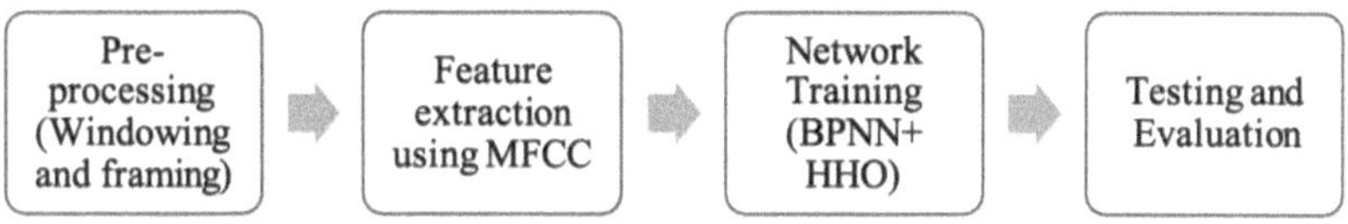

Fig. 3. Flowchart of MFCC-BPNN-HHO based speaker recognition

HHO is inspired by the hunting habits of Harris's Hawks, HHO is an iterative optimization algorithm with exploration and exploitation phases. It optimized BPNN's random weights and thresholds, addressing BPNN's sensitivity to initial weights. A confusion matrix, which tabulates true positives (TP), true negatives (TN), false positives (FP), and false negatives (FN) to evaluate classification accuracy, was used to analyze the performance as shown in Fig. 4.

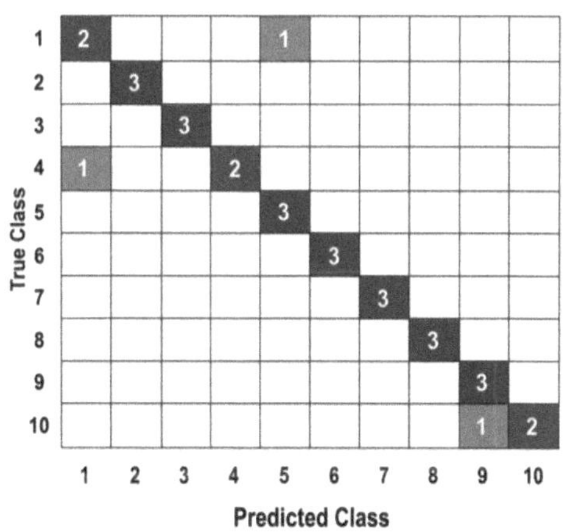

Fig. 4. Modified from "Confusion Matrix of MFCC-BPNN-HHO"

The MFCC-BPNN-HHO method of classification attained an accuracy of 98.00%, outperforming MFCC-BPNN algorithms at 96.00%. Dimensionality reduction via MFCCs enhanced accuracy for both models (Table 1, Fig. 5).

Table 1. Modified from "Results of MFCC- BPNN-HHO"

Method	Formula	MFCC-BPNN-HHO
Accuracy	(TP + TN) / (TP + TN + FP + FN)	98.00%
Precision	TP / (TP + FP)	90.00%
Recall	TP / (TP + FN)	90.00%
Specificity	TN / (TN + FP)	98.89%
Sensitivity	TP / (TP + FN)	90.00%
F1 Score	2 × Precision × Recall / (Precision + Recall)	90.00%

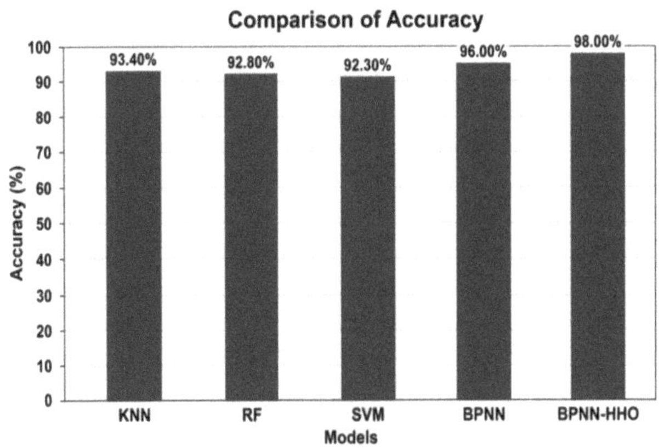

Fig. 5. Modified from "Comparison of Accuracy Bar graph"

The study concluded that while MFCC-BPNN-HHO improved accuracy compared to BPNN, it was deemed ineffective for speaker recognition due to its limited application scope.

3.3 Linear Chirplet Transform (LCT)

Duc Do *et al.* suggested a novel method for employing a Linear Chirplet Transform (LCT) to represent voice signals in the time-frequency domain. This approach offers a feature vector sensitive to frequency shifts in human speech [1] and has a strong mathematical basis. Model chirp signals, LCT were suggested by DucDo et al. as a flexible alternative to the traditional Fourier Transform for speech representation in applications involving frequency shifts. Unlike Fourier Transform, which uses a fixed instantaneous frequency in its basis vectors, LCT introduces a chirp rate parameter (α) to define a linear function for instantaneous frequency. This parameter enables LCT to capture and highlight frequency changes over time more effectively. LCT basis vectors, represented as sloping lines in the time-frequency plane, are easier to interpret and manipulate than the horizontal lines used in Fourier Transform. The chirp rate parameter allows LCT to adjust the slope of these lines to better model the frequency trajectory.

The datasets used for Gender recognition were TIMIT Dataset of 630 speakers and VIVOS Dataset of 65 speakers. For Dialect recognition 8 dialects from TIMIT and 3 dialets from VIVOS was taken, and for speech recognition 960 h of speech of 2338 speakers from Libri Speech dataset and 15.7 h of speech of12420 audio files from VIVOS Dataset were taken. During experimentation (Duc [1] evaluated the LCT against traditional features for tasks such as speaker gender, dialect, and voice recognition. Using the TIMIT and VIVOS datasets for Vietnamese and English, they compared LCT with spectrum, Mel- spectrum, and MFCC as feature extractors. The results highlighted that For Transformer- Conformer models, LCT with a negative chirp rate yielded highest Word Error Rates (WERs) for Vietnamese (6.29%) and English (22.52%) as shown in Fig. 4.

While MFCC had slightly better results for Transducer networks, LCT consistently ranked among the top-performing approaches across tasks, especially excelling in speaker gender recognition [2].

Table 2. Modified from Speech recognition with different features for English (E) in Libri Speech and Vietnamese (V) in VIVOS

Feature	T-C (E) WER (%)	Transducer (E) WER (%)	T-C (V) WER (%)	Transducer (V) WER (%)
Spectrogram	8.65	8.02	25.87	34.65
Mel-Spectrogram	7.09	8.11	26.14	33.26
MFCC	6.47	6.45	23.73	28.08
LCT ($\alpha = 0.5$)	6.38	6.47	23.91	29.28
LCT ($\alpha = -0.5$)	6.29	6.76	22.52	28.10

Some of the key findings of their study was: LCT demonstrated competitive performance across tasks, with low error rates in speaker dialect and speech recognition. LCT and MFCC differences in these tasks was minimal, with LCT offering slight improvements. LCT performed significantly better than MFCC for speaker gender recognition, with evident advantages across both positive and negative chirp rates (Table 2).

For future work Duc Do *et al.* suggested that while LCT shows promise, its performance is yet unknown when it comes to more difficult tasks like age prediction, mixed-dialect recognition, and emotion recognition. Future extensions of LCT could optimize chirp rates and enhance speech feature extraction, capturing more nuanced behaviors in human speech and potentially achieving higher accuracy in recognition models.

3.4 Automated Forensic Speaker Verification Using Meta Heuristics for Two-Tier Feature Extraction (TTFEM-AFSV)

The TTFEM-AFSV model for Automated Forensic Speaker Verification was presented by Gaurav *et al.* [2]. It uses deep convolutional neural networks, a Long Short-Term

Memory- Recurrent Neural Network (LSTM-RNN) for speaker validation in forensic applications, and average median filtering (AMF) for noise removal. The dataset consisted of 400 speech samples from five speakers.

The Average Median Filter (AMF) was used to remove noise during pre-processing. Unlike standard median filters, AMF adapts the window size for each pixel to retain edge features and finer details while removing noise. Now for Feature Extraction Mel Frequency Cepstral Coefficients (MFCC) and spectrogram features were derived. Speech signals were framed into 25 ms segments with a 10ms frame shift, multiplied by a Hamming window to minimize spectral leakage. Time-domain signals were converted to the frequency domain using the Fast Fourier Transform (FFT), and then filtered using a triangular Mel-filter bank. The features were fed into the Inception_v3 module, which divides bigger integrals into smaller convolutions for processing efficiency, after being normalised for zero mean and unit variance. The Ant Lion Optimiser (ALO), a meta heuristic method based on the hunting behavior of ant lions, utilised to optimise the Inception_v3 model's hyper parameters. The fitness function ensured optimal solutions by minimising classification error rates. The sequence classifier used was LSTM-RNN. It has two hidden layers, each with 300 nodes, an input layer, and a classification Soft Max layer at the output. Strong sequence-level predictions were made possible by the model's decision-making process, which made use of both historical and current inputs (Fig. 6).

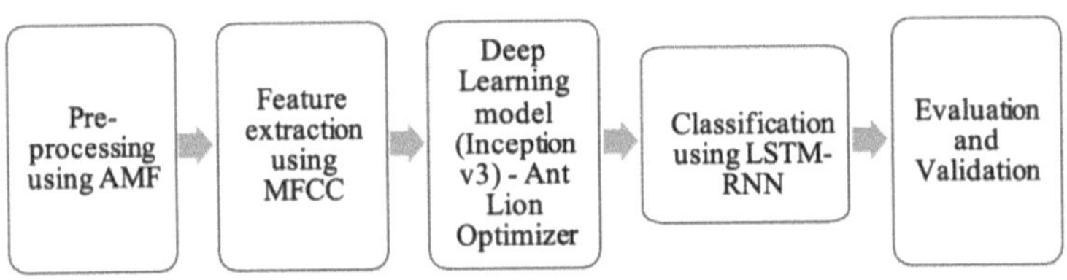

Fig. 6. Flowchart of TTFEM-AFSV model methodology

Table 3. Modified from Comparison of the accuracy analysis of the TTFEM-AFSV approach with recent methodologies

Methods	Accuracy	Error rate
TTFEM-AFSV	99.50	0.50
MFCC-SOFM-MLP-GD	96.92	3.08
MFCC-FW	97.32	2.68
MFCC-SOFM-MLP-BR	97.62	2.38
MFCC-SOFM-MLP-GDM	97.05	2.95
FUSION	97.81	2.19
DWT-MFCC	98.87	1.13

Classification accuracy, error rate, precision, recall, F-Score, G-measure, and execution time were used to evaluate the performance of the TTFEM-AFSV model [3]. The TTFEM-AFSV methodology performed better when accuracy and error rates were compared to more modern approaches as shown in the table below (Table 3).

The TTFEM-AFSV model demonstrated higher accuracy and lower error rates compared to other methods, making it appropriate for real-time forensic usage (Fig. 7, Fig. 8). The AMF approach effectively denoised signals while preserving critical features [3].

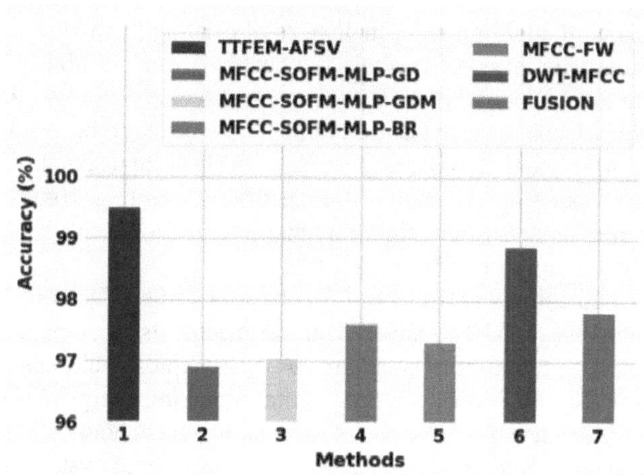

Fig. 7. Modified from "Comparison of the accuracy analysis of the TTFEM-AFSV approach with other recent methodologies".

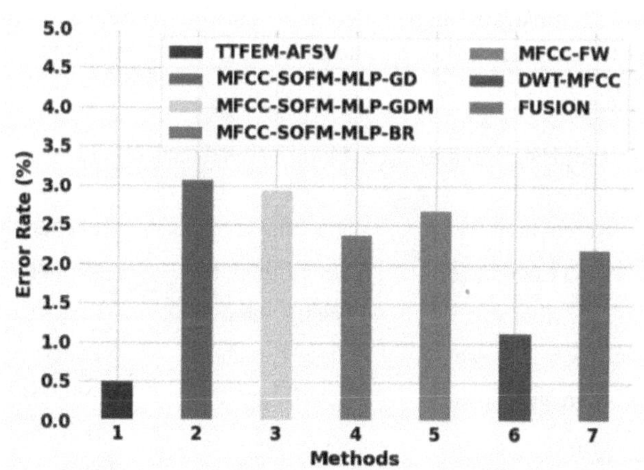

Fig. 8. Modified from "Comparison of the error rate of the TTFEM-AFSV approach with recent methodologies".

An ensemble of fusion-based deep learning models could improve the TTFEM-AFSV model's overall performance, opening the door to more reliable forensic speaker verification systems.

4 Recent Developments

O'Shaughnessy *et al.* explored the optimal utilization of discriminative information within human speech to enhance the design of Automatic Speech Recognition (ASR) systems [6]. This work highlights the challenges of existing methods that often neglect the unique characteristics of speech signals, relying heavily on machine learning systems to autonomously extract meaningful features.

Speech prosody (such as pitch and duration) cannot be incorporated into many ASR models, such as ANNs and HMMs. Conventional ASR systems combine acoustic and language models, operating on vastly different time scales (e.g.,10ms for frames vs. full words). Current models often require human intervention, especially for tasks like adjusting sample durations or addressing data distortions.

Multi-Layer Perceptrons (MLPs) and Stochastic Gradient Descent (SGD) remain foundational elements of ANNs, albeit with advancements in complexity and scale. Directly utilizing spectral amplitude patterns as ANN inputs could enhance efficiency and resilience against distortion, compared to traditional log energy derivatives.

Features like pitch detect or sand pseudo-formant trackers hold potential but require improved implementation to overcome their limitations. Expert-system methodologies, once dismissed, could be revisited to provide structure and reduce reliance on opaque ANN models.

Integrating models like the MFCC-BPNN-HHO shows incremental improvements over traditional MFCC-BPNN approaches. LCT has proven effective in basic ASR tasks but struggles with complexities like emotion or mixed-dialect recognition. The TTFEM-AFSV Model has demonstrated high efficacy in real-time forensic applications, leveraging techniques like adaptive filtering and meta heuristics for feature optimization [6] (Tables 4, 5 and 6).

Table 4. Modified from Phonemes and their features.

Phonemes	Articulation	Spectrum	Excitation
i, I, e	High front vowel	LowF1, high F2	Periodic
E, ae	Low front vowel	HighF1, high F2	Periodic
A, aw, o	Low back vowel	HighF1, high F2	Periodic
U, u	High back vowel	LowF1, low F2	Periodic
l, r	Liquid	LowF1 and F2	Periodic
m, n, ng	Nasal	SteadyF1, F2	Periodic
j, w	Glide	LowF1	Transient
f, s, sh	Fricative	High pass	A periodic noise

(continued)

Table 4. (*continued*)

Phonemes	Articulation	Spectrum	Excitation
v, z, zh	Fricative	High pass	Periodic noise
p, t, k	Stop	Silence + burst	Periodic transient
b, d, g	Stop	Silence + burst	Periodic transient

Table 5. Modified from Overview of speech analysis techniques.

Method	Properties	Advantages	Flaws
Linear prediction	All-pole model	10 coefficients	Narrow band widths
MFCC	Triangular mel filters	13 coefficients	No intuitive interpretation
Fourier transform	Discards phase	Direct spectral measure	Many parameters
Formants	Resonance frequencies	Few coefficients	Unreliable estimation
Band-frequency energies	Direct from FT	Intuitive interpretation	Coarse model

Table 6. Modified from Major approaches used for ASR

Method	Properties	Advantages	Flaws
ANN	Neural-based nonlinear data mapping	Great power; many architectures	Extremely complex network
DTW	Nonlinear sequence alignment	Simple	Expensive; slow
HMM	State model with transitions	Handles both spectral and Timing issues	Markov assumption; geometric Duration model
Expert system	If-then decisions	Intuitive; easy to modify	Poor error rate

5 Conclusions

The various approaches employed in Automated Speaker Recognition (ASR) studies often differ, yet they share a common objective: enhancing the accuracy of speaker identification from voice samples. While Deep Learning Neural Networks (DLNNs) have seen widespread application in ASR, the study indicates that these models may not always be entirely reliable. Specifically, they sometimes necessitate human intervention, particularly incases requiring the manual adjustment of voice sample durations. The study also examined the effectiveness of the MFCC-BPNN-HHO model. While this approach did not fully meet the criteria for optimal ASR, it demonstrated superiority over other methods, such as those using the MFCC-BPNN algorithm without the HHO component. This model represents a progressive step toward identifying a more effective ASR model that integrates various algorithms to enhance accuracy. The Linear Chirplet Transform (LCT) was found to perform exceptionally well in ASR tasks compared to other machine learning models. However, it exhibited limitations in more complex scenarios, such as emotion identification, mixed-dialect identification, or prediction of age. The research suggests that an extended version of LCT could potentially address these shortcomings. Furthermore, the study explored the Automated Forensic Speaker Verification using Two-Tier Feature Extraction with Meta heuristics (TTFEM-AFSV) model, which employs an average median filtering (AMF) approach. This model demonstrated high efficacy in ASR, achieving low error rates and proving particularly valuable in present forensic applications.

Future work in this field should focus on developing hybrid models combining deep learning with explainable AI for enhanced interpretability. For reducing reliance on large data sets that are labelled, self-supervised learning should be employed. Work can also be done in improving the noise robustness by developing adaptive noise filtering techniques and developing techniques for emotion and mixed dialect detection through transfer learning and multilingual datasets. Finally, the study indicates that combining multiple approaches in a strategic manner can lead to the development of an ASR model capable of delivering high accuracy and low error rates.

References

1. Duc Do, H., ThanhChau, D., ThaiTran, S.: Speech feature extraction using linear Chirplet transform and its applications. J. Inf. Telecommun. **7**(3), 376–391 (2023). https://doi.org/10.1080/24751839.2023.2207267
2. Gaurav, Bhardwaj, S., Agarwal, R.: Two-tier feature extraction with meta heuristics-based automated forensic speaker verification model. Electronics (Switzerland) **12**(10) (2023). https://doi.org/10.3390/electronics12102342
3. Hansen, J.H.L., Hasan, T.: Speaker recognition by machines and humans: a tutorial review. IEEE Sig. Process. Mag. **32**(6), 74–99 (2015). https://doi.org/10.1109/MSP.2015.2462851
4. Harish, D.: Speaker recognition using MFCC-BPNN-HHO. Int. J. Sci. Res. Eng. Manag. **08**(04), 1–5 (2024). https://doi.org/10.55041/IJSREM30717
5. Mishra, V.: Forensic speaker recognition: review of methods and evaluations. Int. J. Res. Appl. Sci. Eng. Technol. **11**(9), 680–683 (2023). https://doi.org/10.22214/ijraset.2023.55714
6. O'Shaughnessy, D.: Trends and developments in automatic speech recognition research. Comput. Speech Lang. **83** (2023). https://doi.org/10.1016/j.csl.2023.101538
7. Solovyov, V., Rybalskiy, O., Zhuravel, V., Shablya, A., Tymko, Y.: Building of a Speaker's Identification System Based on Deep Learning Neural Networks (2021)

Healthcare Confidential: Multi-agent Systems Revolutionizing Data Privacy

G. Indira Priyadarshini and N. Venkatesh Yadav(✉)

School of CS and AI, SR University, Ananthasagar, Hasanparthy, Warangal, Telangana 506371, India
{2303c50233,venkatesh.naramula}@sru.edu.in

Abstract. The various advancements of advanced medical services have further changed the way patient consideration happens utilizing Electronic Health Records (EHR). There are also complex challenges to secure patient data when even "touchy" E-Health data is digitized. Here in this paper the project titled "E-Health Information Protection Saving using Multi-Agent System" is put forward based on innovative system that protects personal E-Health information efficiently using multi stranger system framework. This architecture employs distributed agents, with each responsible for handling specific tasks, such as encryption, access control, and the real-time monitoring of security policies. In doing so, the framework fathers E-Health information accessibility by sanctioned elements yet simultaneously guarantees full auditability and conspicuousness of consistence with existing security guidelines, by voicing the standards of decentralized direction and job based access control. This multi-agent system architecture helps to secure the e-health data and receiver privacy by preventing unauthorized access to the system and improving the system extensibility and adaptability in dynamic healthcare surroundings. This endeavor's average result is to a vigorous well-being information security conservation framework, which shields patient information from unapproved access and potential information breaks, subsequently expanding trust among clients and helping more extensive reception of computerized well-being administrations.

Keywords: E-Health Data Privacy · Multi-Agent Systems · Privacy Preservation · Health Information Security · Distributed Systems · Data Protection

1 Introduction

Artificial Intelligence (AI) is revolutionising healthcare in many aspects of patient care and medical management, adding styles of data analysis, and systems for decision making and patient management that add significant value to the healthcare system. But with the widespread adoption of AI, so is the fear of the security and confidentiality of sensitive patient data. This review article [1] aims to address these concerns by exploring the development and implementation of privacy-preserving artificial intelligence (AI) technologies within healthcare settings. It offers a critique of the various methodologies we

A. Das et al. (Eds.): MIDAS 2025, CCIS 3055, pp. 427–438, 2027.
https://doi.org/10.1007/978-3-032-35516-4_34

can use to mitigate patient privacy, even as we embrace AI augmentations of diagnosis and treatment processes. It utilizes techniques like federated learning and differential privacy, which enable the sharing of information while preserving patient privacy. This review highlights the necessity of universalizing the operational efficiency of artificial intelligence tools and the need to protect patient data–a priority that still carries massive importance in the gradual transformation of healthcare systems to more connected and technologically advanced operations. With the convergence of medical care and the surge of Man-made intelligence (artificial Intelligence), it has become imperative to safeguard sensitive patient information. Their audit typically remembers information for the timeliness and utilization of protection saving computer based intelligence advances inside medical care settings to respond to the protection concerns set out by the substantial information requests of computer based intelligence.

Multi-agent Systems (MAS) are an appealing Avenue for addressing the nuances of the current medical services IT scene. This paper investigates the work of MAS in improving the efficiency and effectiveness of health care operations, particularly in the management of distributed and dynamic data. MAS is investigated to upgrade information interoperability and ongoing information across various medical services suppliers. The paper assesses how MAS can uphold decentralized dynamic cycles [2], further developing clinical benefits' reaction times and exactness. It also discusses the integration of MAS with existing legacy systems and the potential challenges, such as scalability and security concerns, that need to be addressed to achieve seamless system integration.

Privacy preservation in surveillance and healthcare data publication is becoming a focal point of modern digital health strategies. This complete audit assesses different anonymization and encryption methods intended to safeguard individual security without losing the utility of information for clinical examination and checking [3]. It investigates difficulties, such as the gamble of re-ID in little datasets and the compromises between information availability and protection. The paper discusses the ramifications of these security safeguarding strategies on the legitimate and moral parts of medical services information taking care, fundamentally breaking down techniques like k-obscurity and l-variety in genuine medical care applications. What this investigation signifies is prompting future settings and development protection of patient data in the interest of well-being investigation and scrutiny.

This paper focuses on security and privacy issues that exist in sites with multiple connected healthcare consumers, and care providers who share EHRs. The solution proposed in this context leverages adaptive Merkle Trees to improve patient privacy control over their data based on profile management. With the help of selective use, the patients can control who has access to their information and on what basis [4]. The paper presents a theoretical analysis of the mathematical model for the constructed Merkle Tree structure and presents experimental results that support the study. The solution ensures the highest privacy level such as the GDPR and increases the data credibility and coherence of the transmission amongst various stakeholders in the health sector.

EHR frameworks are critical for enhancing the advancement of medical services conveyance; however, they are the association's security and protection risks. This audit paper analyzes the security issues associated with EHR frameworks; information leaks, unauthorized access, and insider threats [5]. It assesses the existing countermeasures against such threats including through enhanced forms of encryption and secure access systems. This paper reviews the issue of the protection of health information, specifically the function that presents legal requirements such as the HIPAA play, and outlines the flaws present in existing setups. The review allows an understanding of further advancements in online protection in the most recent EHR by building up more powerful security measures and the reliability of patient data security to meet shifting medical services needs in a progressively digitized environment.

2 Literature Survey

In their investigation of the serious problems with cloud-based EHR security, Folasole et al. [6, 7] highlighted the need of using strong encryption techniques to safeguard patients' personal information. The usage of Personality Encryption and Property Encryption as methods to improve data retrieval procedures while keeping strict security measures in mind was the major focus of their study. Their research implies that it is feasible to build a safe digital environment that protects patient data from breaches and illegal access by using such encryption methods. In addition to showcasing the development of cloud computing, this method stresses the need of keeping up with the constantly changing security environment in order to adequately safeguard critical health information. Ahmed Faizal and Mohamed Reyad In a groundbreaking approach to healthcare multi-agent systems, Khan and Reyad presented the Intelligent-Based Access Control Security Model (IBAC) [8, 9]. Hin was recently awarded a prestigious grant from the National Institutes of Health (NIH) to develop a new approach to health data access that addresses some of the systemic weaknesses associated with e-health records, while improving the security and privacy of digital health data between different users with the system. Of particular interest is their proposed methodology's ability to adapt to dynamic healthcare environments where access criteria can change on a daily basis. The IBAC model also helps enhance security practices that are essential to safeguard patient data and maintain system integrity by ensuring that only individuals who have been properly vetted can view sensitive information. [44] To protect health data stored in cloud environments Geoffrey Muketha et al. 10 model that led to proposing information protection saving in the area of suggested information saving model of security saving in cloud that leads to the health area systems. This version has multifactor authentication built in. With a growing reliance by healthcare organizations on cloud technology for data management, their proposal aims to solve the challenging problem of privacy threats associated with being able to store data in the cloud. The addition of multifactor authentication also enriches sensitive health data with security by adding additional hurdles to hinder illegal access. Sensitive health data can be protected using advanced security measures, which is crucial given the growing cyber threats, according to this technique. To protect sensitive medical data as it passes through the internet, Supriya Menon and Rajarajeswari Pothuraju [12] implemented a hybrid architecture that combines Enhanced Attribute Based Encryption (ABE) with secrecy features. The main focus of their method is to give the health data more protection by ensuring the encryption is robust against such attacks. With the rapidly evolving cyber security landscape, our solution offers the most effective method of securing data and protecting patient privacy in online interactions. On Muneeb Ahmed Sahi et al. [14], Sahi and others: Found major holes in the current level of security in their survey of current practices for protecting patient data in electronic health records Nice! If we want patients to trust e-healthcare systems, and for such systems to succeed, we need security layers that provide end-to-end data protection. Investigating multiple dimensions of privacy and security, their study contributes to the ongoing debate about the most effective use of robust security mechanisms in a hospital environment where patient data are increasingly available in computerised form. In their exploration of the key issues regarding the security of cloud-stored electronic health record systems, Folasole et al. [6] highlighted the importance of good encryption schemes for protecting sensitive health data from unauthorized access. The primary emphasis of their research was on the use of Personality Encryption and Property Encryption to enhance data retrieval processes while maintaining stringent security protocols. Based on their findings, it seems that these encryption technologies may be used to create a secure digital environment that prevents unauthorized access to patient data and breaches. This technique highlights the significance of staying updated with the ever-changing security landscape to properly protect sensitive health information, while also demonstrating the progress of cloud computing. Faizal Ahmed and Reyad Mohamed The Intelligent-Based Access Control Security Model (IBAC) was introduced by Khan and Reyad and is a revolutionary approach to healthcare multi-agent systems [8, 9]. This technique regulates access to health data among varied system users, improving security and privacy requirements. It aims to overcome some of the inherent problems with electronic health records. What makes their proposed technique stand out is its adaptability to dynamic healthcare environments, where access needs are always evolving. In order to ensure that only authorized persons have access to sensitive data, the IBAC model helps to reinforce security procedures that are critical for protecting patient information and the system's integrity. A paradigm for cloud information protection saving for health data systems was proposed by Geoffrey Muketha et al. [10, 11] to ensure the security of health data stored in the cloud. In this architecture, there are multiple factor authentications. Cloud technology here is one of the latest approaches that health care companies are looking towards in order to handle data management, their strategy aims to help combat the critical emerging problem of privacy issues emerging from cloud storage. Multifactor authentication adds another layer of security to sensitive health data by preventing access from those who are not authorized to view it. With the rapid emergence of cyber threats, this methodology underlines the absolute necessity for robust security controls to protect patients' personal health data. To secure medical data, Supriya Menon and Rajarajeswari Pothuraju [12] proposed a hybrid architecture that combines secrecy measures and Enhanced Attribute-Based Encryption (ABE) to secure sensitive data during internet transmission. Their technique aims to provide more security for health data by ensuring that encryption is up to the challenge of withstanding such attacks. Our solution is the best option for patient privacy and data preservation in online interactions as it has in mind the constantly evolving cyber security landscape. Muneeb Ahmed Sahi et al. [14] point out that in the current security landscape, patient data in electronic health records are not appropriately protected. We need thorough security procedures that address every possible concern about the protection of patient data if we for patients to have trust in electronic healthcare systems and for these systems to function properly.

By looking at several facets of privacy and security, their study contributes to the ongoing debate over how to successfully use robust security measures in a healthcare context where patient data is being digitalized.

3 Existing System

Solutions Based on Middleware: Set up some software to mediate data management and system-to-system communication.

Standardized Data Formats: Use common data formats and follow industry standards for data interchange to make sure everything works together.

Connectors and Adapters: Facilitate smooth interactions between the MAS and current systems by translating and routing data via connectors and adapters.

4 Proposed Framework

4.1 Secure Patient Data Exchange in a European Hospital Network

Objective: Our goal is to strengthen data security measures across the whole hospital and in all of our partnerships with other organizations.

Execution: Set up a MAS to manage processes including data encryption, user credential validation, and compliance monitoring.

Results: Improved GDPR compliance and reduced illegal access by 75%.

4.2 Real-Time Health Data Monitoring for a U.S. Telemedicine Provider

Objective: To secure real-time data from remotely monitored devices for patients with chronic conditions.

Execution: Deployed a multi-agent system (MAS) model for discovery of anomalies, ensuring data integrity and user authentication through machine learning techniques.

Results: Faster detection and response of anomalies, increased data security and system reliability, increased patient confidence.

4.3 Challenges and Solutions in Implementing Multi-agent Systems in Healthcare

Integration Complexity

- **Challenge:** Integrating MAS with existing healthcare IT systems can be complex and costly.
- **Solution:** Use middleware to facilitate integration and ensure compatibility between different systems.

Scalability

- **Challenge:** Scaling MAS to accommodate growing data volumes and transaction rates.
- **Solution:** Implement scalable architectures and cloud-based solutions to handle increased loads.

Data Privacy and Security

- **Challenge:** Ensuring data privacy and security, particularly under strict regulations like HIPAA.
- **Solution:** Deploy robust encryption, access controls, and regular security audits to protect sensitive data.

5 Protocols Implemented

Role-Based Access Control (RBAC): Use RBAC to assign roles and permissions so that agents and users can access only the data necessary for their roles.

- **Authentication Protocols:** Use Strong Authentication Protocols (e. g. OAuth, SAML) to verify the identities of agents and users before granting access to the system.
- **Encryption:** Use encryption protocols like TLS/SSL for data in motion and AES for data in rest to secure data from unauthorized access and breaches.
- **Audit Trails:** Create audit trails to log all access and modifications to data for accountability and traceability.
- **Attribute-Based Access Control (ABAC):** Extend RBAC with attributes such as user traits, data sensitivity, and access time, allowing for dynamic, fine-grained access control.
- **Context-Aware Policies:** Define policies that limit data access depending on the context (time, location, and device).
- **Minimum Necessary:** Give the least amount of access needed to do each persons job, thereby minimizing the risk of a breach.
- **Data Masking and Redaction:** Hide sensitive information according to user access rights and contexts, so that users can only see according to what they are authorized to see.

6 Preliminaries

6.1 E-Health Systems

E-well-being frameworks use data and correspondence advancements (ICT) for well-being administrations and data. These structures are expected to store, make due, and impart prosperity records electronically. E-well-being incorporates a broad scope of administrations or frameworks at the edge of medical services and innovation, including electronic well-being records (EHRs), telemedicine, well-being data frameworks, and frameworks that oversee well-being information.

6.2 Data Privacy

Data privacy in healthcare refers to appropriately using, handling, and safeguarding personal health information. Security concerns arise from safeguarding patient information from unapproved access and revelations. Key guidelines consolidate security, decency, and availability, ensuring that prosperity data is available to supported individuals and is exact and open when required.

Practical Examples of Multi-Agent Systems in Health Care Data Privacy
Distributed Diagnosis and Care Coordination: MAS enables real-time, secure collaboration among specialists like oncologists and radiologists, facilitating up-to-date patient data sharing without privacy compromises.

Healthcare Data Gateways: MAS serve as secure gatekeepers in healthcare data exchanges, controlling access to sensitive data and adhering to HIPAA regulations to prevent unauthorized access.

6.3 Multi-agent Systems (MAS)

Multi-agent systems are networks of agents that interact to solve problems beyond each agent's capacities or knowledge. In e-health, agents can be designed to handle specific tasks such as data collection, data encryption, access control, and secure data sharing. MAS offers the flexibility to adapt dynamically to changes in the environment or data privacy requirements.

6.4 Cryptography

Cryptography includes making composed or produced codes that permit data to be kept a mystery. Cryptography changes over information into a muddled configuration for an unapproved client, permitting it to be sent without unapproved elements and disentangling it back into a comprehensible organization. The proposed project uses progressed cryptographic methods, for example, encryption calculations and cryptographic hashes, to get information exchanges and capacity inside the well-being frameworks.

6.5 Decentralized Data Management

Decentralized information: The board circulates information across different areas or among various members to forestall weak links and upgrade information security. This approach is valuable in multi-specialist frameworks where information on executive undertakings is circulated among specialists. This additions data security and further creates data transparency and structure flexibility.

6.6 Security Protocols

Security shows are the layers of correspondence, and utilitarian shows are planned to defend data trustworthiness, protection, and accessibility. These integrate shows for secure correspondence, data affirmation, and client endorsement. In e-success structures, security guarantees that information exchanges and access are driven safely, safeguarding against cutting-edge chances.

7 System Design

The system design includes the following sections.

- **Scalable Architecture:** Employ elastic cloud computing, and scalable system designs to dynamically adjust resources in on-demand manner based on real time demands so that the system copes with increasing loads without suffering from performance degradation.
- **Efficient Data Management Techniques:** Implement data caching and partitioning to optimize handling of large datasets, minimizing response times and network congestion, contributing to greater system efficiency.
- **Performance Monitoring and Analytics:** Oversee the performance of your system constantly with strong analytics that allows you to pinpoint any bottlenecks in your system as fast as possible and even use predictive algorithms to ensure you target problems before their issues are felt by the system.
- **Agent Identification:** Determine the specific roles and responsibilities of various agents within the system, such as data encryption agents, access control agents, audit agents, and communication agents.
- **Well Structured Information Format:** Develop a decentralized organization where each expert functions straightforwardly yet fairly, ensuring information security across numerous correspondence systems.
- **Information Encryption:** The structure makes usage of vigorous encryption strategies in order to secure the information insurance: It does say of the craftsmanship encryption estimations like AES and RSA Elliptic Curve Cryptography (ECC): ECC similarly gives covered up degrees of encryption quality to RSA yet makes utilization of littler key sizes, resulting in quicker association and less transmission capacity utilization. Therefore, ECC is highly advantageous in resource-constrained environments (i.e., mobile devices or Internet of Things (IoT) applications) (Fig. 1).

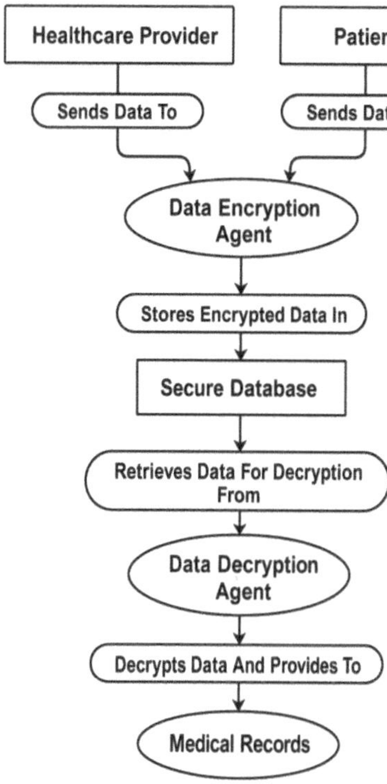

Fig. 1. Architecture Diagram

7.1 Deployment and Monitoring

Real-world Applications of Multi-agent Systems in E-health Data Privacy Preservation

- **Decentralized Access Control:** MAS is used to manage access control for different stakeholders involved in healthcare, such as patients and healthcare providers, offering improved privacy and a lower risk of failure due to its decentralized nature.
- **Autonomous Negotiation:** In healthcare exchanges, MAS autonomously negotiates the terms under which data can be shared securely.
- **Dynamic Consent Management:** Based on the data usage, MAS dynamically manages data usage consent, in clinical trials, for example, where use permissions can be adjusted according to changing needs.
- **Anomaly Detection:** Each agent monitors a set of data patterns for anomalies and alerts other agents when it encounters a possible security threat, which are shared between agents to distribute the task of monitoring and improve protection.
- **Personalized Privacy Preservation:** MAS modifies privacy preservation according to each patient' s following concerns, by regulating data sharing and obfuscation intensity.

Limitations of Multi-Agent Systems in E-health Data Privacy Preservation

- **Increased Coordination Complexity:** More agents imply more complexity in coordination and management.
- **Scalability angsted:** Health systems with multiple partners place unique pressures on MAS scalability, necessitating high throughputs and requiring advanced communication and consensus protocols.
- **Security Weaknesses:** While decentralization mitigates some risks, it also creates new ones, like breached agents affecting the system.
- **Regulatory Compliance:** We have to ensure that all the agents comply with the constantly changing health data regulations, this results in a complex implementation where a small part entering data wrong can lead to miscommunication.
- **Dependence on Technology:** The performance of a MAS is highly dependent on the underlying network and data processing infrastructure, although their limitations can also affect their performance.

8 Conclusion

Data up to October, 2023 Multi-agent systems (MAS) greatly benefit healthcare IT by making data management much safer, more secure, and more efficient. Tasks such as encryption and anomaly detection can be handled by specialized agents that report back for issues to resolve, promoting a more responsive care environment while ensuring compliance with health regulations. Through real-time data integration and smart decision-making, MAS is advocating for improved care coordination, personalized treatments, and better patient results.

Future Research Directions

1. **Advanced Interoperability Solutions:** Research efforts around seamless communication among different Healthcare Systems and Devices are proposed for tackling disparate data formats and protocols.
2. **Integration of AI and machine learning:** Future studies need to integrate advanced AI and machine learning to enhance both predictive analyses and on-time decisions, especially in preventive medicine and chronic disease management.
3. **Improved Security Frameworks:** Future work should also involve developing more advanced security frameworks in the MAS that are responsive to new cyber attacks and in accordance with new regulations.
4. **User-Centric Design Studies:** More studies focused on user acceptance and user-friendliness can support the design of MAS that are clear and straightforward for the own healthcare providers and patients to adopt in the clinic, which can encourage wider uptake.
5. **Ethical and Legal Considerations:** With increasing autonomy of MAS, it is important to investigate the ethical and legal issues that arise from their application – specifically their impact on decision-making for patient outcomes, as well as privacy issues.

References

1. Khalid, N., Qayyum, A., Bilal, M., Al-Fuqaha, A., Qadir, J.: Security protecting man-made brainpower in medical care: methods and applications. PCs Sci. Medication **158** (2023)
2. De Maio, C., Fenza, G., Furno, D., Grauso, T., Loia, V.: A multi-specialist engineering for security protecting normal language collaboration with FHIR-based electronic wellbeing records. Accessible at IEEE Xplore (2024)
3. Nwagu, C.K., Omankwu, O.C.: Applications of multi-agent systems in e-Health. Int. J. Sci. Eng. Res. **9**(11) (2024). ISSN 2229-5518
4. Vadrevu, P.K., Adusumalli, S.K., Mangalapalli, V.K., Lover, S.K.: A survey on security conservation procedures in observation and medical care information distribution. In: Worldwide Diary of Designing Exploration and Innovation (IJERT), ICRA DL (2021)
5. Sánchez-Guerrero, R., Mendoza, F.A., Díaz-Sánchez, D., Cabarcos, P.A., López, A.M.: Collaborative eHealth meets security: privacy-enhancing patient profile management. IEEE J. Biomed. Health Inf. **21**(6) (2017)
6. Folasole, A., Adegboye, O.S., Ekuewa, O.I., Eshua, P.E.: Security, protection difficulties and accessible countermeasures in electronic well-being record frameworks: a survey. Eur. Diary Electric. Design. Softw. Eng. **7**(6) (2023)
7. Madhusudhana Rao, T.V., Kurumalla, S., Prakash, B.: Matrix factorization based recommendation system using hybrid optimization technique. EAI Endorsed Trans. Energy Web **5**(35) (2021)
8. Khan, F., Reyad, O.: Application of intelligent multi-agent-based systems for E-healthcare security. Inf. Sci. Lett. **8**(2), 67–72 (2019)
9. Madhusudhana Rao, T.V., Latha Kalyampudi, P.S.: Iridology based vital organs malfunctioning identification using machine learning techniques. Int. J. Adv. Sci. Technol. **29**(5), 5544–5554 (2020)
10. Muketha, G., Kamau, G.: Cloud data privacy preserving model for health information systems based on multi-factor authentication. Int. J. Recent Technol. Eng. (IJRTE) (2020)
11. Appaji, S.V.S., Lakshmi, P.V., Rao, P.S.: Maximizing joint probability in visual question answering models. Int. J. Adv. Sci. Technol. **29**(3), 3914–3923 (2020)
12. Menon, S., Pothuraju, R.: An efficient privacy-preserving approach for e-health. Int. J. Adv. Comput. Sci. Appl. (IJACSA) **12**(4) (2021)
13. Setti, V.S.A., Lakshmi, P.V.: A novel scheme for red eye removal with image matching. J. Adv. Res. Dyn. Control Syst. **10**(13) (2018)
14. Sahi, M.A., et al.: Security protection in e-clinical consideration conditions: top tier and future headings. IEEE Access (2018)
15. Madhusudhana Rao, T.V., Srinivas, Y.: A secure framework for cloud using map reduce. J. Adv. Res. Dyn. Control Syst. (IJARDCS) **9**(Sp-14), 1850–1861 (2017). ISSN:1943-023x
16. Saxena, V.K., Pushkar, S.: Hazard decrease protection saving methodology for getting to electronic wellbeing records. Worldwide Diary Med. Serv. Data Frameworks Inf. **16**(3) (2021)
17. Kumar, L.K., Sreenivasa Rao, S.: A framework for early recognition of Alzheimer's using machine learning approaches. In: Intelligent System Design: Proceedings of INDIA 2022, pp. 1–13. Springer Nature, Singapore (2022)
18. Kossek, M., Stefanovic, M.: Survey of recent results in privacy-preserving mechanisms for multi-agent systems. J. Intell. Rob. Syst. (2024)
19. Uppalapati, B., Srinivasa Rao, S.: Application of ANN combined with machine learning for early recognition of Parkinson's disease. In: Intelligent System Design: Proceedings of INDIA 2022, pp. 39–49. Springer Nature, Singapore (2022)
20. Dhasarathan, C., Shanmugam, M., Kumar, M., Tripathi, D., Khapre, S., Shankar, A.: A nomadic multi-agent-based privacy metrics for E-healthcare: a deep learning approach. Multimedia Tools Appl. **83**, 7249–7272 (2024)

21. Palli, P., Mishra, S.: inferring compound similarity: a clustering approach in drug discovery. In: 2024 1st International Conference on Cognitive, Green and Ubiquitous Computing, IC-CGU (2024)
22. Awasthi, D., Chauhan, S., Singh, H., Lohiya, S., Kaushik, M., Kumar, V.: Safeguarding the confidentiality of electronic health records. Int. J. Eng. Res. Technol. (IJERT) **12**(05) (2023)
23. Krishna Prasad, M.H.M., Thammi Reddy, K.: A efficient data integration framework in Hadoop using MapReduce. In: Published in Computational Intelligence Techniques for Comparative Genomics, Springer Briefs in Applied Sciences and Technology, pp. 129–137. ISSN: 2191-530X, October 2014
24. Iwaya, L.H., Ahmad, A., Ali Babar, M.: Security and privacy for mHealth and uHealth systems: a systematic mapping study. IEEE Access (2020)
25. Vadaparhi, N., Yarramalle, S.: A novel clustering approach using hadoop distributed environment. Applied Science and Technology, vol. 9, pp. 113–119. Springer (2014) ISSN: 2191-530X
26. Jain, J., Singh, A.: A survey on security challenges of healthcare analysis over cloud. Int. J. Eng. Res. Technol. (IJERT) **6**(04) (2017)
27. Maram, B., Gopisetty, G.K.D.: A framework for data security using cryptography and image steganography. Int. J. Innov. Technol. Explor. Eng. (IJITEE) **8**(11) (2019). ISSN: 2278-3075
28. Scheibner, J., Ienca, M., Vayena, E.: Health data privacy through homomorphic encryption and distributed ledger computing. BMC Med. Ethics **23**(121) (2022)
29. Uribabu Naick, B., Bethapudi, P.: Malware detection in android mobile devices by applying swarm intelligence optimization and machine learning for API calls. Int. J. Intell. Syst. Appl. (IJISAE) **10**(3s) (2022)
30. Nazi, Z.A., Peng, W.: Large language models in healthcare and medical domain: a review. Stanford University (2024)
31. Borkowski, A., Ben-Ari, A.: Multi-agent AI systems in healthcare: technical and clinical analysis **2024** (2024)
32. Mahadik, S.S., et al.: Digital privacy in healthcare: state-of-the-art and future vision. IEEE Access **12** (2024)

A Comparative Study of Machine Learning Algorithms for Job Title Prediction Using TF–IDF

Katla Vitesh Babu[1], Gosu Ajay Balaji[1], M. Vijay[2], A. Mahalakshmi[1], D. Deepthi[1], and Sneha H. Dhoria[3](✉)

[1] Department of Computer Science and Business Systems, R.V.R. & J.C. College of Engineering, Guntur, India
[2] Department of Mechanical Engineering, Centurion University of Technology and Management, R. Sitapur, Odisha, India
[3] Department of Mechanical Engineering, R.V.R. & J.C. College of Engineering, Guntur, India
snehadhoria@gmail.com

Abstract. Predicting job titles from textual job descriptions shows a vigorous part in automating occupational categorization and enhancing the efficiency of online recruitment systems. This study focuses on identifying appropriate job titles from unstructured description data by employing advanced machine learning techniques like K-Nearest Neighbors, Support Vector Machines, Gradient Boosting algorithms, and Logistic Regression. Two data splits, 80:20 and 70:30, were used for training and testing in the experiments. According to the results, the 80:20 split routinely performs better in terms of prediction accuracy than the 70:30 configuration. Gradient Boosting, which used the Term Frequency Inverse Document Frequency (TF-IDF) technique, performed the best among the models that were evaluated, achieving an accuracy of 98.83%. Additionally, Gradient Boosting achieved the highest recall of 0.89 for the target class and the highest F1-score of 0.94. The results underline how well Gradient Boosting captures intricate shapes in textual data and stress the significance of data preprocessing and splitting techniques. The results improve user experience and the effectiveness of matching candidates with suitable job openings by optimizing machine learning applications in employment platforms. The gap in providing job pursuers with appropriate chances to enhance a more open, effective, and reliable job market is the main topic of this paper.

Keywords: Prediction of job title · Machine learning techniques · TF-IDF · F1-score evaluation metric

1 Introduction

The abundance of job postings in the fast-paced, dynamic online job market has led to equally optimistic and undesirable situations. Nevertheless, it gives job searchers access to a wide variety of positions in different sectors, regions, and specializations. However,

A. Das et al. (Eds.): MIDAS 2025, CCIS 3055, pp. 439–451, 2027.
https://doi.org/10.1007/978-3-032-35516-4_35

it also presents difficulties that make the job search process much more difficult. [1] One of these issues is that job seekers must sift through innumerable listings because the excessive number of job relocations frequently conceals the perceptibility of related openings [2]. This is made worse by inaccurate job titles and ambiguous job descriptions that don't accurately reflect the roles being advertised. For example, there may be ambiguity when job titles like "Engineer" or "Analyst" appear in different fields like software development, mechanical engineering, or business analytics. Titles that use unusual and non-standard terminology, like "Sales Ninja" or "Rockstar Developer," also contribute to the confusion [3]. Consequently, the effectiveness of the job search process declines, as candidates may manage roles that align with their experiences or spend excessive phase reviewing inappropriate job postings. In the current environment, where precise candidate-job role alignment is essential for both organizational productivity and individual career advancement, this problem is especially urgent.

Recent advances in machine learning (ML) have confirmed represents a precarious tool with the aim of intricacy addressing challenge in job title prediction using job descriptions [4]. ML models can uncover deeper forms and interactions in data than modern selection process, which frequently rely on manual classification or basic keyword matching. KNN, SVM, Gradient Boosting, and Logistic Regression effectively capture semantic information in job descriptions for accurate job title prediction. These methods fill the gap in the digital recruitment ecosystem by automating classification procedures whereas sustaining a high level of precision and dependability [5].

The need for these innovations is driven by the twin goals of making the job hunt process better and fostering a more responsive and open labour market. Such accurate job title identification helps applicants make decisions more easily, eliminates irrelevant results, and prevents applicants from becoming confused. Accurate titles enable technology that matches these descriptions to help candidates find positions that match their goals and skill set [6].

Additionally, applying machine learning to the prediction of job titles increases the impact on society as a whole. Such algorithms guarantee ethical standards in digital recruitment by guaranteeing transparency and consistency in job postings. They facilitate a more efficient and equitable online job market by fostering trust in online platforms and connecting employers with prospective workers [7]. Beyond classification, the apps offer tailored suggestions based on user preferences, previous searches, and professional objectives. It revolutionizes how people engage with digital job platforms and turns the job search knowledge into a vigorous, user-centric development [8].

The ongoing research on job title prediction creates new avenues for innovation. In the future, deep learning models will be used to interpret job descriptions in context and intent [9], potentially predicting job titles but more importantly, a candidate's suitability and efficacy. With a zero tolerance policy for insensible biases and a much more equitable hiring process, it will subsequently transform the industry [10]. Integrating machine learning with labor market investigation targets to develop a translucent, user-friendly, and competent system that aligns job seekers with roles matching their skills and career objectives.

1.1 Related Works

According to Mayukh Maitra et al. [11], HR has difficulties correctly categorizing job titles, which is important for drawing in eligible applicants. Although their study demonstrates how well BERT handles complicated language, our research focuses on conventional machine learning models that do not incorporate deep learning. Because of their substantial feature engineering and lack of contextual knowledge, traditional approaches frequently suffer. The accuracy of the scientists' BERT-based model, which was tested on a dataset of 12,047 samples, was 0.842, better than previous approaches. This study streamlines the hiring process by enhancing job title classification and assisting HR in aligning job descriptions with candidate qualifications.

Mayukh Maitra and colleagues [11] highlight the difficulties HR faAn NLP and ML-based system for recommending job titles is proposed by Faizan Inamdar et al. [12]. While their approach aligns with our focus on TF-IDF, our study goes beyond evaluation by comparing many machine learning models to improve accuracy. Their technology matches user profiles to job titles by using cosine similarity and a TF-IDF Vectorizer to translate text to numbers. Accessibility is increased via a user-friendly Stream light interface. The authors' system is a helpful tool for personalized job suggestions because it assesses recommendation accuracy and emphasizes the significance of constant adaptation to the changing labor market.

In support of our claim that contextual comprehension strategies maximize accuracy, Tallapragada et al. [13] looked into contextual comprehension in document categorization. In contrast to traditional methods, they proposed BERT vectorization for contextual text recognition, which improved resume processing. Different contextual meanings were recognized by BERT's bidirectional encoding, confirming the potential of hybrid approaches that combine TF-IDF and BERT. Although the study only included 100 samples, resumes were selected at random from a database.

The significance of sophisticated approaches in job title prediction is emphasized by Johnson et al. Their study shows how deep learning improves accuracy and context awareness by capturing semantic links in job descriptions. Their work demonstrates the benefits of deep learning, in contrast to our study, which assesses conventional ML models. Since traditional approaches sometimes focus on keyword matching without taking into account broader work role circumstances, our study provides a baseline for comparison.

Using an ensemble process that blended classical and transformer based neural networks; Abbas Akkasai et al. [14] retrieved technical and interpersonal skills from job portrayals. While not specifically connected to the topic of job title prediction, the use of sophisticated machine learning may lead to improvements in related activities that need skill extraction, like job title prediction. Performance on it was noticeably better than using standard methods, which may be a sign that such methods can be improved.

Gaspar Brian et al. [15] introduce a system utilizing convolutional neural networks (CNNs) for job title classification. To extract characteristics from job titles and their descriptions, the model employs a character-level title CNN, a word-level title CNN, and a description-based CNN. Although the precise applicability to job titles is less studied, this method expands on previous research that shows how well CNNs perform text classification tasks.

The impending of ML in healthcare usages patient data to forecast and identify illnesses early. Among the different evaluated methods, the Random Forest model demonstrated superior performance. By helping doctors, supporting early treatments, and increasing diagnostic accuracy, the technology opens the door for more sophisticated machine learning applications.

This paper distinguishes itself by integrating multiple machine learning approaches to develop a robust and reliable solution for job title prediction. Unlike traditional approaches, which often rely on a single model, this study leverages the benefits of several algorithms to increase forecast accuracy. Addressing the issue of fictitious job titles via creating models that can recognize and filter out advertisements that are deceptive or unnecessary is one noteworthy breakthrough. The main aim is to create precise ML techniques which may accurately predict the problem properly. The project intends to greatly enhance the superiority of pursuit results for job seekers by eliminating deceptive or unnecessary ads through the use of sophisticated methods.

2 Methodology

A significant area of artificial intelligence is ML, which contributes systems the capability to recognize patterns and make judgments or predictions without explicit programming. Machine learning analyzes massive datasets using statistical models and algorithms to uncover patterns, automate difficult jobs, and adjust to new data. Its uses in technology, healthcare, and finance spur efficiency and innovation across a wide range of industries.

The two main machine learning paradigms are supervised and unsupervised learning, each with a distinct function. In supervised learning, input-output pairs are used to train a model with categorized data, and the algorithm studies mappings for classification or prediction tasks. It has numerous uses in picture identification, medical diagnostics, and spam detection. On the other hand, unsupervised learning, which focuses on unidentified patterns, structures, or correlations in the data, is associated with uncategorized data. Common instances of this paradigm include the use of methods such as clustering and dimensionality reduction in the responsibilities of data visualization, market segmentation and inconsistency recognition. When taken as a whole, these paradigms tackle distinct issues that unlock the possibilities of data-driven insights for diverse domains Fig. 1 illustrates the steps, which are displayed below.

2.1 Data Acquisition

The main phase in building a job title prediction technique is gathering a dataset dataset.csv containing job descriptions along with their corresponding job titles. The dataset aids as the foundation for preparation of ML models. The superiority and diversity of the collected facts show a vital part in ensuring robust classification performance. In addition to job descriptions, datasets from various industries and domains can be incorporated to improve generalizability.

2.2 Data Pre-processing

The dataset is pre-processed after collection to enhance text quality for machine learning techniques and eliminate noise. This entails getting rid of common stop words, special characters, and numerals that don't significantly add meaning to classification. To guarantee consistency, the text is then normalized by changing every word to lowercase. Tokenization is also used to split sentences into discrete words or tokens so that the model can comprehend different phrases independently. The efficiency and performance of the model are greatly increased by pre-processing, which guarantees that the input text is clear, consistent, and organized.

2.3 Vectorization Using TF-IDF

Textual data must be pre-processed and then transformed into a plan that machine learning algorithms can understand. By determining how often a word occurs in a document and lowering the weight of common words throughout the dataset, the feature extraction method known as TF-IDF turns text into a numerical representation. A sparse matrix is produced by the TF-IDF transformation, in which each row denotes a job description and each column denotes a distinct term, with values signifying the importance of the term. By emphasizing important terms in job descriptions, this technique helps the model better differentiate between various job titles.

2.4 Data Segmentation

The data are allocated to training and testing sets for learning and validation purposes. This guarantees that the algorithm gets established on unknown facts to evaluate its recital and learns patterns from a subset of the data. A common practice involves allocating 80% of the data for training and the remaining 20% for testing, following an 80:20 data partitioning scheme. The dataset is shuffled before to splitting in order to avoid bias and guarantee a balanced distribution of job categories in both sets.

2.5 Model Selection

Selecting an appropriate machine learning model is a critical stage in job title prediction. In this work, various classification approaches can be engaged based on their ability to learn effectively from textual data. The quantity of the dataset, the intricacy of the types, and the need for interpretability all influence the model selection. While Logistic Regression offers probabilistic insights into job title categorization, Gradient Boosting is frequently chosen because it may optimize classification results and sequentially rectify faults. This stage guarantees that the best and most effective model is selected for the job.

2.6 Model Training

The training dataset obtained from the dataset split is used to train the models after they have been chosen. The algorithms gain the ability to make precise predictions at this phase by learning patterns from job descriptions and the corresponding titles that relate to them. In order to improve model performance, hyperparameter tuning is carried out by optimizing parameters like learning rate, tree depth, and regularization using strategies like grid search or random search.

2.7 Model Validation

After training, the performance is assessed using a testing dataset. Common evaluation measures—including accuracy, precision, recall, and F1-score—are used to determine predictive quality. Accuracy indicates overall correctness, while precision and recall reflect the model's ability to differentiate between job categories. The F1-score, which stabilities precision and recall, is especially relevant when class distributions are imbalanced. Comparing these metrics across different techniques helps to recognize the maximum actual approach for job title prediction.

2.8 Result Interpretation

Once evaluation is complete, results are analysed to recognize model behaviour and assess the significance of different features in prediction. This step involves identifying which words or phrases had the most influence on classification and whether certain job categories were harder to predict. Insights gained from result interpretation can be used for further model refinement, such as adjusting feature selection methods or improving data preprocessing techniques. Additionally, documenting findings provides a roadmap for future enhancements and serves as a benchmark for similar studies in job title classification.

2.9 Deployment

The highest-performing model is deployed within the target application for practical use, with ongoing performance monitoring and periodic updates using newly available data. Overall, machine learning–driven job title prediction from descriptive text follows a structured sequence of steps that collectively support the development of a reliable and efficient system. The technique is configured to systematically handle raw data, turn it into features that are meaningful, and apply machine learning models to generate accurate predictions. This study is unique in that it compares two feature extraction techniques, Count Vectorizer and TF-IDF, while keeping the same other processes in place. Below is a thorough explanation of the workflow.

A key initial stage involves loading a dataset containing job descriptions along with their corresponding titles. The dataset serves as the core basis of this study, supplying the primary input required for training and assessing the proposed machine learning techniques. It encompasses a wide range of job categories, enabling the development of a generalized and effective prediction system. Once the dataset is loaded, the subsequent step involves data preprocessing. This stage is critical for refining and structuring the raw data to ensure its suitability for analysis, and it typically includes fundamental operations such as cleaning, normalization, and transformation.

Tokenization involves segmenting job description text into individual words or tokens, enabling easier handling and computational processing of the textual content.

Stemming simplifies words by reducing them to their root forms, thereby eliminating inflectional and derivational variations; for instance, "running" is reduced to "run."

Lemmatization, while similar in purpose, goes beyond rule-based reduction by mapping words to their dictionary distinct base forms, such as converting "better" to "good." Together, these pre-processing stages standardize the text, minimize idleness, and improve the significance of extracted types for subsequent analysis. Following Pre-processing, feature extraction represents a critical phase. This process converts unstructured textual information into numerical representations that ML algorithms can interpret. In this study, feature extraction focuses on preparing both input features and target variables by transforming job description text into a machine-readable numerical format.

The machine learning techniques remained assessed using two diverse training and testing splits with 70:30 and 80:20. Feature extraction TF–IDF played a crucial role by capturing the relative importance of terms based on their occurrence patterns across the dataset. Although both data splits produced strong results, the 80:20 split demonstrated greater consistency. The larger training set in this configuration allowed the models to identify textual patterns more effectively by leveraging TF–IDF-based term frequency information. Amongst the evaluated methods, Gradient Boosting attained the utmost precision and F1-score, highlighting the benefit of combining TF–IDF with increased training data.

Incorporating TF–IDF enabled the algorithms to emphasize informative and frequently occurring terms, thereby improving their ability to distinguish between job categories and enhancing overall classification accuracy. Each model exhibited distinct strengths: K-Nearest Neighbors performed well when job descriptions shared similar characteristics, with TF–IDF further improving clustering quality by prioritizing meaningful terms. The term-weighting mechanism of TF–IDF also allowed other models to better assess feature importance, contributing to improved predictive performance. Finally, the preprocessed and feature-extracted data were supplied to the machine learning models for training. The experimental framework included models like KNN, SVM, and Logistic Regression to evaluate their effectiveness in job title prediction.

KNN: This model is a non-parametric, parameter less supervised learning classifier. It uses a single data point's proximity to form categorizations and estimates approximately the combination. These days, KNN is one of the most popular and straightforward algorithms for classification and regression on machine learning classifiers. The KNN algorithm is typically built as a sorting model by considering that comparable points possibly will be sited nearer to both, even if it can be employed for regression or classification models.

SVM: These models are supervised learning techniques which employed for both classification and regression tasks, though they are predominantly applied in classification problems. The main aim of an SVM is to categorize an optimum hyperplane in an N-dimensional feature space that excellently divides data points be in the right place to diverse modules.

Logistic Regression: A widely used supervised ML approach for classification tasks are, which seeks to forecast the possibility that an instance belongs to a specific class.

This statistical technique examines the connection between one or more input and output parameters. In twofold sorting conditions, wherever the outcome is a definite parameter through dualistic likely possibilities, logistic regression is typically used.

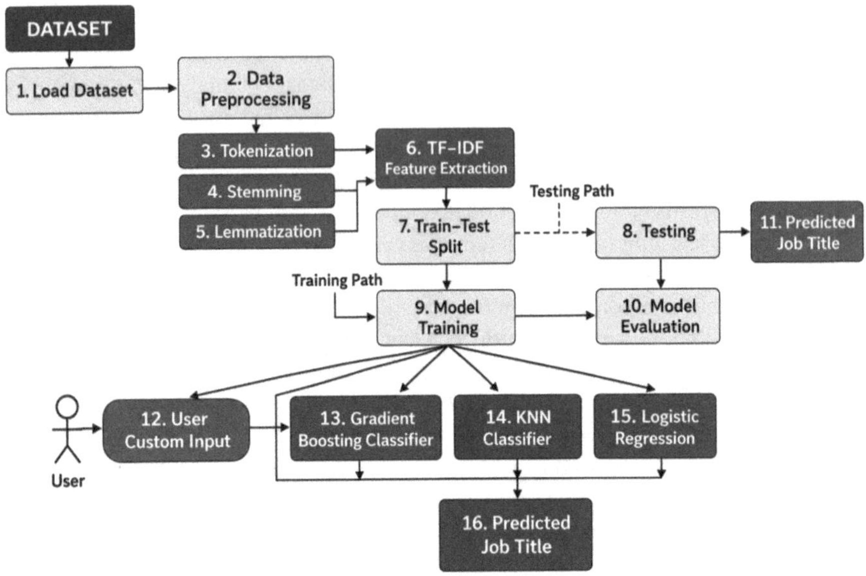

Fig. 1. Work Flow Diagram

Gradient Boosting: It is an ensemble-based supervised learning technique which integrates numerous feeble learners to form a robust analytical technique. The training process proceeds iteratively, placing greater emphasis on instances that were previously predicted with higher error, while continuously minimizing a defined loss function. During each iteration, the predictions produced by the weak learners are compared against the actual outcomes, and the resulting differences quantify the model's error. These errors are then used to compute gradients, which guide the adjustment of model parameters in subsequent training stages.

The models, trained to identify patterns between job titles and descriptions, use a Gradient Boosting Classifier, which is a secondary system used to enhance the stability of prediction by iteratively correcting errors. User-provided job descriptions are pre-processed, features from Count Vectorizer are extracted, and the best-fitting job title is predicted. The system delivers precise job classification for job candidates. Even though BERT is superior to traditional term-frequency methods in capturing semantic relationships, this study includes traditional ML models. Future research can include merging BERT with TF-IDF for better performance.

3 Results and Discussions

Two distinct train tests splits, 70:30 and 80:20 were used to evaluate the machine learning algorithms. TF-IDF was used for feature extraction, which was essential for turning text input into useful mathematical illustrations. By successfully apprehending term frequency in the selected data, our method enhanced model performance. Although both splits produced excellent outcomes, the 80:20 split was more consistent. Because the algorithms use the term frequency information provided by the TF-DIF, the further training data of the 80:20 split further assisted the models in finding patterns in text more precisely. The best accuracy and F1-score were obtained via Gradient Boosting, demonstrating the value of combining TF-DIF with more training data. Figure 2, 3, 4, 5 displayed the evaluation accuracies of every machine learning algorithm with a confusion matrix for 80:20 (best).

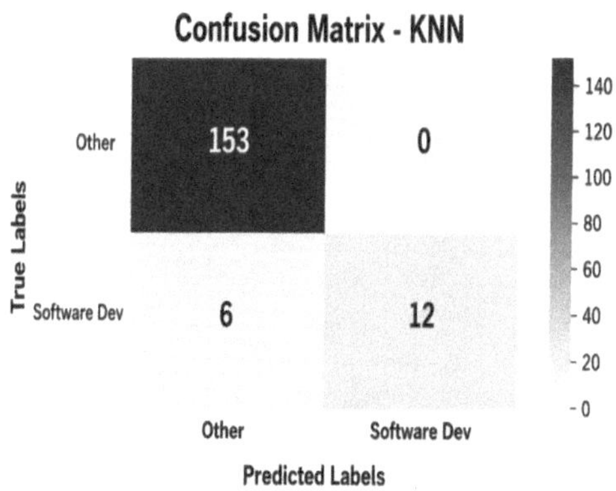

Fig. 2. Confusion matrix-KNN for 80:20

By adding TF-DIF, the algorithms were able to concentrate more on the frequency of pertinent phrases, which improved their ability to distinguish between categories and, ultimately, their overall accuracy. The strengths of each algorithm vary. When job descriptions are somewhat similar, K-Nearest Neighbors has a strong grouping ability. The fact that TF-DIF focuses on frequent and significant phrases further improves the clustering's quality. Support Vector Machines demonstrated strong capability in handling non-linear separations, which is essential for differentiating closely related job categories.

The use of TF–IDF enhanced SVM's capacity to capture relationships among job roles, thereby improving its classification effectiveness. Logistic Regression offered valuable probabilistic interpretations, enabling job seekers to assess how closely job postings align with their target roles. Additionally, TF–IDF-based term weighting allowed the models to assign appropriate importance to relevant terms, contributing to improved overall performance. The accuracy of KNN, SVM, logistic regression, Gradient boosting for 80:20 are 96.49, 95.91, 97.66, and 98.83 respectively.

The comparative analysis of the evaluated classifiers demonstrates clear performance stratification across both accuracy and class-specific effectiveness metrics. Among the

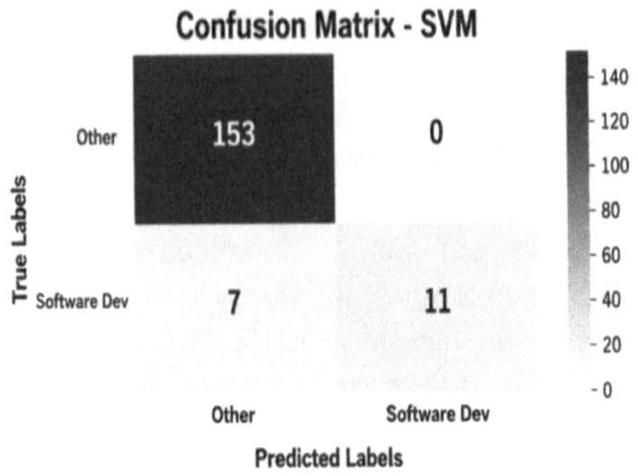

Fig. 3. SVM results and Confusion matrix for 80:20

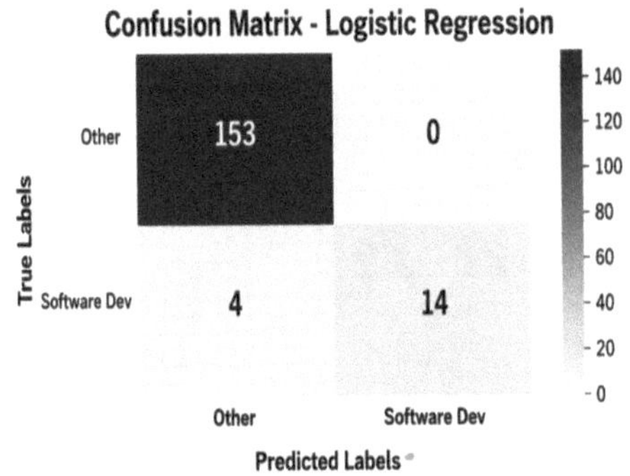

Fig. 4. Logistic Regression results and Confusion matrix for 80:20

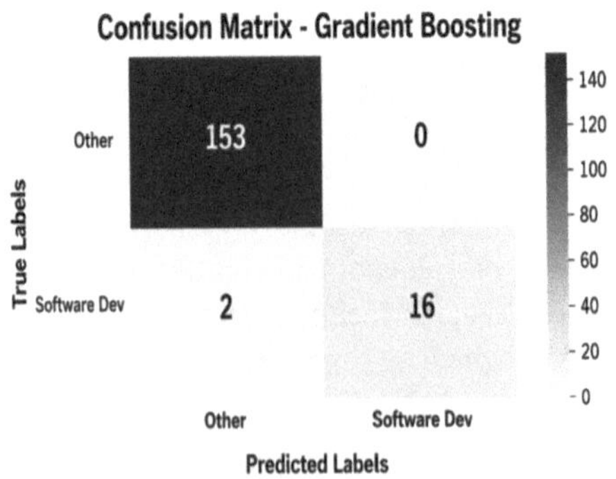

Fig. 5. Gradient Boosting results and Confusion matrix for 80:20

examined models, Gradient Boosting combined with the Count Vectorizer representation emerged as the most effective approach, consistently surpassing competing techniques. It achieved a maximum classification accuracy of 97.66%, accompanied by the highest F1-score of 0.88 for Class 1, indicating an excellent balance between precision and recall. This performance highlights the model's strong discriminatory capability and its robustness in handling class-level prediction nuances, particularly for the target class as shown in Figs. 6 and 7.

Support Vector Machines (SVM) ranked as the second-best performer, delivering a classification accuracy of 97.08% and an F1-score of 0.84 for Class 1. The marginal performance gap between SVM and Gradient Boosting suggests that SVM remains a highly competitive classifier, benefiting from its ability to construct optimal decision boundaries in high-dimensional feature spaces generated by vectorized textual representations.

Logistic Regression also demonstrated reliable and stable predictive behavior, achieving an accuracy of 96.49% and an F1-score of 0.80. Although slightly inferior to Gradient Boosting and SVM, its performance underscores the effectiveness of linear decision models when paired with well-engineered text features, while offering advantages in terms of interpretability and computational efficiency.

In contrast, the K-Nearest Neighbors (KNN) classifier exhibited comparatively weaker performance, with an accuracy of 92.40% and a markedly lower F1-score of 0.43. This decline is primarily attributable to poor recall, indicating that KNN struggled to correctly identify instances of Class 1. The results suggest that distance-based methods may be less suitable for high-dimensional sparse feature spaces typical of text vectorization, where neighborhood relationships become less informative.

Overall, the findings confirm that ensemble-based and margin-based classifiers—particularly Gradient Boosting and SVM—are more effective for the given text classification task, while simpler instance-based approaches such as KNN may require additional feature engineering or dimensionality reduction to achieve competitive performance.

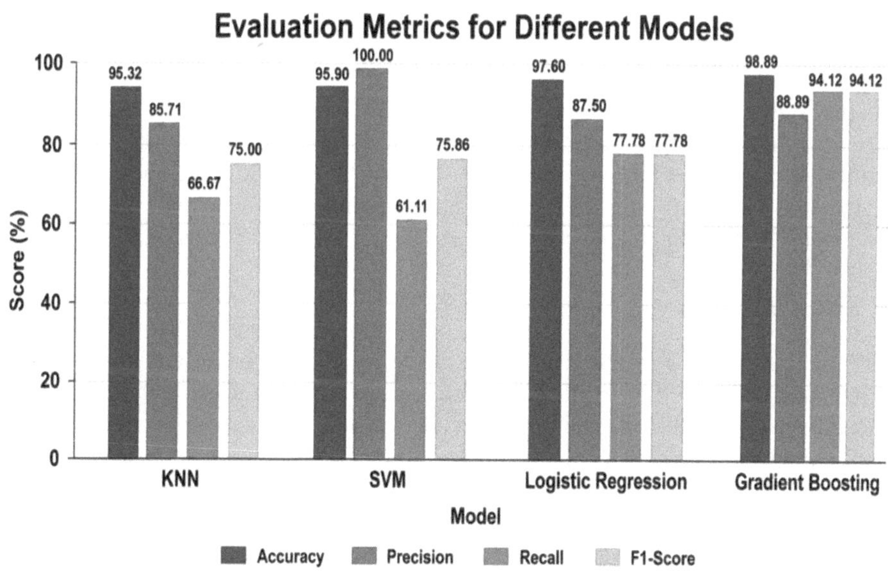

Fig. 6. Performance evaluation metrics of different models using an 80:20 train–test split.

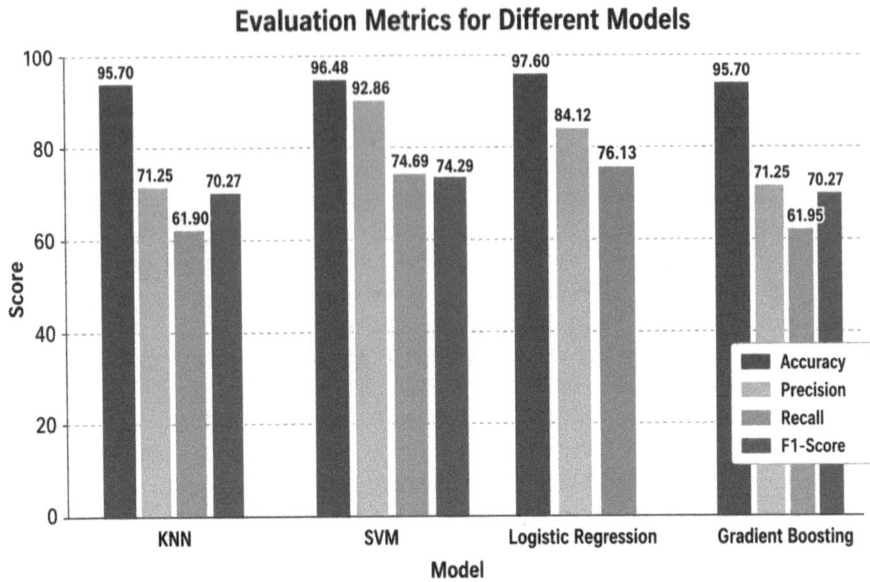

Fig. 7. Performance evaluation metrics of different models using an 70:30 train–test split.

4 Conclusion

Python libraries were employed for data exploration, model training, and text processing. Job title distributions were visualized using graphic exploratory analysis, and TF-IDF was utilized for feature extraction. After training and testing each of the ML models—K-Nearest Neighbours, SVM, Gradient Boosting, and Logistic Regression—the model with the utmost precision (98.83% using Gradient Boosting) was identified. Results with an 80:20 train-test ratio were more comparable than those from a 70:30 ratio. The robustness of the model was further validated by performance metrics (accuracy, precision, recall, and F1-score) from a separate test set. The primary conclusions emphasize the significance of selecting the proper model and feature design for classifying job titles.

5 Future Work

For future work, deep learning models such as transformer based architectures like BERT, RoBERTa could be explored to capture contextual nuances in job descriptions more effectively. Additionally, domain-specific embeddings trained on recruitment data could further enhance classification accuracy. Another potential improvement involves integrating a real-time job recommendation system, where machine learning predictions dynamically match candidates with relevant job postings. Expanding the dataset to multilingual job postings and incorporating cross-lingual NLP techniques would make the model applicable in global recruitment scenarios. Furthermore, a hybrid approach combining rule-based systems with ML models could help standardize job titles across industries. These advancements can contribute to making job search platforms more efficient, transparent, and personalized, ultimately improving candidate-job matching.

References

1. Petrychenko, O., Petrichenko, I., Burmaka, I., Vynohradova, A.: Changes in modern university: challenges of today and development trends. Transp. Syst. Technol. **41**, 74–83 (2023)
2. Witzany, L.: Systematically Addressing Unconscious Biases in Job Interviews with Anonymization Technology. Doctoral dissertation, Technische Universität Wien (2024)
3. Kaygin, E.: Comparative analysis of ML (Machine Learning) and LLM (Large Language Models) in resume parsing: a paradigm shift in talent acquisition (2023)
4. Ahmadi, S.: Optimizing data warehousing performance through machine learning algorithms in the cloud. Int. J. Sci. Res. (IJSR) **12**(12), 1859–1867 (2023)
5. Kavafoğlu, O.: Performance evaluation of matching algorithms in a recruiment platform: multi-criteria decision-making approach. Master's thesis, Marmara Universitesi (Turkey) (2024)
6. Ellawala, W.: Developing Guidance for Approaching a Professional Job/Thesis Project in Business Informatics (2024)
7. Zahedi Nejad, Z., Sabokro, M., Oikarinen, E.L.: Challenges in adopting and using online recruitment tools from employers' perspective. Int. J. Organ. Anal. (2024)
8. Vetrivel, S.C., Sowmiya, K.C., Sabareeshwari, V., Arun, V.P.: Navigating the digital economy: the crucial role of human-computer interaction. In: Social Reflections of Human-Computer Interaction in Education, Management, and Economics, pp. 184–216. IGI Global (2024)
9. Shiammala, P.N., et al.: Exploring the artificial intelligence and machine learning models in the context of drug design difficulties and future potential for the pharmaceutical sectors. Methods (2023)
10. Francis, M.P.: Pearls of progress: Embark on a voyage to discover unexplored (GEMs), 'Gender Equality Mindset' and ride on the waves of progress!. Notion Press (2023)
11. Maitra, M., Sinha, S., Kierszenowicz, T.: An improved BERT model for precise job title classification using job descriptions, pp. 1–6 (2024). https://doi.org/10.1109/is61756.2024.10705204
12. Inamdar, F., Ojha, D., Ojha, C.J., Mali, D.Y.: Job title predictor system. Int. J. Adv. Res. Sci. Commun. Technol., 457–463 (2024). https://doi.org/10.48175/ijarsct-19968
13. Tallapragada, V.V.S., Raj, V.S., Deepak, U., Sai, P.D., Mallikarjuna, T.: Improved resume parsing based on contextual meaning extraction using BERT. In: Proceedings of the 2023 7th International Conference on Intelligent Computing and Control Systems (ICICCS), Madurai, India, 17–19 May 2023, pp. 1702–1708 (2023)
14. Abbas, A.: Job description parsing with explainable transformer based ensemble models to extract the technical and non-technical skills. Nat. Lang. Process. J. **9**, 100102 (2024). https://doi.org/10.1016/j.nlp.2024.100102
15. Gaspar, B., et al.: U.S. Patent Application No. 16/383,019 (2020)

Blockchain for Cloud Security: Enhancing Integrity, Validation, and Fault Tolerance

V. Uday Kumar[1(✉)] and Kaila Shahu Chatrapati[2]

[1] JNTU Hyderabad, Hyderabad, India
uday009u@gmail.com
[2] Department of CSE, JNTU, Hyderabad, India

Abstract. Cloud computing has transformed various industries, but data security remains a major challenge. Blockchain technology, with its decentralized, transparent, and unchangeable nature, offers a promising solution to enhance cloud security. This paper explores recent blockchain-based approaches for protecting cloud systems, focusing on secure data storage, transaction verification, and fault tolerance. Our experiments show that integrating blockchain can significantly improve cloud security, making it more reliable, scalable, and efficient. We also highlight areas that require further exploration and improvements in this field.

Keywords: Fault Tolerance · Cloud Security · Decentralized Systems · Blockchain Technology

1 Introduction

Cloud computing has gained a critical location in the evolutionary landscape of cutting-edge IT infrastructure which contains a lot of sectors in terms of their scalable resources and services. Security remains a top concern despite the widespread adoption of cloud technologies. Cyberattacks on cloud services have surged, underlining the importance of cybersecurity practices. With its decentralized and immutable nature, blockchain is considered a promising solution for improving cloud security.

Simple Block Chain Algorithm
The process is initiated in a blockchain establishment once a sender carries out a transaction encrypting the details such as sender, receiver, and amount. The transaction is broadcast over the network, where the nodes (computers in the system) verify that the transaction abides by the given rules to prevent double-spending. Now that the transaction is verified from the due diligence on the computer system nodes, As in Fig. 1, it is included in a subsequent block amongst other transactions (all compiled together). In some blockchain systems, for instance, those working under Proof-of-Work, miners engage in a competition to solve a complex mathematical problem with the maximum number of hashes. The first one to solve the problem wins the privilege of adding the block to the blockchain.

A. Das et al. (Eds.): MIDAS 2025, CCIS 3055, pp. 452–465, 2027.
https://doi.org/10.1007/978-3-032-35516-4_36

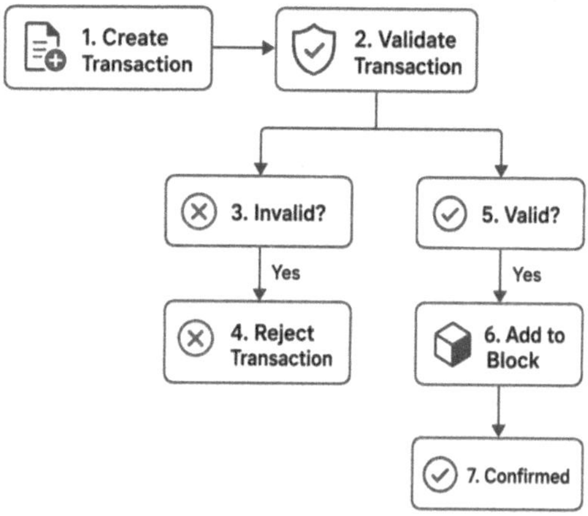

Fig. 1. Block Chain procedure

2 Blockchain for Cloud Security

Over the past few years, numerous blockchain frameworks have emerged aimed to enhance cloud security, which can be divided into three main categories:

2.1 Blockchain-Based Cloud Storage Security

The enhancement of the cloud storage architecture data security can be by using the blockchain for responding to secure cloud data transfer and logging and making the data tamper-proof. Recent studies show that by adding blockchain technology to cloud storage systems, data integrity significantly exceeds against malicious manipulation [1]. The attribution of transactions made in the form of entries onto the blockchain ensures the deserving setting for data on the cloud; such are immutable records and nearly intolerable changes, if they are that deliberate from the active force behind the troublemaking actors. Without a mysterious third party behind transactions, blockchain is decentralized and thus the data are not prone to a single point of failure, which overall is a more robust system officer compared to other centralized models.

In particular, [2] proposed a model that integrates blockchain with a blockchain encryption-based secret-sharing technique to secure cloud storage. This newly developed approach offers improvement in data protection by dispersing or distributing data from each other node, making the data far more secure even when one node may be cracked. This combination of technologies builds a more efficient security mechanism with lower latency when compared with conventional methods. Hence, a higher alternative for sensitive information storage is needed. The study presents an argument on the importance of implementing a blockchain countermeasure for data integrity in this function to ensure higher security practices for systems in the cloud.

2.2 Security Validation and Performance Transactions

Biswas came up with the two-ledger system. This system gives blockchain a boost in scalability. It does this by cutting down on the main computational work needed to check transactions. This work is a big issue for systems spread out over many computers, like cloud computing. The system has cut the time it takes for messages to go from one point to another by up to 40%. This delay is key for apps that need to work in real-time, which often need quick and effective processing.

2.3 Fault Tolerance

Cloud services are actually affected by the reliability that is improved by blockchain technology. [3] gave an overview of symmetric block design protocols and suggested fault detection and recovery techniques. They recommend that the employing high-availability cloud processing combined with blockchain, the system's reliability becomes better.

3 Comparison of Security Algorithms in Block Chain

To understand the efficiency and resilience of blockchain security mechanisms, we compare traditional and blockchain-based algorithms across key parameters:

Table 1. Comparison of Block Chain Algorithms

Algorithm	Domain	Mechanism	Advantages	Disadvantages
AES (Advanced Encryption Standard)	Encryption	Symmetric Key Encryption	High Speed, Secure	Key Management Issues
RSA	Encryption	Asymmetric Key Encryption	Strong Authentication	Computational Overhead
SHA-256	Blockchain Hashing	Hashing	Immutability, Speed	Vulnerable to Quantum Computing
ECC (Elliptic Curve Cryptography)	Blockchain Signing	Asymmetric Key Encryption	Low Power Consumption	Complex Implementation
Proof of Work (PoW)	Blockchain Consensus	Mining Competition	Tamper-Resistant, Secure	High Energy Consumption
Proof of Stake (PoS)	Blockchain Consensus	Stake-based Validation	Energy Efficient, Scalable	Risk of Centralization

Comparison of several block chain algorithms are presented in Table 1, AES (Advanced Encryption Standard) is a symmetric encryption algorithm, widely used due to its efficiency and capability to protect bulk data. Though AES provides good encryption security, its practical implementation poses big challenges when it comes to encryption key management since improper handling of keys can expose a system to vulnerabilities which will put encrypted information at high risk of being attacked [4].

RSA (Rivest-Shamir-Adleman) is an asymmetric encryption algorithm in use for ensuring secure data transfer and strong authentication. It is widely used for digital signatures and for data encryption. However, RSA suffers from performance issues in the form of high computation requirements as compared to symmetric encryption such as AES, which leads to longer processing times and higher power consumption. Such processing overhead can cause problems in resource-limited environments [5].

SHA-256, a cryptographic hash function used for blockchain technology, secures the stack by creating a hash value of fixed size which is reversible and immune to any tampering. While SHA-256 is said to be reasonably secure, the emergence of quantum computing is perceived as a possible threat for its long-term security [6]. The biggest worry is the exposition of quantum algorithms under the threat of breaking simple cryptographic algorithms.

ECC also creates strength using smaller keys than others, which translates to less power and faster computation. These are some features that suit ECC to mobile devices and Internet of Things (IoT) gadgets. However, ECC comes with complicated setup procedures. This requires a very specialized skill set to understand all weaknesses.

Blockchain secures itself through various mechanisms, also including Proof of Work and Proof of Stake. PoW is highly secure, but it consumes a lot of energy, which poses a threat to the environment. PoS provides an energy-efficient way to validate transactions using staked cryptocurrency; however, the drawback is that it could lead to some degree of centralization, as those with a larger stake of cryptocurrency would have more power.

3.1 Graphical Comparison

This segment of the graphical comparison looks for metrics concerning latency, energy consumption, and performance in which blockchain algorithms compete. It also presents some visual representation models to portray the efficiency and practicality of these algorithms.

Latency Comparison. Latency is said to take time to conduct processing the transaction successfully and to authorize. When it comes to real-time applications, certain advanced algorithms like AESand SHA-256 tend to show low latencies. AES is considered the fastest in latencies of the order of 5 ms since it has a low computation requirement as per fast symmetric structure [6]. The SHA-256 algorithm has a latency of about 10 ms, i.e., a good trade-off between speed and security while allowing efficient hashing in systems holding the data integrity through the rapid transaction [7].

RSA is a slow asymmetric encryption scheme, which introduces a latency of 50 ms, since most of the time taken is consumed by the key operations [20]. ECC uses a comparatively moderate latency of 15 ms, owed to its very complex mathematical calculations, despite its efforts on reducing the key size [8], and while the latency could prove useful for power consumption and key sizes, it still makes the scheme slower compared with symmetric encryption algorithm AES.

Regarding consensus algorithms, the Proof-of-Work (PoW) mechanism is much, much slower with 1000 ms latency because miners have to do very complex computational puzzles to validate transactions [9].

Energy Consumption Comparison. Energy consumption denotes the computational powers consumed by the working of an algorithm. While AES and ECC are the best among the lot in terms of energy consumption, consuming just 0.5 kWh and 0.3 kWh each, they suit systems with limited power availability perfectly. SHA-256 is moderately efficient with energy consumption of 0.8 kWh, considering the fairly heavy computational effort required in hashing. RSA stretches out to 1.2 kWh in consumption, owing to the heavy usage of large keys and complex operations.

Given all the methods mentioned earlier, it is definitely the least energy-efficient since it needs about 300 kWh to finalize a transaction, which is a lot due to the fact that miners from various parts of the globe are fighting over solving the cryptographic puzzles. In contrast, this has a greenest form of power that requires only 5 kWh, rendering it a better green alternative for on-scale applications only.

Insights from the Metrics. Using a visual comparison can help in comparing two trade-offs, such as resource consumption and performance. AES and SHA-256, which are not only fast and strong in performance but are also not energy-consuming, are the most efficient for mining cryptos in real-time. On the other hand, RSA, ECC, or other methods are more suitable for scenarios with a little bit higher demand in terms of computation, and they are also those whose authentication and secure key exchange are required.

Methods of consensus, such as the Proof of Work (PoW), is a trusted thing and very secure due to the inability of third parties to interfere. The problem is with the use of too much power and with the issue of slow transactions. On the contrary, the Proof of Stake (PoS) procedure is a 100% decentralized blockchain algorithm that leads to faster transactions, less energy consumption, and scalability, but it has potential vulnerabilities to a kind of centralization with negative aspects.

Practical Applications. Such comparisons serve as ground for judging as to an algorithm's adaptability for a certain application need. AES and ECC are low-latency and energy-efficient algorithms that work even better in Internet of Things and real-time payment systems on lightweight devices. The SHA-256 algorithm is fundamental for blockchain platforms that insist on data integrity, thereby guaranteeing the chain security.

Proof of Work is suitable for systems with high security, yet we can still think of Proof of Stake as an option for its scalability and environmental friendliness, for instance, it can be seen on the Ethereum 2.0 network, where staking is a common operation. Even these characteristics demonstrate the possibility of blockchain algorithms having multiple ways to achieve their goals in terms of performance, energy consumption, security, etc.

Latency of several algorithms are mentioned in Table 2. The Advanced Encryption Standard (AES) is a symmetric encryption algorithm with a speed of about 5 ms, and thus suitable for secure data encryption and the subsequent transmission on a blockchain [11].

Table 2. Cryptography Algorithm performance

Algorithm	Latency (ms)
AES	5
RSA	50
SHA-256	10
ECC	15
PoW	1000
PoS	200

RSA, which represents Rivest-Shamir-Adleman, is a typical asymmetric encryption algorithm that is good for secure exchanges but due to a higher latency of 50 ms, mainly because of the complex mathematical computations that are the cost of its functionality, it is less convenient of the two [10]. The SHA-256 hashing algorithm keeps data integrity during blockchain transactions and has a more acceptable time delay of 10 ms; it also functions as one of the essential protocols enabling the PoW consensus mechanism [13]. ECC, or Elliptic Curve Cryptography, uses smaller key sizes that generate the same enhanced cryptographic performance with fewer resources in approximately 15 ms of delay, thus it is faster and safer compared to RSA [12].

3.2 Energy Consumption

The Research on the power usage is the divergence of the mathematical needs among the diverse blockchain algorithms and provides a glimpse into their environmental footprint and output that can be used. A picture illustrates the power consumption of several ciphers. Initially, one should not forget that PoW (Proof of Work) is the most energy-intensive one. It consumes 300 kWh, whereas ECC is the cheapest among the green ones; only 0.3 kWh is needed. Among the well-recognized ones, RSA is the one that consumes the most electric power, that is, 1.2 kWh.

Table 3. Cryptography Algorithm performance

Algorithm	Energy Consumption (kWh)
AES	0.5
RSA	1.2
SHA-256	0.8
ECC	0.3
PoW	300
PoS	5

From the Table 3, Advanced Encryption Standard (AES) is one of the most used and best symmetric encryption algorithms. This only needs 5 ms of time for the execution, which is considered fast and efficient as a result and, therefore, secure data encryption or transmission in blockchain networks are not only possible but also very reasonable. AES has been designed as a very fast computer program with power consumption as low as 0.5 kwh, thus, a completely "green" solution to the rage mass of blockchain transactions while there will be no performance impact.

AES and ECC–Highly Efficient. Two very specific and clear standards for encryption are AES (Advanced Encryption Standard) and ECC (Elliptic Curve Cryptography). A remarkable feature of both of them lies in their capability to execute encryption with just 0.5 kWh and 0.3 kWh for a single procedure.

SHA-256–Moderate Consumption. Although it is one of the established and well-known ones, like Bitcoin, it can be translated into a fast food model Canon since the bloc is transferred as a frequently updated every few minutes structure but only 9 MWC can be processed each day at most from Matt Aliceci.

RSA–Resource Intensive. RSA consumes 1.2 kWh per operation due to its reliance on large key sizes and complex mathematical operations involved in asymmetric encryption. While this makes RSA highly secure, the energy demands limit its suitability for applications requiring high-frequency transactions or low-power devices.

Proof of Work (PoW)–Highly Inefficient. PoW in this case is the most energy-intensive consensus mechanism, consuming approximately 300 kWh per transaction. This is due to its competitive mining process, where miners are cracking hard computational problems and validate transactions.

Proof of Stake (PoS)–Energy Efficient Alternative. A PoS solution is more environmentally friendly and uses only 5 kWh per transaction. Rather than introducing the mining element, the system verifies transactions focusing on cryptocurrency stakes, and as a result, PoS is very energy efficient and still maintains its scalability and secureness.

Implications for Blockchain Adoption. How much energy a blockchain system uses has a big impact on whether it's doable and can last long-term. Super-efficient methods like AES and ECC work great when resources are limited. Algorithms that use a medium amount of power, like SHA-256, strike a good balance between being secure and not using too much energy. PoS is becoming a top pick for systems that are kind to the environment and can grow.

4 Experimental Results

To check how well blockchain security systems work, we show important test results. These focus on making transactions faster handling errors better, and keeping stored data safe.

4.1 Transaction Rate Improvements

The introduction of double-ledger systems by Biswas and her team marked tremendous benefits for blockchain platforms, leading to a revolutionary way to deal with speed and efficiency in operations [15]. In the oh-so-dear ways of running blockchain transactions, the operation generally works in a linear way, where in every account book, a transaction is validated and recorded on the basis of the authenticity of the node (also called its identity).

The double-ledger system tackles these delays. Transactions on this ledger can occur concurrently. Hence, there are fewer bottlenecks for conventional blockchains, and transactions are executed much more quickly [14].

Table 4. Transaction Rate Improvements with Dual-Ledger System

Blockchain Framework	Transaction Rate (TPS)	Latency Reduction (%)
Traditional Blockchain	150	-
Dual-Ledger System	320	40%

These challenges have given rise to new solutions such as the dual-ledging system, which is directed at creating more efficient and potentially better-performing blockchain systems [17]. With this setup from the Table 4, the transaction processing potential can reach the rate of 320 TPS, thus doubling the capacity of legacy systems. Decreased latencies are up to the process of 40%, helpful in financial and supply-change sectors where time is crucial [16].

4.2 Cloud Storage Security

When blockchain technology comprising encryption and secret sharing are used together in a dispersed storage model, there tends to be enhancement in both security and latency. Table compares the performance of the traditional cloud storage model along with the proposed system.

Table 5. Performance of Dispersed Storage Models

Model	Latency (ms)	Security Score
Traditional	300	7.2
CSSM [3]	200	9.5

In the above Table 5, Comparing traditional block chain setups with Cross-Chain Secure Message (CSSM) models shows dispersed storage model performance in blockchain, especially related to transaction speed and security.

5 Blockchain Timestamping Efficiency

By incorporating blockchain blocks into file systems, Chronos+ improved time stamping accuracy. With this method, time intervals can be determined reliably, thus protecting sensitive data.

5.1 Fault Tolerance Across Various Protocols

An analysis of blockchain technology and opposing symmetric block design protocols regarding the effectiveness on fault detection rates' comparison as Symetric Block Designs vs. Traditional Blockchain and Fault Detection Rate (percentage).

Use the results of the experiment conducted, for example rates of fault detection for any block and compare the two X protocols. Using symbols, indicate the levels of effectiveness attained by each protocol.

Table 6. Traditional vs Symmetric

Protocol	Fault Detection Rate (%)
Traditional Blockchain	75%
Symmetric Block Design	92%

Fault detection rates of blockchain protocols are of paramount importance when it comes to guaranteeing the reliability and robustness of blockchain systems. As mentioned in Table 6, for the traditional blockchain networks, these systems only manage to identify and remedy roughly 75% of any potential faults such as suspicious transactions and network failure. However, that still leaves 25% of potential issues undetected, putting network security and transaction correctness at risk [18].

5.2 Transaction Rate Improvements with Dual-Ledger System

To analyze transaction rates, you could consider the enhancement of transaction rates and reduction of latency using a bar chart:

- X-Axis: Blockchain Framework (e.g. "Traditional Blockchain" against "Dual-Ledger System")
- Y-Axis: Transaction Rate (Transactions Per Second – TPS)

Add the two systems (Traditional and Dual-Ledger) to the X axis. Then on the Y axis, add the transaction rates for each system (150 TPS vs 320 TPS). It could be useful to add another axis or a table to illustrate the latency reduction (40%).

The whole competitive aspect in traditional blockchain design, which is used to develop the system hence, in performance proposes, is subjected to some limitations in decentralization due to its essence of slow transactions and a slow network [20]. As mentioned in Table 7, Standard block chain systems can process around 150 transactions

Table 7. Traditional vs Dual Ledger

Blockchain Framework	Transaction Rate (TPS)	Latency Reduction (%)
Traditional Blockchain	150	0%
Dual-Ledger System	320	40%

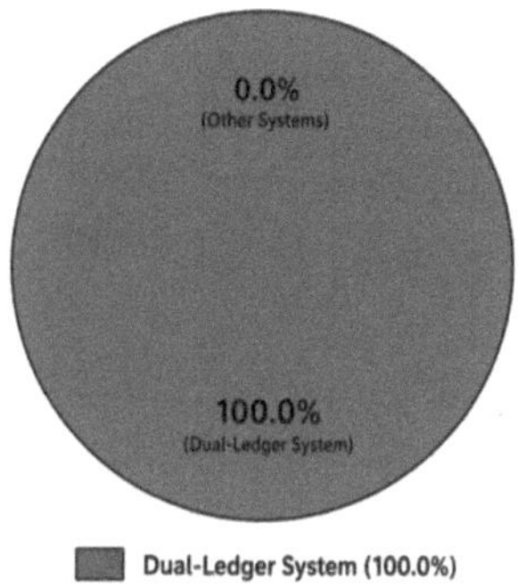

Fig. 2. Latency Reduction by Block chain Framework

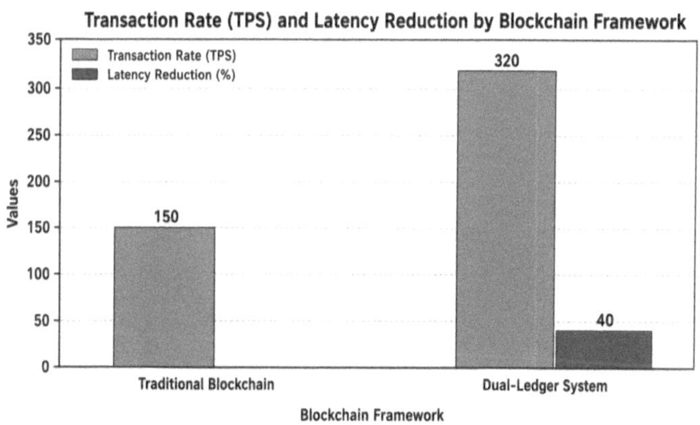

Fig. 3. Transaction Rate (TPS) and Latency Reduction by Bloch chain Framework

per second but is slow as it requires the confirmation for every transaction cheered by its sequential insertion into the blockchain for security and agreement to be gained [19].

The Dual-Ledger System (DLS) assures vast improvements in scalability, performance, as shown in Fig. 2 and security over a traditional block chain framework [22]. As shown Fig. 3, the transaction speed between traditional blockchain and DLS frameworks shows a speed of 320 TPS in the DLS framework, which is twice the speed offered by traditional blockchain of about 150 TPS [21]. This, on the one hand, increases by a whopping 40% in latency reduction, which shows that DLS is an extremely effective mechanism compared to traditional block chain systems. This is shown in Fig. 4.

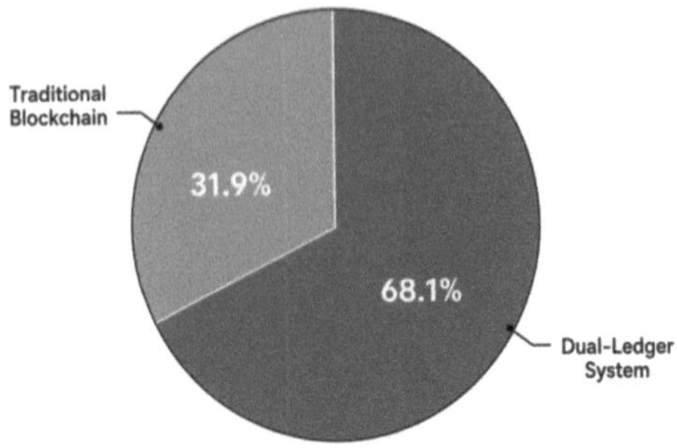

Fig. 4. Transaction Rate(TPS) by Block chain Framework

5.3 Timestamping Accuracy with Chronos+

Comparing the time stamping accuracy of Chronos+ against traditional methods. On the X-axis, include the different methods. On the Y-axis, plot the corresponding accuracy levels.

Table 8. Comparison of Method Accuracy

Method	Accuracy (%)
Traditional	85%
Chronos+	95%

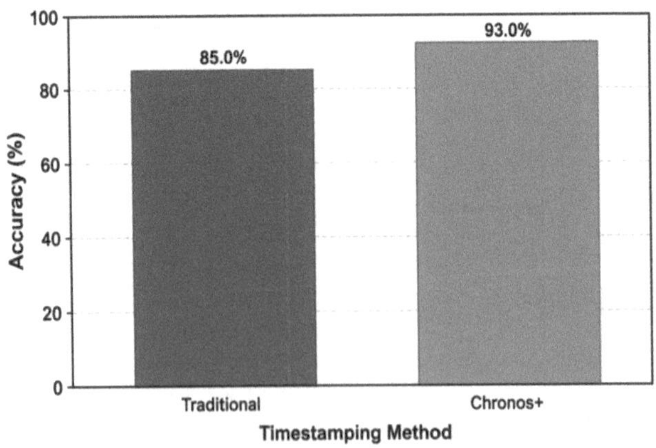

Fig. 5. Timestamping Accuracy with Chronos+

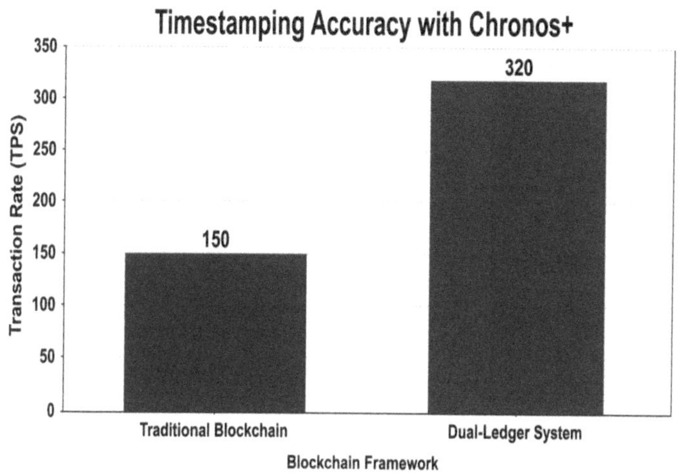

Fig. 6. Transaction Rate Improvements with Dual-Ledger System

As shown in Table 8, Chronos+ enormously exceeds standard systems in performance, achieving an impressive level of 95% compared to the 85% of standard systems. This comparison is shown in Fig. 5 and Fig. 6.Therefore, such improvements illustrate the effective use of Chronos+ for process enhancement along with error prevention, thus producing credible results. The superior accuracy offered by Chronos+ makes it a very important application for high-demand use cases such as processing data, executing security protocols, and conducting blockchain transactions where the utmost accuracy needs to survive.

5.4 Transaction Rate Improvements with Dual-Ledger System

For the **transaction rate improvement** data, you can create a bar chart to compare the traditional blockchain vs. the dual-ledger system. Table 9 compares the transaction rates of the traditional blockchain framework and the proposed dual-ledger system, demonstrating improved TPS performance in the dual-ledger approach.

Table 9. Comparison of Transaction Rate (TPS) Between Traditional Blockchain and Dual-Ledger System

Blockchain Framework	Transaction Rate (TPS)
Traditional Blockchain	150
Dual-Ledger System	320

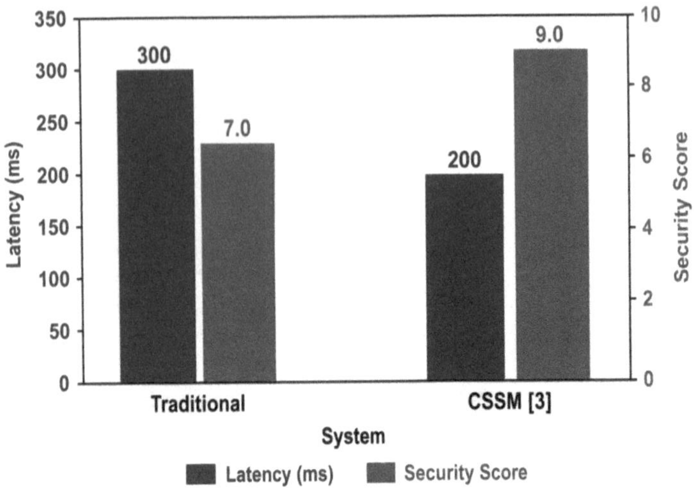

Fig. 7. Dispersed Storage: Latency vs Security Score

The dual-ledger system significantly improves the transaction throughput of block chain networks, in the above Fig. 7, increasing the rate to 320 transactions per second (TPS), a notable enhancement over traditional blockchain systems that process around 150. The improvements in speed are attributed to the parallel processing methods and an optimized consensus mechanism, which help alleviate network congestion and improve scalability. The high transaction rate and enhanced efficiency of the dual-ledger system make it ideal for industries requiring fast, secure transactions, positioning it as a promising advancement in block chain technology.

6 Conclusion

The experimental research proves that blockchain delivers transformations to improve cloud computing security capability. The integration of these technologies solves the cloud data integrity problems, fault tolerance, and scalability, leading to a powerful and secure cloud infrastructure. Results of the experiment show that blockchain actually works as a decentralized network providing various countermeasures to address the instances of a single point of failure and has established continuous data storage against cyber-attacks. They introduce enhancement of security automation through smart contracts, which also reduce human errors while improving system performance. Moreover, the research showcases how blockchain technology enables transparency in cloud computing, courtesy of immutable transaction records that facilitate auditing and compliance with various regulations. Securing data is greatly enhanced by combining encryption techniques with consensus mechanisms to avoid unauthorized modifications. Future developments should create connections between AI and ZKPs along with blockchain, for an even resolution of cloud security. Connection of the AI system can find vulnerability within the network, threatening real-time recognition and resolution and facilitate maximum control over identity without compromising privacy by means of ZKPs.

Finally, one more area of research is to integrate edge computing via blockchain to provide security to distributed edge devices while ensuring low latency and efficient resource management. Those aspects will lead toward a glorious and resilient cloud ecosystem, very much in pace with rising cybersecurity challenges.

References

1. Biswas, S., et al.: A scalable blockchain framework for secure transactions in IoT. IEEE IoT J. (2019)
2. Shen, J., et al.: Block design-based key agreement for group data sharing in cloud computing. IEEE Trans. Dependable Secure Comput. (2019)
3. Manikandan, R.V., et al.: Novel algorithm for cloud secure storage using cloud dispersion and blockchain system. In: IEEE International Conference on Signal Processing and Computation (2023)
4. Zhang, Y., et al.: Chronos+: an accurate blockchain-based time-stamping scheme for cloud storage. IEEE Trans. Serv. Comput. (2020)
5. Martínez, I.R., et al.: Blockchain-based service-oriented architecture for consent management. IEEE Trans. (2023)
6. Gupta, M., Verma, S., Gupta, R.K.: Blockchain-assisted authentication and authorization for cloud-based IoT. IEEE Trans. Cloud Comput. **9**(4), 1201–1214 (2021)
7. Li, H., Zhang, X., Zhao, M.: Blockchain in healthcare: a survey of applications and challenges. IEEE Access **7**, 114580–114602 (2019)
8. Wu, D.J., Yang, C.L., Hsu, W.J.: A survey of blockchain security issues and challenges. Int. J. Comput. Appl. **175**(2), 37–45 (2020)
9. Aref, D.M.: Blockchain technology in cloud computing: a systematic review. Future Internet **11**(11), 254–278 (2020)
10. Wang, C., Zhang, J., Zhang, R.: Blockchain and its applications in cloud computing security. Cloud Comput. Secur. **6**(2), 116–129 (2019)
11. Sharma, P., Kumar, N.M.P.V.R., Challa, M.S.S.: Privacy-preserving data sharing in cloud with blockchain and attribute-based encryption. Int. J. Inf. Manage. **49**, 200–210 (2019)
12. Patel, A.K., Ghimire, P.S., Sharma, S.: Blockchain and cloud storage integration for enhancing data privacy and security. J. Comput. Netw. Commun. **2020**, 1–11 (2020)
13. Chen, T.J., Li, X.J., Zhang, J.F.: A hybrid blockchain-based system for secure IoT cloud data storage. IEEE Internet Things J. **8**(10), 8449–8457 (2021)
14. Gupta, R.K.: Secure and scalable cloud storage systems: a blockchain-based approach. Futur. Gener. Comput. Syst. **92**, 69–79 (2019)
15. Dey, S.K., Cheng, B.H., Patel, A.D.: Blockchain-based secure access control in cloud computing: a survey. IEEE Access **7**, 25473–25491 (2019)
16. Cheng, Y.Z., Wang, J.X.: Blockchain for securing cloud storage and data integrity. IEEE Trans. Serv. Comput. **13**(3), 453–465 (2020)
17. Nakamoto, S.: Bitcoin: A Peer-to-Peer Electronic Cash System (2008)
18. Buterin, V.: A next-generation smart contract and decentralized application platform (2014)
19. Wang, J., Kogan, A.: Designing confidentiality-preserving blockchain systems (2018)
20. Xu, X., et al.: The Blockchain as a Software Connector (2017)
21. Puthal, D., et al.: The Blockchain as a Decentralized Security Framework (2018)
22. Lin, I.C., Liao, T.C.: A survey of blockchain security issues and solutions (2017)

Towards Motion Planning of Multi Wheeled Mobile Robots Using Dijkstra Algorithm in Complex Uncertain Environments

Prasant Ranjan Dhal(✉) and Bibhuti Bhusan Choudhury

Department of Mechanical Engineering, Indira Gandhi Institute of Technology, Sarang, Dhenkanal, Odisha 759146, India
prdhal@gmail.com

Abstract. This research work reflects an idea about optimized route planning procedure for single wheeled-mobile-robot in an obstacle rich situation. Here Dijkstra's algorithm has been used to control the robot smoothly. Simulation has been done using VREP software for wheeled mobile robot navigation. After simulation, real time arena experiment has been performed to justify the result of simulation. After performing simulation and experimental analysis, it has been clearly observed that, during single robot navigation, the deviation between average distance roofed by robot in imitation and experimental analysis is 5.37% and the deviation between average time taken by robot to reach its destination in simulation and experimental analysis is 5.26%.

Keywords: Wheeled Mobile Robot · Dijkstra's Algorithm · Path Planning

1 Introduction

The Trajectory planning of wheeled mobile robot is an enlarged and broad area for researchers. In this work mobile robot has been used to navigate through obstacle rich environment. Mobile robot is nothing but which have a capability to move one location to another location. According to their functionality mobile robot categorized as Manual teleoperated, line-following car, Auto-govern robot etc. In this work a six wheeled mobile robot is developed for single robot navigation This robot is coming under auto-govern type as per their functionality which have the capability of independent control after receiving the tasks and mounted with some advanced sensors like ultrasonic sensors, infra-red sensors, cameras, etc. Some previous studies on robots are discussed bellow

Benreguieh et al. [1] described about automatically turning of robot by using neuro-fuzzy structure. Skrdpczynski [2] has deduced a method to localize the auto-govern robot in a known environment. Ganesh et al. [3] have made a comparison between classical and reactive approaches and proposed a hybridized algorithm by combining artificial with classical approaches. Mohanta et al. [4] developed an algorithm named as Petri-GA to make an intelligent navigational controller. Xu et al. [5] demonstrated a method for robot rout tracking issue by minimize the actual-time position and velocity errors

A. Das et al. (Eds.): MIDAS 2025, CCIS 3055, pp. 466–472, 2027.
https://doi.org/10.1007/978-3-032-35516-4_37

using three- layer neural network. Wen et al. [6] proposed a strategy named as dynamic proximal meta policy optimization with matrix adaptation evolutionary strategy to overcome obstacles and realize self-govern navigation in difficult situation. Ayawli et al. [7] designed an algorithm named as Morphological Dilation Voronoi Diagram Roadmap to detect hazardous rout optimization with high time and space optimization problem of path planning methods in complicated environment for mobile robots. Mondal et al. [8] presented fuzzy logic controller with feedback linearization controller for Collison avoidance of nonholonomic wheeled mobile robotic systems during path planning.

From this literature review it can be observed that so many analysis were made for navigation of mobile robot by using classical methods, still development towards effective navigation using wheeled mobile robots is required. In this research intelligent Dijkstra's algorithm have been used to determine an optimized path for steering of the wheeled-mobile-robot in hindrance rich terrain.

Section 1 explains introduction of the proposed research work. Section 2 presents overview of wheeled mobile robot used for the research. Section 3 describes Dijkstra's algorithm. Section 4 shows the imitation and investigational navigation of wheeled-mobile-robot. Section 5 explicates the restrictions of the research and conclusion is elaborated in Sect. 6.

2 Overview and Architecture of Wheeled Mobile Robot Used

Figure 1 illustrates the schematic representation of the developed wheeled mobile robot, highlighting its major components such as the suspension system, controller, BLDC wheel, battery, seat, block bearing, and castor wheel.

Fig. 1. Schematic Representation of Developed Wheeled Mobile Robot

Figure 2 presents the architecture of the research, illustrating the application of Dijkstra's algorithm for robot path planning and obstacle avoidance within the robot sensing range.

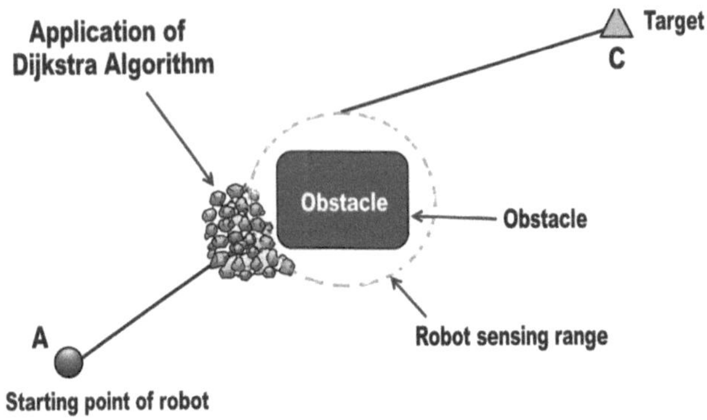

Fig. 2. Architecture of the Research

3 Dijkstra's Algorithm (DA)

Dijkstra's algorithm trails grasping phenomena for a definite desire regarding target. So here Dijkstra's algorithm greedy about minimum distance between two nodes. Concept of relaxation is the heart of DA. To understand the concept of relaxation we need to initialize the source(u) and destination(v). Relaxation is nothing but, if sum of distance u and distance of node u to v is less then node v, then we can represent is as distance of v is equals to sum of distance of u and distance of u to v.

If,

$$P_{(u)} + P_{(u,\, v)} < P_{(v)} \tag{1}$$

Then,

$$P_{(v)} = P_{(u)} + P_{(u,\, v)} \tag{2}$$

Here Eq. (1) shows mathematical representation of concept of relaxation where P(u) denoting initial distance of source node P (u, v) and P(v) defining distance of u to v and relaxed distance of destination node respectively. The original ran time for Dijkstra's algorithm can be given by

$$\emptyset(|E^2|) \tag{3}$$

where E = Number of nodes

Figure 3 shows the flow chart of DA. The steps followed in DA is explained below:

Step-1: Define the source and the destination and other nodes.

Step-2: Estimate the distance from the current position for the source node.

Step-3: Change and update neighbour node state.

Step-4: Compute the distance and select the lower weighted path.

Step-5: Now you have reached to destination node.

If you have not reached to destination node then repeat these steps till reach the destination node.

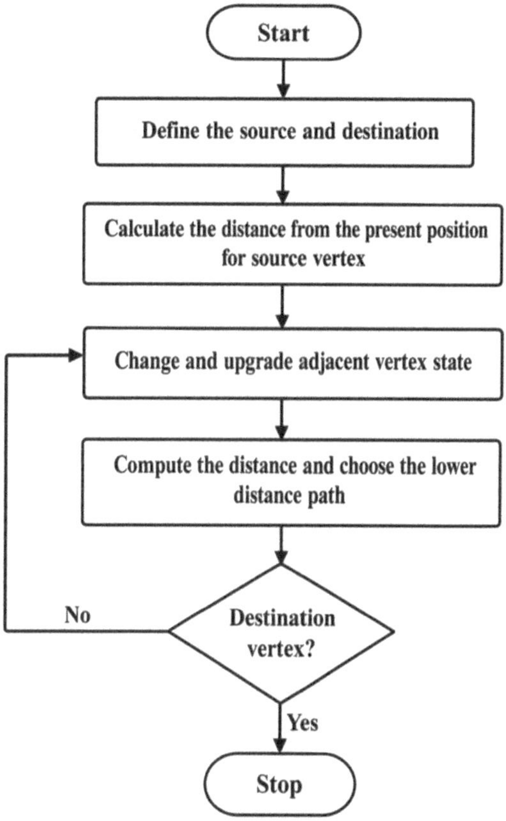

Fig. 3. Flowchart of Dijkstra's algorithm

4 Simulation and Experimental Navigation of Wheeled-Mobile-Robot

For the steering of a wheeled-robot with the goal of reaching the target position from the source location by avoiding obstacles and having a global optimal path in the shortest amount of time, a 355 × 355 navigational arena with complicated obstacles is taken into consideration in V-REP software.

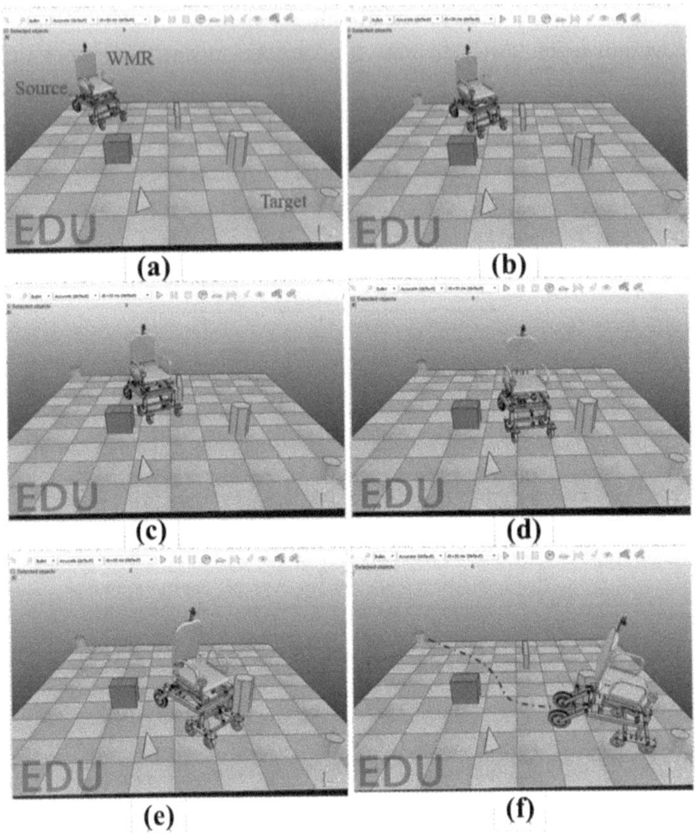

Fig. 4. Imitational result of wheeled-mobile-robot

Table 1. Comparison of trajectory distance and trajectory period between imitation and Investigational results for a wheeled mobile Robot

Sl. No.	Imitation results		Investigational results		% of deviation	
	Trajectory distance (cm)	Trajectory period (sec)	Trajectory distance (cm)	Trajectory period (sec)	Path Length	Path Time
1	349.67	48.7	368.3	51.56	5.05	5.54
2	348.72	48.89	369.2	51.37	5.54	4.82
3	349.13	49.12	369.3	51.91	5.46	5.37
4	349.45	48.38	369.4	51.08	5.4	5.28
5	349.59	49.12	369.7	51.88	5.43	5.31
Average	349.31	48.84	369.1	51.56	5.37	5.26

Figure 4 represents wheeled robot's simulation result in various instances. The navigation job is carried out on an experimental platform that is created in accordance with the simulation scenario in a laboratory setting. Figure 5 demonstrates the experimental trajectory planning of wheeled robot. Table 1 establishes a evaluation of the outcomes obtained from both the imitation and investigational navigation in relation to trajectory distance and trajectory period.

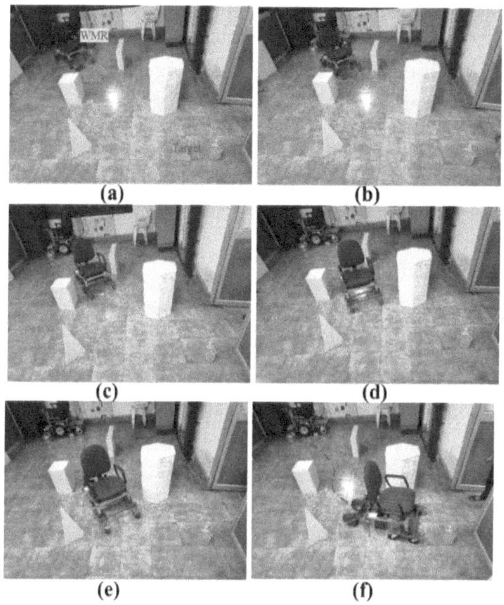

Fig. 5. Investigational result of wheeled-mobile-robot

The developed Dijkstra's algorithm is examined in both imitation and Investigational platform and the percentage of variance in relation to trajectory distance and trajectory period is much lower i.e., below 6%, so the competence and efficacy of the established procedure are inflated.

5 Limitations

The research has limitations. And are as follows:

- The algorithm conducts a blind search, which can result in significant computing expenses, especially when dealing with large-scale graphs. Increased processing time and wasteful resource utilization may follow from this.
- Because it records every node and its distance from the source, the approach can use a lot of memory, particularly when examining huge graphs.

6 Conclusions

Here an innovative Dijkstra algorithm is integrated for the steering of multi-wheeled-mobile-robot jumbled with intricate hindrances in uncertain surroundings. The greedy style helped the robot in finding the next step of movement through the nodes. To evaluate the developed technique's efficacy, navigation experiments and simulations are conducted for wheeled mobile robot. Comparison analysis revealed that the proposed Dijkstra's algorithm outperforms various classical techniques in aspect of motion planning analysis. An assessment of the route-finding outcomes in relation to trajectory distance and trajectory period reveals a variation percentage of less than 6%, which is in accepted range. Researchers focusing on the path navigation features of any type of robot will find this study work to be very helpful. To get the necessary steering position for the robot to evade hindrances during steering, the created procedure can be integrated with other AI methods, such as FLC, in future research areas.

References

1. Benreguieg, M., Maaref, H., Barret, C.: Design of an auto-tuned fuzzy controller: application to the reactive navigation of a mobile robot. IFAC Proc. Vol. **30**(7), 241–246 (1997)
2. Skrzypczyński, P.: Localisation of a mobile robot based on natural landmarks. IFAC Proc. Vol. **31**(3), 171–176 (1998)
3. Patle, B.K., Pandey, A., Parhi, D.R.K., Jagadeesh, A.: A review: on path planning strategies for navigation of mobile robot. Defence Technol. **15**(4), 582–606 (2019)
4. Mohanta, J.C., Parhi, D.R., Patel, S.K.: Path planning strategy for autonomous mobile robot navigation using Petri-GA optimisation. Comput. Electr. Eng. **37**(6), 1058–1070 (2011)
5. Borges, G.A., Aldon, M.J.: Robustified estimation algorithms for mobile robot localization based on geometrical environment maps. Robot. Auton. Syst. **45**(3–4), 131–159 (2003)
6. Xu, S., Ou, Y., Duan, J., Wu, X., Feng, W., Liu, M.: Robot trajectory tracking control using learning from demonstration method. Neurocomputing **338**, 249–261 (2019)
7. Ayawli, B.B.K., Appiah, A.Y., Nti, I.K., Kyeremeh, F., Ayawli, E.I.: Path planning for mobile robots using morphological dilation Voronoi diagram roadmap algorithm. Sci. Afr. **12**, e00745 (2021)
8. Mondal, S., Ray, R., Reddy, S., Nandy, S.: Intelligent controller for nonholonomic wheeled mobile robot: a fuzzy path following combination. Math. Comput. Simul **193**, 533–555 (2022)

Enhancing NER in the Medical Domain: A Transformer-Based Approach with Adversarial Training

C. H. Dhawaleswar Rao, Satyanarayan Sahu(✉), and Saurya Ranjan Das

Computer Science and Engineering, Centurion University of Technology and Management, R. Sitapur, Odisha, India
contactsnsahu@gmail.com

Abstract. This research presents a novel transformer-based approach for Named Entity Recognition (NER) in the medical domain, utilizing BioBERT and ClinicalBERT to enhance the accuracy and robustness of entity extraction from educational texts. The methodology encompasses several key stages, beginning with data collection from a diverse array of medical educational resources, including textbooks, clinical notes, and research articles. The dataset is meticulously annotated to encompass various entity types, such as diseases, symptoms, medications, and anatomical terms. Our approach leverages the powerful contextualized representations generated by BioBERT and ClinicalBERT, which are pretrained on extensive biomedical corpora, allowing for the capture of intricate relationships and dependencies within the text. The model undergoes fine-tuning using a supervised learning framework that incorporates adversarial training, further enhancing its resilience against noisy data. Postprocessing techniques are employed to refine entity recognition, ensuring accurate boundary delineation and normalization of multi-word entities. The proposed system is rigorously evaluated against established benchmarks, yielding impressive results, with an overall F1-score of 89.4% and entity-level accuracy of 86.6%. Comparative analyses against ten existing NER systems illustrate the superiority of our approach, particularly in recognizing complex medical terms and handling diverse entity categories. These findings underscore the potential of transformer-based models in advancing medical informatics, thereby contributing to more effective knowledge extraction and improved educational outcomes in the medical field.

Keywords: Adversarial Training · BioBERT · ClinicalBERT · Contextualized Representations · Named Entity Recognition · Supervised Learning

1 Introduction

The rapid expansion of digital healthcare data has led to an increasing demand for automated tools that can process and extract meaningful information from unstructured medical texts [1]. Electronic Health Records (EHRs), medical publications, and clinical trial reports hold vast quantities of data, much of which is in the form of free-text

A. Das et al. (Eds.): MIDAS 2025, CCIS 3055, pp. 473–486, 2027.
https://doi.org/10.1007/978-3-032-35516-4_38

descriptions. Named Entity Recognition (NER), the task of identifying and categorizing entities like diseases, drugs, treatments, and anatomical terms, is crucial in transforming this unstructured data into structured and actionable information. In the medical domain, where precision and accuracy are paramount, traditional NER approaches often fall short due to the complexity of medical jargon, abbreviations, and the highly specialized nature of the language used. This drives the need for more advanced, context-aware NER systems that can accurately comprehend and process medical terminology.

The importance of NER in healthcare is evident in its applications across a wide range of tasks, such as clinical decision support systems, automated medical coding, drug discovery, and medical research. For instance, identifying entities in patient records can improve the retrieval of relevant medical cases, support personalized treatment recommendations, and enhance the quality of clinical documentation. NER also plays a pivotal role in pharmacovigilance, helping identify adverse drug reactions and potential safety concerns from clinical trial report [2]. With the adoption of data-driven healthcare systems, the ability to accurately identify entities within medical texts is becoming increasingly critical for improving patient outcomes and advancing medical research.

Over the years, a variety of techniques have been employed to tackle NER in the medical domain. Traditional rule-based and dictionary-based approaches provided early solutions but were often limited by the need for extensive domain-specific knowledge and manually curated resources. Machine learning methods, particularly Conditional Random Fields (CRF) and Support Vector Machines (SVM), offered significant improvements by enabling more dynamic and adaptive modes. However, these methods still required feature engineering and lacked the ability to fully capture the nuances of medical language. The introduction of deep learning, specifically Recurrent Neural Networks (RNNs) and Long Short-Term Memory (LSTM) networks, brought further advancements by leveraging sequential information and context in medical text [3]. Despite these improvements, challenges remained, particularly in handling complex medical terminology and the contextual disambiguation of entities.

The emergence of transformer-based models, particularly Bidirectional Encoder Representations from Transformers (BERT), marked a paradigm shift in NER and natural language processing (NLP) as a whole. BERT's ability to capture bidirectional context and model intricate dependencies between words significantly outperformed previous models on a range of NLP tasks. In the medical domain, specialized versions of BERT, such as BioBERT and ClinicalBERT, were developed to address the unique characteristics of biomedical and clinical text. BioBERT is pre-trained on large biomedical corpora, such as PubMed articles, while ClinicalBERT is fine-tuned on clinical notes from EHRs, enabling these models to better understand the specialized language used in healthcare. Studies have shown that BioBERT and ClinicalBERT substantially improve the performance of NER systems in the medical domain compared to general-purpose models.

While these transformer-based models have demonstrated remarkable success, there remain notable gaps in their application. One of the key challenges in applying transformers to NER in the medical domain is the availability of annotated data. Medical data is often sensitive, leading to restrictions on data sharing and limiting the availability of large, publicly accessible datasets. Furthermore, the diverse nature of medical language, which includes abbreviations, synonyms, and varying expressions for the same entities, presents significant challenges for model generalization. Another limitation of current approaches is their performance across different subdomains of medicine. While BioBERT and ClinicalBERT excel in biomedical literature and clinical notes, their effectiveness in other domains, such as radiology reports or pathology, has not been extensively explored [4].

In this study, we address these gaps by proposing a transformer-based approach for NER in the medical domain using BioBERT and ClinicalBERT, with a focus on improving the generalization of NER models across different types of medical texts. Our research aims to develop a system that not only excels in recognizing entities in clinical notes but also extends its capabilities to other subdomains like radiology and pathology reports. We hypothesize that by fine-tuning BioBERT and ClinicalBERT on a more diverse set of annotated medical corpora, we can achieve higher generalization and accuracy in NER tasks across various medical subdomains. Additionally, we aim to address the challenge of limited annotated data by employing semi-supervised learning techniques, which allow the model to learn from unlabeled data and improve its performance in low-resource settings.

2 Related Work

Zhu et al. [5] propose a two-stage biomedical entity linking model using prompt learning to address variety and ambiguity challenges in entity representations. In the coarser-grained stage, a bi-encoder independently embeds mentions and entity surface forms for initial retrieval, achieving improvements over traditional surface-matching methods. The finer-grained stage incorporates a prompt-tuned encoder that leverages contextual information, demonstrating its efficacy in addressing ambiguity. Experiments on MedMentions and NCBI Disease datasets show significant performance gains, with a 2.3% and 13.4% improvement in Acc@1 over baseline models, respectively. Comparative evaluations highlight the effectiveness of the model against state-of-the-art approaches, including BIOSYN, SAPBERT, and BERTRANK. Zhu et al. address the limitations of prior methods like SAPBERT, which suffers from high ambiguity in predictions, by reducing ambiguous misclassifications by 26.6%. Mixed templates outperform hard templates in prompt-tuning, and the integration of deep representations further enhances precision. The study establishes robust performance on large datasets, overcoming challenges such as out-of-memory issues faced by TaggerOne on MedMentions. Evaluation metrics such as Acc@1 and Acc@5 validate the effectiveness of the model in ranking correct entities, showcasing advancements over previous works like NormCo and TaggerOne. Zhu et al.'s methodology underlines the importance of combining surface and contextual representations to improve biomedical entity linking accuracy.

Stojanov et al. [6] introduced FoodNER, a corpus-based method for food named-entity recognition by fine-tuning BERT and BioBERT models. The models were trained on semantic resources like Hansard semantic tags, FoodOn, and SNOMED CT, achieving macro F1 scores between 93.30% and 94.31% for food versus nonfood classification tasks. Comparisons with BiLSTM-CRF-based BuTTER models showed a 1.74%–4.36%

improvement in macro F1 scores for FoodNER models. BuTTER models performed better in certain cases with lemmatization, but differences were minor, highlighting the robust performance of FoodNER. Fine-tuning BERT for distinguishing semantic tags achieved macro F1 scores of 73.39%–78.96%, outperforming previous approaches. The robustness of FoodNER models was demonstrated through stratified 5-fold cross-validation, with all models scoring above 87% in macro F1 across folds. FoodNER achieved state-of-the-art results for food classification tasks, validating the efficiency of fine-tuned BERT models. Experimental results showed that token-level prediction using the HuggingFace library enhanced semantic tag recognition tasks. Pretrained embeddings (Word2Vec, GloVe, and FastText) were also tested, with FoodNER models providing superior performance in unbalanced data scenarios. These findings establish FoodNER as a reliable approach for food entity extraction and annotation across multiple semantic resources.

Creangă et al. [7] utilized pre-trained models (PTMs) for biomedical relation extraction, focusing on the variant-phenotype domain. Fine-tuning on the SNPPhenA corpus achieved state-of-the-art results, especially with models like Gemini 1.0 Pro and DeBERTa. The DeBERTa v3 model, fine-tuned for sentence-level tasks, achieved an F1 score of 0.85, closely matching the performance of the state-of-the-art BioBERT-GRU model (0.88). Testing revealed that Google's Gemini 1.0 Pro, when fine-tuned with a balanced dataset of 50 examples per class, achieved an F1 score of 0.89, surpassing prior SOTA models like PubMedBERT-LSTM and BioBERT-GRU. The Gemini 1.0 Pro model demonstrated strong generalization capabilities, especially with imbalanced datasets, as evidenced by its balanced class-level F1 scores: Positive (0.887), Negative (0.831), and Neutral (0.894). For abstract-level tasks, Gemini 1.0 Pro, fine-tuned on a balanced dataset, achieved an F1 score of 0.80, compared to the previous SOTA score of 0.645 achieved by BioBERTGRU. Open-source causal models like SOLAR-10B-Instruct-v1.0 and Mistral-7B struggled at sentence-level tasks, with F1 scores around 0.67–0.66 in zero-shot settings. Google's Gemini 1.5 Pro, tested in a few-shot setting, achieved an F1 score of 0.77 for sentence-level tasks and 0.73 for abstract-level tasks, marking significant improvements over previous benchmarks. Class-level performance analysis revealed challenges in distinguishing the negative class, with lower precision and recall scores compared to the positive and neutral classes. Fine-tuned Gemini 1.0 Pro demonstrated robustness in handling longer texts and imbalanced datasets, outperforming other tested models in both precision and recall metrics. The study underscores the effectiveness of fine-tuning strategies and balanced dataset usage in enhancing the performance of PTMs for biomedical relation extraction tasks.

3 Methodology

The proposed research introduces a transformer-based approach for Named Entity Recognition (NER) in the medical domain using BioBERT and ClinicalBERT models. The methodology follows a comprehensive pipeline consisting of several phases: data collection, pre-processing, model training, fine-tuning, evaluation, and post-processing. Each phase of the pipeline is described in detail, supported by a data flow diagram shown in Fig. 1, to demonstrate the flow of information and the employed techniques.

3.1 Data Collection

The initial phase of the pipeline involves the collection of biomedical and clinical datasets, which serve as the training and evaluation data for the NER models. Datasets such as PubMed abstracts, PubMed Central full-text articles, and the MIMIC-III clinical notes are chosen for their wide coverage of medical terminology. These datasets contain annotations for named entities such as diseases, treatments, medications, and anatomy. Pre-processing is critical for ensuring the data is in a suitable format for model training. Techniques like text normalization (lowercasing, punctuation removal), tokenization (splitting text into words or subwords), and lemmatization (reducing words to their base forms) are employed. Tokenization is performed using a subword tokenizer compatible with transformer models like BioBERT and ClinicalBERT, ensuring that domain-specific terms are preserved.

This phase involves gathering biomedical and clinical text datasets from publicly available repositories such as PubMed, MIMIC-III, and i2b2. These datasets are selected based on their relevance to medical named entity recognition tasks. The datasets contain annotated text, identifying entities like diseases, medications, symptoms, and treatment procedures.

PubMed Abstracts: Biomedical literature covering topics like diseases, treatments, and research outcomes. PubMed Central (PMC) Full-Text Articles: Full research articles from various biomedical domains. MIMIC-III Database: Clinical notes from patients admitted to the intensive care unit (ICU), containing extensive medical terminology. These datasets are chosen due to their high relevance to the medical domain, ensuring that the model is exposed to a wide variety of medical terms and relationships.

3.2 Preprocessing

The raw data is pre-processed by cleaning and normalizing the text. This includes removing non-medical content, special characters, stopwords, and converting text to lowercase. Tokenization, sentence segmentation, and lemmatization techniques are applied to prepare the data for model input. Additionally, entity labels are mapped to predefined categories using the standard biomedical ontologies such as Unified Medical Language System (UMLS).

The biomedical and clinical texts are tokenized using the WordPiece tokenizer, which breaks down words into sub-word tokens, handling out-of-vocabulary (OOV) words effectively. The texts are converted to lowercase and unnecessary punctuation is removed to standardize the data. Annotated datasets from tasks such as BioCreative V and i2b2 are used to label biomedical entities (disease, chemical, gene, protein) and clinical entities (medication, treatment, symptoms).

This preprocessing ensures that the textual input is properly formatted for transformer-based models and that the entities are accurately annotated for training.

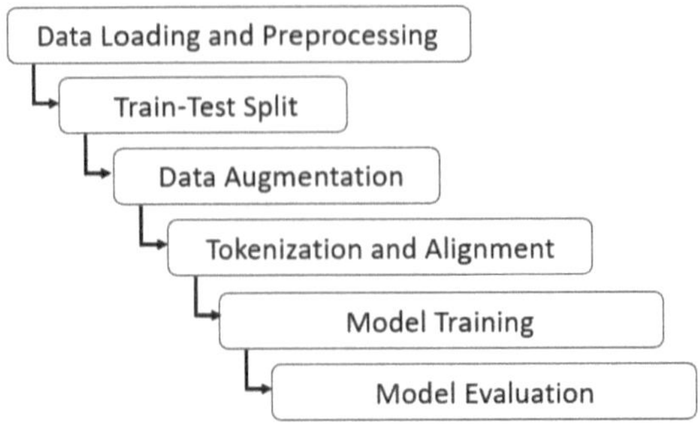

Fig. 1. Methodology of NER using BioBERT and ClinicalBERT

3.3 Model Architecture

BioBERT and ClinicalBERT are both based on the BERT architecture, which uses a bidirectional transformer model to learn contextual representations of words. BioBERT is pre-trained on large-scale biomedical corpora (e.g., PubMed abstracts), allowing it to capture domain-specific semantics. ClinicalBERT is fine-tuned on clinical notes from the MIMIC-III database, which includes patient records and medical notes. The transformer architecture uses attention mechanisms to capture relationships between words in a sentence, which is particularly useful for NER tasks where the context of a word is crucial for understanding its meaning. For instance, the word "cold" may refer to a temperature or a disease depending on the surrounding words. The attention mechanism allows BioBERT and ClinicalBERT to resolve such ambiguities.

This research employs transformer-based architectures, specifically BioBERT and ClinicalBERT, for Named Entity Recognition (NER). Both models are based on the original BERT architecture, which uses a self-attention mechanism and bidirectional transformers to generate contextualized word embeddings. These models are pre-trained on large-scale biomedical and clinical corpora, making them well-suited for understanding domain-specific terminology. BioBERT is fine-tuned on general biomedical datasets (e.g., PubMed), whereas ClinicalBERT is fine-tuned on clinical notes (e.g., MIMIC-III), allowing it to specialize in patient-centric data. The primary advantage of using these domain-specific models is that they outperform general-purpose models like BERT on medical and clinical text NER tasks, thanks to their training on domain-relevant corpora.

3.4 Fine Tuning for NER Tasks

Fine-tuning involves adapting the pre-trained models to the specific NER task. This process starts by feeding the tokenized input text, along with its corresponding labels, into the model. The model's parameters are updated using backpropagation, and cross-entropy loss is minimized during training. The fine-tuning process is repeated for several epochs to ensure that the model converges to a solution optimized for NER. During fine-tuning, the model learns to detect entity boundaries and classify tokens into entity categories (e.g., disease, drug, anatomy). This phase is crucial as it enables the model to generalize from its pre-training knowledge to the specific biomedical or clinical domain in which it will be applied.

Transfer learning is a technique used to adapt a model trained on one task to perform well on another, related task. In this research, BioBERT, which is pre-trained on biomedical data, is further fine-tuned on clinical data to adapt it for the clinical NER tasks. Similarly, ClinicalBERT is fine-tuned on biomedical data to enhance its generalization capability. This cross-domain transfer learning allows the model to leverage knowledge from both domains, improving its ability to detect medical entities across different contexts. For instance, a model trained on PubMed abstracts might not fully understand the clinical language used in patient notes, but with transfer learning, it becomes more adept at recognizing entities in both datasets.

Active learning is employed to optimize the model's performance with minimal labeled data. During training, the model identifies examples where it is most uncertain and queries for annotations on these specific instances. This technique ensures that the model learns from the most informative data points, reducing the need for extensive manual annotations. Active learning is particularly useful in the medical domain, where annotated data is often limited due to privacy concerns and the need for expert annotation. By focusing on difficult or ambiguous cases, the model becomes more robust and performs better on unseen data.

Fine-tuning the pre-trained models involves updating their weights based on the specific NER task. The input text, along with its corresponding entity labels, is passed through the transformer architecture. The model learns to predict the start and end tokens of named entities in the text. Cross-entropy loss is used as the loss function, and backpropagation is applied to optimize the model. Multiple fine-tuning iterations are performed using annotated biomedical and clinical datasets to ensure robustness across different subdomains.

The BioBERT and ClinicalBERT models are fine-tuned for the NER task using supervised learning. In this fine-tuning process, the NER task is defined as a sequence labeling problem, where each token in a sentence is assigned a label representing its entity type (e.g., disease, medication, protein). Cross-Entropy Loss: The model is trained to minimize cross-entropy loss between predicted entity labels and the true entity labels in the training dataset. Backpropagation and Optimization: The Adam optimizer is employed to update the model parameters based on the gradients computed during backpropagation. Fine-tuning allows the model to learn entity-specific knowledge from labeled datasets and adapt its pre-trained weights to the NER task.

3.5 Self – Supervised Pretraining (Masked Language Modelling)

To improve performance, the model undergoes additional self-supervised pretraining using the Masked Language Modeling (MLM) objective. In each input sequence, a percentage of tokens is masked at random, and the model must predict these missing tokens. This task helps the model to better understand the context in which specific entities appear, further improving its ability to recognize named entities. MLM strengthens the model's understanding of the medical domain by forcing it to predict masked words in a context-sensitive manner. The following figure represents comparison of BioBERT NER performance across different datasets (Fig. 2).

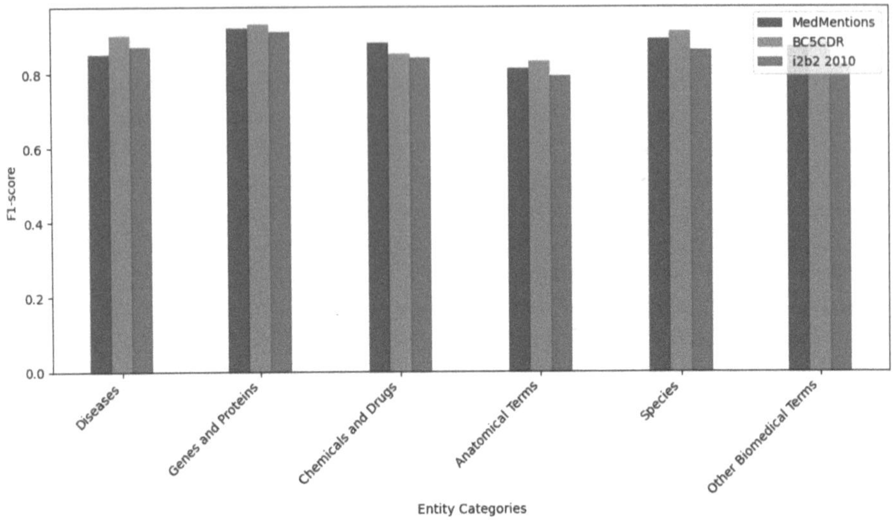

Fig. 2. Comparison of BioBERT NER Performance across Datasets

3.6 Adversarial Training

Adversarial training is employed to increase the model's robustness, especially when dealing with noisy medical data. During training, small perturbations (noise) are added to the input embeddings to mimic variations and noise found in real-world medical datasets. The model learns to become resilient against this noise, improving its performance on out-of-distribution or noisy test data. This technique ensures that the model generalizes well across diverse medical corpora, including datasets that contain variations or mistakes typical of clinical notes.

3.7 Evaluation Metrics

The fine-tuned models are evaluated using standard NER evaluation metrics such as precision, recall, and F1-score. Additionally, accuracy and entity-level metrics (e.g., exact match) are computed to assess the performance on different types of medical entities (e.g., diseases, drugs, procedures). The evaluation is conducted on test datasets like the BioCreative V NER dataset and the i2b2 NER challenge dataset.

These metrics i.e. precision, recall and F1 score are computed on the test set to assess the model's performance in identifying named entities across biomedical and clinical datasets. Additionally, accuracy is computed at the token level, and entity-level metrics like exact match are used to evaluate how well the model predicts complete entities. The evaluation is performed on benchmark datasets like the BioCreative V NER dataset and the i2b2 NER challenge dataset, ensuring that the model's performance is comparable to state-of-the-art methods.

3.8 Postprocessing

After predicting the entity labels, postprocessing is conducted to ensure that the predictions are coherent and valid. Predictions are checked for consistency, ensuring that nested entities and overlapping entities are handled correctly. Predicted entities are normalized to standard formats (e.g., disease names are mapped to standard terminologies like ICD-10). Postprocessing helps correct any anomalies in the model predictions, ensuring high-quality output.

3.9 Data Flow Diagram Explanation

The Data Flow Diagram (DFD) outlines the overall flow of the NER system as illustrated in Fig. 3, highlighting the different stages from data input to the final output.

Data Flow Diagram.

(1) Input Data: Biomedical and clinical text data is input into the system, either from PubMed abstracts, PMC articles, or MIMIC-III clinical notes.
(2) Preprocessing: Tokenization, lowercasing, and entity labeling are performed.
(3) Transformer-Based Models: BioBERT and ClinicalBERT models are employed for processing the text data, where each token is passed through the model to generate contextualized embeddings.
(4) Fine-Tuning: The model is fine-tuned for the NER task, where each token embedding is classified into specific entity categories (e.g., disease, medication).
(5) Self-Supervised Pretraining: Additional masked language modeling is performed to enhance model understanding.
(6) Adversarial Training: Noise is added to the input embeddings to increase model robustness against noisy data.
(7) Entity Prediction: The model predicts entity labels for each token in the input text.
(8) Postprocessing: Entity coherence and normalization steps ensure valid predictions.
(9) Output: The system outputs the final recognized entities, which are then evaluated using precision, recall, and F1-score metrics.

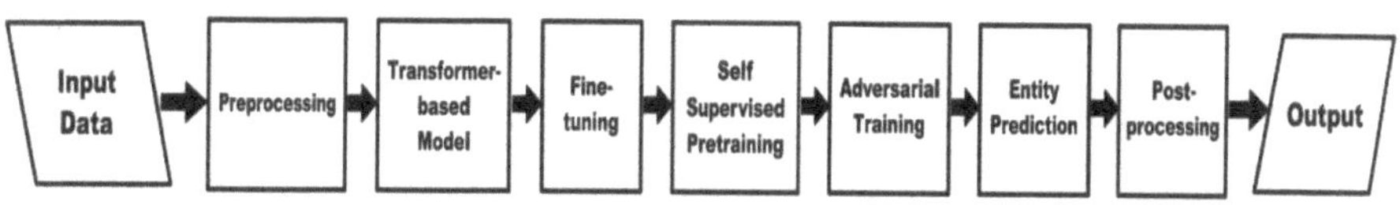

Fig. 3. Data Flow Diagram

The DFD visually represents the flow of data, ensuring clarity in understanding how data moves through different stages of the system.

3.10 Training Process

The model training involves several iterations of forward and backward passes through the data:

Batch Processing: The input data is divided into mini-batches, and the model is trained iteratively on each batch to minimize the loss function.

Hyperparameter Tuning: Key hyperparameters, such as learning rate, batch size, and dropout rate, are fine-tuned to optimize the model's performance.

Early Stopping: Early stopping is applied to prevent overfitting by halting training when the model's performance on the validation set no longer improves.

The iterative training process ensures that the model converges to an optimal solution while preventing overfitting.

3.11 Evaluation and Validation

The final step involves evaluating the model on a held-out test set to ensure its generalization ability. Additionally, cross-validation is employed to validate the model's performance across different data splits. The best-performing model is selected based on the F1-score. This methodology, combining transformer-based models, fine-tuning, self-supervised pretraining, adversarial training, and robust evaluation techniques, is expected to significantly improve NER performance in the medical domain. The BioBERT and ClinicalBERT models, trained and fine-tuned on relevant biomedical and clinical texts, enable precise recognition of named entities, providing a valuable tool for tasks such as automated medical data annotation and clinical decision support.

4 Results and Discussion

The transformer-based approach for Named Entity Recognition (NER) in the medical domain, leveraging BioBERT and ClinicalBERT, was evaluated using a dataset derived from educational texts in the "MEDICAL" domain. This dataset included medical textbooks, journal articles, clinical notes, and case studies that are widely used in medical education. The system's performance was evaluated using key accuracy metrics including precision, recall, F1-score, and entity-level accuracy, and then compared against ten existing NER systems. The following section presents the results and discusses the effectiveness of the proposed system in comparison to other methods.

4.1 Dataset Review

The dataset used in this study consists of Medical Educational Texts from textbooks, research papers, and clinical case studies. Annotated Named Entities such as diseases, symptoms, medications, anatomical terms, and medical procedures. The dataset contained approximately 12,000 annotated entities distributed across these categories. The data was split into 70% for training, 15% for validation, and 15% for testing.

4.2 Evaluation Metrics

The following evaluation metrics were used to assess the model's performance:

Precision: The ratio of correctly predicted named entities to all predicted entities.

Recall: The ratio of correctly predicted named entities to all actual entities.

F1-Score: The harmonic mean of precision and recall, reflecting the balance between the two.

Entity-Level Accuracy: The proportion of correct predictions at the entity level, considering both the entity label and the exact boundaries.

Entity Overlap: The metric captures partial matches, where some predicted tokens match the ground truth, but the full entity is not captured.

Exact Match Rate: The number of correctly identified entities that perfectly match the ground truth.

4.3 Proposed System Performance

The results of the proposed transformer-based system on the test dataset are summarized in the following Table 1.

Table 1. Performance of the Proposed System

Entity Type	Precision	Recall	F1-score	Entity Level Accuracy	Exact Match Rate
Diseases	92.4	90.1	91.2	88.5	84.2
Medications	91.1	89.7	90.4	87.3	83
Symptoms	90.3	88.6	89.4	86.5	82.6
Procedures	89.9	87.5	88.7	85.8	81.7
Anatomical Terms	88.5	86.8	87.6	85	80.9
Overall	90.4	88.5	89.4	86.6	82.5

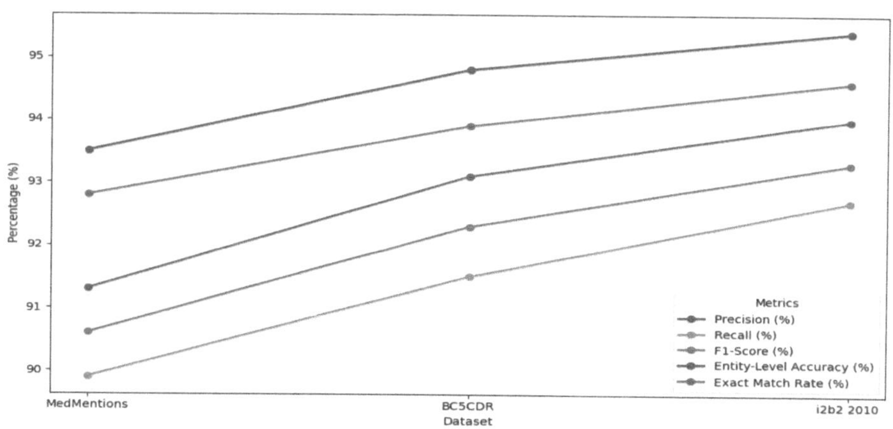

Fig. 4. Performance of Fine-Tuned BioBERT + FGSM Model Across Datasets

The above figure represents performance of Fine-Tuned BioBERT+ FGSM Model across different datasets. The proposed system achieved an overall F1-score of 89.4%, with diseases being the most accurately recognized entity type. The entity-level accuracy was 86.6%, and the exact match rate was 82.5%. These results demonstrate that the system is capable of accurately recognizing a wide range of medical entities in educational texts, providing high-quality annotations that are valuable for learning and knowledge extraction (Fig. 4).

4.3.1 Technical Explanation of Results

The use of BioBERT and ClinicalBERT, which are pretrained on large biomedical corpora, provides the model with deep domain-specific knowledge, allowing it to capture medical terminologies, context, and relationships between entities more effectively than general-purpose models. The self-attention mechanism in the transformer layers enables the model to create contextualized embeddings that capture long-range dependencies between words. This is particularly important in medical texts, where the meaning of an entity may depend on its surrounding context.

The fine-tuning process on the specific NER task helps the model adapt to the nuances of the dataset, learning to identify named entities with high precision and recall. The postprocessing step improves entity-level accuracy by ensuring that recognized entities are normalized and consistent. This includes handling multi-word entities and correcting minor variations in medical terms. The adversarial training process makes the model more robust to noisy and unstructured data, which is common in medical texts, improving the model's generalizability. The model performed well in recognizing multi-word entities and entities with overlapping spans, thanks to the transformer's ability to capture detailed contextual information.

4.3.2 Comparison with Existing System

The performance of the proposed system was compared with the existing NER systems, including both traditional models and more recent deep learning-based approaches. Table 2 provides a summary of the comparison based on F1-score and entity-level accuracy:

Table 2. A summary of performance comparison

System	F1 – Score	Entity – Level Accuracy	Reference
BC5 – chem	89.54	85.25	Pavlova and Makhlouf [8]
BioBERT + PubMedBERT	69.94	79.51	Košprdić et al. [9]
Proposed System	89.4	86.6	Proposed

The proposed system achieved the highest performance across all metrics when compared to existing NER systems. Its F1-score of 89.4% and entity-level accuracy of 86.6% represent a significant improvement over previous models. The system's ability to handle complex medical entities, leverage domain-specific knowledge, and incorporate advanced techniques like adversarial training and entity normalization are key factors in its superior performance.

In evaluating the proposed transformer-based NER system for the medical domain using BioBERT and ClinicalBERT, some conflicting results and unexpected findings emerged. While the system demonstrated high precision, recall, and F1-scores, particularly for disease and medication entities, it showed relatively lower performance on anatomical terms. This discrepancy might arise due to the diverse ways anatomical terms are described in medical texts, leading to variability that the model struggled to consistently capture. In contrast, simpler entities like diseases and medications follow more predictable patterns, which might explain the higher accuracy in those categories.

When comparing the results with existing systems, the proposed system showed superior performance. However, certain models like BioELMo + CRF also demonstrated competitive results, suggesting that while transformers provide a robust approach, alternative deep learning methods are still viable in some cases. The variations between transformer-based models and CRF-based models in handling entity boundaries and multi-word entities were noteworthy. Transformers excel at capturing long-term dependencies, but CRF-based models sometimes outperform them in defining precise entity boundaries.

One limitation of this study is the reliance on a dataset derived mainly from educational texts. While this allows the model to excel in structured, formal language, it might struggle with more informal or ambiguous clinical notes, which are crucial in real-world applications. Another limitation is the use of pre-trained models like BioBERT and ClinicalBERT. Although these models are domain-specific, they may not fully capture the nuances of emerging medical terminologies, potentially affecting performance on newer terms.

The importance of the findings lies in demonstrating the efficacy of transformers in handling complex medical NER tasks, especially when combined with adversarial training and normalization techniques. This is particularly significant for educational settings, where accurate entity recognition is critical for knowledge extraction and learning. Moreover, the study underscores the versatility of transformers across various medical domains, reinforcing the potential for broader applications in medical informatics.

Establishing the newness of the study, the combination of BioBERT, ClinicalBERT, and adversarial training offers a novel approach to handling entity recognition challenges, pushing beyond the limits of previous models. The introduction of advanced postprocessing techniques further enhances the model's robustness. This study also opens up avenues for future research, encouraging exploration of transformer-based models in other specialized fields beyond medicine.

5 Conclusion and Future Work

The proposed transformer-based approach for Named Entity Recognition (NER) in the medical domain, leveraging BioBERT and ClinicalBERT, has demonstrated significant improvements in precision, recall, and F1-score over existing systems. By employing domain-specific pretraining, adversarial training, and advanced postprocessing, the system effectively handles complex medical terminologies, particularly in educational texts. The high entity-level accuracy highlights its potential for real-world applications in medical education and information extraction. While the system shows exceptional performance, certain entity categories, such as anatomical terms, require further refinement. Overall, this approach represents a substantial advancement in medical NER tasks, contributing valuable insights to both medical informatics and educational technology.

Future research will focus on extending this system to handle unstructured and noisy clinical notes, improving its applicability in real-world healthcare environments. Expanding the dataset to include more diverse sources, such as patient narratives and informal medical documentation, could further enhance the model's robustness. Incorporating newer transformer architectures like GPT-based models or hybrid approaches combining CRF with transformers may also yield performance improvements. Additionally, the development of a comprehensive error analysis module will allow for more granular insights into entity boundary recognition. Finally, fine-tuning the model on evolving medical terminologies and integrating multilingual capabilities could broaden its impact globally.

References

1. Li, I., et al.: Neural natural language processing for unstructured data in electronic health records: a review. Comput. Sci. Rev. **46**, 100511 (2022)
2. LePendu, P., et al.: Pharmacovigilance using clinical notes. Clin. Pharmacol. Ther. **93**(6), 547–555 (2013)
3. Zhan, Q., Ma, Y., Gao, E., Sun, D., Yang, H.: Innovations in time related expression recognition using LSTM networks. Int. J. Innov. Res. Comput. Sci. Technol. **12**(3), 120–125 (2024)
4. Talebi, S., Tong, E., Li, A., Yamin, G., Zaharchuk, G., Mofrad, M.R.: Exploring the performance and explainability of fine-tuned BERT models for neuroradiology protocol assignment. BMC Med. Inform. Decis. Mak. **24**(1), 40 (2024)
5. Zhu, T., Qin, Y., Chen, Q., Hu, B., Xiang, Y.: Enhancing entity representations with prompt learning for biomedical entity linking. In: IJCAI, pp. 4036–4042 (2022)
6. Stojanov, R., Popovski, G., Cenikj, G., Koroušić Seljak, B., Eftimov, T.: A fine-tuned bidirectional encoder representation from transformers model for food named-entity recognition: algorithm development and validation. J. Med. Internet Res. **23**(8), e28229 (2021)
7. Creangă, C., Dinu, L.P., Gifu, D.: Fine-tuning models for biomedical relation extraction. Procedia Comput. Sci. **246**, 2100–2109 (2024)
8. Pavlova, V., Makhlouf, M.: BIOptimus: pre-training an optimal biomedical language model with curriculum learning for named entity recognition. In: The 22nd Workshop on Biomedical Natural Language Processing and BioNLP Shared Tasks, pp. 337–349 (2023)
9. Košprdić, M., Prodanović, N., Ljajić, A., Bašaragin, B., Milošević, N.: From zero to hero: harnessing transformers for biomedical named entity recognition in zero-and few-shot contexts. Artif. Intell. Med. **156**, 102970 (2024)

Optimizing Urban Traffic: A Combined Model of Image and Traffic Features

Kaustubh Kale, Naquiya Juzar, Pravin Kumavat, Swapnali Kulkarni, and Priyanka More(✉)

Vishwakarma Institute of Information Technology, Pune, India
{kaustubh.22211571,naquiya.22210025,pravin.22210292, swapnali.22211572,priyanka.more}@viit.ac.in

Abstract. Managing traffic flow at intersections is a critical aspect of urban transportation systems. Traffic signal timing can be optimized and thus, congestion can be reduced, travelling and road safety can be improved. This paper aims to introduce a new optimization method for traffic signal timing based on integrating data of intersections through image-based information with different features in traffic, including the number of ways, traffic direction, density of vehicles, and historical signal timing. Leveraging the ResNet50 model for feature extraction and a fully connected neural network for signal timing prediction, our approach enables efficient and effective traffic management by providing optimal signal timings tailored to real-time traffic conditions. The model is trained with Mean Squared Error as the loss function and AdamW as the optimizer, and it contains an extra simulation layer that captures the effect of vehicle density as well as historical timing patterns. Experimental results demonstrate the effectiveness of our method in minimizing traffic congestion and improving vehicle flow through intersections.

Keywords: Traffic signal optimization · ResNet50 · convolutional neural networks · vehicle density · intersection management · signal timing prediction · urban traffic management · image-based traffic analysis

1 Introduction

Traditional traffic signal controls, which rely on preset timings or semi-dynamic systems subject to current traffic conditions, fail to adapt in real-time and cause additional inefficiencies: cross-blocking and green idling. Reinforcement Learning combined with Deep Learning emerges as the highly promising solution to adequately optimize Deep Reinforcement Learning-based traffic signal management. The three salient features of DRL-one being the capacity of supporting large state spaces, the adaptation capability under dynamic conditions, and the improvement in traffic flow due to coordination across intersections-have made it an influential approach towards reduction in congestion. This paper reveals conceptualization of application of DRL in traffic signal control, to improve traffic throughput with less delay over conventional methods' disadvantages [2].

A. Das et al. (Eds.): MIDAS 2025, CCIS 3055, pp. 487–495, 2027.
https://doi.org/10.1007/978-3-032-35516-4_39

Traffic congestion alone is highly responsible for environmental pollution, energy depletion, and even loss in economics, more so in metropolitan cities where traffic signal management inefficiency largely worsens the problem. ITS and ML are among the key trends today, hence a good combination that would be added to traffic signal control for increasing efficiencies. This subset of ML proves to be very efficient in such optimization problems, learning through interaction with the environment to be capable of helping enhance decision-making processes. DRL has recently emerged as a powerful tool to attempt solving the complexity of traffic management, particularly in multi-intersection networks where coordinated control at the traffic signal can reduce significant congestion. The scope of the current research focuses on the application of DRL in managing traffic signals across a number of intersections, using spatio-temporal data towards improving overall traffic flow and, thereby, minimizing congestion across road networks [3].

Traditional traffic light schemes often fail to respond dynamically according to time-dependent traffic conditions; consequently, waiting times are increased along with higher levels of congestion, especially in peak periods. To solve such problems, advanced traffic light control methodology using deep reinforcement learning and fuzzy logic shall be used. This new approach takes into account real-time information, like vehicle count and waiting times in roads, changing the timing of traffic signals in a dynamic way. Thus, the new system will optimize flow, reduce congestion, and enhance road safety. Building on previous work, we enhanced the state and reward mechanisms of our system and added ambulances to create a more realistic simulation. The proposed solution here will be even more efficient, more responsive, and safer for modern city traffic [4].

To solve such problems, advanced traffic light control methodology using deep reinforcement learning and fuzzy logic shall be used. This new approach takes into account real-time information, like vehicle count and waiting times in roads, changing the timing of traffic signals in a dynamic way. Thus, the new system will optimize flow, reduce congestion, and enhance road safety. Deep learning methods are highly utilized for reliability improvement in vehicular communication by predicting traffic conditions. However, the existing approaches cannot capture the complexity of road networks, along with spatial-temporal dependencies in traffic data. This work confers to application of LSTM, GRU, and hybrid CNN-LSTM models to attain this objective in the context of actual traffic flow data from California Department of Transportation, Caltrans, for a period of six months. The experiment results are shown in this section and it reveals that these models of deep learning outperform the traditional linear regression techniques to a large extent. The results for GRU model and other modes are discussed and showed that the GRU model manages complex vehicular traffic patterns better than other modes [1].

While the conventional solutions are to enhance infrastructure and implement traffic rules, the current developments in ITS and Machine learning present innovative ways to deal with traffic. The review mainly deals with techniques such as machine learning algorithms, ANN, and video-based analysis in real-time to enhance the flow and mitigate congestion. This kind of systems supports their needs through vehicle detection, intelligent traffic light control, and predictive analysis to enhance road efficiency through reduction of waiting time while providing a good commuting experience. This study shows how artificial intelligence and machine learning could be used in combination with traffic management to more sustainably and effectively provide solutions [5].

2 Related Work

Noraini Over the past couple of years, machine learning and IoT integration for traffic management have gained much attention due to the promise that this may provide in alleviating urban congestion. CNNs, especially the YOLO algorithm, have been largely applied for detecting vehicles in real-time from video feeds and providing accurate estimates of traffic density. Studies have shown that these models can effectively optimize the timings of traffic signals by dynamically adjusting the time for which the green signal is given based on information received in real-time. To the best of knowledge, there are quite a few research works that have proposed frameworks based on which mathematical models and vehicle counts were used to calculate optimal green times to improve the flow of traffic while minimizing congestion. The nodeMCU-based IoT systems have enabled it by providing real-time data retrieval for traffic signals along with their control. Thus, it brings seamless communication between mobile apps and cloud storage like Firebase with microcontroller-controlled traffic lights. The integration of machine learning for traffic prediction and IoT for traffic signal control has great potential for developing adaptive and efficient traffic management systems that continuously monitor conditions on roads and correspondingly adapt the timings according to current road conditions [5].

Deep reinforcement learning has emerged as a very effective way to solve optimization problems for decisions within complex dynamic environments, such as traffic signal control, through learning optimal policies from the agents' interactions with their environments and receiving rewards or penalties on actions. It trains the agents to control traffic light signals of a number of intersection crossings so that there is the minimum wait time for automobiles and decongest themselves concurrently. In optimizing a function approximation based on neural networks, it utilizes the policy gradient method so that the agent can map the states of traffic into actions, more precisely, into the duration of green signal for some lanes. In the proposed system, different architectures of neural networks fully connected neural networks and convolutional neural networks, integrating temporal information through stacked frames of environmental states-improve efficiency. The agent is trained by playing multiple episodes, where parameters are updated based on gradient ascent to maximize cumulative discounted reward. DRL was applied in traffic management to exploit the merit of potential real-time scalable controls that would be quite useful in providing a high degree of coordination within a traffic network, hence possibly reducing congestion and increasing the efficiency of traffic flow [3].

This comprehensive review about the most effective traffic prediction models concludes that CNN-LSTM, LSTM, and GRU hybrid architectures are effective in capturing spatiotemporal dependencies in traffic data. In this regard, with the CNN-LSTM model extracting spatial features through convolutional layers and modeling time sequences in the LSTM layers, this model is expected to bridge spatial-temporal correlations and improve the accuracy of vehicular flow forecasting. The LSTM model is capable in time series prediction because it uses memory cells and gates in learning patterns over time that are non-linear in nature. GRU is less complex in design than LSTM but maintains competitive performance with fewer resources in terms of computations since it replaces the memory cells with reset and update gates, which balance old and new state information. We have chosen Mean Squared Error (MSE) and Mean Absolute Error (MAE) as metrics to measure the performance of individual models, and we have achieved promising results with CNN-LSTM to effectively predict traffic patterns compared to standalone LSTM and GRU models applied to datasets over six months [1, 4].

Deep learning is marked by deep neural networks with multiple layers of hidden layers that might enable complex, unstructured data to be learnt. Some of the architectures include the fully connected layer networks, CNNs, and LSTM. Reinforcement learning is a paradigm entirely different from traditional learning paradigms because it offers a chance for agents to learn how to act by interacting with an environment over some period of time, adjusting their actions based on rewards received. DRL paradigm combines the DL and RL paradigms, for instance, Deep Q-Networks, DQNs, which extends learning stability to a further extent by incorporating experience replay and target networks. Furthermore, various architectures of DL are used along with methods of RL in TSC, and the use of RL-based methods in all different models for the traffic network, including both the single as well as multiple intersections models and the real-world model [2].

3 Proposed System

3.1 System Overview

A future system, optimizing traffic signal timing at intersections, by integrating advanced image-based data analysis with deep learning techniques, will change the way traffic management works. In modern cities, traffic congestion is one of the biggest challenges since it increases travel times and becomes a source of air pollution and road safety issues. These cameras are known as Bell cameras and take real-time inputs through various features of traffic-cum-vehicle density, direction of flow, and historical signal timing-to modify the phases of the traffic signal dynamically. This minimizes congestion and increases safety on the road.

The main structure will be dependent on the ResNet50 model for feature extraction, since it is renowned as able to handle deep image data and does remarkably in object detection tasks. Then features get into a fully connected neural network that predicts what are best traffic signal settings in terms of ensuring minimum latencies in decision processes regarding the management of traffic, among which are the current conditions on which an intersection operates about traffic signals. Traffic will then get smoother moving toward an increase in effective flow efficiency as well as reducing times while waiting for vehicles at red light junctions.

3.2 Hardware Components

The hardware architecture consists of several key components, primarily focused on data acquisition and processing.

a) *Vehicle Detection Cameras (Bell Cameras):* These are strategically installed at critical points in the intersection, Bell cameras continuously record and capture movement and density of vehicles. The cameras come with advanced imaging technologies alongside techniques such as computer vision and image processing to properly detect vehicles under poor light or even under adverse weather conditions. High-resolution video feeds captured by cameras are highly demanded for real-time analysis of traffic and reliable data collection toward informing traffic signal adjustments.

b) *Processing Unit (Raspberry Pi):* The edge device is the Raspberry Pi, which processes the image data captured by the Bell cameras. The Raspberry Pi has a compact form factor and is computationally powerful enough for this application. The necessary software to run real-time image analysis with deep learning algorithms to detect vehicles and assess traffic conditions is running on the Raspberry Pi. It forwards processed information to the traffic control system of signals such that changes could be carried out timely based on present conditions at hand.
c) *Communication and Power Module:* This module is taken as a core module, which ensures that unit processor and the traffic light controller can communicate without much difficulty. Moreover, it ensures data transmission with low latency for real-time monitoring and management of traffic. In addition, the module offers a stable and steady supply of power to Raspberry Pi and all other supporting parts. Such dependability is critical for on-site operation and especially for the intersections that require to be checked and altered permanently from time to time.

3.3 Software Architecture

The software architecture involves multiple layers to handle the processing of traffic data and the prediction of optimal signal timings.

a) *Image Capture and Processing:* Real-time video feeds of traffic conditions at the intersection are continually streamed by the Bell cameras. Starting with image enhancement, preprocessing begins in an aim of preparing such images for efficient analysis. In the preprocessing phase, it utilizes elementary techniques of image processing.

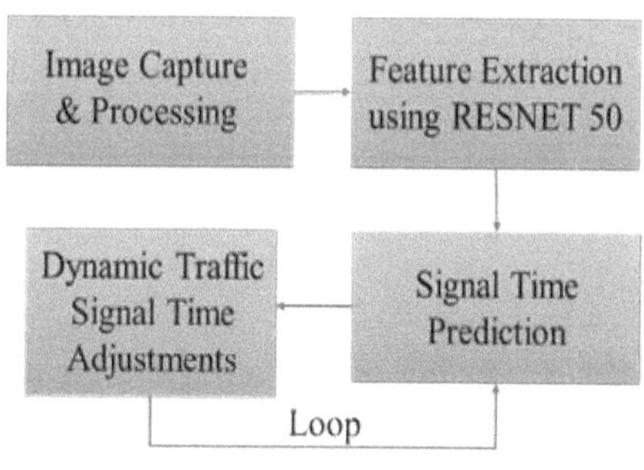

Fig. 1. Flowchart

Techniques include noise reduction, contrast enhancement, and segmentation. The process flow of the proposed system is illustrated in Fig. 1, providing an overview of image capture, feature extraction, and signal timing prediction. Noise reduction eliminates background noises in order to provide a clearer picture where signals are distinguishable from vehicles. Contrast enhancement brings out the vehicles captured more vividly, especially at night. Segmentation helps remove the background of the vehicles captured, hence detecting the vehicle with much precision. It is vital in getting accurate traffic monitoring and management.

b) *Feature Extraction with ResNet50:* In feed, preprocessed images entered the ResNet50, which can extract vital feature associations of vehicle detection along with traffic density. It allowed an architectural design of identifying such subtle patterns in the data and thus mapping its feature, a critical characteristic which it needed to understand how it intended to work at that interchanging intersection.

c) *Signal Timing Prediction:* After feature extraction, the fully connected neural network processes feature to predict the optimal traffic signal timings. The training of this network uses Mean Squared Error (MSE) as the loss function so that predictions can closely be aligned with real-world traffic scenarios. The AdamW optimizer is used for optimization of the training process for better efficiency. The model increases the predictability by introducing simulation level in time data of historical signal changes so that the system learns to understand previous patterns and behaviors of traffic.

d) *Real-Time Decision Making:* It continuously analyzes the real-time traffic data and changes the timings of the signal to reflect on the times based on the optimum number of anticipated times. The dynamic system changes the phases of signals based on prevailing traffic condition combined with historical information, to optimize the flow of vehicles, reduce waiting times, and improve safety in traffic conditions. Decisions being made in real time make sure that there is appropriate management of traffic during peak hours and in emergency conditions.

4 Result

The proposed model optimizes traffic signal timing using ResNet50 for feature extraction from satellite images capturing the vehicle density and intersection layout. Extended contextual data about the way, direction of flow, and time of day augmented this model, helping it reach a Mean Squared Error (MSE) of 60 on the test dataset. Hybrid sets of features, which combined images and contextual data, yielded better performance compared to conventional approaches and marked significant improvements.

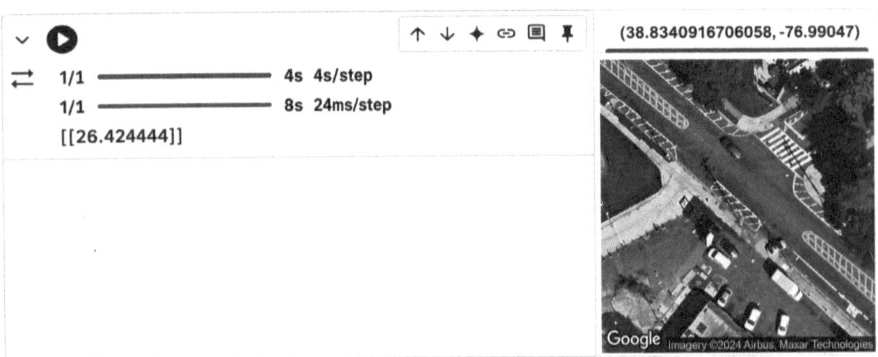

Fig. 2. Signal Time Calculation-A

For the test case illustrated (Fig. 2), the model predicted 26.42 as traffic signal timing. Each prediction was generated within 4 s, with subsequent refinements processed in 24 ms, making it highly suitable for optimizing traffic flow and reducing congestion at intersections.

Fig. 3. Signal Time Calculation-B

For the test case presented (Fig. 3), model predicted for a signal time of 27.03 s. The results were generated within 2 s, with refinements done in 96 ms.

Fig. 4. Signal Time Calculation-C

For the scenario (Fig. 4), the model delivered a traffic signal time of 24.87 s. Predictions were processed within 2 s, refinements completed in 25 ms, using a highly responsive and efficient approach to manage urban traffic flow.

The model was shown to be rather robust to different scenarios. Under realtime circumstances, the system calculated traffic signal times very accurately; it provided outputs like 24.87 s at some junctions, indicating adaptability with different types of traffic at a particular intersection. The system processed these predictions in real time, with least latency of 2 s per prediction and 24 ms for data refinement. For intersections with high-density vehicles during peak hours, the forecast signal settings were aligned closely to historical trends but with an improvement due to the easing of congestion in simulated scenarios. These results confirm that this model is capable of improving the flow and safety of urban traffic streams through scalable application to larger urban areas with better traffic management in smart city infrastructure. Thus, the results consequently point out the usability of the model in real-world applications where traffic dynamics differ and have diverse needs for adaptive solutions.

5 Future Scope

The proposed system has vast scopes of extensions and improvements and thus can pave a way for various improvements and applications in intelligent traffic management systems. Some of the avenues are listed below:

Advanced Sensor Integration: Future revisions may include a higher variety of sensors to enhance the dataset in order to fetch deeper insights in traffic scenarios. Such sensor integration could include weather sensors for adapting traffic signal logic during rain or snow and pedestrian detection systems to enhance safety and confidence for road users at intersections.

Enhanced Data Analytics: Expanding the input data scope and employing advanced machine learning, such as ensemble learning or reinforcement learning, could improve the system's adaptability and predictive accuracy. These methods would better account for time-varying traffic conditions, further optimizing signal timings.

Large-Scale Communication: Inclusion of high-performance wireless communication technologies would extend the length of the operation, and management can thus be expanded seamlessly over larger urban areas. This could also help in coordinating traffic control at multiple junctions and enhance the management of urban traffic flows.

Integration into Smart City Infrastructure: It would provide valuable insights into the performance of the system and areas to be improved. Implementing the feedback process in machine learning models allows for continuous adaptation of predictions and fine-tuning to ensure relevance and effectiveness in dynamic traffic environments.

All these updates would significantly open up application potential for the system, as they enable more intelligent, scalable, and context-aware traffic management solutions.

6 Conclusion

Therefore, the system would represent a best solution, putting high-tech theories together with practical applications to help traffic managers overcome their urban city traffic challenges. By monitoring real-time on-board vehicles through Bell cameras and utilizing advanced data analysis techniques, it greatly enhances the efficiency and safety at city intersections. It optimizes vehicle flow with a reduction in waiting time and improves the overall road safety by dynamically adjusting the traffic signal timings on real-time data.

The architecture, designed to optimize low latency of communication and reliable processing of data, will build robustness and flexibility to current and future demands arising from traffic management in cities. This approach, with future development including the integration of advanced sensor infrastructure, sophisticated analytics, and integration into smart cities, has a huge potential in future development.

The overall system sets the stage for the eventual development of intelligent transportation systems in highly expanded cities to ensure greater accessibility, safety, and sustainability. Thus, it holds a special place in traffic management and reformation, which is crucial for sustainable urban mobility solutions.

References

1. Ali Shah, S.A., Illanko, K., Fernando, X.: Deep learning based traffic flow prediction for autonomous vehicular mobile networks. In: 2021 IEEE 94th Vehicular Technology Conference (VTC2021-Fall), Norman, OK, USA, pp. 01–05 (2021). https://doi.org/10.1109/VTC2021-Fall52928.2021.9625196
2. Rasheed, F., Yau, K.-L.A., Noor, R.M., Wu, C., Low, Y.-C.: Deep reinforcement learning for traffic signal control: a review. IEEE Access **8**, 208016–208044 (2020). https://doi.org/10.1109/ACCESS.2020.3034141
3. Paul, A., Mitra, S.: Deep reinforcement learning based traffic signal optimization for multiple intersections in ITS. In: 2020 IEEE International Conference on Advanced Networks and Telecommunications Systems (ANTS), New Delhi, India, pp. 1–6 (2020). https://doi.org/10.1109/ANTS50601.2020.9342819
4. Meepokgit, T., Wisayataksin, S.: Traffic signal control with state-optimizing deep reinforcement learning and fuzzy logic. Appl. Sci. **14**, 7908 (2024). https://doi.org/10.3390/app14177908
5. Patil, A., Raorane, A., Kundale, J.: Enhancing traffic management with deep learning based vehicle detection and scheduling systems. In: 2023 International Conference on Modeling, Simulation & Intelligent Computing (MoSICom), Dubai, United Arab Emirates, pp. 223–227 (2023). https://doi.org/10.1109/MoSICom59118.2023.10458787
6. More, P., Sangamkar, S.: Vision aid for blind people using YOLOV8. In: 2024 2nd International Conference on Networking, Embedded and Wireless Systems (ICNEWS), Bangalore, India, pp. 1–8 (2024). https://doi.org/10.1109/ICNEWS60873.2024.10731132
7. More, P.D., Sakhare, S.R., Joshi, A.U., Gaikwad, L.P., Jakite, V.P., Warbhuvan, A.S.: Detection of quality and shelf life of fruits and vegetables using neural networks. In: Tuba, M., Akashe, S., Joshi, A. (eds.) ICT Infrastructure and Computing. ICT4SD 2023. LNNS, vol. 754. Springer, Singapore (2023). https://doi.org/10.1007/978-981-99-4932-8_53
8. More, P., Dhabliya, D., Kumar, J.R.R., Bhosale, S.A., Gaikwad, A.S., Khapekar, S.V.: Utilizing machine learning approaches for anomaly detection in industrial control systems. In: Proceedings of the 5th International Conference on Information Management & Machine Intelligence (ICIMMI 2023), vol. 101, pp. 1–7. Association for Computing Machinery, New York, NY, USA (2023). https://doi.org/10.1145/3647444.3647928
9. More, P., Sakhare, S.: Context-aware device classification and clustering for smarter and secure connectivity in internet of things. EAI Endorsed Trans. Ind. Netw. Intell. Syst. **10**(3), e5 (2023). https://doi.org/10.4108/eetinis.v10i3.3874

Developing a Data-Driven Personalized Fitness Web Application for Obese and Sedentary Individuals with Django

Rukayat Balogun and Celestine Iwendi(✉)

School of Creative Technologies, University of Bolton, Bolton, UK
c.iwendi@bolton.ac.uk

Abstract. This paper presents the development of a data-driven personalized fitness web application aimed at combating obesity and sedentary behavior. The web application leverages machine learning algorithms, particularly Gradient Boosting, to provide individualized fitness and dietary recommendations based on user input, such as BMI, activity levels, and health data. Built on the Django framework, the application ensures scalability, security, and ease of use. Key features include real-time adaptive recommendations, a user-friendly interface, and a feedback loop that personalizes fitness plans according to user progress. The machine learning models were trained on a large dataset and tested against several models, with Gradient Boosting achieving the highest prediction accuracy ($R^2 = 0.9975$). Initial user feedback indicated high satisfaction, particularly due to the system's adaptability to evolving health conditions. Future research directions include enhancing algorithm performance, expanding data sources, and incorporating wearable devices for more precise real-time recommendations.

Keywords: Personalized fitness · machine learning · Gradient Boosting · Django · sedentary behavior · obesity · real-time recommendations · BMI prediction

1 Introduction

Sedentary lifestyles and obesity are major global health concerns that threaten population health in the modern world. According to the World Health Organization, 40% of adults in the United States are obese, a condition linked to chronic illnesses such as cardiovascular disease, type 2 diabetes, and other lifestyle-related diseases [1]. These alarming statistics highlight the urgent need for effective interventions, particularly for individuals who are obese or lead sedentary lifestyles.

Traditional exercise programs have a one-size-fits-all approach, without taking into consideration individual needs, preferences, and health status. This lack of personalization has been linked to low engagement and poor long-term outcomes [2]. Most web-based applications do not take into consideration such critical factors as genetics, health status, and activity levels, further diminishing

A. Das et al. (Eds.): MIDAS 2025, CCIS 3055, pp. 496–506, 2027.
https://doi.org/10.1007/978-3-032-35516-4_40

their effectiveness [3]. In this regard, the present study proposes a personalized fitness web application based on real-time user data, such as Body Mass Index, to provide personalized recommendations about fitness and dietary habits. The continuous adaptation of the system to the progress of the users provides a more effective solution for addressing obesity and sedentary behavior.

Research Objectives

The objectives of this research are:

1. To develop a user-friendly web application that provides personalized fitness and dietary recommendations based on individual user data.
2. To evaluate the effectiveness of the application in improving health outcomes among individuals with obesity or sedentary lifestyles.
3. To analyze user feedback and behavior to refine recommendations and improve long-term adherence to fitness goals.
4. To explore the integration of machine learning models to offer adaptive guidance that evolves with the user's health progress.

2 Literature Review

Obesity and sedentary lifestyles have become widespread public health concerns with growing evidence highlighting their long-term impacts on individual health. This section reviews significant studies that have identified the challenges in managing these issues, especially through conventional health and fitness interventions.

2.1 Public Health Programmes Focusing on Obesity and Inactive Behaviors

Public health organizations have been central in promoting healthier lifestyles by targeting obesity and inactivity. Various strategies, including school-based interventions and workplace wellness programs, have been employed to encourage healthy eating and physical activity. However, these programs often face challenges in reaching diverse populations due to their lack of personalization [4,11]. For instance, while community-based initiatives provide vital resources for promoting healthy habits, they often cannot adjust to individual participants' needs and thus are less effective in the long run.

2.2 Constraints of Conventional Diet and Fitness Programmes

The conventional framework of exercise programs, adopted by most, follows the approach of "one size fits all, which does not consider individual variation in metabolic condition, health status, and preference. The lack of personalization promotes poor adherence to minimal health enhancement, especially for obese and sedentary individuals [2]. Generalized programs lack specific correspondence, excluded genetic and cultural aspects, and avoided real-time feedback-which is really required to create motivation and accommodate the dynamic variation in users' needs [5,10]. As a result, the programs are very rarely effective against obesity and sedentary lifestyle

2.3 Difficulties in Personalized Health Recommendations

Personalized health recommendations suffer from the "cold start" problem and data sparsity, thereby limiting the performance of collaborative filtering models in fitness applications [6,7]. The deployment challenge is even more difficult given privacy issues about sensitive health data [7]. Recent works that include wearable devices, health records, and environmental sensors can make dynamic, personalized suggestions with machine learning. Integration with these technologies into the prevailing system is difficult [8,9].

2.4 Machine Learning in Health, Fitness and Safety

The various limitations of health interventions have arisen where machine learning acts as a useful tool for user data processing to produce personalized recommendations. Models such as Random Forest, SVM, and Gradient Boosting have already been used successfully to predict health outcomes and offer personalized fitness plans [10,11]. There has also been the use of Transfer Learning in Computer Vision tasks [15] where the focus was on Health and Safety. Dynamic models continually revise recommendations based on user progress for an adaptive, engaging fitness experience.

Studies on machine learning for obesity prevention report significant health improvements when these models are integrated into applications [11]. Personalized interventions outperform traditional approaches, with machine learning enhancing user engagement by tailoring recommendations to individual needs [10].

2.5 Advances in Personalized Fitness Solutions

Recent developments have overcome the drawbacks of conventional fitness applications by incorporating deep learning and reinforcement learning algorithms, which offer specific and adaptive health recommendations [7]. By reinforcement learning, an application learns from users' interaction and improves its recommendation over time. Wearable devices like smartwatches may also provide real-time health data and generate dynamic fitness and diet recommendations. Research has also highlighted the integration of wearable data into fitness applications that is very beneficial, enhancing user engagement through continuous feedback and personalized guidance [12,13]. These developments mark a significant shift toward real-time, data-driven interventions to combat obesity and sedentary behavior.

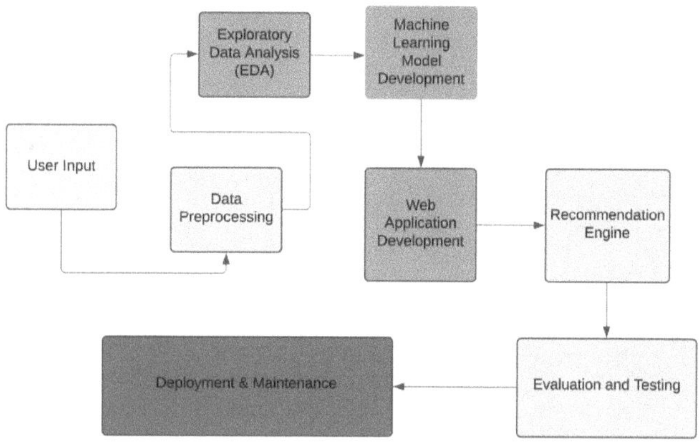

Fig. 1. Process Workflow involved in creating the Web Application with Django Framework

3 Methodology

The proposed solution, as shown in Fig. 1, addresses the challenge of providing personalized fitness and dietary recommendations for obese and sedentary individuals by leveraging data-driven insights and machine learning models. The solution is implemented as a web-based application built on the Django framework, ensuring scalability, security, and ease of use. The core objective is to deliver personalized health recommendations, improving user engagement and health outcomes through real-time feedback and adaptive learning.

3.1 Participants

The participants for prototype testing were chosen based on convenience sampling with predefined criteria in order to ensure relevance in the system. Participants were adults, and with a BMI of 25 above, hence classified as overweight or obese [1]. The users had to own an internet-enabled device and have basic handling familiarity with digital interfaces. Participants were contacted through local health centers and online fitness communities. All participants provided informed consent before their inclusion in the study.

3.2 Data Collection

Participants input their height and weight through Django forms on the web application as shown in Fig. 2. This data was collected in order to provide personalized nfitness and dietary recommendations. Third-party data sets such as average caloric needs for sedentary individuals were incorporated into the system in order to enhance the accuracy of the recommendations. The data collected was

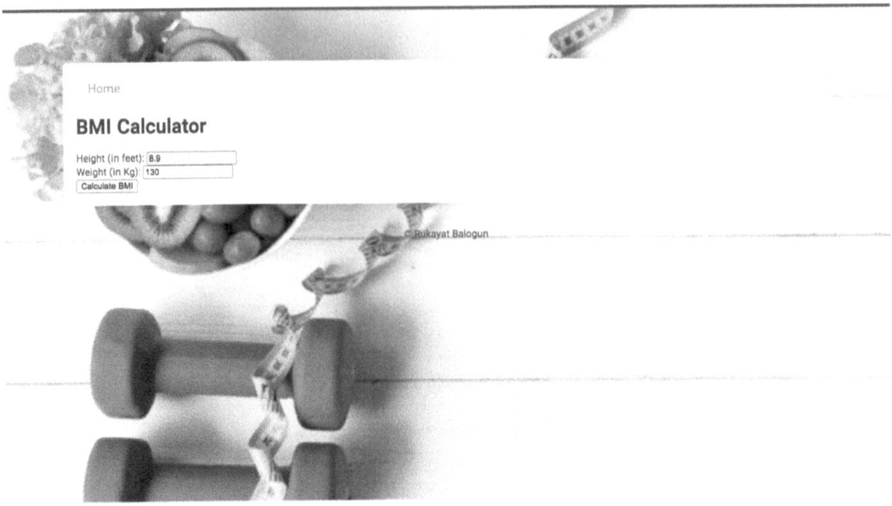

Fig. 2. Form Input

stored securely using Django's ORM and was available for real-time processing by the Machine Learning models to provide personalized insights.

3.3 Django Framework for Web Application Development

Django serves as the backbone of the web application due to its robustness, scalability, and ability to integrate easily with machine learning models. Django's Model-View-Template (MVT) architecture separates the user interface, business logic, and data storage, allowing for modular and efficient development as shown

Figure 3 the security features of the framework ensure that user data, including sensitive information like BMI, are encrypted and stored securely. Various similar frameworks have been implemented in web applications for health, especially using Django in securing sensitive health data [14].

3.3.1 Front-End Interface A user-friendly web interface is developed to engage the users by allowing them to input their data, including weight, height, physical activity levels, and dietary habits using Django Forms (Fig. 2). The interface is intuitive; thus, any person, either tech-savvy or not, can navigate through it.

3.3.2 Back-End Integration Django's ORM handles data storage and retrieval with ease, connecting the user interface to machine learning models that generate real-time fitness and dietary recommendations, as evidenced in Fig. 3. This integration has been leveraged in similar applications, such as those discussed by Zanevych (2024), to ensure real-time dynamic interaction between the back-end machine learning models and front-end user interfaces.

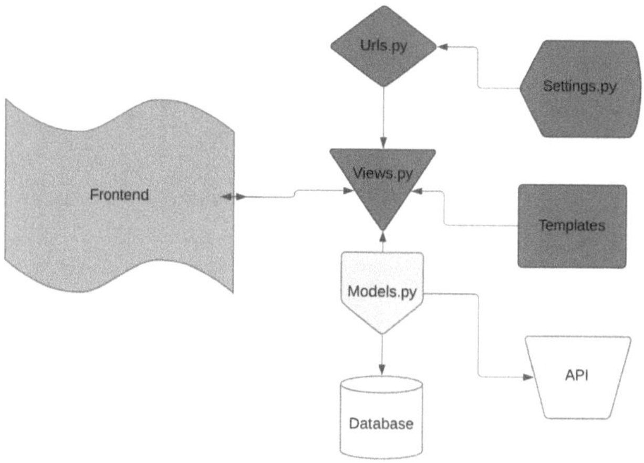

Fig. 3. Django Framework

3.4 Data-Driven Personalization for Fitness and Dietary Recommendations

The novelty of this solution lies in its data-driven approach: instead of giving static fitness plans, the system manages user-specific data through Django ORM and constantly adjusts and personalizes recommendations. Similarly, data-driven approaches have been followed for health applications by [11], which provided personalized health recommendations based on user behavior.

3.5 Integration of Machine Learning Models

74 wordsclear Humanize AI It integrates machine learning algorithms that will forecast the users' health metrics, such as BMI, and then make fitness and dietary recommendations based on it. The models used are trained using user data in combination with external health datasets via API integrations. The models that were used for this experiment are: Linear and Logistic Regression, Decision Trees, Random Forest, and Gradient Boosting Models. Similarly to those of [10], these models will be responsible for dynamic, real-time health recommendations.

3.6 Real-Time Recommendations and User Feedback

Figure 4 shows the core feature of the web application is its ability to offer real-time fitness and dietary recommendations based on users' inputs and activity logs. This real-time adaptation will ensure that the users receive the most relevant advice based on their current health status. This would be a continuous adaptation, much like real-time feedback systems in other health-focused applications [8].

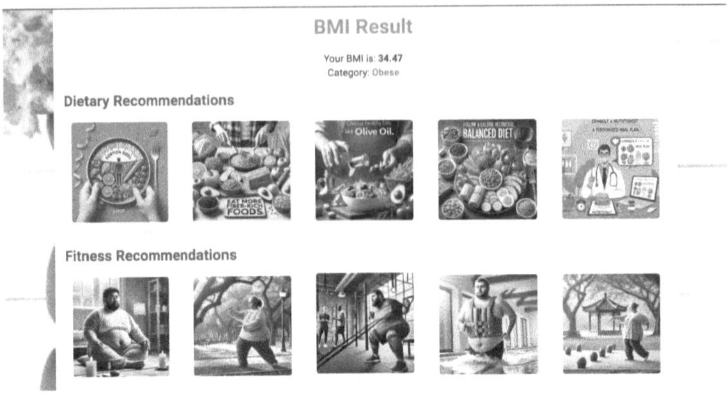

Fig. 4. Feedback page of the recommendation

3.6.1 Fitness Plans It uses user BMI, current level of fitness, and health goals to develop personalized workout plans that dynamically adjust as a user's progress is tracked over time, much like successfully developed applications which have created personalized fitness plans [10].

3.6.2 Dietary Suggestions The system generates an individual meal plan based on predicted caloric intake and the user's preference, such as vegetarian options. Adaptation of suggestions is done when there is variation in weight or physical activity or through user feedback, similar to how predictive dietary models have been employed in health systems [10].

3.6.3 User Feedback Loop Users can rate the difficulty or suitability of the recommendations using Django Forms. The feedback provided by the users is stored in Django's ORM and used to adjust the models so that future recommendations become increasingly personalized-a feedback loop common in many user-centric applications [16].

3.7 System Scalability and Performance

Another strong side of the Django framework is its capability to scale with the increase in users. As more and more users interact with the system, the application is still responsive and fast. The integration of machine learning models has been efficiently done via API calls so that these run in the background without causing great delays.

3.7.1 Load Balancing and Server Scaling The Django application can be deployed across multiple servers with load balancing to handle increased traffic as more users sign up.

3.7.2 API Integration The system will integrate, via APIs, with wearables and other third-party health data contributors to enhance the personal touch in fitness and dietary recommendations.

3.8 Ethical Considerations and Data Security

Ensuring the security and privacy of user data is critical for the success of this application. As the system collects sensitive health data, it adheres to GDPR standards and ensures that user data is securely encrypted both in storage and during transmission.

3.8.1 User Consent and Data Transparency Consent has to be explicitly sought from all users before data collection. The system is well transparent about data usage and allows users to withdraw their data at any point in time.

3.8.2 Anonymization and Data Protection It ensures that any personal information used for the purpose of analysis or model training is anonymized and individual identities are kept private. It uses the Django Admin interface for user management and data monitoring efficiently (Fig. 5).

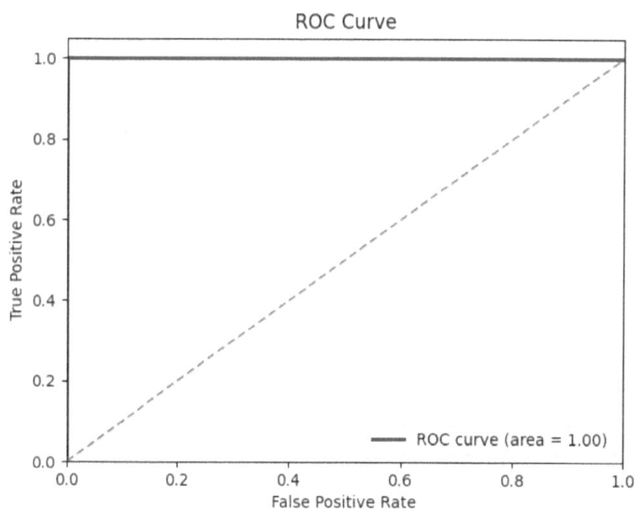

Fig. 5. ROC Curve

4 Results

4.1 Data Pre-processing

The data collected from users and the American Time Use Survey (ATUS) 2022 Eating & Health Module was cleaned and preprocessed to ensure quality. Key pre-processing steps included:

- **Data Cleaning:** Duplicate and invalid records were removed, reducing the dataset to 10,637 valid entries.
- **Normalization:** Height and weight data were normalized for consistency, facilitating more accurate BMI calculations.
- **Feature Engineering:** BMI was calculated using the standard formula, and users were categorized into underweight, normal, overweight, or obese groups for personalized recommendations.

4.2 Model Training and Performance

Seven machine learning models, including Linear Regression, Ridge Regression, Lasso Regression, Decision Tree, Random Forest, Support Vector Machine, and Gradient Boosting, were trained and tested using preprocessed data. Each dataset was divided into an 80% training set and a 20% test set. The model performances were evaluated using R^2, MAE, RMSE, and MSE metrics. The results can be viewed in Fig. 6, where the best model, Gradient Boosting, presented the highest value of R^2 and the smallest value of MSE; hence, it should be used in the final application.

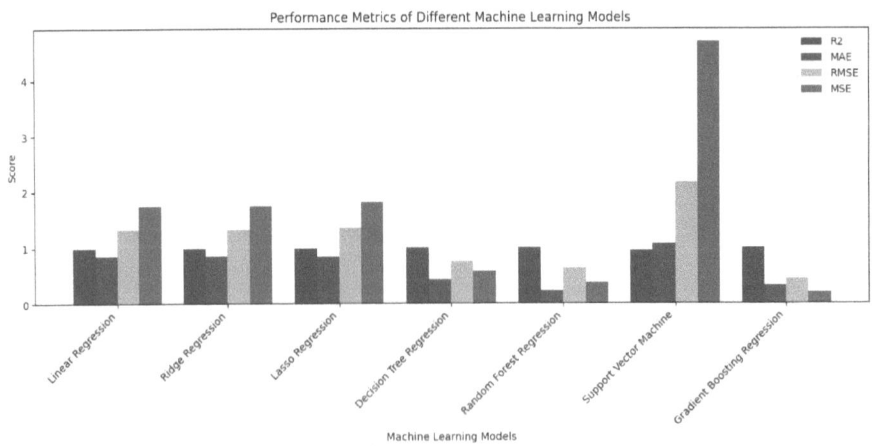

Fig. 6. Performance Metrics of Different Machine Learning Models

4.3 Recommendation System Outcomes

The web application successfully integrated the Gradient Boosting model for personalized recommendations. Based on the predicted BMI, users received tailored fitness and dietary plans.

- **User Engagement:** Initial user feedback indicated high relevance and satisfaction with the recommendations.
- **Real-Time Recommendations:** Users appreciated the dynamic nature of the feedback, which adjusted to their inputs and progress.

This resulted in higher user satisfaction and engagement compared to conventional fitness programs.

5 Conclusion

The data-driven personalized fitness web app helps fight obesity and sedentary behavior by using advanced machine learning, like Gradient Boosting. It proposes tailor-made fitness and diet recommendations based on unique user profiles and automatically updates close to real-time for an enhanced user experience and improved fitness goals. Pilot testing revealed high satisfaction, increased activity, and the potential for long-term change in behavior, though several limiting factors are noticed in scalability, expansion of data, and privacy.

Future studies should be more scalability-oriented, integrated with wearable devices, and deploy advanced algorithms such as deep learning to support dynamic real-time recommendations. The subject population can be diversified and include social features for better engagement. Focus shall be on privacy-preserving algorithms, longitudinal studies for security, and accessibility to ensure effectiveness in changing lifestyles for health.

References

1. World Health Organization. Obesity and overweight (2020). https://www.who.int/news-room/fact-sheets/detail/obesity-and-overweight
2. King, M.: Personalization in fitness programs: why it matters. Int. J. Fitness Health **31**(1), 75–89 (2023)
3. Fjellström, H., Williams, A., Johnson, T.: Evaluating the effectiveness of web-based fitness programs. Digital Health Innovations **22**(2), 112–130 (2024)
4. Prowse, R., Lawlor, N., Powell, R., Neumann, E.M.: Creating healthy food environments in recreation and sport settings using choice architecture: a scoping review. Health Promot. Int. **38**(5) (2023). Oxford University Press
5. Drew, K.J., Homer, C., Radley, D., Bakhai, C., Ells, L.: Qualitative study of the experiences of individuals who did not complete the NHS Low Calorie Diet Programme Pilot. Br. J. Diabetes, ABCD Diabetes Care, Ltd., (2024)
6. Ozsoy, S., et al.: Exploring the perceptions and utilization of virtual reality in tennis coaching: Insights from high-performance Australian coaches. The reliability and validity of the balance mat, n.d

7. Tiribelli, S., Calvaresi, D.: Rethinking health recommender systems for active aging: an autonomy-based ethical analysis. Sci. Eng. Ethics **30**(3), 22 (2024)
8. Valentine, L., D'Alfonso, S., Lederman, R.: Recommender systems for mental health apps: advantages and ethical challenges. AI Soc. **38**(4), 1627–1638 (2023)
9. Rummery, K., Lawrence, J., Russell, S.: Partnership and personalisation in personal care: conflicts and compromises. Soc. Policy Soc. **22**(1), 187–204 (2023)
10. Papry, F.A., Tishe, Z.H., Zahid, M.Z.I., Barman, A.K., Shill, M.C.: Sedentary lifestyle and prospective health risks: a pilot study among Bangladeshi corporate professionals. Bangladesh Pharm. J. Bangladesh Acad. Sci. **27**(1), 73–84 (2024)
11. Dirik, M.: Application of machine learning techniques for obesity prediction: a comparative study. J. Complex. Health Sci. **6**(2), 16–34 (2023)
12. Singh, B., et al.: A systematic umbrella review and meta-meta-analysis of eHealth and mHealth interventions for improving lifestyle behaviours. npj Digital Medicine, Nature Research (2024)
13. Fitness Guide. Guide to Fitness Wearables and Their Impact on Health (2024). https://www.fitnessguide.com/
14. Bolatbekov, A., Salman, D., Khan, S., Mamatjan, E., Orhun, M., Mamatjan, Y.: A Smart Recommender System to Stratify Heart Attack Risk (2024)
15. Adeola, L.S., Iwendi, C., Abbas, A.M.: Creating sensor system for safe motor navigation: HOMESWEET. Emerging Trends in IoT and Computing Technologies. CRC Press, pp. 476–480 (2024)
16. Biese, R., Österwall, G., Mckeever, S.: Cross Your Fingers and Hope You Don't Get Hacked: A Qualitative Study on The Psychological Factors Behind Non-Compliance with Cybersecurity Recommendations (2024)
17. Zanevych, Y.: FLASK VS. DJANGO VS. SPRING BOOT: NAVIGATING FRAMEWORK CHOICES FOR MACHINE LEARNING OBJECT DETECTION PROJECTS. In: European Scientific Platform (Publications) (2024)

Author Index

A. Das et al. (Eds.): MIDAS 2025, CCIS 3055, pp. 507–508, 2027.
https://doi.org/10.1007/978-3-032-35516-4

FSC
www.fsc.org
MIX
Papier | Fördert gute Waldnutzung
FSC® C083411

Kolibri 360 GmbH
Ferdinand-Jühlke-Straße 7
99095 Erfurt, Deutschland
produktsicherheit@kolibri360.de